# Handbook of Pressure-Sensitive Adhesive Technology

# Handbook of Pressure-Sensitive Adhesive Technology

Edited by

**Donatas Satas**

VNR VAN NOSTRAND REINHOLD COMPANY
NEW YORK CINCINNATI TORONTO LONDON MELBOURNE

Library of Congress Catalog Card Number: 81-10455
ISBN: 0-442-25724-4



Manufactured in the United States of America

Published by Van Nostrand Reinhold Company Inc.
135 West 50th Street, New York, N.Y. 10020

Van Nostrand Reinhold Publishing
1410 Birchmount Road
Scarborough, Ontario M1P 2E7, Canada

Van Nostrand Reinhold Australia Pty. Ltd.
17 Queen Street
Mitcham, Victoria 3132, Australia

Van Nostrand Reinhold Company Limited
Molly Millars Lane
Wokingham, Berkshire, England

15 14 13 12 11 10 9 8 7 6 5 4 3 2 1

**Library of Congress Cataloging in Publication Data**

Main entry under title:

Handbook of pressure-sensitive adhesive technology.

Includes indexes.
1. Adhesives. I. Satas, Donatas.
TP968.H36 668'.3 81-10455
ISBN 0-442-25724-4 AACR2

# Contributors

Charles Bartell. B.S.Ch.E. (1942) from Cooper Union Engineering, M.S. (1949) and Ph.D. (1951) from Ohio State University. Was associated with Rohm and Haas Co., Permacel, W. R. Grace and Co. Currently research director at Mystik Tape, Northfield, Illinois.

Erhard Braeunling. Doctorate in chemistry. Project leader at Beiersdorf AG, Hamburg, Germany, engaged in the development of electrical tapes.

G. L. Butler. B.Sc. in chemistry and natural philosophy at University of St. Andrews, Scotland. Development Manager of DRG Sellotape Products, Borehamwood, England, responsible for the development and quality control of pressure-sensitive products.

Carl A. Dahlquist. Now retired and a part-time consultant, was a research scientist of 3M Co., St. Paul, Minnesota. Main field of interest adhesion, rheology of adhesive materials, mechanical properties of polymers. Author of many papers and patents.

Norman P. DeBastiani. B.S. in chemistry from American International College (1960). Was associated with Plastic Coating Corp. Currently in charge of development and quality control of photographic, graphic arts, and pressure-sensitive products at Chartpak, Leeds, Massachusetts.

R. Dowbenko. Ph.D. in organic chemistry. Involved during his career in the synthetic aspects of coating materials. Currently manager of polymer research at PPG Industries C&R Division, R&D Center, Allison Park, Pennsylvania.

Earle E. Ewins, Jr. Technical associate at Shell Development Company, Houston, Texas. Responsible for compounding principles and associated technologies for pressure-sensitive adhesives.

Mary Daniels Fey. B.S. in chemistry from University of Michigan (1974). Currently development representative and group leader in the Technical Service of Dow Corning Corp. in Midland, Michigan.

Keiji Fukuzawa. Doctor of Engineering from Yokohama National University, Japan (1954). Chemical consultant, Tokyo, Japan. Specializes in pressure-sensitive adhesives, tapes, labels, surgical products. Member Japan Consulting Engineers' Association.

Whiteford D. Grimes. Chemical engineer (Virginia Polytechnic Institute). Was associated with DuPont, Sun Chemical. Twenty-five years experience in film processing, laminating, coating technology. Currently marketing manager of automotive products in the Plastics Division of Stauffer Chemical Co., Westport, Connecticut.

Fred H. Hammond, Jr. Graduated from Colby College and studied physical chemistry at Massachusetts Institute of Technology. Early research involved studying the surface chemistry and polymer rheology of pressure-sensitive adhesives and development of Polyken probe tack tester. He is presently information services specialist at Kendall Co., Lexington, Massachusetts.

George M. Harris. B.S.Ch.E. from Illinois Institute of Technology (1950). Was associated with the Kendall Co. in R&D, technical service and marketing functions of corrosion protective tapes. Currently vice president at Tapecoat Company, Evanston, Illinois.

John J. Higgins. B.S. from Iona College and M.S. in chemistry from Fordham University. Currently sales representative with Exxon Chemical Company USA, Florham Park, New Jersey.

Dirk de Jager. Research chemist, Koninklijke/Shell Laboratorium, Amsterdam, Holland. Engaged in adhesive technology involving thermoplastic rubbers.

Frank C. Jagisch. B.E. and M.S. from Stevens Institute of Technology. Currently a product applications and customer technical service specialist with Exxon Chemical Company, Baton Rouge, Louisiana.

David N. Kendall. B.A. and M.A. in chemistry from Wesleyan University (1938 and 1939) and Ph.D. in chemical physics from Johns Hopkins (1943). Consultant since 1953, founder and owner of Kendall Infrared Laboratories (Plainfield, New Jersey). Previous experience at American Cyanamid Co. and National Lead Co.

Jim Komerska. A.B. University of Chicago (1951). Experience in R&D and other technical activities in reprographics, identification, and labeling pressure-sensitive adhesives. Currently manager of product commercialization and quality assurance at Paper Manufacturers Company, Southampton, Pennsylvania.

William H. Korcz. Staff research engineer responsible for coordination of Shell Development Co. effort in adhesives, sealants and coatings at Westhollow Research Center, Houston, Texas.

Duane F. Merrill. Technical marketing specialist in the Fluids, Resins and Specialties, Silicone Products Department, General Electric Company, Waterford, New York. He has been engaged for 30 years in research on silicone resins and related products, varnishes, paints and pressure-sensitive adhesives.

Helmut W. J. Mueller. Graduated as a chemical engineer, Berlin 1958. Currently in charge of the pressure-sensitive adhesive work team at BASF, Ludwigshafen, Germany. Previous experience in the automotive industry.

Valentinas Rajeckas. Doctor of Technical Sciences, professor and head of department of Leather and Textile Products Technology at Kaunas Polytechnical Institute, Kaunas, Lithuania, USSR. Specializes in theoretical principles and instrumental methods in prediction of adhesive joint strength.

Donatas Satas. Chemical engineer (B.S.Ch.E., Illinois Institute of Technology, 1953). Independent consultant since 1975 (Satas & Associates, 99 Shenandoah Rd., Warwick, Rhode Island). Specializes in adhesives, coating, and laminating technology.

David J. St. Clair. Staff research engineer, Shell Development Co., Houston, Texas. Specializes in new polymer and adhesive development.

James A. Schlademan. B.S. in chemistry from Carleton College (1964), M.S. in organic chemistry from South Dakota School of Mines (1966) and Ph.D. in organic chemistry from Clarkson College of Technology (1975). He was associated with Neville Chemical Co. in charge of adhesives and rubber application research. Presently research chemist in the New Products Development Group, ARCO Polymers, Inc., Newtown Square, Pennsylvania.

N. Eugene Stucker. B.S.Ch.E., University of Kansas. Currently a polymerization specialist in the plant technical department with Exxon Chemical Company USA, Baytown, Texas.

Christopher Watson. Chemist, B.S. from Kings College (1968). Technical operations manager at Kleenstick Laminating Division, Ajax, Ontario, Canada. Previous experience with Zimmer Paper Co., Riegel Paper Co., Lamco Consultants in the area of hot melt technology.

John E. Wilson. B.S. in chemistry from Cleveland State University (1966) and M.S. in chemistry from Case Western Reserve University (1968). Currently a paper industry specialist in Paper Coatings, Technical Service and Development at Dow Corning Corporation, Midland, Michigan.

# Preface

Pressure-sensitive adhesive products are used in most industries as well as at home, in the office, and in hospitals. These products constitute one of the fastest growing segments of the adhesives industry, which in itself is growing at above the average rate. Such an active area deserves a compilation of technical information. The absence of an extended treatment so far perhaps could be explained by the lack of published information in some segments of this industry. Patent disclosures are still the best source of up-to-date information in many areas of the pressure-sensitive adhesive technology. The situation is gradually improving, mainly due to the publishing effort of raw material suppliers interested in providing better organized information to their customers. The pressure-sensitive adhesive products industry has not been eager to disseminate technical information beyond the description of the products.

The book is aimed mainly at the technologist in the pressure-sensitive adhesive industry. While an average worker might have considerable experience in some areas, surpassing the depth and detail presented in the book, he usually has not had the opportunity to be exposed to all aspects of this technology. The book should contribute in increasing the breadth of his outlook. It should also be of interest to the technologist in related areas of other adhesives, sealants, and coatings industries. Much of the technology is interrelated and transferable. A businessman engaged in sales or manufacturing of pressure-sensitive tapes, labels, or other products should benefit by a better insight into the technical problems. While a portion of the material deals with technical details of interest to a specialist only, a large part should be of sufficiently general interest. A better understanding of technical tasks should improve the communication with the technical personnel.

An attempt is made to cover most of the aspects of the pressure-sensitive technology. This leads into various disciplines: mechanical details of the equipment, polymer chemistry and physics, business and marketing. In some areas, a highly technical discussion is required; in others, fairly simple descriptive presentation is sufficient. Such is the nature of this technology.

My gratitude is expressed to all contributors as well as to numerous authors whose work we have discussed. My appreciation goes to Susan Munger and Denis Riney of Van Nostrand Reinhold for their help in processing the manuscript. I am also grateful to my son Paul V. for preparation of many drawings, my daughter Audrone M. for her help in typing, and also to my wife Saule for her assistance in the preparation of indexes.

D. Satas

Warwick, R. I.

# CONTENTS

**Contributors/v**
**Preface/ix**

**1. PRESSURE-SENSITIVE ADHESIVE PRODUCTS IN THE UNITED STATES, Donatas Satas/1**

Production Volume/2 History/3 Tapes/4 Labels/19 Other Products/20 Manufacturers/20 Adhesive Compounding/21 Adhesive Raw Materials/21

**2. PRESSURE-SENSITIVE BUSINESS IN JAPAN AND ASIA, Keiji Fukuzawa/25**

Japan/25 Korea/28 Taiwan/30 China/30 Southeast Asia/31

**3. TACK, Fred H. Hammond, Jr./32**

Rolling Ball Tack Test/33 Peel Tests for Tack/35 Probe Tack Tests/36 Surface Chemistry and Tack/39 Rheology and Tack/42 Compounding and Tack/43 Conclusion/47

**4. PEEL, Donatas Satas/50**

Testing/51 Transition from Cohesive to Adhesive Failure/53 Transition to Oscillating Failure/56 Effect of Temperature/61 Effect of Adhesive Thickness/63 Effect of Backing/65 Effect of Other Variables/68 Stress Distribution/70 Analysis of Peel/73

**5. CREEP, Carl A. Dahlquist/78**

Short-Term Creep and Tack/79 Long-Term Creep/83 Long-Term Creep and Steady State Viscosity/84 Creep and Non-Newtonian

Viscosity/88 Chemical Cross Linking and Creep Resistance/90 Physical Crosslinks/92 Degradation and Creep/94 Mechanical Property Measurements/94 Summary/97

6. ELECTRICAL PROPERTIES, Erhard Braeunling/99

Dielectric Strength/100 Electrolytic Corrosion/102 Insulation Resistance, Volume Resistivity/103 Dielectric Constant, Dielectric Dissipation Factor, Loss Index/103 Temperature-Time Limits/104 Other Important Test Methods/106 International Specifications/107

7 BOND STRENGTH AND ITS PROGNOSIS, Valentinas Rajeckas/109

Strength Concept and Failure Mechanism/109 Bond Strength/111 Stress Concentration in Adhesive Bonds/113 The Effect of Mechanical Properties of the Adhesive and the Substrate/118 The Influence of Scale Factor/121 The Influence of Temperature and Loading Regime/122 The Evaluation of Long-Term Strength at Variable Loads/130 The Influence of Temperature and the Deformation Rate/137 Heat Resistance/139 The Application of Analogies for Strength Prognosis/141 Evaluation of Complex Strength/148

8. AGING PROPERTIES, Keiji Fukuzawa/152

Effect of Tackifying Resins/152 Effect of Antioxidants/155 The Correlation Between Accelerated Aging Tests and the Shelf-Life/157

9. ANALYTICAL TECHNIQUES FOR IDENTIFICATION AND CHARACTERIZATION, David N. Kendall/168

Infrared Methods/169 Multiple Internal Reflection/171 Analysis of Backing Material/172 Identification of Component Parts/175 Characterization of Pressure-Sensitive Adhesives/179

10. NATURAL RUBBER ADHESIVES, G. L. Butler/189

Composition of Adhesives/190 Manufacture/197 Mechanisms of Tack and Adhesion/198 Specially Modified Adhesives/203

Emulsion Adhesives/210 Major Applications/211 Summary and Future Trends/218

**11. BLOCK COPOLYMERS, William H. Korcz, David J. St. Clair, Earle E. Ewins, Jr. and Dirk de Jager/220**

Nature of the Basic Molecule/220 General Formulating Principles/224 Application of Formulating Principles/241 Formulating to Meet Constraints/262

**12. BUTYL RUBBER AND POLYISOBUTYLENE POLYMERS, J. J. Higgins, F. C. Jagisch and N. E. Stucker/276**

Basic Properties/277 Adhesives Compounding/281 Solvents and Solution Preparation Procedures/285 Applications and Formulations/289

**13. ACRYLIC ADHESIVES, Donatas Satas/298**

Monomers/299 Copolymerization/305 Molecular Weight/307 Crosslinking/311 Secondary Bonding/315 Ionic Binding/319 Functional Groups/322 Compounding/324 Coating Methods/326

**14. VINYL ETHER POLYMERS, Helmut W. J. Mueller/331**

Monomers/332 Production of Polymers/332 Commercial Products/333 Properties/334 Stabilization/335 Application/337

**15. SILICONE PRESSURE-SENSITIVE ADHESIVES, Duane F. Merrill/344**

Manufacturing and Products/344 Crosslinking/348 Test Methods/349 Primers/350 Tapes/350 Lamination/351

**16. TACKIFIER RESINS, James S. Schlademan/353**

Rosin and Rosin Derivatives/354 Hydrocarbon Tackifier Resins/356 Evaluating Tackifier Resins/364

**17. RELEASE COATINGS, Donatas Satas/370**

Polymer Coatings/371 Waxes/373 Silicones/373 Long Chain Branched Polymers/376 Amines/378 Chromium Complexes/379 Miscellaneous/380 Testing of Release Quality/380

**18. SILICONE RELEASE COATINGS, Mary D. Fey and John E. Wilson/384**

Physical Properties/385 Chemistry/388 Coating Types/391 Solvent-Borne Release Coatings/391 Water-Borne Systems/394 Solventless Silicone Coatings/395 Coating Application/397 Coating Evaluation/400 Summary/403

**19. SATURATED PAPER AND SATURATED PAPER TAPES, Charles Bartell/404**

Physical Properties/404 Saturation/407 Raw Paper/408 Saturated Papers/409 Paper Tapes/413

**20. HOSPITAL AND FIRST AID PRODUCTS, Donatas Satas/419**

Tape Uses/420 Skin Irritation/421 Nonocclusive Tapes/422 Other Products/423 Adhesives/423

**21. PACKAGING TAPES, Keiji Fukuzawa and Donatas Satas/426**

Polypropylene Tapes/428 Unplasticized Vinyl Film Tapes/429 Kraft Paper Tapes/431 Cloth Tapes/432 Saturated Paper Tapes/433 Japanese Paper Tape/433 Reinforced Tapes/434 Other Film Tapes/434 Tape Dispensing/435 Tape Testing/436

**22. AUTOMOTIVE APPLICATIONS, Whiteford D. Grimes/438**

Paint/439 Bright Metals/440 Plastics/441 Applications/441 Exterior Woodgrain Films/442 Striping Films/446 Interior Woodgrains/447 Informational Decals/448 Exterior Automotive Tops/448 Adhesive System Tapes/449

**23. PLASTIC TAPE PIPELINE COATINGS, G. M. Harris/450**

History/450 Basic Composition/451 Types of Tape Coatings/452 Long Line Tape Coatings/453 Polyethylene Tape Plant Coating Systems/456 Specification/457 In-Service Performance/459

**24. GRAPHIC ART APPLICATIONS, Norman DeBastiani/463**

Graphic Tapes/464 Graphic Films/470 Die Cut Pressure-Sensitive Products/475 Dry Transfer Lettering/476

**25. LABELS AND DECALS, Jim Komerska/478**

Adhesives Composition/479 Methods of Application/480 Release Liners/481 Liner Backings/483 Release Coatings/484 Release Levels/486 Delayed Action Adhesives/486 Labels/488 Nameplates/495 Protective and Decorative Sheets/496 Decals/496 Embossable-Imagible Tapes/497

**26. COATING, Donatas Satas/498**

Machineability/501 Coating Methods/503 Knife and Blade Coaters/504 Analysis of Knife and Blade Coating/510 Air Knife Coating/513 Bar Coaters/516 Reverse Roll Coaters/518 Mechanism of Roll Coating/520 Other Roll Coaters/522 Gravure Coating/525 Calendering/527

**27. DRYING, Donatas Satas/533**

Mass Transfer Mechanism/534 Drying Curves/538 Solvent Retention/540 Drying of Latexes/542 Heat Transfer/544 Infrared Drying/546 Coating Imperfections/546 Curling/547 Equipment/548 Convection Dryers/549 Floater Dryers/552

**28. HOT MELT APPLICATION, Christopher Watson and Donatas Satas/558**

Products/558 Adhesives/559 Equipment Selection/562 Slot Orifice Coaters/563 Roll Coaters/565 Extrusion Coaters/568 Coater Operation/569

**29. POLYMERIZATION, Donatas Satas/574**

Solution Polymerization/575 Emulsion Polymerization Mechanism/576 Emulsifiers/578 Initiators/580 Equipment/582

**30. RADIATION CURING, R. Dowbenko/586**

Ultraviolet and Electron Beam Polymerization/587 Free Radical Polymerization/588 Photoinitiators/589 Radiation Cure Equipment/591 Polymerizable Compositions for Radiation Pressure-Sensitive Adhesives/593 Adhesive Syrups/595 Adhesive Polymers for Radiation Cure/598 Conclusion/601

**Name Index/605**
**Subject Index/611**

# Handbook of Pressure-Sensitive Adhesive Technology

# Chapter 1

# Pressure-Sensitive Adhesive Products in the United States

**Donatas Satas**

*Satas & Associates*
*Warwick, Rhode Island*

Pressure-sensitive adhesives are used for many products in many different ways. It is convenient to subdivide the products into the following categories: tapes, labels, and miscellaneous products.

All of these are combinations of two components: backing and the adhesive. The relative importance of the two can vary greatly. In many cases, the adhesive must take a secondary position as a means to secure the flexible backing to another surface, while the backing as a tape or label performs the main function. In the electrical insulation tapes, the most important property is the dielectric strength of the backing while the adhesive serves only as a convenient way to secure the backing in place. In the case of labels, the printed paper or film surface carries the message, while the adhesive again only keeps the label in place.

While pressure-sensitive adhesives clearly take a position of secondary importance in most products, the successful applications of these products and their position in the market would not have been possible without the pressure-sensitive adhesives. Perhaps the relative importance of these two constituents, flexible web and adhesive, can be best measured by their relative cost. The cost of the adhesive, including its application, release, and prime coatings, often surpasses the price of the backing many times. Clearly, the cost of the adhesive, its application and that of the release paper is much higher than the cost of coated paper used in the pressure-sensitive labels. The same is true with most pressure-sensitive tapes. These products are primarily pressure-sensitive adhesive oriented and constitute the bulk of this business. There are many other products where the cost of the pressure-sensitive adhesive is minor. Floor tiles with pressure-sensitive adhesive for an easy application, decorative (especially woodgrained) vinyls, and foam tapes constitute products where the cost of the backing surpasses that of the adhesive, although the adhesive performance might be very important for the successful use of such products.

## PRODUCTION VOLUME

Various estimates of the pressure-sensitive adhesive business have been published. Frost and Sullivan, Inc. estimates the U. S. volume of pressure-sensitive products as follows.[1]

- Tapes $1.1 billion
- Labels $640 million

Other products, such as floor tiles, automotive woodgrained vinyls, and other decorative vinyls, constitute a large additional dollar volume, but the value added by the adhesive is relatively small and inclusion of such products would misleadingly increase the importance of pressure-sensitive products. Another use of pressure-sensitive adhesives is lamination of difficult to adhere films and plastics such as polyolefins.

The U. S. consumption of pressure-sensitive adhesives has been estimated at 337 million lb. in 1980 by Frost and Sullivan[1] and at 360 million lb. in 1979 by H. S. Holappa and Associates and Springborn Laboratories, Inc.[2]

Natural rubber adhesives still have the largest volume, but acrylates and styrene–butadiene block copolymers have grown fast and contest the supremacy of natural rubber. The following breakdown of various elastomers used for pressure-sensitive adhesives is reported.[2]

| | Million lb. |
|---|---|
| Natural rubber | 63 |
| SBR random polymers | 44 |
| SIS block copolymers | 30 |
| Polyacrylates | 53 |
| Polyisoprene | 2 |

It should be remembered that the elastomers are compounded with resins and other ingredients, thus increasing the weight of adhesive by approximately a factor of two. Polyacrylates are either not compounded, or compounded very little.

In addition to the above elastomers, butyl rubber is used in large quantities for corrosion protective pipe wrap tapes. A smaller quantity of atactic polypropylene is consumed for low price hot melt adhesives, such as those used for floor tiles.

## HISTORY

The beginning of the pressure-sensitive adhesive industry is in the medical application of adhesive tapes and plasters. The use of adhesive masses and plasters can be traced from the beginnings of recorded medical history.[3] These were not pressure-sensitive adhesive products and it is not entirely clear at which point these masses evolved into products sufficiently resembling pressure-sensitive adhesives of today. During the mid-19th century, natural rubber was added to the adhesive formula that previously contained resins and beeswax. The invention of rubber pressure-sensitive adhesives is attributed to Dr. Henry Day.[4] Such adhesive consisted of India rubber, spirits of turpentine, turpentine extract of cayenne pepper, litharge, pine gum, and other ingredients. A U. S. patent to Shecut and Day was issued in 1845.[5] A German patent was issued in 1882 to a druggist P. Beiersdorf for a plaster based on gutta-percha. In 1899, natural rubber-based, zinc oxide containing adhesive has been developed. The adhesive plasters were produced industrially in larger quantities by the turn of the century. Wood rosin was the weak link for some time. The resin was easily oxidized and the adhesive lost tack. The problem was solved by modifying wood rosin to improve its stability.

For quite sometime, pressure-sensitive adhesive applications were limited to first aid uses. The first electrical insulating tapes were produced during the period 1920–1930.[6] This was not strictly a pressure-sensitive tape, but a friction tape. The adhesive mass was designed to adhere to itself. Its cohesive strength was low and the adhesive failed cohesively during unwind of this tape.

An important utilization of pressure-sensitive tape was for paint masking applications. It is said that a manufacturer of hospital tapes has noticed that the tape sales of one of the distributors was increasing at an unusually fast rate. A further investigation disclosed that the distributor was selling the tape to the automobile industry to hold paper sheets employed as paint masks. Saturated paper tapes have been developed for this application and the product grew into the largest single tape used for many different uses besides paint masking.[7]

The first transparent synthetic film tape that found many uses in offices and households was cellophane film tape.[8,9] Further development of various tapes was fabulous. Every conceivable flexible backing has been used for pressure-sensitive tape applications. Tapes found use in every industry for a tremendous variety of purposes.

The shortage of natural rubber during World War II prompted a look for substitutes. Polyisobutylene was among the first polymers to replace natural rubber.[10] Many other elastomers have been used as the basis for pressure-sensitive adhesives, but the dominance of natural rubber has been threatened only recently by polyacrylates and by block copolymers which are especially useful for hot melt adhesives.

## TAPES

Tapes are the most important of the pressure-sensitive products. Their volume is the largest, and construction can be quite difficult and demanding. The variety of tapes is claimed to be 200 by some,[4] 400 by others. It all depends on the definition of what constitutes a different tape and what is merely a variation. The important point is that industry requires such a large number of different tapes. This indicates the wide acceptance of the pressure-sensitive adhesive tapes and other products.

The tape must be constructed so that the adhesive remains on one side of the backing when it is unwound. The adhesive must not split or transfer to the other side, even if aged at higher temperatures and at other unfavorable conditions and under considerable pressure, which can be exerted on the adhesive in a tightly wound roll. The unwind force must be sufficiently low, but not too low. In order to meet these requirements, the properties of the adhesive must be carefully chosen. Release, backsize, and prime coatings may be used. In especially difficult cases, it might be necessary to use a release-coated interlayer that, of course, detracts from the easy use of a pressure-sensitive tape.

The tapes may be classified according to construction or according to use. Classification according to both criteria produces some repetition, but it also allows coverage of various aspects of pressure-sensitive tapes.

Tapes may be subdivided according to their construction into the following categories.

- Fabric tapes
- Paper tapes
- Film tapes

- Nonwoven fabric tapes
- Foil tapes
- Reinforced tapes
- Foam tapes
- Two-faced tapes
- Transfer tapes

As is obvious from the above list, the backing, rather than the adhesive, provides the main distinction among various tapes, although classifications like acrylic adhesive tape or hot melt adhesive tape might be heard occasionally.

Tapes may be classified according to their function. Bemmels,[4] in describing various tapes, uses classifications such as holding or bundling, masking, sealing, protecting, reinforcing, splicing, stenciling, identifying, packaging, insulating, mounting, etc. We will try to classify the tapes according to a more general area of application which has to do with both the function of the tape and the industry for which the tape is intended. Thus the following tapes will be discussed.

- Hospital and first aid tapes
- Office and graphic art tapes
- Packaging and surface protection
- Building industry products
- Electrical tapes
- Automotive industry products
- Shoe industry tapes
- Appliance industry products
- Splicing tapes
- Corrosion protective tapes
- Miscellaneous tapes and pressure-sensitive products

## Fabric Tapes

Woven fabric is the oldest backing material used for pressure-sensitive adhesive tapes. Hospital tapes were, and many still are, made with a woven fabric backing. Fabric has some unique characteristics which are difficult to obtain with other backings, and therefore fabrics are still used despite their higher cost. The combination of high tensile strength with a high flexibility are not obtained in other materials. A low tear strength can be incorporated into a fabric backing without sacrificing the tensile strength. The tape could be made to tear easily across or lengthwise removing the need for cutting. This property has been useful for hospital tapes and other applications.

Fabric backing can be heavily filled and coated with starch, nitrocellulose, vinyl, or acrylic coatings. Fabrics can also be used without any backcoating for tape backings. The rough surface of a woven cloth provides sufficient release if the adhesive does not flow into the fabric. While cotton and rayon are the most common fibers used, polyester fabric backings are used for many applications, including high tensile packaging uses. Nylon fabric tapes have found some uses in the shoe industry. Glass fiber fabrics are used for their high tensile strength and especially for their resistance to high temperatures. Such tapes with silicone adhesive are useful over a temperature range −54–232°C. Glass also does not support combustion.

## Paper Tapes

Paper is the least expensive material available in the web form and its dominance has been only recently threatened by some low cost polymeric films.

The paint masking tape for the automotive industry, an important application for a low cost tape, prompted the development of a family of saturated paper tapes. These tapes proliferated beyond the use for paint masking and became general-purpose tapes serving many diverse applications. Two main types of saturated paper backings are used: creped paper and flatback paper. The saturation process and various tape products are discussed in detail by Bartell in a separate chapter.

In addition to the saturated paper tapes, unsaturated kraft tapes have been developed more recently. They are especially popular in Japan. The main disadvantage is poor delamination resistance. Because of it, efficient silicone-based release coatings are used. These tapes exhibit a poor adhesion to their backing and have a tendency to flag.

Various specialty papers are also used: rope fiber paper for higher tensile strength, and nylon high-temperature papers for increased temperature resistance in electrical applications.

## Film Tapes

Polymeric films have found numerous applications as tape and label backings. The properties inherent in polymeric films (impermeability, thinness, smooth surface, good dielectric properties, inertness) are desired for many electrical, packaging, and decorative applications. Some films are also inex-

**Table 1-1. Physical properties of some film tapes.**

| FILM | TENSILE STRENGTH (kg/cm) | ELONGATION (%) | THICKNESS (mm) |
|---|---|---|---|
| Clear cellophane | 5,7 | 19 | 0.07 |
| Clear acetate | 3.6 | 20 | 0.07 |
| Etched acetate | 3.6 | 25 | 0.08 |
| Acetate fiber tape | 8.6 | 3 | 0.15 |
| Polyester transparent | 4.5 | 110 | 0.05 |
| Transparent vinyl | 3.6 | 16 | 0.07 |
| Carton brown vinyl | 6.3 | 40 | 0.07 |
| Polypropylene clear | 3.2 | 14 | 0.06 |
| Polypropylene clear | 6.8 | 73 | 0.06 |

pensive on the basis of surface area. They are competing with paper for applications where price is the determining factor, and are replacing paper in many cases. Some properties of typical film tapes are shown in Table 1-1. The table lists only the most commonly used film tapes. There are many more varieties of film tapes produced.

Cellophane is the oldest transparent film tape. It was and still is widely used as an office and general-purpose household tape. Cellophane tape is being replaced by other film tapes. Matte cellulose acetate film tape replaced cellophane in some offices, paper mending and general-purpose applications, despite the fact that cellulose acetate film is more expensive than cellophane. Cellophane is hygroscopic, becoming quite brittle when dry and quite soft in a humid atmosphere. These deficiences are absent in acetate tapes. Furthermore, a matte acetate film surface does not reflect light and is therefore inconspicuous. It also accepts writing.

Poly (vinyl chloride) films are widely used as tape backings. Plasticized vinyl film tapes are predominant in electrical insulation applications, while rigid vinyl film tapes are important for packaging and general-purpose uses. In Europe, unplasticized vinyl films have occupied a dominant position among the packaging and general-purpose film tapes. In the United States, unplasticized vinyl films are not popular; polyester films have been preferred for some applications and biaxially-oriented polypropylene film tapes are taking over the packaging tape area. Plasticized vinyl tapes are used for protecting racks immersed in electroplating solutions, masking metal surfaces to be plated, sealing cans, protecting metal window frames, and fabricating corrosion protective tapes for underground pipes and printable tapes for general labeling. Colored vinyl tapes are used for identification and floor markings.

Polyester film tapes are used in electrical applications and in various uses where high tensile strength and high tear are important. Oriented polypropylene tapes have replaced polyester tapes in some applications because of their lower price.

Polyethylene film tapes have not been as successful for general use as some of the other films. It is used for special applications, some of which are large volume uses. It is used for some medical tapes, electrical insulating, duct insulating, and carpeting. One large application is corrosion protective tapes for underground gas and oil lines.

Tetrafluoroethylene (TFE) tapes offer a high dimensional stability, resistance to elevated temperatures, chemical inertness and a low coefficient of friction. Because of these properties, it has found many applications despite its high cost. In combination with a silicone adhesive, TFE tapes are good for continuous use at 250°C. TFE tapes are useful in many electrical applications such as coil winding, transformers, cables, relays, condensers, resistors, etc. The tape is useful to wrap various rollers to prevent sticking of adhesives and coatings, to line conveyors, chutes, guide rails, and hoppers in order to eliminate friction. It is useful as a dry lubricant to prevent abrasion, rubbing, and squeaking. It provides a nonstick surface in heat sealing equipment, and glue dispensers. Many tapes are also available with a TFE coating over a fabric, usually a glass fiber cloth.

Polyimide film tapes are used for specialty electrical applications for high-temperature insulation, flat cable construction and masking in soldering operations.

### Nonwoven Fabric Tapes

Nonwoven fabrics do not find many uses as tape backings. There is a successful hospital tape with a nonwoven fabric backing. These fabrics are used for some electrical tapes where the porous nonwoven fabric helps to hold the resinous impregnant.

### Foil Tapes

Aluminum, lead and some copper foil tapes are used in the industry. Aluminum is by far the most important foil tape and it is used in applications where a moisture seal is required. Foil tapes are also used in electrical applications as a static bleedoff on metallic structures, in heating insulation and air conditioning ducts as a reflective heat shielding, in high-temperature masking operations (as experienced in the manufacture of printed circuit boards), in repairing sheet metal, in blocking radiation on X-ray plates, and other applications. Dead soft aluminum foil is most commonly used for

**Table 1-2. Properties of foil tapes.**

| BACKING | ADHESIVE | TOTAL THICKNESS (mm) | TENSILE STRENGTH (kg/cm) | PEEL ADHESION (g/cm) | ELONGATION (%) | USEFUL TEMPERATURE RANGE (°C) |
|---|---|---|---|---|---|---|
| Aluminum | Acrylic | 0.09 | 3.6 | 730 | 5 | −65–177 |
| Aluminum | Acrylic | 0.13 | 5.4 | 450 | 5 | −65–177 |
| Aluminum | Acrylic | 0.13 | 5.4 | 730 | 5 | −65–177 |
| Aluminum | Acrylic | 0.29 | 9.0 | 730 | 13 | −65–177 |
| Aluminum | Silicone | 0.09 | 3.2 | 220 | 5 | −65–316 |
| Aluminum foil glass cloth | Silicone | 0.15 | 13.4 | — | — | −54–316 |
| Lead | Acrylic | 0.13 | 3.0 | 450 | 15 | −65–177 |
| Lead | Rubber | 0.15 | 3.0 | 390 | 15 | −60–106 |
| Lead | Rubber | 0.19 | 3.6 | 390 | 15 | −60–106 |

tape backings. Aluminum foil laminated to cloth, foam, and other bulkier materials is used for vibrational and acoustical damping. Properties of some aluminum and lead foil tapes are given in Table 1-2.

Lead foil tape is used for its chemical resistance, especially in plating operations, for its high density as a radiation shield and acoustical dampening tape, and for its low water vapor transmission as a sealing tape. Many foil tapes, especially lead foil, are sold with a release liner, because of the difficulty to provide an adequate release or prime coating on lead. Lambert and Smith describe a primer for lead foil which removes the need to use an interliner.[11]

Copper foil pressure-sensitive adhesive tapes do not find many applications. In flat cables, which is a major application for copper foil, heat-activated adhesives have found a greater acceptance than pressure-sensitive ones.

Slitting and winding of foil tapes present special problems. Score slitting deforms the foil, leaving a bead at the edge. Also, the razor blades are dulled easily. Shear slitting might deform the edge, unless the knives are set very shallowly. Rider rolls are generally required to keep the foil smooth and to iron out any wrinkles or deformations.

## Reinforced Tapes

Packaging operations such as bundling of heavy materials, i.e., steel rods, require high tensile tough tapes. The toughness, as measured by the area under stress-strain curve, is perhaps a better index of the performance of these tapes in heavy-duty applications than just the tensile strength.

The first reinforced tapes were constructed by laminating acetate fiber rope to acetate film. These filaments are embedded in the adhesive mass. They can move and distribute the load evenly over the fibers, resulting in high tensile strength tapes. Some of these tapes are listed in Table 1-1. Other yarns, especially glass and rayon, and other films, polyester and polypropylene, are also used for construction of reinforced tapes. High tensile strength polyester fiber and glass fiber fabrics have been also used as reinforcements.

The reinforced tapes are used for heavy-duty bundling, closing and reinforcing fiberboard cartons, palletizing, as a tear tape for cardboard packages, as an edging for photo albums and documents. Some reinforced tapes are printable, or the film is printed in reverse.

## Foam Tapes

Foam tapes are used for several different functions and therefore products of several types are on the market. Foam tapes are used for sealing and gasketing. Lower density foams are generally used for these applications. Foam tapes made from various foam/foil and foam/film laminates are used for thermal and acoustical insulation. An important application for two-sided, adhesive-coated foam tapes is mounting of various objects: reflectors for automobiles, trucks and trailers, side trim, mounting of balancing weights to automobile wheels, bulletin boards, telephone pen caddies, and bathroom fixtures to walls, medallions to wine and liquor bottles, and electrocardiogram terminals to patients and many other mounting applications. Foam tapes are suitable for such applications because the foam can conform to the uneven surfaces. Foam distributes the load and prevents stress concentration that causes failure in such applications.

Polyethylene, vinyl, urethane, and polychloroprene foams are used for tape applications. Closed cell radiation crosslinked polyethylene foams manufactured by Sekisui Chemical Company are becoming very important for tape construction. The foam is available in several formulations: cell size, bulk density and thickness. Polyurethane foams have been important for mounting applications. High-density foams with a well-adhered skin are useful for this purpose. Low-density polyurethane foams are used for sealing and insulation purposes and also for miscellaneous general-purpose uses where the low cost of such foams becomes an important advantage. Vinyl foams are useful for general-purpose applications. Proper adhesives must be chosen which are not affected by plasticizer migration. Polychloroprene foams are used because of their solvent resistance in automotive and other applications.

The properties of mounting foams are illustrated in Table 1-3 and of

**Table 1-3. Properties of pressure-sensitive foam mounting tapes.**

| FOAM | ADHESIVE | DENSITY (kg/m$^3$) | THICKNESS (mm) | TENSILE STRENGTH (kg/cm) | ELONGATION (%) | PEEL ADHESION (gm/cm) | TEMPERATURE LIMITS (°C) |
|---|---|---|---|---|---|---|---|
| PVC [a] | Acrylic | 256 | 0.79 | 0.68 | 150 | 500 | −18–71 |
| PVC | Acrylic | 256 | 1.59 | 1.16 | 150 | 500 | −18–71 |
| PVC | Acrylic | 256 | 3.12 | 2.06 | 150 | 500 | −18–71 |
| PU [b] | Acrylic | 400 | 0.79 | 0.50 | 350 | 710 | −35–93 |
| PU | Acrylic | 400 | 1.59 | 1.34 | 350 | 710 | −35–93 |
| PU | Acrylic | 400 | 3.12 | 2.06 | 350 | 710 | −35–93 |
| PU | Rubber | 400 | 0.79 | 0.50 | 350 | 535 | −35–93 |
| PU | Rubber | 400 | 1.59 | 1.34 | 350 | 535 | −35–93 |
| PU | Rubber | 400 | 3.12 | 2.05 | 350 | 535 | −35–93 |
| PE [c] | Rubber | 96 | 0.70 | 0.89 | 250 | 557 | −29–82 |
| PE | Rubber | 96 | 1.59 | 1.25 | 250 | 557 | −29–82 |
| PE | Rubber | 96 | 3.12 | 2.32 | 250 | 557 | −29–82 |

[a] PVC. Polyvinyl chloride.
[b] PU. Polyurethane.
[c] PE. Polyethylene.

some foam sealants in Table 1-4. Foam tapes are sold faced with silicone release-coated paper.

## Two-faced Tapes

Two-faced tapes can be subdivided into two groups: thin tapes, usually coated polyester or polypropylene film (0.013 mm thick or heavier), or lightweight paper and heavy two-faced tapes, usually coated foam or fabric. Heavy tapes are used for adhesion to a rough surface where the bulk is

**Table 1-4. Properties of pressure-sensitive adhesive-coated foam sealants.**

| TYPE OF FOAM | DENSITY (kg/m$^3$) | MAXIMUM TEMPERATURE (°C) |
|---|---|---|
| PVC [a] closed cell | 256 | 82 |
| PVC closed cell | 176 | 82 |
| PVC closed cell | 96 | 71 |
| PU[b] open cell | 32 | 135 |
| PU[b] open cell | 32 | 121 |
| Polychloroprene closed cell | 192 | 71 |

[a] PVC: Polyvinyl chloride.
[b] PU: Polyurethane.

required for the conformability, or a heavier adhesive layer is required for distributing the force over a larger area. Many mounting applications require heavy two-faced tapes and the foam tapes are the most popular. Holding down carpets, and securing lithographic printing plates to the printing cylinder are two examples of applications of heavy two-faced tapes.

Lightweight tapes are used for applications such as splicing, lamination, business forms, disposable sandpaper pads, nameplates, and tags. Table 1-5 lists various carrier webs used for two-faced tapes, illustrating the variety of constructions that are needed.

Most two-faced tapes require the use of two-sided coated interliner. A two-faced tape has been produced which is wound on itself without an interliner. Synthetic adhesives used are sufficiently incompatible to prevent the diffusion of polymer from one side to the other, and the tape can be unwound quite easily.[12] Tapes with different adhesion levels on each side have been produced. A product with a permanent adhesive on one side and a removable one on the other is of interest for packaging applications.

## Transfer Tapes

Transfer tapes consist of an unsupported adhesive coating applied over a silicone release-coated paper. The release paper used must have a sufficient differential release in order to retain the adhesive on the side to which it was originally coated. The release level, however, must be sufficiently low to allow a good adhesive transfer.

**Table 1-5. Backings used for two-faced tapes.**

| BACKING | BACKING THICKNESS (mm) | TOTAL THICKNESS (mm) |
|---|---|---|
| Polyester film | 0.013 | 0.157 |
| Polyester film | — | 0.127 |
| Polyester film | 0.025 | 0.097 |
| Polyester film | 0.025 | 0.178 |
| Biaxially-oriented polypropylene film | 0.030 | 0.150 |
| Polyethylene film | 0.076 | 0.178 |
| Vinyl film | 0.076 | 0.216 |
| Vinyl carpet tape | — | 0.127 |
| Semiopaque tissue paper | 0.076 | 0.140 |
| Tissue paper | — | 0.114 |
| Repulpable tissue paper | — | 0.089 |
| Rope tissue | 0.076 | 0.140 |
| Rope tissue | 0.051 | 0.147 |
| Cloth | — | 0.381 |
| Cloth, flame retardant | — | 0.330 |

Transfer tapes, like thin film two-faced tapes, are used for various lamination purposes and adhesion of name plates. They are especially useful in providing the adhesive in the exact location for later use. Removal of the release paper exposes the adhesive and makes it available for lamination.

Most of the transfer tapes consist of unsupported adhesive coating. The adhesive used is often acrylic, but rubber-resin adhesives are also widely used. There is a product on the market which has free fibers dispersed in the adhesive as a reinforcement. The fibers increase the cohesive strength, tensile strength and tear resistance of the adhesive coating[13] without sacrificing thinness or transparency. Silicone pressure-sensitive adhesive tape on silicone-coated release paper is also on the market. This product is used for its resistance to high temperature and for its good adhesion to low energy surfaces, such as unetched TFE and polypropylene.

Most of the transfer tapes have the adhesive coating up to the edge. Some have a free, uncoated edge that is helpful in facilitating the removal of the release paper. The free edge tapes are made by pattern coating and slitting in the uncoated area. Slitting of the transfer tapes presents some problems unique to these tapes. The adhesive may easily transfer to the equipment during handling. Also, the adhesive may be pushed by the knife over the edge of the release backing and adhere to the next adhesive layer. This causes legging during unwind and may cause lifting of the adhesive in the area of contact. The release paper and the slitting method should be such as not to produce free fibers at the slit edge. Fibers may adhere to the adhesive coating and also cause legging and lifting. Adhesive coatings which flow excessively are not suitable for transfer tapes where the adhesive extends to the edge. The release lining should be sufficiently thick to prevent contact between adjacent adhesive layers.

## Hospital and First Aid Tapes

Hospital tapes were the first application of pressure-sensitive products. This application has grown to a sizable business, estimated by Frost and Sullivan at $30 million.[1] In addition to this sizable volume, the same products are consumed at physicians' offices and at home for first aid uses. Similar products are also used for foot care and for athletic protection. These products are covered in a separate chapter by Satas.

## Office and Graphic Art Tapes

Clear cellophane film tape was the first product for this application. These applications have developed into a large business requiring a diverse assortment of tape and various coated film products. This is discussed in a separate chapter by DeBastiani.

## Packing and Surface Protection

The competition between the paper tapes and film-backed tapes is most visible in the area of packaging. Packaging applications are quite sensitive to price, and film tapes are replacing more expensive saturated paper backings. While the unplasticized vinyl films do not find a wide acceptance in the U.S. market, biaxially-oriented polypropylene film tapes are becoming the most frequently used tapes for packaging. Saturated paper is still maintaining its dominant position in the surface protective tape and sheet products. Packaging tapes are discussed in a separate chapter by Fukuzawa and Satas and saturated paper is discussed in another chapter by Bartell.

## Building Industry Products

Pressure-sensitive tapes do not find many applications in the construction industry, although the volume of adhesives and sealants consumed by this industry is very large.

Tapes are used for temporary applications such as paint masking. Two-sided coated foam tapes are used for temporary attachment of gypsum panels to interior surfaces until the permanent adhesives develop the full strength. Tapes are used to keep buildings sealed during cold weather by securing plastic film sheets. Similarly, dust is kept out by film partitions held by tapes. Expensive finishes and plastic or metal surfaces may be protected by pressure-sensitive coated sheets.

Permanent uses of pressure-sensitive tapes include employing foam tapes as sealants, sealing of floor to wall joints to eliminate draft, and bridging narrow cracks using a lightweight tape which is overpainted. Two-faced tapes of heavier construction are used to secure the jacketing over pipe insulation. Pressure-sensitive polyethylene coated fabric, aluminum foil, and other construction tapes are used for insulation of heating and air conditioning ducts. Water vapor transmission is important in many sealing applications. Table 1-6 shows the water vapor transmission rates of some of the tapes used in building construction. Polyvinyl fluoride film pressure-sensitive adhesive tapes have been used for sealing corrugated cement-asbestos siding and an experimental roofing system.[14]

Electrical work on the construction side requires large quantities of electricians' tape. Door frames and moldings may be manufactured from sheet metal, particle board, or low quality wood. These can be covered with woodgrain printed vinyl film with pressure-sensitive adhesive coating.

**Table 1-6. Water vapor transmission of some tapes used in building construction**

| TAPE | WATER VAPOR TRANSMISSION ($g/m^2$, 24 hr.) |
|---|---|
| Foil tapes | 0 |
| Polyethylene film | 2.3–3.9 |
| Polyethylene coated fabric | 11–14 |
| Polyester film | 19–23 |
| Vinyl film | 31–47 |
| Vinyl coated fabric | 23–47 |
| Cotton fabric | 31–93 |
| Glass fabric | 62–93 |

Very lightly metallized polyester film with a pressure-sensitive adhesive is used to control solar radiation by applying the film over window glass. Acrylic adhesives are used because of their clarity and stability in the presence of light. The product must be of a high quality optically and it requires careful coating. Various other coated and embossed films are used to restrict light passage through the windows.

Printed vinyl film with a pressure-sensitive adhesive is used for wall covering, shelf inserts, and similar decorative uses. The applications for wall covering are potentially very large. The pressure-sensitive products have not made a large penetration mainly because of the poor repositionability of the adhesive-coated wall coverings. Many ingenious ways have been suggested. One method involves placing glass beads on the surface of the adhesive, to prevent contact between the adhesive and the wall, until pressure is applied which either crushes or drives the glass beads into the adhesive coating. Incorporation of encapsulated activating agents has also been suggested, allowing the activation of the adhesive on application of pressure.

Adhesive-coated floor tiles are successfully used for a fast installation of tiles. Hot melt adhesives based on block copolymers and on amorphous polypropylene are used. A heavy adhesive coating is required and the hot melt application was especially suitable.

Applications of pressure-sensitive products for building construction have been discussed by Gerstel,[16] Wallace,[15] and Wilson.[14]

## Electrical Tapes

The electrical tapes can be subdivided into two categories: tapes intended for the original equipment manufacturers (OEMs) and electricians' tapes used for electrical insulation during installation. The electrical properties of such tapes and their testing is discussed by Braeunling in a separate chapter. The chemical purity, because of the required noncorrosiveness, and the resistance to deterioration at high temperatures are the important requirements of electrical tapes. Natural rubber and synthetic elastomer-resin pressure-sensitive adhesives are used for continuous performance up to 130°C. Acrylic adhesives are considered to be suitable up to 155°C and silicone resin-elastomer adhesives for continuous use up to 180°C.

The electricians' black friction tape is not a pressure-sensitive product; the adhesive is very soft and it splits on the unwind. This product was the prototype for the development of electricians' vinyl tapes and also for many other industrial tapes. The current electricians' tape has either a plasticized vinyl film backing, or a polyethylene film backing. These tapes are conformable and are required to meet nonflammability specifications.

A large variety of tapes is used for OEM applications. In addition to electrical insulation, the tapes are used to hold, reinforce, and protect the electrical wires, to provide a matrix for varnish impregnation, to identify wires in the electrical circuitry and many other uses. Some tapes are used for temporary purposes to help in manufacturing or in the assembly. For example, printed circuit board manufacturers use pressure-sensitive tapes to protect the terminals during manufacturing.

Many different backings are used for OEM tapes as shown in Table 1-7. Cloth tapes made from cotton, acetate taffeta, or glass fibers are used for electrical tapes. Cloth tapes offer more protection against physical damage because of their bulk. Their surface is rough and absorbent and better suited for varnish impregnation. Cloth tapes are heavier than film tapes and therefore not as suitable where space is limited.

In addition to vinyl film tapes, which are standard electricians' tapes, polyester film tapes are widely used for OEM applications. Films from 0.013 to 0.05 mm thick are used for tape backings. Some less common films such as polyimide or tetrafluoroethylene find uses for OEM applications. TFE tapes exhibit a high dielectric strength, and good arc resistance. TFE vaporizes under sustained arcing, but does not leave a carbonaceous path. Properties of TFE-coated glass cloth tapes with silicone pressure-sensitive adhesive are shown in Table 1-8.

Electrical coils are often encapsulated in resins for protection from the environment: moisture, dirt, mechanical abuse, and thermal shock. Film tapes do not form a good bond with the impregnating varnish (usually epoxy resin). The bond formation has been improved by using polyester web–polyester film laminates. Polyester webs alone are also used as a matrix for varnish.

Paper tapes are used for OEM applications. Nonwoven fabrics and

**Table 1-7. Properties of some electrical tapes.**

| BACKING | BACKING THICKNESS (mm) | TOTAL THICKNESS (mm) | TENSILE STRENGTH (kg/cm) | ELONGATION (%) | DIELECTRIC STRENGTH (volts) | INSULATION CLASS (°C) |
|---|---|---|---|---|---|---|
| Films | | | | | | |
| Polyester | 0.013 | 0.025 | 2.3 | 100 | 3800 | 130 |
| Polyester | 0.025 | 0.063 | 4.5 | 100 | 5000 | 130 |
| Polyester | 0.050 | 0.089 | 8.9 | 100 | 6500 | 130 |
| Vinyl | — | 0.178 | 3.6 | 200 | 9500 | 80 |
| Vinyl | — | 0.254 | 5.3 | 190 | 11,500 | 80 |
| Vinyl | — | 0.381 | 10.6 | 250 | 15,000 | 80 |
| Polyimide | 0.025 | 0.075 | 5.3 | 60 | 7500 | 180 |
| Polyimide | 0.050 | 0.100 | 9.8 | 80 | 11,500 | 180 |
| Polypropylene | 0.032 | 0.063 | 6.8 | 75 | 8000 | 85 |
| Polyvinyl fluoride | 0.050 | 0.089 | 2.5 | 45 | 6000 | 130 |
| Fluorohalocarbon | 0.050 | 0.089 | 2.3 | 50 | 8500 | 130 |
| Cloth | | | | | | |
| Cotton | — | 0.267 | 8.9 | 5 | 3000 | 105 |
| Acetate taffeta | — | 0.203 | 8.0 | 10 | 3000 | 105 |
| Glass (rubber adhesive) | — | 0.191 | 30.3 | 6 | 2500 | 130 |
| Glass (silicone adhesive) | — | 0.191 | 32.1 | 7 | 3500 | 180 |
| Paper | | | | | | |
| Kraft, creped | — | 0.254 | 3.6 | 10 | 2500 | 105 |
| Kraft, microcreped | — | 0.152 | 8.0 | 10 | 2500 | 105 |
| Nomex, flatback | — | 0.089 | 3.9 | 8 | 2800 | 155 |
| Nomex, creped | — | 0.381 | 5.3 | 50 | 2600 | 155 |
| Laminates | | | | | | |
| Aluminum foil, glass cloth | — | 0.165 | 17.8 | 5 | — | 180 |
| Polyester film, paper | — | 0.127 | 6.2 | 3 | 4500 | 105 |
| Polyester film, nonwoven | — | 0.152 | 7.1 | 35 | 5000 | 130 |
| Acetate film, acetate cloth | — | 0.203 | 8.9 | 10 | 5500 | 105 |

**Table 1-8. Properties of TFE-coated glass cloth tape with silicone adhesive.**

| BACKING THICKNESS (mm) | TOTAL THICKNESS (mm) | TENSILE STRENGTH (kg/cm) | ELONGATION (%) | DIELECTRIC STRENGTH (volts) | TEMPERATURE RANGE (°C) |
|---|---|---|---|---|---|
| 0.076 | 0.127 | 8.9 | < 5 | 4000 | −73 to 260 |
| 0.076 | 0.127 | 13.4 | < 5 | 3000 | −73 to 260 |
| 0.127 | 0.178 | 17.9 | < 5 | 4000 | −73 to 260 |
| 0.152 | 0.203 | 17.9 | < 5 | 5000 | −73 to 260 |
| 0.254 | 0.305 | 38.8 | < 5 | 6500 | −73 to 260 |

paper made from unusual fibers such as Nomex find uses for higher temperature applications.

Aluminum foil and foil-paper or foil-fabric laminates find uses as antistatic shields and radiation reflectors. Copper foil tapes are used as electrostatic shields for coils, microwave components, printed circuit boards, and flat conductor harnesses. Electrically conductive tapes, including conductive adhesives, are used for various shielding applications.

Flexible epoxy film tapes or coatings over glass cloth are used as insulating tapes for higher temperatures than vinyl tapes. They are good for use at 150°C instead of 105°C for plasticized vinyl. Shrinkable polyester film for encapsulation of small electrical coils, silicone rubber or silicone rubber-coated glass fiber fabric tapes are used for various electrical applications.

## Automotive Industry Products

A large volume of electrical tapes are used in the electrical system of a car. These products are similar to the OEM tapes discussed before. Other automotive pressure-sensitive products are discussed by Grimes in a separate chapter. A general discussion of such products has been published in *Chemical Week*.[17]

## Shoe Industry Tapes

In the shoe industry, the pressure-sensitive tapes are used to cover the backseam. The tape reinforces the seam and reduces pressure spots. Most of the tapes for this application are cotton fabric-backed, although some paper tapes and polyester and nylon fabric tapes are also used. Pressure-sensitive tapes are also employed as binding tapes and reinforcing tapes in various shoe construction applications.

### Appliance Industry Products

Decorative vinyl strips, especially woodgrain, backed with pressure-sensitive adhesive are widely used in the appliance industry. Nameplates are attached by transfer tapes or two-faced tapes. Foam gasketing with pressure-sensitive adhesive coating for easy installation is often used, and foam pads for sound insulation are often secured with pressure-sensitive adhesives.

### Splicing Tapes

Tapes are widely used to splice various webs during conversion or manufacturing operations. The paper industry requires the use of repulpable tapes, so that waste paper could be returned back to the beater and the adhesive would not gum up the equipment. Various water-dispersable polymers have been compounded into pressure-sensitive adhesives for this application. Two-faced tapes and transfer tapes are used for lap splices; film and paper tapes, for butt splicing. Splicing of silicone-coated release paper requires silicone adhesive tapes.

### Corrosion Protective Tapes

This is a very large application for pressure-sensitive polyethylene tapes. These products are discussed in detail by Harris in a separate chapter.

## LABELS

The pressure-sensitive label industry is a separate business different from the tape manufacturing in several aspects. In the case of labels the backing, its printability, flatness, ease of die cutting, and properties of the release paper are important properties often overshadowing the pressure-sensitive adhesive. The firms experienced in these areas have established themselves in the pressure-sensitive label business, rather than the tape manufacturers expanding into the label area.

A label stock manufacturer sells his product faced with a release paper either in rolls or in sheets for later printing and die cutting by a converter. The conversion operation can be a part of the same organization. The adhesive is usually applied over the release paper and dried. Then the label stock is laminated at the end of the line. Larger operations might also apply the silicone release coating in line.

Labels are discussed in detail by Komerska in a separate chapter.

## OTHER PRODUCTS

Pressure-sensitive adhesives are used in many applications which do not fit into the categories of tapes and labels. For example, such important applications as pressure-sensitive adhesive-coated floor tiles, wall covering, automotive woodgrained films, and various decorative sheets are outside of these categories. This area of miscellaneous products is growing as pressure-sensitive adhesives continue to find new applications.

## MANUFACTURERS

The pressure-sensitive tape and label manufacturing is dominated by several large companies. The largest tape manufacturers are:

- 3M Co.
- Johnson and Johnson, including Permacel
- Nashua Corporation
- Tuck Industries
- Mystik Tape
- Shuford Mills, Inc.
- Polyken Division, Kendall Co.
- Arno Adhesive Tapes, Inc.
- Anchor Continental

In addition, there is a large number of smaller tape manufacturers. Many smaller manufacturers specialize in some product line and they may be important suppliers to a particular segment of the industry. An additional 80 to 100 manufacturers are coating and processing pressure-sensitive tapes.

The label manufacturing is concentrated in several companies:

- Avery International
- Morgan Adhesives Co. (Bemis)
- Dennison Manufacturing Co.
- Fitchburg Coated Products Division, Litton Industries
- Coated Products, Inc. (subsidiary of Essex Chemical Corp.)

The label industry is subdivided into the coaters of label stock and the printers and converters. In both of these categories, there is a large number of smaller manufacturers that successfully compete for the label business. The tape, and especially label industry, is showing signs of dispersion. Many paper, vinyl film, and other substrate manufacturers are entering the pressure-sensitive coating market and many converters are attempting to expand vertically by manufacturing their own label stock.

## ADHESIVE COMPOUNDING

Traditionally, the pressure-sensitive adhesive manufacturers compounded their own adhesives and the demand for such adhesives was very small. The advent of acrylic pressure-sensitive adhesives has changed this situation. The in-house manufacturing of acrylic adhesives is done by larger tape and label manufacturers, but the technology has not been so easily available to the smaller coaters. A market for ready-to-coat pressure-sensitive adhesives has been created.

This ready availability of the adhesives has induced many coaters to produce pressure-sensitive products. It became unnecessary to acquire the know-how in the adhesive compounding, coating capability became sufficient to produce pressure-sensitive products, especially labels and other release paper-faced products. On the other hand, many larger adhesive users are tempted to decrease their raw material costs by vertical integration, and also to protect their products from competition by designing special properties into their proprietory adhesives.

## ADHESIVE RAW MATERIALS

Pressure-sensitive adhesives may be subdivided into two major classes: adhesives which are compounded to pressure-sensitive products by blending an elastomer with tackifying resins, plasticizers and few other ingredients and adhesives which consist of polymers that are inherently pressure-sensitive and require little or no compounding. Polyacrylates are the most important adhesives of this type. Their properties are described in a separate chapter by Satas. Polyvinyl ether adhesives can also be used without tackifying resins. These polymers have been discussed by Mueller in a separate chapter.

The most important elastomer for pressure-sensitive adhesives still is natural rubber. It is discussed in a separate chapter by Butler.

Random styrene-butadiene copolymers are also widely used for formulation of pressure-sensitive adhesives either as a single elastomer or in combination with natural rubber and other elastomers. Korpman[18] describes the use of SBR as a single elastomer. Low styrene content copolymers are used for pressure-sensitive adhesives. SBR is not compatible with polyterpene tackifying resins which are the preferred resins for natural rubber-based adhesives, but it is compatible with rosin type tackifiers. Lanolin improves the compatibility of SBR with polyterpene resins. Formulation of a curable adhesive with SBR as the sole elastomer is given below.

| | |
|---|---|
| SBR elastomer (SBR 4502 Goodrich Chemical Co.) | 100 |
| Tackifier (Picco L-30-3, Piccodiene 2050, or Picco N-10-5 by Hercules) | 150 |
| Heat-reactive phenol formaldehyde resin (Schenectady SP-1056) | 15 |
| Accelerator (zinc resinate) | 5 |
| Antioxidant (Agerite Superlite by Vanderbilt) | 2 |

The adhesive is dissolved in toluene.

SBR elastomers are often used with natural rubber. A formulation of such adhesive is shown below.[19]

| | |
|---|---|
| Pale crepe rubber | 50 |
| Styrene-butadiene copolymer (29/71, such as GRS 1022) | 50 |
| Zinc oxide | 85 |
| Polyterpene resin (m.p. 115°C) | 30 |
| Glycerol ester of hydrogenated rosin | 40 |
| Zinc resinate | 10 |
| Octyl phenol formaldehyde resin | 10 |
| 2,5-di-tert-amyl hydroquinone | 2 |

This adhesive is suitable for electrical tapes.

Block copolymers of styrene-butadiene and styrene-isoprene copolymers are especially important for pressure-sensitive hot melts. The polymers are covered in a separate chapter.

Butyl rubber and polyisobutylene, although less important, have found large application in pressure-sensitive corrosion protective tapes. These elastomers are covered in a separate chapter by Higgins, Jagisch, and Stucker. The corrosion protective tapes are covered by Harris.

Butadiene-acrylonitrile rubbers have been used for vinyl film tapes after tackifying with a resin.[20] These polymers are of interest in improving the solvent resistance of tapes. Adhesives for electrical applications based on butadiene-acrylonitrile copolymers have been discussed by Hendricks and Dahlquist.[21] A simple example of such adhesive is given below.

| | |
|---|---|
| Butadiene-acrylonitrile copolymer (70/30 such as Hycar OR-25 by Goodrich Chemical Co.) | 100 |
| Diphenyl cresyl phosphate plasticizer | 50 |
| Resin (pine wood resin, rosin substituted phenol resin, or other suitable tackifier) | 40 |

Polychloroprene latex has been proposed for pressure-sensitive applications.[22] Emulsified tackifying resins are added. Heat-reactive resin, such as liquid hexamethoxymethylmelamine (Cymel 301 by American Cyanamid Co.), which reacts with polychloroprene, may be added. Good antioxidant is needed.

Polyurethanes have been compounded to pressure-sensitive adhesives of good temperature and solvent resistance. Reaction products of polyols with aromatic polyisocyanates compounded with tackifying resins and plasticizers can be used for high-temperature resistant masking tapes.[23] Webber[24] describes a transparent low tack, low self-adhesion adhesive prepared by reacting a polyol with diisocyanate and using α-pinene resin as tackifyer. The tack level was controlled by the ratio of diisocyanate to polyol. The adhesive is suitable to manufacture two-faced tapes that can be wound on itself without an interliner. Polyols react with diisocyanates, forming an adhesive *in situ.*[25,26]

Polyvinylpyrrolidone and vinylpyrrolidone copolymers are suitable for preparation of water soluble or water dispersable pressure-sensitive adhesives[27] which are useful for splicing tapes used in paper mills.

Ethylene vinyl acetate copolymers of a high vinyl acetate content are inherently pressure-sensitive and useful for lamination applications where the pressure-sensitive properties help to adhere to difficult to bond surfaces such as polyolefin plastics. Ethylene vinyl acetate copolymers have been also compounded to pressure-sensitive hot melt adhesives. Amorphous polypropylene is also used for hot melts. These adhesives are briefly discussed in the chapter on hot melt application.

## REFERENCES


1. *Pressure-Sensitive Products and Adhesives,* Frost and Sullivan, Inc. Report No. 879, 1981.
2. *Water-borne Adhesives, The Rapid Growth & Maturing Global Markets, 1979–1989.* H. S. Holappa and Associates and Springborn Laboratories, Inc., April 1980.
3. *Professional Uses of Adhesive Tape.* Third edition. Johnson and Johnson, New Brunswick, N. J., 1972.
4. Bemmels, C. W. Pressure-Sensitive Tapes and Labels. *Handbook of Adhesives* (I. Skeist, ed.). Van Nostrand Reinhold Co., New York, 1977.
5. Shecut, W. H., and Day, H. H. U.S. Patent 3,965 (1845).
6. Levine, I. W. U.S. Patent 1,573,978 (1926).
7. Drew, R. G. U.S. Patent 1,760,820 (1930), Reissue 19,128 (1934) (assigned to Minnesota Mining and Manufacturing Co.).
8. Hendricks, O. J., and Dahlquist, C. A. "Pressure-Sensitive Adhesive Tapes." *Adhesion and Adhesives,* Vol. 2, pp. 387–408 (R. Houwink and G. Salomon, eds.). Amsterdam: Elsevier Publishing Co., 1967.
9. Drew, R. G. U.S. Patent 2,177,627 (1939).

10. Tierney, H. Canadian Patent 514,402 (1940) (assigned to Minnesota Mining and Manufacturing Co.).
11. Lambert, R. J. and Smith, A. G. U.S. Patent 3,205,088 (1965) (assigned to Minnesota Mining and Manufacturing Co.).
12. Kalleberg, M. O. U.S. Patent 2,889,038 (1959) (assigned to Minnesota Mining and Manufacturing Co.).
13. Kallenberg, M. O., and Turner, C. E. U.S. Patent 3,062,683 (1962) (assigned to Minnesota Mining and Manufacturing Co.).
14. Wilson, P. H., in *Adhesives in Building*. Building Research Inst. Natl. Academy of Sci.–National Res. Council. Publication No. 979. Washington, D.C. 1962.
15. Wallace, D. A. in *Adhesives in Building*. Building Research Inst. Natl. Academy of Sci.–National Res. Council. Publication No. 979. Washington, D.C. 1962.
16. Gerstel, M.H. *Adhesives and Sealants in Building*. Building Research Inst. Natl. Academy of Sci.–National Res. Council. Publication No. 577, Washington, D.C. 1958.
17. *Chemical Week*, October 16, 1974.
18. Korpman, R. U.S. Patent 3,535,153 (1970) (assigned to Johnson and Johnson).
19. Lavanchy, P. U.S. Patent 3,051,588 (1962) (assigned to Johnson and Johnson).
20. Waldman, H. L. U.S. Patent 3,374,134 (1968) (assigned to Johnson and Johnson).
21. Hendricks, J. D., and Dahlquist, C. A. U.S. Patent 2,601,016 (1952) (assigned to Minnesota Mining and Manufacturing Co.).
22. Fitch, J.C., and Snow, A. M. Jr. *Adhesives Age* **20**(10):23–26(1977).
23. Dahl, R. U.S. Patent 3,437,622 (1969) (assigned to Anchor Continental).
24. Webber, C. S. U.S. Patent 3,246,049 (1966) (assigned to Norton Co.).
25. Dahl, R. U.S. Patent 3,802,988 (1974) (assigned to Continental Tapes, Inc.).
26. Kest, D. O. U.S. Patent 3,879,248 (1975).
27. De Groot von Arx, E. U.S. Patent 3,096,202 (1963) (assigned to Johnson and Johnson).

# Chapter 2

# Pressure-Sensitive Adhesive Business in Japan and Asia

**Keiji Fukuzawa**
*Consulting Chemical Engineer*
*Tokyo, Japan*

## JAPAN

The pressure-sensitive industry is divided into three areas: tape, label stock, and surgical product manufacturing. The pressure-sensitive tape manufacturers are organized into the Japan Adhesive Tape Makers Association, which was established in 1961 and currently has 26 members as shown in Table 2-1.

The largest producers from the members of the Association are Nitto, Nichiban, Sekisui, and Teraoka. There are some 20 tape manufacturers who are not members of the Japan Adhesive Tape Makers Association. Oji Tape Co., Ltd., Bostik Japan Co., Ltd., and Hitachi Chemical Co., Ltd. are the largest tape manufacturers among the nonmember companies.
The Association reports statistical data on production of the member companies annually. These data are shown in Tables 2-2 and 2-3.

Table 2-3 shows the value of the manufactured tape in U. S. dollars converted at a rate of \$1 = 40 yen. These data show the effect of inflation.

The pressure-sensitive business shows a rapid increase until 1973. The recession caused a sharp decrease in production, and the recovery has not yet been complete.

The most important tape is a kraft paper tape used for various packaging applications. The demand for this tape amounts to 24 million $m^2$ per month and the largest manufacturer of this tape is Oji Tape Co., Ltd. It has

**Table 2-1. Membership of the Japan Adhesive Tape Makers Association.**

| COMPANY | MAIN PRODUCTS |
|---|---|
| Betto Industrial Co., Ltd. | Japanese paper tape, kraft tape. |
| Ebisu Kako Co., Ltd. | Protective sheet. |
| Fujikura Rubber Works, Ltd. | Rubber tape |
| Joto Seisakusho Co., Ltd. | Insulating tape. |
| Kamoi Kakoshi Co., Ltd. | Japanese paper tape. |
| Kikusui Tape Co., Ltd. | Kraft tape. |
| Kuramoto Sangyo Co. | Label stock. |
| Kyowa Limited | Cellophane tape, vinyl tape. |
| Modern Plastic Industries Co., Ltd | Label stock, double-faced tape. |
| Nichiban Co., Ltd. | Stationary tape, surgical tape, industrial tapes. |
| Nissei Co., Ltd | Cellophane tape. |
| Nitto Electric Industrial Co., Ltd. | Various industrial tapes, surgical tape. |
| Okamoto Riken Gomu Co., Ltd. | Cloth tape. |
| Rinrei Tape Co., Ltd. | Japanese paper tape, cloth tape. |
| Sayama Kako Tack Paper Co., Ltd. | Label stock. |
| Sekisui Chemical Co., Ltd. | Cellophane tape, kraft tape, polypropylene tape. |
| Semedaine Co., Ltd. | Polyester film tape. |
| Sohken Paper Conversion Co., Ltd. | Label stock, kraft tape. |
| Sony Chemicals Corp. | Double-faced tape, label stock. |
| Soshin Industry Co., Ltd. | Protective sheet. |
| Sugawara Industrial Co., Ltd. | Cloth tape, filament reinforced tape. |
| Sumitomo 3M Limited | Various industrial tapes. |
| Teraoka Seisakusho Co., Ltd. | Various industrial tapes. |
| Toyo Chemical Co., Ltd. | Vinyl tape. |
| Uni Industrial Corp. | Cellophane tape. |
| Union Kogyo Inc. | Kraft tape, Japanese paper tape. |

about 25% of the market share. As noted before, Oji is not a member of the Japan Adhesive Tape Makers Association, and its production is not included in Tables 2-2 and 2-3. Of the other nonmembers, Bostik Japan Ltd. is

**Table 2-2. Production of various tapes by the member companies of the Japan Adhesive Tape Makers Association.[a]**

| Product | PRODUCTION LEVEL (MILLION $M^2$) | | | | | | | | | |
|---|---|---|---|---|---|---|---|---|---|---|
| | 1969 | 1970 | 1971 | 1972 | 1973 | 1974 | 1975 | 1976 | 1977 | 1978 |
| Paper tapes | 138 | 169 | 198 | 242 | 303 | 238 | 205 | 252 | 235 | 241 |
| Cloth tapes | 49 | 55 | 54 | 58 | 58 | 34 | 39 | 44 | 42 | 47 |
| Film tapes | 115 | 130 | 141 | 151 | 193 | 132 | 149 | 169 | 163 | 173 |
| Specialty tapes | 4.0 | 5.7 | 7.0 | 10.1 | 13.8 | 9.5 | 12.0 | 15.4 | 15.9 | 18.9 |
| Label stock | 43 | 51 | 60 | 70 | 90 | 68 | 74 | 92 | 85 | 76 |

[a] Data of Japan Adhesive Tape Makers Association.

important in double-faced tape production and Hitachi Chemical Co., Ltd. manufactures electrical and protective tapes.

Table 2-4 shows the estimated total production of tapes in Japan.

Pressure-sensitive label stock is manufactured by these paper companies: Oji Paper Co., Ltd., Kanzaki Paper Mfg. Co., Ltd., Sohken Paper Conversion Co., Ltd. Other label stock manufacturers are: Fuji Shiko Co., Ltd., and Nichiban Co., Ltd. The production of label stock is shown in Table 2-5.

Accurate statistical data on hospital pressure-sensitive adhesive products are not available. It is estimated that the annual production of surgical adhesive tape is $10.4 million, finger bandages and other unit dressings $83.3 million, and plaster $50 million. A pressure-sensitive sheet intended to catch cockroaches was selling well at a level of $42 million annually. Some 20 pharmaceutical producers were engaged in manufacturing this product. The production of this product has lately been decreasing.

**Table 2-3. Value of tapes produced by the member companies of the Japan Adhesive Tape Makers Association.[a]**

| Product | VALUE OF PRODUCTION ($MM) | | | | | | | | | |
|---|---|---|---|---|---|---|---|---|---|---|
| | 1969 | 1970 | 1971 | 1972 | 1973 | 1974 | 1975 | 1976 | 1977 | 1978 |
| Paper tapes | 29.8 | 35.9 | 40.7 | 48.6 | 85.7 | 86.4 | 63.4 | 79.0 | 73.7 | 74.9 |
| Cloth tapes | 26.6 | 30.1 | 30.5 | 32.5 | 54.3 | 38.8 | 42.0 | 46.0 | 42.7 | 45.9 |
| Film tapes | 53.9 | 57.1 | 60.8 | 66.1 | 104 | 90.9 | 93.7 | 107 | 101 | 106 |
| Specialty tapes | 7.5 | 10.4 | 12.9 | 17.8 | 30.3 | 26.2 | 30.6 | 40.3 | 41.7 | 47.5 |
| Label stock | 16.9 | 19.6 | 24.4 | 29.4 | 45.4 | 46.4 | 47.8 | 58.8 | 55.2 | 48.4 |

[a] Data by Japan Adhesive Tape Makers Association.

**Table 2-4. Production of industrial tapes in Japan.**[a]

| PRODUCT | 1977 $MM | 1977 MILLION $m^2$ | 1980 $MM | 1980 MILLION $m^2$ |
|---|---|---|---|---|
| Paper tapes | | | | |
| Kraft paper | 72.3 | 273.0 | 77.5 | 299.0 |
| Japanese paper | 29.4 | 61.0 | 34.3 | 71.0 |
| Crepe paper | 3.2 | 1.9 | 4.5 | 2.5 |
| Other | 1.4 | 1.5 | 1.1 | 1.4 |
| Cloth tapes | | | | |
| Rayon fabric | 45.7 | 45.1 | 45.4 | 46.6 |
| Cotton fabric | 1.1 | 3.8 | 1.2 | 4.4 |
| Other | 0.07 | 0.05 | 0.08 | 0.05 |
| Film tapes | | | | |
| Cellophane | 38.6 | 91.2 | 35.6 | 84.2 |
| Vinyl | 40.9 | 55.7 | 41.6 | 60.9 |
| Polypropylene | 13.8 | 28.8 | 13.2 | 30.9 |
| Polyester | 11.9 | 9.3 | 13.0 | 10.5 |
| Acetate | 3.2 | 1.7 | 5.7 | 3.3 |
| Teflon | 1.9 | 0.07 | 2.8 | 0.09 |
| Other films | 0.4 | 0.07 | 0.25 | 0.14 |
| Foam tapes | 2.3 | 1.0 | 3.0 | 1.4 |
| Metal foil tapes | 1.4 | 0.5 | 2.9 | 0.9 |
| Rubber tapes | 0.3 | 0.08 | 0.6 | 0.1 |
| Filament reinforced tapes | 2.0 | 1.0 | 2.5 | 1.2 |
| Double-faced tapes | | | | |
| Paper and nonwoven | 27.8 | 11.5 | 31.2 | 13.3 |
| Fabric | 6.8 | 2.7 | 6.5 | 2.6 |
| Plastic | 1.7 | 0.4 | 5.1 | 1.2 |
| Foam | 4.3 | 0.4 | 8.6 | 0.8 |
| Total | 310.47 | 587.37 | 336.33 | 632.48 |

[a] Data by Sogo-hosoushupan Co., Ltd.

The current trend is towards solvent-free systems of which hot melts are of major interest. Most of the labels are made using aqueous acrylic adhesives.

## KOREA

There are about 30 pressure-sensitive adhesive product manufacturers in Korea. The major companies are shown in Table 2-6.

**Table 2-5. Production of pressure-sensitive label stock and sheeting.[a]**

| PRODUCT | PRODUCTION LEVEL $MM | MILLION $m^2$ |
|---|---|---|
| With release sheet | | |
| Labels, stickers | 152 | 210 |
| Decorative labels | 8.5 | 4.7 |
| Reflective labels and signs | 18.7 | 1.2 |
| Without release sheet (mainly protective sheeting) | | |
| Kraft and other paper | 1.4 | 5.6 |
| Film | 43.3 | 54.7 |
| Total | 223.9 | 276.2 |

[a] 1976 data by Sogo-hosoushupan Co., Ltd.

The largest tape manufacturer in Korea is Daeil Chemical Co., Ltd. The main packaging tapes are made from oriented polypropylene film and cellophane film backings. Tapes are very widely distributed in Korea. At many street stalls there, tape is cut to size for the customer.

**Table 2-6. Pressure-sensitive product manufacturers in Korea.**

| COMPANY | MAIN PRODUCTS |
|---|---|
| Daeil Chemical Co., Ltd. | Industrial tapes, surgical products, gummed tape. |
| STC Chemicals, Inc. | Cellophane tape, oriented polypropylene film tape. |
| Hae Sung Tape Ind. Co., Ltd. | Cellophane tape, oriented polypropylene film tape, polyester tape. |
| Hyup Sung Tape Industrial Co., Ltd. | Oriented polypropylene tape, polyester tape, masking tape. |
| Je Il Tape Industrial Co., Ltd. | Kraft paper tape, oriented polypropylene tape. |
| Hyundai Chemical Ind. Co., Ltd. | Oriented polypropylene tape. |
| Lucky Tape Co. | Vinyl tape. |
| Shin Poong Works Co., Ltd. | Labels cellophane tape. |
| Nan Mi Industrial Co. | Labels. |
| Sea Lim Label Co. | Labels. |
| Sam Hyup Label Co. | Labels. |

## TAIWAN

Some 30 companies are manufacturing pressure-sensitive tapes in Taiwan. The major manufacturers are listed below.

- Four Pillars Enterprise Co., Ltd.
- Asia Chemical Corp.
- Globe Industries Corp.
- Tung Tai Paper Corp.
- Chi Ming Industrial Co., Ltd.
- Polychemicals Co., Ltd.
- Chung Jih Plastics Co., Ltd.
- K. K. Converter Co., Ltd.
- Fu Ta Physical and Chemical Co., Ltd.
- 3M Taiwan Ltd.
- Teh Fong Industrial Co., Ltd.
- Chung I Tape Co., Ltd.
- Jih Ta Physical and Chemical Industrial Co., Ltd.

The Adhesive Manufacturers' Association (Taiwan) has been recently organized. Four Pillars Enterprises Co., Limited is the largest adhesive tape manufacturer producing various industrial tapes. Asia Chemical Corporation is manufacturing oriented polypropylene film tape with an acrylic adhesive. Globe Industries Corporation is making an embossed soft poly (vinyl chloride) film tape for packaging applications. This tape can be torn easily and cleanly by hand. Taiwan excels in production of plasticized vinyl film tapes and exports these products all over the world, including the United States. Three vinyl tape manufacturers had their products approved by Underwriters' Laboratories, Inc. Taiwan Plastics Corporation is manufacturing the resin and Nan Ya Plastics Corporation is making the film for the tape companies.

## CHINA

There are some 30 tape manufacturers in China. They are manufacturing the following products: cellophane tape, kraft paper tape, polyester film tape, double-faced tape, plasticized vinyl film tape, gummed tape, and surgical products. Raw materials made locally are used only. Rosin and polyterpene tackifying resins are the only ones available. The industry needs much modernization in this country.

## SOUTHEAST ASIA

There are four tape manufacturers in the Phillipines: Hope Adhesive Paper Products, Inc. (HAPPI), which makes masking tape, kraft paper tape, double-faced tape, and labels; Globe Tape Inc., which manufactures masking tape, oriented polypropylene film tape and cellophane tape, 3M Phillipines Inc., which makes masking tape, cellophane tape, and reflective products, and Sibesco Mfg. Co., which makes label stock and vinyl tape.

Loytape Berhad in Malaysia is the largest pressure-sensitive tape manufacturer in Southeast Asia. The company has established the well-known trade name "Loytape." Loytape Berhad produces gummed tape, cellophane tape, rigid and plasticized vinyl film tapes, and masking tape. Their products not only dominate the Malaysian market but also have substantial market shares in Singapore, Thailand, Australia, Indonesia, and other countries in the region.

Central Industrial Corporation Sendirian Berhad (ICI) produces a wide range of label stock and tapes. Loytape Berhad and Central Industrial Corporation both belong to the Central Securities Group. Further growth of the pressure-sensitive tape manufacturing is expected in this area.

# Chapter 3

# Tack

**Fred H. Hammond, Jr.**
*Kendall Co.*
*Lexington, Massachusetts*

Pressure-sensitive tack is the adhesive property related to bond formation. It is the property which enables the adhesive to form a bond with the surface of another material upon brief contact under light pressure. Pressure-sensitive tack is differentiated from the tire makers' tack of elastomer-to-elastomer bonding, and the printers' tack of resistance-to-separation by a viscous fluid between two surfaces. Implicit in a definition of tack is the assumption that the adhesive separates from the adherend cleanly, that is, without leaving visible residue. The ASTM definition[1] of pressure-sensitive tack requires the bond established to be of measurable strength. In this chapter, the word tack will mean pressure-sensitive tack.

Each individual's impression of tack is the sensation experienced in bringing the thumb or finger lightly into contact for a short time with a pressure-sensitive adhesive and then quickly withdrawing it. By varying the pressure and time of contact and noting the difficulty of pulling the thumb from the adhesive, one perceives how easily, quickly, and strongly the adhesive can form a bond. As a test for tack, the thumb test has the faults of being subjective, nonquantifiable, uncertain in differentiation among adhesives except for tests made at the same time, and indicative only of the adhesive/skin bonding which may or may not relate to adhesive/other-adherend bonding. Even so, the thumb test made by a trained and experienced person is often as good an indication of tack as any mechanical test.

Tests for pressure-sensitive tack can be classified into three groups:

1. Rolling ball,[9,16,37,42] rolling cylinder[47] and rotating[8] drum tests in which distance traveled, energy lost, or force observed in the interaction of these objects with an adhesive (tape) is measured.

2. 90 or 180 degree peel tests[9,10,37,48] involving low external contact pressure and short application time, in which the peel force is taken as the tack value.
3. Probe tests[2,3,17,20,32,49] in which the butt tensile strength of the bond formed between the tip of a probe and the adhesive after a short time at low pressures is taken as the tack level.

## ROLLING BALL TACK TESTS

The rolling ball tack test is the oldest and most widely used mechanical test. It has been used for at least 50 years. Early literature references[9,16] cite the Douglas Aircraft test. Rolling ball tack is the Pressure-Sensitive Tape Council's Test Method No. 6.[37]

In the most common form of the rolling ball tack test,[37] a stainless steel ball 1.1 cm (7/16 in.) diameter is released at an elevation on an inclined track so as to roll down and at the bottom come into contact with the horizontal, upward-facing adhesive. The distance the ball travels out along the tape is measured as tack. This roll out distance measure of tack gives an inverse scale of tack: the greater the distance, the less tacky is the adhesive. One variation of the rolling ball test is to place a section of adhesive (tape) on the incline some distance below the ball's release point and then find the diameter of the ball that just fails to traverse the adhesive.

Attempts have been made to create scaling of rolling ball tack values which would group combinations of release heights and roll out distances for equivalent tack levels. Table 3-1 was created 40 or more years ago at Bauer and Black, then part of the Kendall Company. Such scaling tables have limited utility because of the high frequency of exceptions to them that are observed.

The rolling ball tack test was developed when most pressure-sensitive adhesives were compounded from natural rubber and tackifying resins. With such adhesives, there is an approximate relation between rolling ball and thumb tack. For adhesives compounded from synthetic elastomers or single component adhesives, however, the correspondence of rolling ball to thumb tack is less than good. Many functional adhesives synthesized from acrylates and other monomers do not stop the ball in the maximum prescribed length of tape and thus have no apparent rolling ball tack.

The ready availability of balls of different size, composition, surface texture, density, weight, moment of inertia, etc., invites the investigation of the effects of these parameters on apparent rolling ball tack.[9,16] Such investigations probably have been made in several laboratories but results have not been published. Gillespie[18] made an extensive study using stroboscopic photography to determine the time/position of the ball and thus its velocity

**Table 3-1. Tack level.**

| ROLL OUT, INCHES | RELEASE HEIGHTS, INCHES | | | | | |
|---|---|---|---|---|---|---|
| | 1 | 2 | 3 | 4 | 5 | 6 |
| 1/4 | — | — | — | — | — | 7.0 |
| 1/2 | — | — | — | — | — | 6.9 |
| 3/4 | — | — | — | — | — | 6.3 |
| 1 | 2.0 | 3.5 | 4.4 | 5.0 | 5.4 | 5.9 |
| 1 1/4 | 1.8 | 3.3 | 4.2 | 4.8 | 5.2 | 5.7 |
| 1 1/2 | 1.5 | 3.0 | 3.9 | 4.5 | 5.0 | 5.4 |
| 1 3/4 | 1.3 | 2.8 | 3.7 | 4.3 | 4.7 | 5.2 |
| 2 | 1.0 | 2.5 | 3.4 | 4.0 | 4.5 | 4.9 |
| 3 | 0.8 | 2.0 | 2.8 | 3.2 | 3.9 | 4.3 |
| 4 | 0.5 | 1.6 | 2.5 | 3.0 | 3.4 | 3.9 |
| 5 | — | 1.2 | 2.1 | 2.5 | 3.1 | 3.5 |
| 6 | — | 1.0 | 1.8 | 2.3 | 2.9 | 3.2 |
| 7 | — | 0.9 | 1.5 | 2.1 | 2.6 | 3.1 |
| 8 | — | 0.8 | 1.4 | 2.0 | 2.4 | 2.9 |
| 9 | — | 0.7 | 1.2 | 1.8 | 2.3 | 2.6 |
| 10 | — | 0.6 | 1.1 | 1.6 | 2.1 | 2.5 |
| 11 | — | 0.5 | 1.0 | 1.5 | 2.0 | 2.4 |
| 12 | — | 0.5 | 1.0 | 1.3 | 1.8 | 2.3 |
| 13 | — | — | 0.9 | 1.2 | 1.7 | 2.2 |
| 14 | — | — | 0.8 | 1.1 | 1.6 | 2.1 |
| 15 | — | — | 0.8 | 1.0 | 1.5 | 2.0 |
| 16 | — | — | 0.7 | 1.0 | 1.4 | 1.9 |
| 17 | — | — | 0.7 | 0.9 | 1.3 | 1.8 |
| 18 | — | — | 0.6 | 0.8 | 1.3 | 1.7 |
| 19 | — | — | 0.6 | 0.7 | 1.2 | 1.6 |

and energy. Table 3-2 presents his analysis of the mechanics of the rolling ball test. Gillespie found instances where balls slowed down as if they were acted on by a constant force while on the adhesive and lost energy in proportion to distance traveled. In these instances, the ball rolls out a distance proportional to the first power of the release height. He found instances in which balls slowed down as if energy was absorbed from them in proportion to the time of contact with the tape in which balls roll out distances proportional to the 3/2 power of the release height. Investigating filled compounded commercial adhesives, Gillespie found more instances where energy was lost in proportion to time of contact than in proportion to distance traveled. He found that heavy, dense balls tended to lose energy in proportion to time and light Ping-Pong balls in proportion to distance. He found some adhesives which changed from one mode to the other as the release height was changed. Polytetrafluoroethylene balls tended to lose energy with time in the beginning portion of a roll out and with distance in the final portion. He was unable to summarize the data by a unified theory.

**Table 3-2. Rolling ball tack test mechanics analysis.**

| | | |
|---|---|---|
| Prior to release: ball has potential energy (P.E.) = gmh = gmdcos$\theta$ | | |
| On incline: ball's P.E. is converted to kinetic energy (K.E.) = $mv^2 + Iv^2/2r^2$ | | |
| | for solid balls $I = 2mr^{2/5}$ | |
| | for thin shell balls $I = 2mr^{2/3}$ | |
| At bottom of incline, ball has: | | |
| | velocity $(V) = (gh/0.7)^{1/2}$ | |
| | kinetic energy (K.E.) = $0.7mv^2$ | |
| While rolling on tape: | | |
| | If ball loses energy proportional to distance | If ball loses energy proportional to time |
| Deceleration: | $-dv/dt = a$ | $-dV/dt = b/v$ |
| Velocity: | $v = Vt - at$ | |
| | $v^2 = V^2 - 2ax$ | $v^2 = V^2 - 2bt$ |
| Distance at time $t$: | $x = Vt + at^2/2$ | $x = v^2(1 + (2bt/v^2 - 1)^{3/2})/3b$ |
| Time to stop: | $T = V/a$ | $T = V^2/2b$ |
| Roll out distance | $X = v^2/2a$ | $X = (gh)^{3/2}/1.76b$ |
| | $X = gh/1.4a$ | |
| where | | |
| | $g$ = acceleration due to gravity | $I$ = moment of inertia |
| | $m$ = mass of ball | $V$ = velocity of ball at bottom |
| | $h$ = height of release | $T$ = time to stop |
| | $d$ = position on incline | $X$ = roll out distance |
| | $\theta$ = angle of incline | |
| | $v$ = translational velocity | |
| | $a$ = deceleration constant | |
| | $b$ = deceleration constant | |
| | $x$ = distance along tape | |

Gillespie found roll out distances and the area of contact between the ball and adhesive to vary with how well the tape was held down. He constructed a vacuum hold-down system which gave more consistent results than the common practice of hand tensioning of the backing.

The ultimate simplicity, low cost, ease of execution, and generally good reproducibility of the rolling ball tack test cause it to be used in adhesive tape quality control. Its short (0.001–0.01 sec), indeterminate contact time, dependence on degree tape hold-down, lack of applicability to all adhesives, inverse scaling, and emphasis on adhesive softness limits its use in theoretical studies or as a will-it-function-in-application test.

## PEEL TESTS FOR TACK

Peel tests for tack are logical extensions of standard adhesion peel tests for bond strength. They are adhesion tests made at short contact times and low contact pressures before the bond is completely formed. The usual concept is to consider the force to be indicative of completeness of bond formation and for the peel force to be taken directly as the tack measure. A less common concept of tack measure is to use the ratio of peel force after a short contact time at low pressure to peel force at standard conditions (longer time and higher pressure).

In the Pressure-Sensitive Tape Council's "Quick-Stick" test,[37] the tape is laid on the test surface with no contact pressure beyond that of its own weight. The tape is then pulled from the surface at 90 degrees at 30 cm/min (12 in./min) with the peel force required to break the bond taken as the tack value. In the 1950's, a similar procedure referred to as the Kreck test using a loop of adhesive tape was proposed to the ASTM but was not made a standard. Apparatus for carrying out peel tack tests are common constant rate of extension tensile testers and simple test fixtures.

As long as end-use application conditions of a tape are closely similar to those in the test, tape peel tests offer a fairly good method of comparing tapes. The more application conditions differ from the situation in which the tape is merely laid down on a substrate and removed right away, the less useful they are. Peel tack is highly influenced by the backing on which the adhesive is coated. In peel tests, there is a force couple consisting of the measured peel force acting on one side of the peel line and a resultant force acting on the opposite side, pressing the adhesive against the adherend surface. Plastic film, stiff paper, or metal foil tape backings transfer force across the line of peel into a compressive contact force, causing even an initially low peel force to cascade rapidly into a higher force, and consequently a higher tack value. On the other hand, cloth or soft paper backings cannot effectively transfer peel forces to compressive forces, so the peel force and tack value remain at low levels.

## PROBE TACK TESTS

Probe tack tests are mechanical simulations of thumb or finger tack tests. In them, the tip of a probe is brought into contact with a supported adhesive under low contact pressures for a short time and then pulled away at a fixed rate, during which the peak force of separation is measured. Wetzel[49,50] pioneered the procedure using an Instron Tensile Tester and special fixtures. Other probe tack designs[5,7,32,36] are reported in the literature. Hammond[20] developed the self-contained instrument in which the individual factors involved in bond formation can be varied one at a time. It became the commercially available Polyken Probe Tack Tester.

Probe tack tests can examine separately each of the factors involved in adhesive bond formation. The first requirement of bond formation is that the adhesive and the adherend be brought into intimate contact. Seldom are both adhesive and adherend ideally smooth planar surfaces. Adhesives on tape from a roll have the surface texture of the backing with which it has been in contact. Adherends have a very wide range of surface textures from smooth glass to rough oxidized metal. In bond formation, it is invariably the adhesive which must physically conform to the irregularities of the adherend.

Bond formation and probe tack increase with increased contact pressure. As shown in Fig. 3-1, probe tack of smooth surface adhesives is high even at low contact pressures and rises quickly to an asymptotic value at high pressures while probe tack of rough surface adhesives is low at low contact pressures and rises slowly as pressure is increased.

Bond formation and probe tack values increase with increased contact time. As shown in Fig. 3-2, probe tack of smooth surface adhesives is high even at short contact times, while probe tack of rough surface adhesives is low at short contact times and increases gradually as contact time is increased.

While bond formation increases with increasing temperatures, probe tack values most often does not. As shown in Fig. 3-3, probe tack of smooth

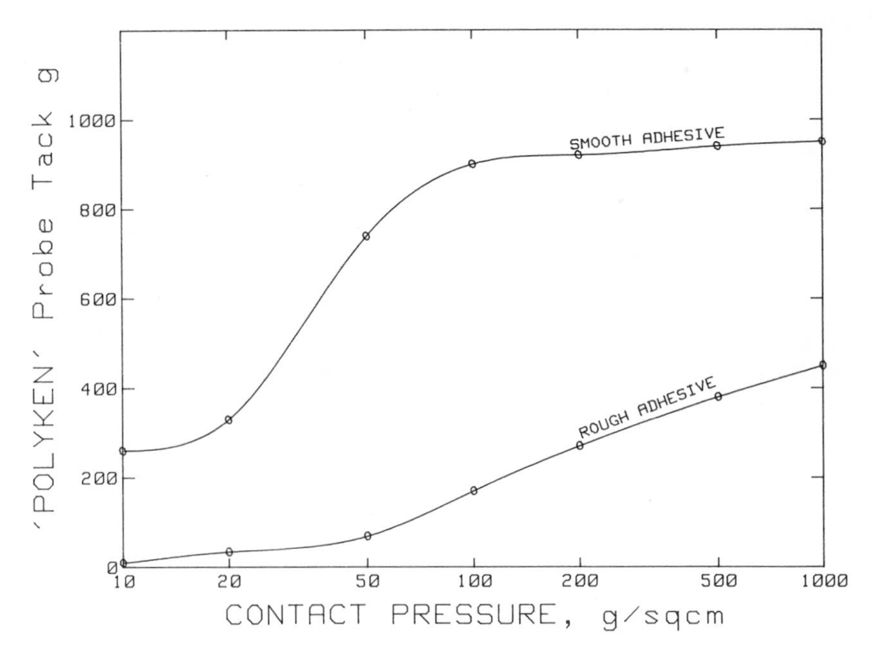


Fig. 3-1. Probe tack of smooth (film backed) and rough (cloth backed) pressure-sensitive adhesive on a standard 1/2 cm diameter stainless-steel probe, 1 second contact time, 1 cm/sec separation rate at room temperature.

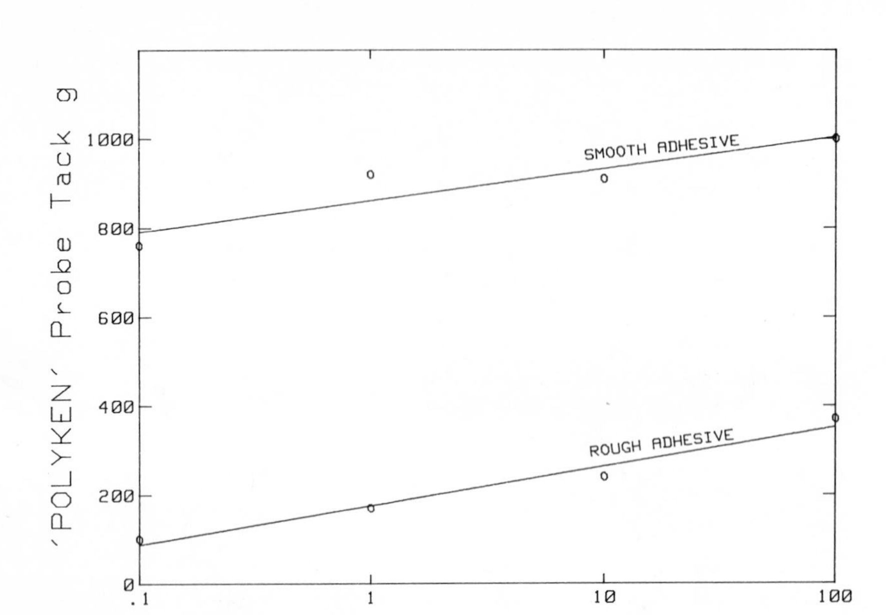


Fig. 3-2. Probe tack of smooth (film backed) and rough (cloth backed) pressure-sensitive adhesive on a standard 1/2 cm diameter stainless-steel probe, 100 g/cm$^2$ contact pressure, 1 cm/sec separation rate at room temperature.

surface adhesives decreases with increasing temperature. This decrease in probe tack is the consequence of the rate at which bond strength decreases with temperature being greater than the rate at which the nearly complete contact increases. Probe tack of the rough surface adhesive remains almost constant over this range of temperature because the rate of increase in bond formation is approximately balanced by the decrease in bond strength.

Bond strength and probe tack depend on the chemical composition of the adherend.[3,20,42] Probe tack values for three commercial adhesive tapes as measured on stainless steel and various plastic probes using the Polyken instrument are shown in Fig. 3-4. The three tapes decrease in tack with decreasing surface energy of the probe. It is also seen that the tack values and relative ranking of the adhesives are different on different composition probes.

In most instances, apparent bond strength and probe tack are greater when the rate at which the bond is broken is greater. Probe tack values typically increase with the rate of separation of the probe from the adhesive as illustrated in Fig. 3-5.

No single tack value will fully characterize an adhesive because of the differences in effect from one adhesive to another, on bonds formed under

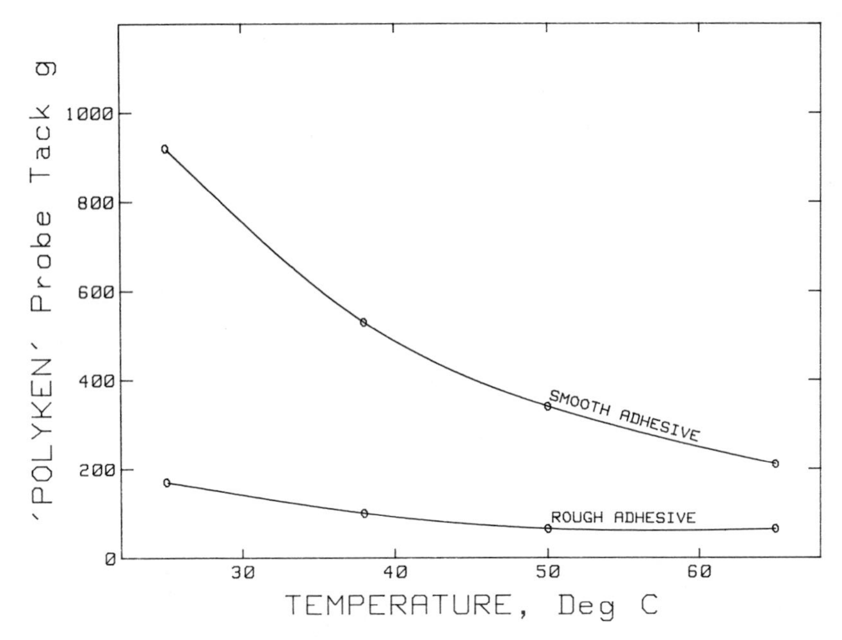


Fig. 3-3. Probe tack of smooth (film backed) and rough (cloth backed) pressure-sensitive adhesive on a standard 1/2 cm diameter stainless-steel probe, 100 g/cm$^2$ contact pressure, 1 second contact time, 1 cm/sec separation rate.

different pressures, times, adherend surface roughnesses and compositions, etc. Nor is there only one probe tack value, since the several factors involved in formation can be varied to approximate those in a particular adhesive application. There is, however, advantage in establishing test conditions that are generally acceptable for use in comparing adhesives. The ASTM D14 "Method of Test for Tack Using an Inverted Probe Machine"[2] prescribes a stainless steel probe with a surface roughness of 2 microin. rms, a contact pressure of 100 gm/cm$^2$, a contact time of 1 sec, and a rate of separation of 1 cm/sec as standard test conditions.

Irrespective of how pressure-sensitive adhesive tack is defined or measured, it is ultimately the outward manifestation of the surface chemical and bulk physical properties of the adhesive. With a substantial degree of understanding, rolling ball, peel, and especially probe tack can be interpreted in terms of surface energy, contact angle, wetting, modulus, loss tangent, etc.

## SURFACE CHEMISTRY AND TACK

Pressure-sensitive tack, and bond formation in general, ultimately involves molecular interactions at the adhesive/adherend interface. The major driv-

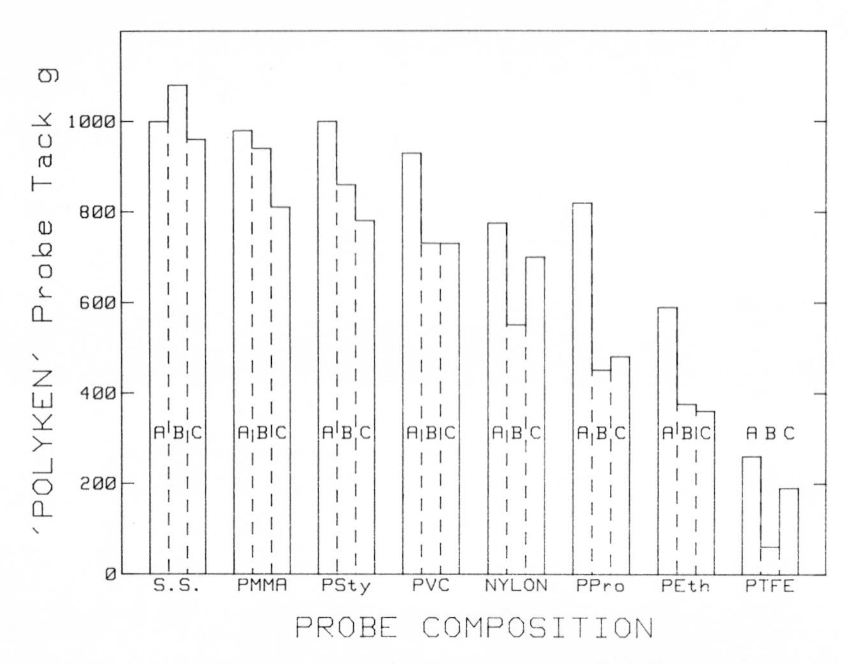


Fig. 3-4. Probe tack of three commercial adhesives tapes—A, B, and C—on 1/2 cm diameter probe, 100 g/cm$^2$ contact pressure, 1 second contact time at room temperature.

ing force towards bond formation is the surface chemist's "work of adhesion," which can be related to surface energies of the adhesive and adherend. Countering this "work of adhesion" and preventing immediate and complete bond formation are geometric micro and macroscopic surface roughnesses and the resistance to deformation (rheological) properties of the adhesive and adherend. Thanks to the contribution of a number of persons, there is now a substantial theoretical understanding of tack.[6,11,12,14,24,28,33,42,44,49,51]

To have good pressure-sensitive tack, an adhesive should have low surface energy. Zisman's[51] concept of critical surface tension for a solid provides a basis for estimating the surface energy of an adhesive and also understanding why low surface energy adhesives are desirable. Small drops of liquids placed on solid surfaces make characteristic angles of contact with the solid that depend on the liquid's surface tension. Plots of observed contact angle versus liquid surface tension, especially in the region of low contact angles, are most frequently straight lines. Zisman suggested, and it has been borne out since, that the liquid surface tensions obtained by extrapolating these lines to zero contact angle are characteristic of the solid's surface. Gardon[19] showed how critical surface tensions may be re-

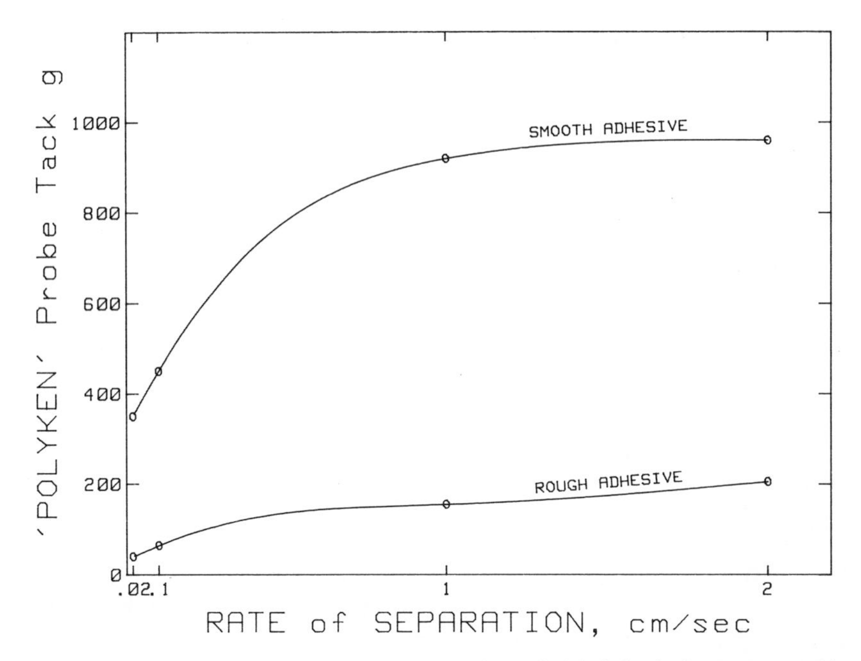


Fig. 3-5. Probe tack of smooth (film backed) and rough (cloth backed) pressure-sensitive adhesive on a standard 1/2 cm diameter stainless-steel probe, 100 g/cm$^2$ contact pressure, 1 second contact time at room temperature.

lated to the solid's chemical composition through Hildebrand's solubility parameter and cohesive energy density.

The critical surface energies of pressure-sensitive adhesives and adherends can be measured following the methods of Zisman and co-workers. Care must be used in selecting test liquids to avoid complications of solubility, specific chemical reactions, and time dependency of contact angle.[14] Often, the critical surface tension can be well enough estimated by reference to published tables[11,14,45] for low energy plastic surfaces, and then using the one for the composition most closely resembling the adhesive or adherend. Table 3-3 contains a brief listing of polymers and their critical tensions as found in the literature.

The extent of contact between adhesive and an adherend can depend on their relative surface energies.[20,43] A simple liquid with a surface tension less than the critical surface tension will make a zero contact angle and spontaneously wet and spread over a solid's surface. Adhesives with surface tensions less than the adherend's critical surface tension are also expected to wet and spread, but this requires some period of time because of the adhesive's viscosity. Sharpe and Schonhorn[38,39] first recognized the significance

**Table 3-3. Critical surface tensions of various polymers.**

| POLYMER | CRITICAL SURFACE TENSION (dyne/cm) |
|---|---|
| Polytetrafluoroethylene | 18.5 |
| Polypropylene | 29 |
| Polyethylene | 31 |
| Polystyrene | 33 |
| Polyvinylacetate | 37 |
| Polyvinyl chloride | 39 |
| Polyethylene terephthalate | 43 |
| Polyhexamethyleneadipamide | 46 |

to strong bond formation of the (fluid) adhesive having a lower surface energy than the adherend. Adhesives of hydrocarbon composition (therefore having low surface energy) generally exhibit good tack to the widest variety of adherends.

The surface chemistry that gives the greatest tack may not always have a critical surface tension less than the adherend but may be some higher value. Zisman pointed out the maximum "work of adhesion" between simple liquids and solids was for liquids with surface tensions greater than the solid's critical surface tension. This circumstance should be true of adhesives and adherends as well. Huntsberger[24,25,26,27] has shown that complete wetting of solid surfaces does not require zero contact angle but can occur for liquids making contact angles up to 90 degrees. Huntsberger also reasoned the greatest rate of wetting would be achieved by adhesives having surface tensions somewhat greater than the solid's critical surface tension. Toyama[43] and others have observed experimentally a maximum in probe tack when the surface energy of the adhesive was slightly greater than that by the adherend.

## RHEOLOGY AND TACK

While adhesive and adherend surface chemistries determine interaction energy at the molecular level, their physical properties also determine the rate and extent of contact (and bond strength) between them.[29] Pressure-sensitive adhesives differ from other materials primarily because of the unique rheological properties they possess. Rheological properties are the major factors in the phenomenon of pressure-sensitive tack.

For good pressure-sensitive tack and rapid bond formation, the adhesive should be easily deformed in a time span on the order of a fraction of a second. Dahlquist found the 1 sec compliance of a typical pressure-sensitive adhesive having good probe tack to be $10^{-6}$ $cm^2/dyne$. Toyama, Kraus,[32,33,34,35] and others found comparable values. Adhesives with good rolling ball tack need to be easily deformed in time spans of about 0.001 sec. Tackifier resins, plasticizers, solvents, and temperature all impart tack to elastomers by their increasing the ease of deformation of the adhesive in short time spans.

For good tack, the adhesive deformation in the bonding stage should be in large part viscous. Energy-absorbing fluidlike flow is desirable to relax the stress put on the adhesive as it is made to conform to the irregularities of the adherend. If bonding deformation were only elastic, the recoverable stored stress would assist external stress in the rupture of the bond. Pressure-sensitive adhesive tack is favored by a high ratio of viscous to elastic properties at time spans comparable to bonding times. In keeping with this, pressure-sensitive adhesives are characteristically very good adsorbers of mechanical energy at audio frequencies.

Since tack is measured as the force or energy to break, adhesives should have a high modulus at the strain rates and magnitudes imposed on them during rupturing of the bond. Tackifiers raise the glass transition temperature of elastomers so that the adhesive mixture has a high modulus at high strain rates and normal ambient temperatures. (The difference between a tackifier and a plasticizer can be that a tackifier increases the modulus at low temperature, short time, and high frequencies but decreases the modulus at high temperature, long times, and low frequencies, whereas a plasticizer decreases the modulus at all temperatures, times, and frequencies.) Elastomers that increase in modulus greatly at some particular elongation also make high tack adhesives. Natural rubber is the best example.

Because bond strength, and apparent tack, is greater when more of the adhesive is actively resisting bond rupture at any instant, an adhesive should have the ability to distribute stress throughout its volume. High tack is associated with adhesives that would have high elongations to break in simple tensile strength tests. Tacky adhesives form long "legs" during bond strength tests. High molecular weight and flexible backbone polymers provide this stress distribution property.

## COMPOUNDING AND TACK

Much of the art of compounding or synthesizing pressure-sensitive adhesives is centered on juggling physical properties to give a balance of tack, peel adhesion, and cohesive strength (shear creep resistance). It is not at all necessary to measure the rheological properties of the mixture or composition, even though such measurements can be extremely helpful. Empirical relations between tack and composition are readily established in simple "make it and test it" adhesive experiments. Probe tack tests are particularly well suited to empirical adhesive composition studies.

The probe tack of mixtures of an elastomer and a tackifying resin is simply related to composition. Wetzel[49] showed this in his original publication. Hammond,[20] Toyama[42] and others using the Polyken Probe Tack Tester have reported this to be generally true. The probe tack of rubber-resin mixtures as a function of percent resin concentration gives the smooth curves in Fig. 3-6. Hammond found that plotting probe tack as a function of resin content expressed as parts per hundred gave curves with two straight line sections as in Fig. 3-7. In the region of low resin concentration, tack is unchanged from that of the elastomer. Above a characteristic point tack increases with resin concentration; the rate of increase depends on the chemical type and melting point of the resin. At some critical concentration, also dependent on resin chemical type and melting point, tack values become erratic and abruptly drop to zero. Hock[22,23] observed that phase changes in the mixtures paralleled changes in tack. Marginal compatibility between elastomer and resin was postulated as the cause of tack.[30,31] However, fully compatible systems having tack have been observed. This postu-

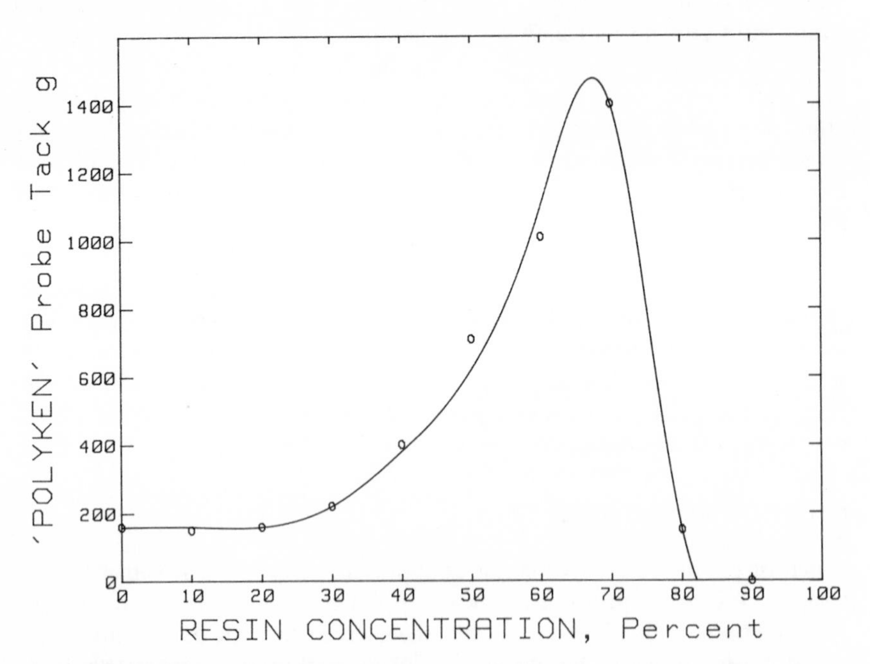


Fig. 3-6. Probe tack of natural rubber/Staybelite Ester No. 10 adhesives on a standard 1/2 cm diameter probe, 100 g/cm$^2$ contact pressure, 1 second contact time, 1 cm/sec separation rate at room temperature.

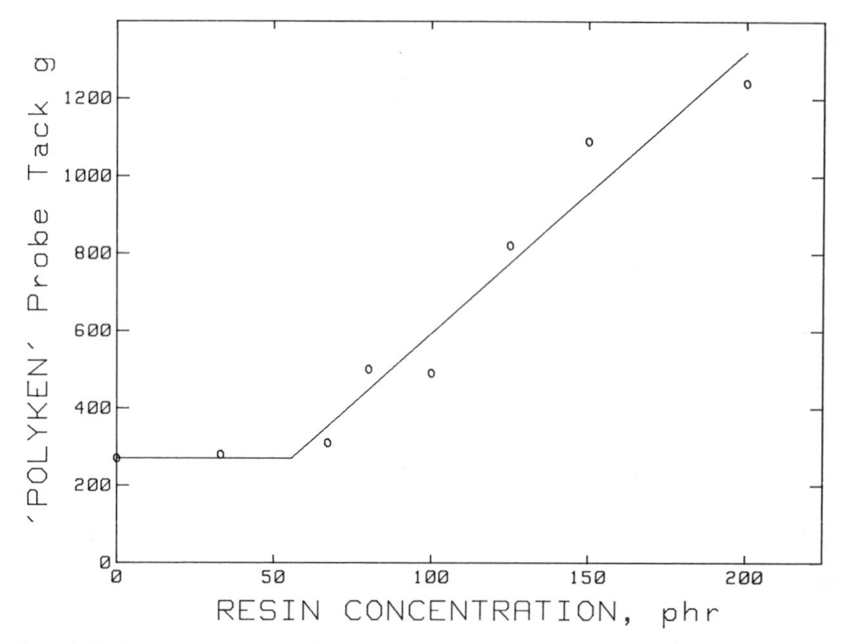


Fig. 3-7. Probe tack of natural rubber/Staybelite Ester No. 10 adhesives on a standard 1/2 cm diameter probe, 100 g/cm$^2$ contact pressure, 1 second contact time, 1 cm/sec separation rate at room temperature.

late, then, applies only to some systems. More fundamental to tack than phase changes are probably the changes in rheological properties of the adhesive.

The form of the relation between probe tack and composition illustrated in Fig. 3-7 is quite typical of two component mixtures of elastomer and resin.[15,48] Fortunately, tack/resin concentration relations determined in two-component systems carry over to three-component systems.[21] As one example, if an adhesive is made from two elastomers and a resin, both are tackified by the resin, and the tack of three-component mixtures will be the average of the two-component tacks at the same resin concentration. Such a system is natural rubber, SBR-1011 and Staybelite Ester No. 10, whose tack is summarized in Fig. 3-8. As a second example, if an adhesive is made from two elastomers and a resin which tackifier only one of them, the tack of a three-component adhesive will be simply that of the two-component system of the tackified elastomer at the resin concentration calculated as if the second elastomer were not present. Such a system is natural rubber, SBR-1011 and a polyterpene resin with a high melting point, for which tack as a function of concentration is shown in Fig. 3-9.

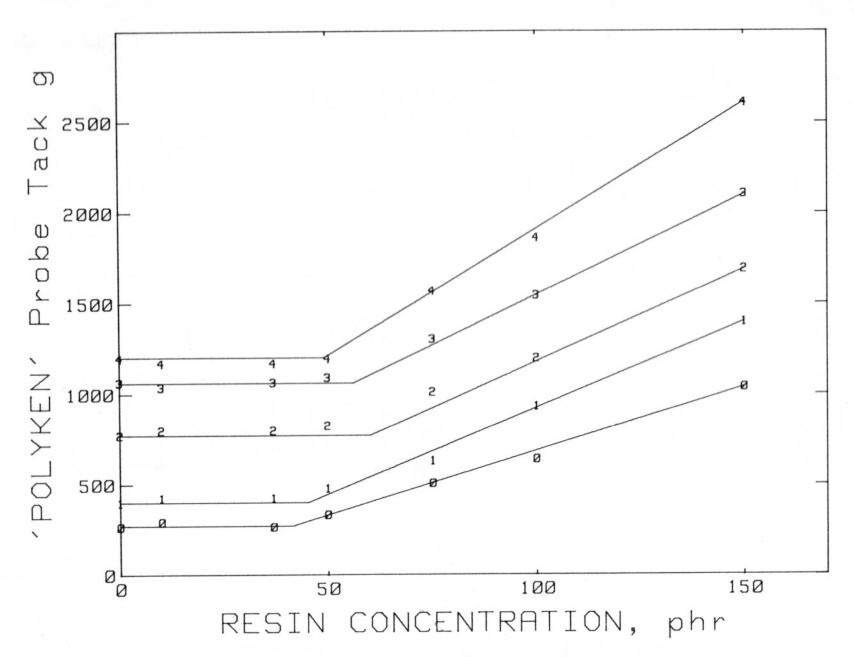


Fig. 3-8. Probe tack of natural rubber/SBR-1011/Staybelite Ester No. 10 adhesives. (0) NR only, (1) NR/SBR = 3/1, (2) NR/SBR = 1/1, (3) NR/SBR = 1/3 and (4) SBR only.

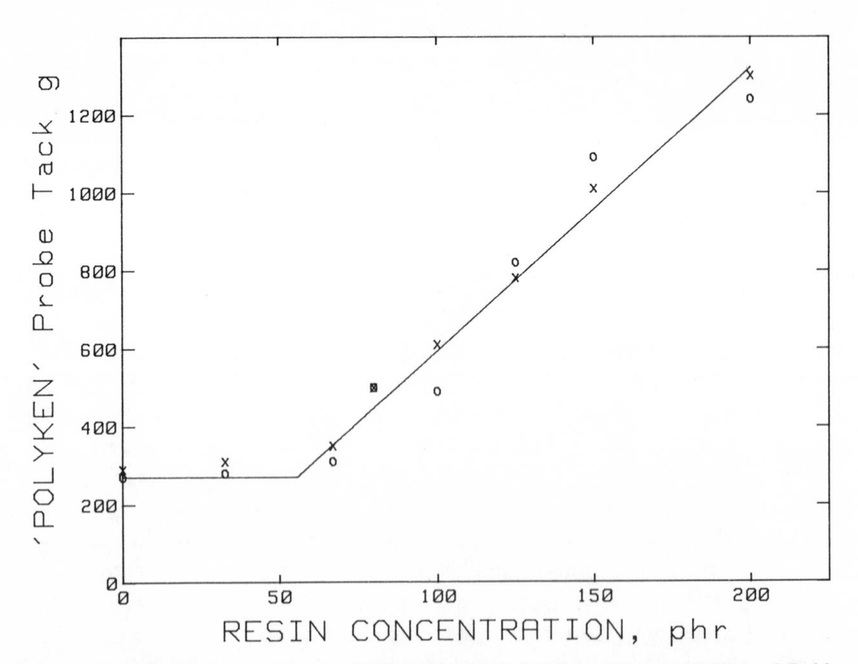


Fig. 3-9. Probe tack of natural rubber/SBR-1011/Piccolyte S115 adhesives. (0) Natural Rubber only (x) NR/SBR = 1/1.

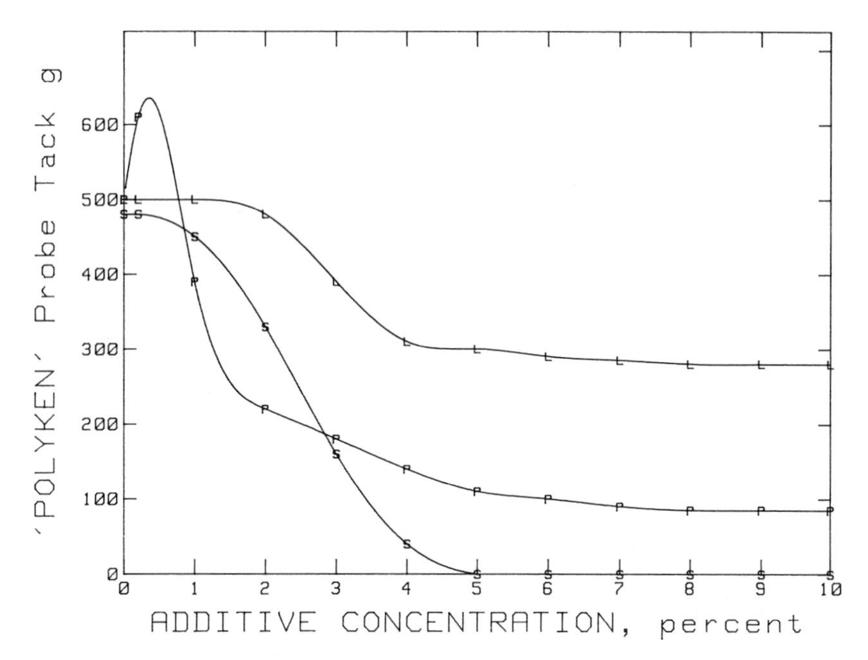


Fig. 3-10. Probe tack of natural rubber/Piccolyte S115 adhesives (100 phr P S115) with (S) stearic acid, (P) paraffin (sealing wax) and (L) Lanolin.

Tack sometimes can be greatly affected by small amounts of certain materials.[21] Stearic acid, lanolin, or paraffin wax added to a natural rubber/rosin ester adhesive decreases the probe tack as shown in Fig. 3-10. These small quantities of additives do not change to any extent the rheological properties of the adhesives; they modify the surface so as to make it appear there was a physically weak layer present.

## CONCLUSION

In the three decades from 1950 to 1980, pressure-sensitive tack has become a science of surface chemistry polymer rheology. The adhesive scientist now knows the chemistry that enables a bond to form and the viscoelastic properties that allow the bond to form quickly and yet be strong. The formulator of commercial pressure-sensitive adhesives still must look at a number of compositions, but he has insight into what materials to use, how much, and why.

## REFERENCES


1. *Compilation of ASTM Standard Definitions*, third edition, ASTM: 630 (1976).
2. *Pressure Sensitive Tack of Adhesives using an Inverted Probe Machine*, ASTM, D2979-71.
3. Bates, R. *J. Appl. Polym. Sci.* **20**: 2941–2954 (1976).
4. Bauer, R. F. *J. Polym. Sci.* **A-2**(10): 541–48 (1972).
5. Beatty, J. R. *Rub. Chem. and Tech.* **42**: 1040–1053 (1969).
6. Bikerman, J. J. *J. Coll. Sci.* **2:** 163–175 (1947).
7. Brunt, N. *Rheol. Acta*, Band No. 2: 242–277 (1958).
8. Bull, R. F., Martin, C. N., and Vale, R. L., *Adhesives Age* 20–24, (May 1968).
9. Chang, F.S.C. *Rub. Chem. and Tech.* **20**: 847–853 (1957).
10. Chang, F.S.C. *Adhesives Age*, **1**: 32–39, (November 1958).
11. Crocker, G. J. *Rub. Chem. and Tech.* **42**: 30–69 (1969).
12. Dahlquist, C. A. "Tack" in *Adhesion, Fundamentals and Practice*, Ministry of Technology, Maclaren, London, p. 143 (1969).
13. Dahlquist, C. A. in *Aspects of Adhesion* Vol. 5 (D. J. Alner, ed.), Univ. of London Press, p. 183 (1969).
14. Dann, J. R. *J. Colloid and Interface Science* **32**(2): 302–331 (February 1970).
15. DeWalt, C. *Adhesives Age*, 38–45 (March 1970).
16. Dow, J. *Proc. Inst. Rub. Ind.* **1**: 105 (1954).
17. Forbes, W.G., and McLeod, L. A. *Trans. Inst. Rubber Ind.* **35**(4): 154–174 (August 1958); *Rubber Age* **81**: 97 (1957); *Rubber World* **136**: 7676 (1957); *Rubber Chem. and Tech.* **32**: 48 (1959).
18. Gardon, J. L. *J. Phys. Chem.* **67**: 1935 (1962).
19. Gillespie, R. H. "Rolling Ball Tack," unpublished Kendall Co. Report (1962).
20. Hammond, F.H. Jr. *Polyken Probe Tack Tester*, ASTM Spec. Pub. 360 (1963).
21. Hammond, F.H. Jr. "Pressure-sensitive Tack Measurements," unpublished Kendall Co. Report (1971).
22. Hock, C. W. *J. Polym. Sci.* **Part C 3**: 139–149 (1963).
23. Hock, C. W. *Adhesives Age* **7**(3) 21–25 (1964).
24. Huntsberger, J. R. *Chemistry and Engineering News*, 82–86, (November 1964).
25. Huntsberger, J. R. *J. Paint Tech.* **39**(507) April 1967.
26. Huntsberger, J. R. *J. Adhesion* **7**: 289–299 (1976).
27. Huntsberger, J. R. *Adhesives Age* (December 1978).
28. Iyengar, Y., and Erickson D. E. *J. Appl. Polym. Sci.* **11**: 2311–2324 (1967).
29. Kaelble, D. H. *Rub. Chem. and Tech.* **45**: 1604–1622.
30. Kamagata, K. *J. Adhesion* **2**: 279 (1970).
31. Kamagata, K., Kosaka, H., Hino, K., and Toyama, M. *J. Appl. Polym. Sci.* **15**: 483–500 (1971).
32. Kambe, H., and Kamagata, K. *J. Appl. Polym. Sci.* **13**: 493–504 (1969).
33. Kraus, G., and Rollmann, K. W. *J. Appl. Polym. Sci.* **21**: 3311–3318 (1977).
34. Kraus, G., Jones, F. B., Marrs, O. L., and Rollmann, K. W. *J. Adhesion* **8**: 235–258 (1977).
35. Kraus, G., Rollmann, K. W., and Gray, R. A. *J. Adhesion* **10**: 221–236 (1979).
36. Parry, S. A., and Ritchie, P. F. *Adhesives Age* 28–33 (November 1966).
37. *Test Methods for Pressure-Sensitive Adhesives*, Sixth Edition Pressure-Sensitive Tape Council.
38. Sharpe, L. H. and Schonhorn, H. *International Science and Technology*, p. 26 (April 1964).
39. Sharpe, L. H., and Schonhorn, H. in *Contact Angle, Wettability and Adhesion*, ACS Advances in Chemistry Series No. 43 (1964).
40. Sherriff, M., Knibbs, R. W., and Langley P. G. *J. Appl. Polym. Sci.* **17** 3423–3438 (1973).

41. Schonhorn, H. *J. Appl. Phys.* **37**: 4967 (1966).
42. Toyama, M., and Ito, T. *Polymer-Plast. Tech. Eng.* **2**(2): 161–229 (1973).
43. Toyama, M., Ito, T., and Moriguchi, H. *J. Appl. Polym Sci.* **14**: 2039–2048 (1970).
44. Toyama, M., Ito, T., and Moriguchi, H. *J. Appl. Polym. Sci.* **14**: 2295–2303 (1970).
45. Toyama, M., Ito, T., Nukatsuka H., and Tikeda, M. *J. Appl. Polym. Sci.* **17**: 3495–3502 (1973).
46. Uffner, M. W. *Adhesives Age* 30–33 (December 1967).
47. Voit, A., and Geffken, C. F. *Ind. and Eng. Chem.* **43**, 1614 (1951).
48. Wake, W. C. *J. Inst. Rub. Ind.* (April 1972).
49. Wetzel, F. *Characterization of Pressure-Sensitive Adhesives.* ASTM Bulletin No. 221, 64–68 (1957).
50. Wetzel, F. Function of Resin Derivatives in Pressure-Sensitive Adhesives. *Am. Chem. Soc. Div. Paint, Plastics Ink Chem.*, Papers presented at the Miami Meeting No. 1, 225–232 (1957).
51. Zisman, W. A. in *Contact Angle, Wettability and Adhesion*, ACS Advances in Chemistry Series, p. 1 (1964).

# Chapter 4

# Peel

**Donatas Satas**
*Satas & Associates*
*Warwick, Rhode Island*

Peel adhesion is one of the important characteristics of pressure-sensitive adhesives. This property is evaluated by measuring the tensional force required to remove the adhesive tape.

The term peel adhesion is used rather loosely. Adhesion denotes a state in which two surfaces are held together. Resistance to peel or peel force denotes the tensional force required to separate the adherend in a peel test. The peel force might be dependent on the adhesion, but it also depends on many other factors: viscoelastic properties of the adhesive, stiffness of the adherend, rate of separation, temperature, etc. Yet the terms peel adhesion, peel force, and resistance to peel are used interchangeably, disregarding the important differences between adhesion and the other terms.

The standard peel tests are carried out at a constant peel rate and it is expected that a pressure-sensitive adhesive tape will strip off cleanly from the adherend, leaving no visually noticeable residue. This type of failure is called adhesive failure and it occurs at or near the adhesive-adherend interface. Many pressure-sensitive adhesives exhibit a transition from cohesive to adhesive failure at low peel rates or at elevated temperatures. In case of crosslinked or very high molecular weight adhesives, this transition might not be observed.

Besides the cohesive-adhesive failure transition, another one from a relatively steady to an oscillating peel force is observed at higher peel rates. Figure 4-1 shows a generalized peel force vs. peel rate curve. The dotted line at low peel rates represents the cohesive failure, the solid line an adhesive failure and the striped line at high peel rates represents the oscillating peel force. The peel force shows a decrease immediately after the transition to

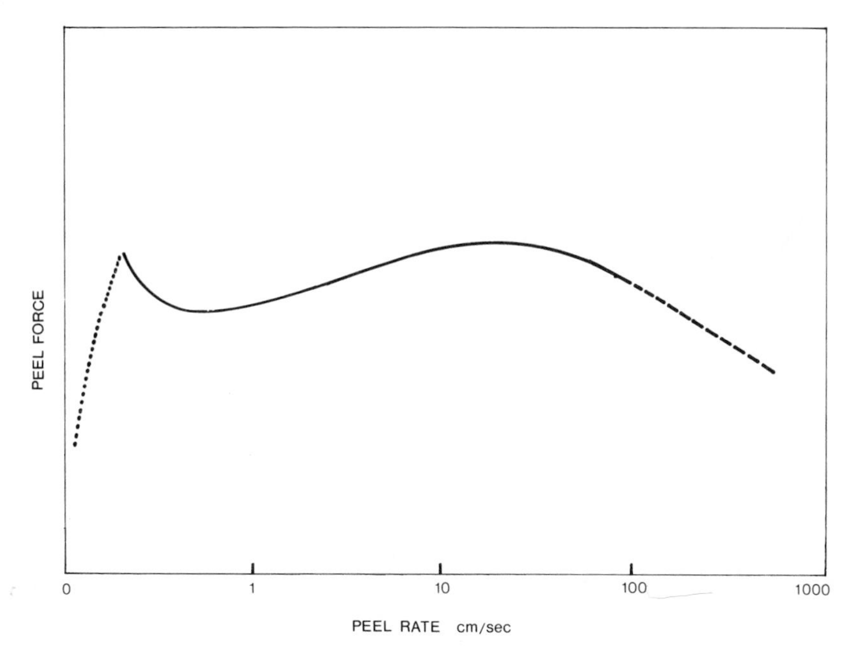


Fig. 4-1. Generalized peel force vs. peel rate curve.

adhesive failure, then a steady increase with increasing peel rate and again a gradual decrease just before the onset of oscillations. The oscillations of peel force observed are of nonrandom nature. Hendricks and Dahlquist[1] call this second maximum the "yield point" above which the stripping becomes raspy and jerky and is generally referred to as "stick-slip" peel.

## TESTING

There are several geometrically different combinations to test resistance to peel of pressure-sensitive adhesive tapes. The most common arrangements are shown in Fig. 4-2. The test shown in Fig. 4-2a is the most frequently used for peel of pressure-sensitive tapes, the 180° peel adhesion test. The 90° peel test is shown in Fig. 4-2b. This test is used less frequently. In both of these tests, the tape is applied over a rigid steel panel. The drum test (Fig. 4-2c) is rarely used, but if the drum is replaced by a roll of tape, then such a test measures the unwind force, which is an important property of a pressure-sensitive adhesive tape. The T-peel test (Fig. 4-2d) measures the force required to separate an adhesive-to-adhesive bond. While it is of interest for some specialty tapes, this test is not used for the common characterization of the tapes. It is a useful quick test to check the anchorage

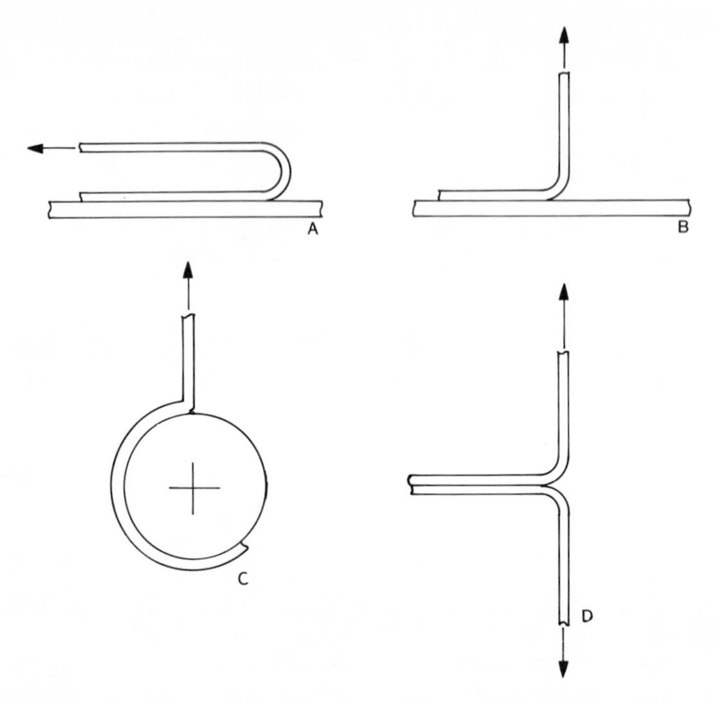


Fig. 4-2. Schematic diagram of various peel tests: (a) 180° peel, (b) 90° peel; (c) Drum peel (tape unwind); (d) T-peel.

of the adhesive mass to the backing. The T-peel test can also be used to measure the force to separate the tape from a flexible adherend.

The basis for most of the tapes of pressure-sensitive adhesive tapes is ASTM D1000-66 or later versions.[2] These tests have been specifically designed for electrical tapes, but most of them are also suitable for other tapes. Method D1000 contains 19 test methods and about half are of interest for all pressure-sensitive tapes. The procedures of interest for peel testing include sample preparation, adhesion to steel plate (180° peel test), drum adhesion (90° peel test), and other tests. The testing is specified to be performed at 0.5 cm/sec (12 in./min). While the driven jaw is moving at 0.5 cm/sec, the actual peel rate, because of the geometry in the 180° peel test, is 0.25 cm/sec. Stainless steel plates finished to a surface roughness of 0.25–0.50 $\mu$m are used. The test results are reported in grams per centimeter or in ounces per inch width.

The T-peel test is described by ASTM D1876-61T and is used to check adhesive-to-adhesive bonds.[3]

The Pressure-Sensitive Tape Council (PSTC), a manufacturers' trade association, has several tests for peeling: Peel Adhesion for Single Coated

Tapes at 180° Angle (PSTC-1), Peel Adhesion for Double Coated Tapes at 180° Angle (PSTC-3), Adhesion to Liner of Pressure-Strength Tapes at 180° Angle (PSTC-4), Unwind Adhestion (PSTC-8), High-Speed Unwind Adhesion (PSTC-13), and Adhesion to Fiberboard at 90° Angle and Constant Stress (PSTC-14). These tests are similar to ASTM tests and have been adopted from ASTM procedures. Unwind adhesion (at 12 in./min) and high speed unwind adhesion (at 200 ft/min) are important to determine the behavior of the tape in roll form. Adhesion to fiberboard is designed mainly for evaluation of the performance of carton closure tape. A constant force is applied and the time to failure is measured.

Federal Test Method Standards 101 (Preservation, Packaging and Packaging Materials) and 147 (Tapes, Pressure-Sensitive and Gummed, Methods of Inspection, Sampling, and Testing) are tests that involve measurement of peel force. The National Electrical Manufacturers' Association (NEMA) has an interest in pressure-sensitive tape testing, mainly for tape applications in original equipment. In addition, the Tag and Label Manufacturers Institute (TLMI) has tests for 180° and 90° peel adhesion similar to the ones described.

The 180° peel test is used most often to characterize the pressure-sensitive tapes. It is easier to carry out than the 90° test, the data show slightly less scattering, and it is more sensitive to variations in tape construction. Usually, one measurement of peel force at specified conditions is given. This is often not enough to form an opinion about the tape performance under use conditions. A much better judgment about the adhesive can be formed it a peel force vs. peel rate or peel force vs temperature curve is given. This has been discussed by Satas.[4]

Various special peel tests have been devised. Diefenbach[5] describes an apparatus used to measure the peel force of low adhesion pressure-sensitive tapes, such as used for securing the postage stamp coils. A peel force of about 30 g/cm is expected of such tapes. An interesting way to measure the peel force has been described by Šalkauskas and Paulavičius.[6] A triangular cut is peeled at 90° by applying a dead load. The peel rate decreases because the width of the peeled tape or coating increases until equilibrium is reached and the peeling ceases. The bond strength is measured by dividing the dead load applied by the width of the tape at equilibrium point. The values so obtained are somewhat lower than the values obtained by a regular 90° peel test.

## TRANSITION FROM COHESIVE TO ADHESIVE FAILURE

This phenomenon observed at low peel rates or at elevated temperatures has been a subject of many investigations. The generalized peel curve in Fig.

4-1 shows such a transition at the low peel rate. The response of the pressure-sensitive adhesive to the stress is of a viscoelastic nature. At low rates of force application or at elevated temperatures, the response is predominantly viscous and the result is a cohesive failure. At higher rates of force application, the response becomes predominantly elastic and the failure becomes adhesive, or at least appears to take place at the adhesive-adherend interface. The high peel force in the cohesive failure region can be important in the performance of an adhesive tape. If the rate of force application is slow, or if the force is applied over a short distance, this property may help to retain the tape in place. Adhesives applied over the skin, as in hospital tapes or finger bandages, benefit from this phenomenon. Such adhesives are discussed in the chapter on Hospital and First Aid Products by Satas. Flagging in electrical and other tapes is eliminated by an adhesive exhibiting a possibility of cohesive failure at low peel rates.

This transition from cohesive to adhesive failure has been discussed by many researchers. McLaren and Seiler[7] have shown that low molecular weight polyvinyl acetate adhesives fail cohesively in a peel test, while higher molecular weight materials exhibit adhesive failure. Similar behavior was also observed at elevated temperatures. The peel force showed a continuous increase with increasing temperature until the mode of failure changed from interfacial to cohesive. At that point, the peel force started to decrease with increasing temperature. Huntsberger[8] has shown similar behavior for poly(*n*-butyl methacrylate) adhesive. When the temperature decreases, the relaxation rate becomes slower than the rate of force application, and the stress concentration occurs at the edges, resulting in an interfacial failure. Further lowering of temperature causes a decrease in peel force, because of a higher stress concentration.

The most often cited author describing this transition in pressure-sensitive adhesives is Bright.[9] Figure 4-3 shows some of this data. Segment C of the curve represents clean adhesive failure. Segment B represents the transitional period which takes place in a narrow temperature range, and segment A represents a cohesive failure within the adhesive mass.

The analogy between the peel rate and temperature in pressure-sensitive tapes has been discussed by many authors. Hendricks and Dahlquist[1] show a construction of a master curve for a given temperature which allows prediction of the force vs. time relation over a wide time range from measurements obtained at a convenient time range and at several temperatures. This time-temperature superposition is also discussed in detail by Rajeckas in Chapter 7.

It is generally agreed that the transition from cohesive to adhesive failure is associated with the transition from liquidlike viscous to rubberlike elastic behavior.[10,11] Other possible mechanisms have been discussed by

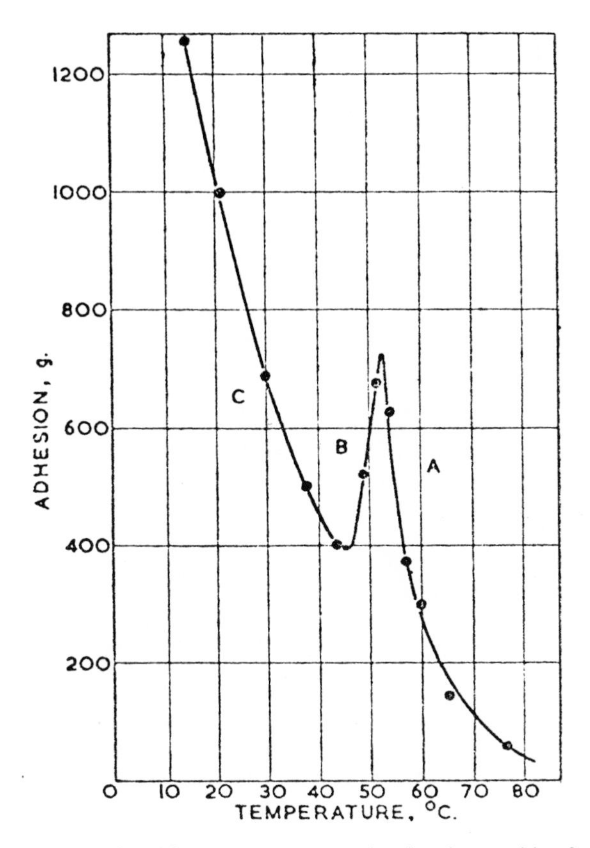


Fig. 4-3. The dependence of peel force on temperature showing the transition from adhesive to cohesive failure.

Gent and Petrich.[11] It has been proposed that the transition is caused by cavitation within the adhesive. The critical negative hydrostatic pressure to cause cavitation in rubbery materials is 7–10 $kg/cm^2$. Peel tests were conducted under an imposed hydrostatic pressure above these values. The imposed hydrostatic pressure was more than sufficient to prevent cavitation from occurring. The values of peel force were the same with and without the imposed hydrostatic pressure, leading to the conclusion that the failure mechanism does not include cavitation. That the failure mode transition is due to a change from a viscous to an elastic response is also supported by the fact that crosslinked adhesives, which do not exhibit liquid flow, do not fail cohesively at low peel rates. Gent and Petrich also point out that the change from cohesive to adhesive failure is not an inevitable feature of this transition at low peel rates. The liquid response is not absolutely necessary.

Similar performance can be shown by the materials which yield and draw. High adhesive strength is obtained from the materials which have large deformation energies within the stress limit set by the interfacial bond strength.

Gent and Petrich[11] discuss the quantitative treatment of peel at low peel rates. The peel force $P$ is given approximately by the sum of tensile stresses $f$ in the elements of the adhesive layer up to the point of failure.

$$P = \int_0^{y_m} f\,dy = h\int_0^{e_m} f\,de,$$

where

$y$ = distance through which an element of the adhesive is stretched
$e$ = corresponding fractional extension
$y_m$ and $e_m$ = maximum values obtained
$h$ = thickness of the adhesive layer.

Equation (1) is based on two main assumptions: the substrate film is completely flexible, and the elements of the adhesive can be treated as independent.

Equation (1) can be rewritten in terms of stress limits

$$P = h\int_0^{f_m} f\,de, \tag{2}$$

where $f_m$ = maximum stress attained.

Equation (2) predicts that the peel strength depends on the deformation energy of the polymer up to the point at which the tensile stress is large enough either to break the adhesive bond or to rupture the polymer.

## TRANSITION TO OSCILLATING FAILURE

At high peel rates, a second transition is observed from a smooth adhesive failure to a failure in which the peel force is oscillating in a nonrandom manner. Random fluctuations are occasionally observed at various peel rates and are usually caused by surface irregularities, imperfections in the tape or other extraneous factors. The oscillations of the peel force observed after the transition period at high peel rates are clearly of nonrandom nature. The peel force vs. peel rate curve has a negative slope in that region as shown in Figure 4-1.

This behavior has been observed and discussed by many researchers. In pressure-sensitive adhesives, it has been discussed by Hendricks and Dahlquist.[1] Greensmith[12,13] in the studies of tear properties of rubber has reported steady and also nonrandomly oscillating stick-slip forces. Similar behavior in peel tests of adhesive coatings has been reported by Dannenberg,[14] Gardon,[15], and Lannus and Zdanowski.[16] This transition has been attributed to a change from elastic to glassy response, to localized plastic yielding,[17] or to storage of elastic energy in the adhesive and in the substrate.

This peel behavior at high peel rates is often observed during the unwind of pressure-sensitive tapes from a self-wound roll. Such peel force data can be obtained by unwinding the tape at various rates. The test is essentially a 90° peel test of the tape from its backing. A schematic diagram of the test is shown in Fig. 4-4.

Two different types of force vs. time curves in the region of random adhesive failure have been observed and are shown by curves "a" and "b" in Fig. 4-5. Curve "a" shows a randomly oscillating force. Curve "b" shows a relatively continuous peel force free of oscillations. This type of unwind can be observed at various unwind rates and the variations in the peel force can be due to surface imperfections or other extraneous reasons. Peel of the type shown by curves "a" and "b" is considered to belong to the region of steady adhesive failure.

In the region of nonrandomly oscillating adhesive failure, two slightly different patterns of force vs. time curves are observed. These are shown in Fig. 4-5 as curves "c" and "d". Both curves have a definite nonrandomly oscillating pattern, except in curve "d", a random oscillation appears to be superimposed over the longer, nonrandom wavelength pattern. Most of the tapes studied exhibited a transition from an "a" type curve to "b" to "c" (or "d"). Type "a" failure is associated with the increasing slope portion of the curve in Fig. 4-1, "b" with the constant force portion and "c" or "d" with decreasing slope shown by a discontinuous line in Fig. 4-1.

It is well known that a electrostatic charge is generated during unwinding of tapes. It was observed that during the steady unwind force the

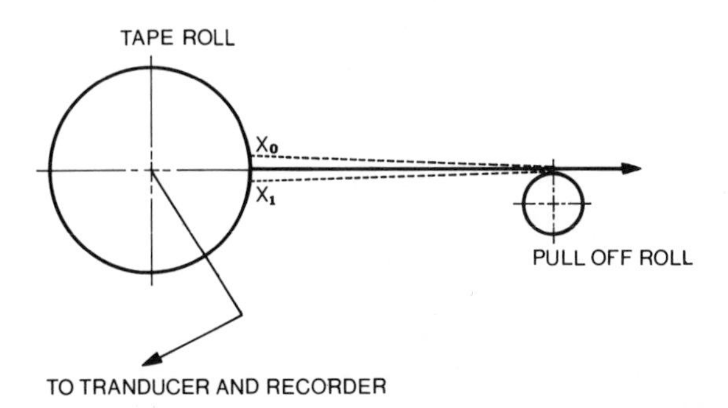


Fig. 4-4. Schematic diagram of tape unwind test.

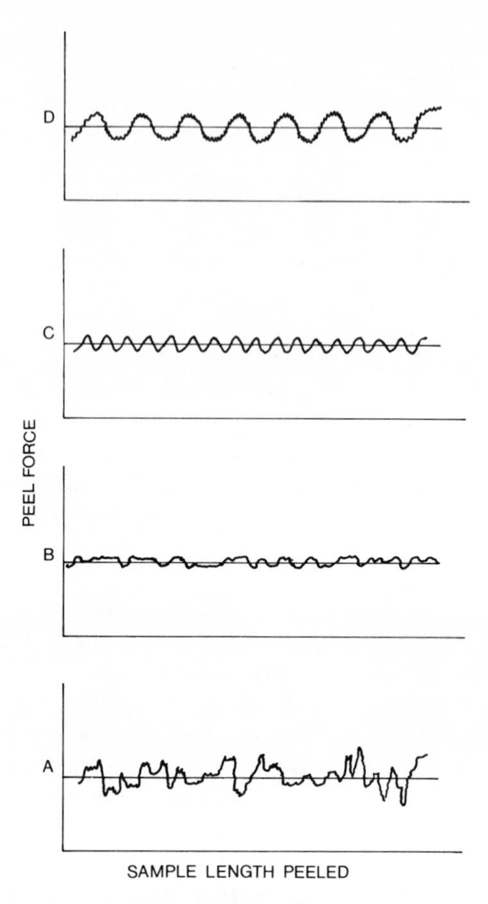


Fig. 4-5. Various peel force oscillations observed during the tape unwind.

electrostatic charge developed is maintained at a constant level; during a nonrandomly oscillating force, the electrostatic force also oscillates. Table 4-1 shows the observed data.

Deryagin[18,19] attributes a much more important role to the electrostatic charge observed during the separation of an adhesive bond, than just a by-product that is generally observed during the separation of surfaces. He attributes the rate dependence of the peeling force to the electrical double layer produced during peeling.

Oscillations are generally connected with energy storage and the nonrandomly oscillating peel force is explained as a result of storage of energy elastically until it is released by a sudden failure.[20,21] Oscillations of the peel force correspond to the oscillations of fracture points that travel be-

**Table 4-1. Electrostatic charges developed during tape unwind.**

| UNWIND RATE (cm/sec) | PEEL FORCE (g/cm width) | TYPE OF OSCILLATION | ELECTROSTATIC CHARGE (volts) |
|---|---|---|---|
| 8 | 106 | Random | 0 |
| 13 | 98 | Random | 500 |
| 21 | 89 | Random | 1000 |
| 32 | 80 | Nonrandom | 0–1000 |
| 43 | 89 | Nonrandom | 0–2000 |

tween points $x_0$ and $x_1$ in Fig. 4-4. When the fracture point travels from $x_0$ to $x_1$, the elastic energy is stored and the peel force increases. The elastic energy is released during the reverse cycle, the force decreases and reaches a minimum at point $x_0$. Adhesive coating applied over an elastic backing, such as rubber sheeting, exhibits the nonrandomly oscillating failure at much lower peel rates than if applied over a rigid substrate.

Pressure-sensitive adhesives are viscoelastic materials with a decreasing viscous component at higher rates of deformation. Therefore, it is expected that softer, less elastic adhesives should exhibit the transition to nonrandomly oscillating peel force at higher peel rates than harder, crosslinked

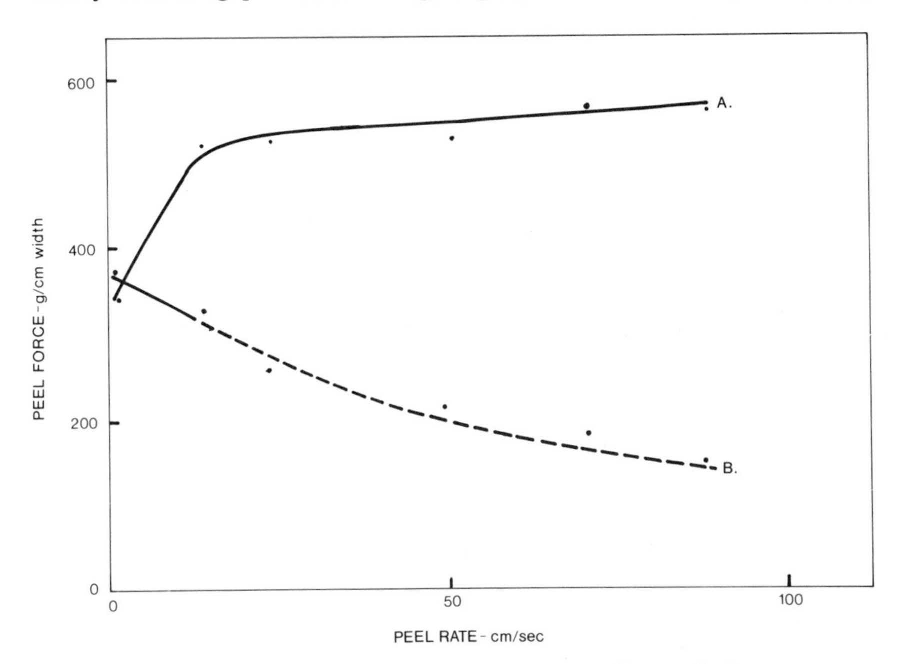


Fig. 4-6. Peel force of aluminum foil tapes with two different adhesives.

and more elastic ones. Figures 4-6 through 4-9 illustrate the effect of adhesive properties on the peel force.

Figure 4-6 shows peel force data of two 0.1 mm thick soft aluminum foil tapes, backcoated with polyethylene. Tape A has a soft uncrosslinked acrylic polymer-based adhesive, while tape B has an adhesive of similar composition, but crosslinked and considerably more elastic. It is obvious that these two curves represent different sections of the generalized peel force vs. peel rate curve shown in Fig. 4-1. Curve A is the steady peel and curve B is the nonrandomly oscillating peel force portion of the generalized curve. The solid line denotes steady peel force and the discontinuous line denotes oscillating peel force.

Figure 4-7 shows the unwind of polyethylene coated fabric tapes. Tape C has a soft natural rubber-resin adhesive; tape D, a crosslinked natural rubber-resin adhesive; tape E, an acrylic polymer-based adhesive. The data illustrate the shift of the transition point to the higher peel rates with decreasing elasticity of the adhesive.

The effect of hydrogen bonding on the properties of pressure-sensitive acrylic polymers is discussed in Chapter 13. It has been shown that the introduction of hydrogen bonding via acrylonitrile comonomer increases polymer stiffness. Figure 13-9 shows the effect of acrylonitrile on the penetrometer compliance, illustrating its stiffening effect. The peel behavior of

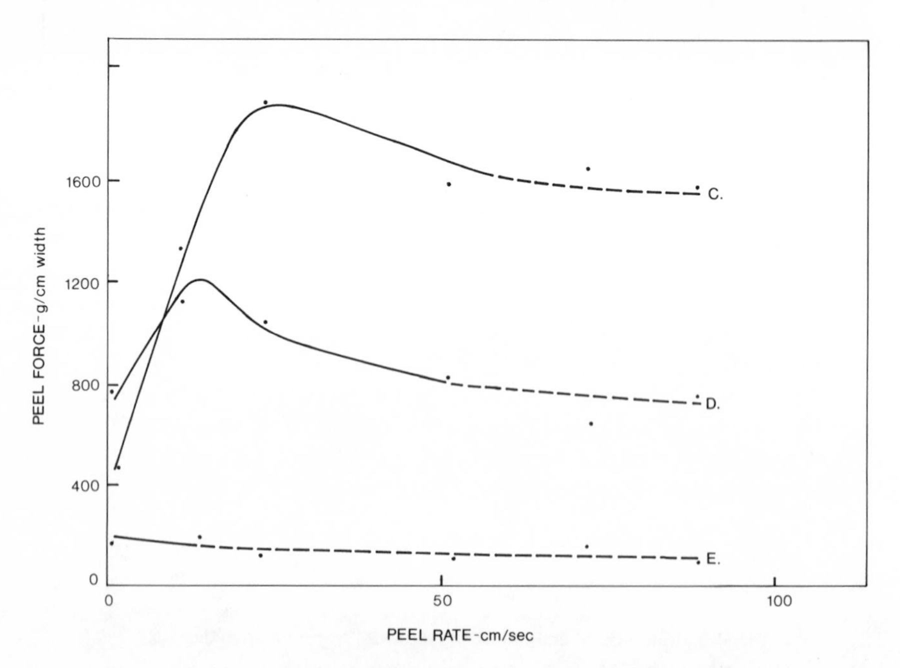


Fig. 4-7. Peel force of polyethylene coated fabric tapes with different adhesives.

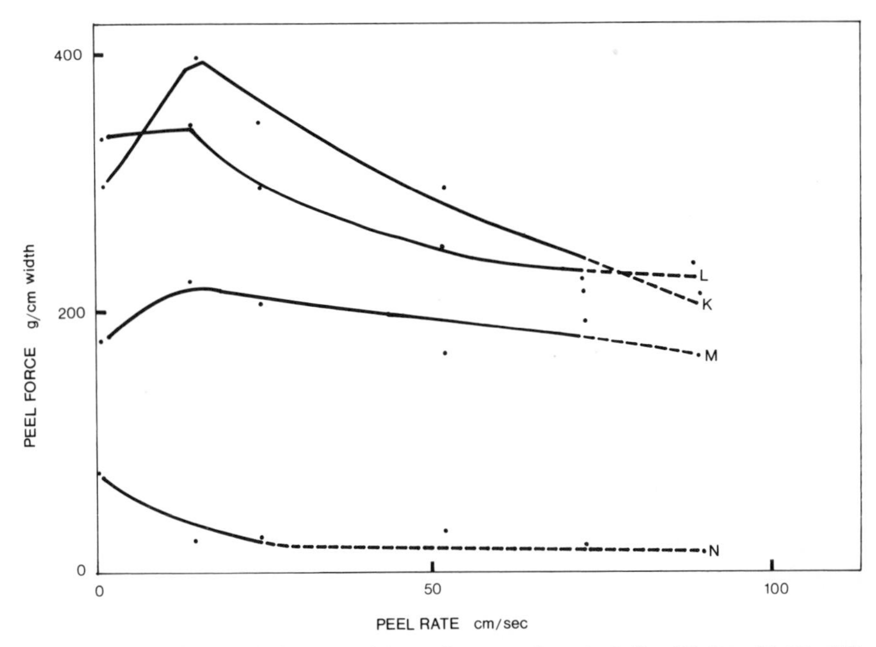


Fig. 4-8. Effect of acrylonitrile on peel force. Amount of acrylonitrile: (K) 0%; (L) 1%; (M) 2.5%; (N) 10%.

these polymers is shown in Fig. 4-8. The decrease of the peel force and the shift of the transition point is affected by increasing acrylonitrile content.

The peel behavior of pressure-sensitive adhesive tapes depends not only on the adhesive, but also on the properties of the backing. It has been reported that roughening of cellophane film surface in a cellophane film-acrylic binder system caused a change of the mode of failure from nonrandomly oscillating to steady peel.[15] A nonrandomly oscillating force is usually not observed when the adhesive is peeled off a rough, fibrous textile surface. Apparently, a relatively smooth surface is needed to propagate the fracture. The backing thickness, while affecting the peel force level, did not appear to affect the transition point significantly.

In designing pressure-sensitive adhesive tapes, release coating is often used to decrease the unwind. Figure 4-9 shows the unwind of aluminum foil tapes with various release coatings. The peel force level and the transition point from steady to oscillating peel have been affected by the type of the release coating used.

## EFFECT OF TEMPERATURE

The investigation of peel behavior as a function of temperature is an interesting tool to obtain information about pressure-sensitive adhesives.

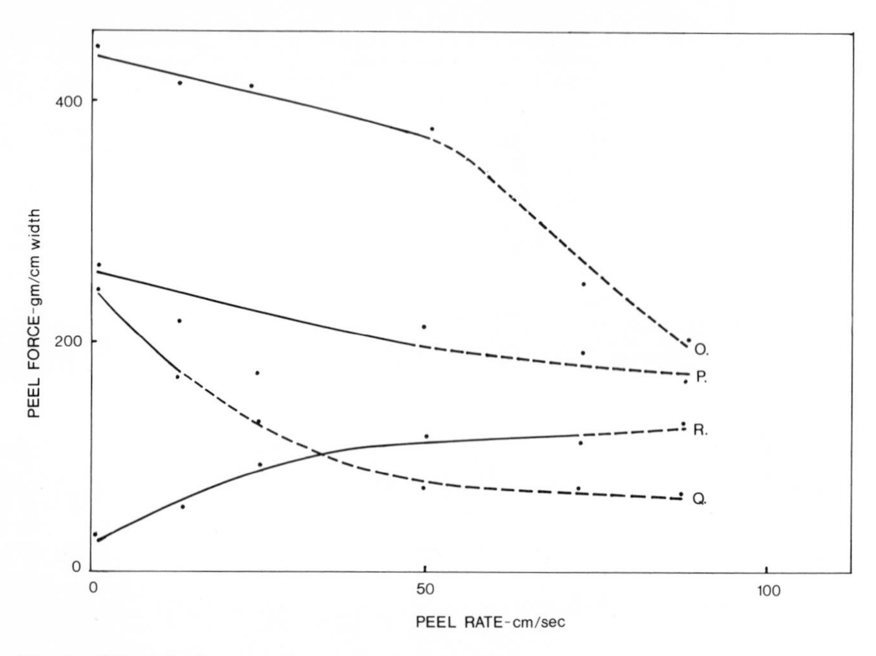


Fig. 4-9. Effect of release coatings on peel: (O) acrylic release coating; (P) Quilon C release; (Q) Polyethylene release coating; (R) Silicone release coating.

Such tests can be carried out by employing a bar over which a temperature gradient is established. One end of the bar is heated and the other is cooled. Heating is accomplished by insertion of a heater into a hole drilled through the bar. Cooling is done by circulating cold water through a hole provided for that purpose at the other end of the bar. Stainless steel plates used for peel tests are attached to the bar by clamps. The peel test is started at the cooler portion of the bar and continued towards the increasing temperature. A schematic diagram of such a bar is shown in Fig. 4-10. The temperature at which the transfer from adhesive to cohesive failure takes place is called the splitting temperature and it is a useful measure of the adhesive behavior. Table 13-3 in Chapter 13 shows a group of pressure-sensitive acrylic polymers of various molecular weight. The splitting temperature clearly follows the molecular weight variations as well as the creep resistance of the adhesives.

The peel force vs. temperature curves are different for polymers with different degree of crosslinking, or hydrogen and other secondary bonding. An increasing degree of crosslinking increases the splitting temperature. Acrylic polymers containing carboxylic groups were treated with Na ions in order to affect ionic binding.[22] Figure 4-11 shows the peel force vs. temper-

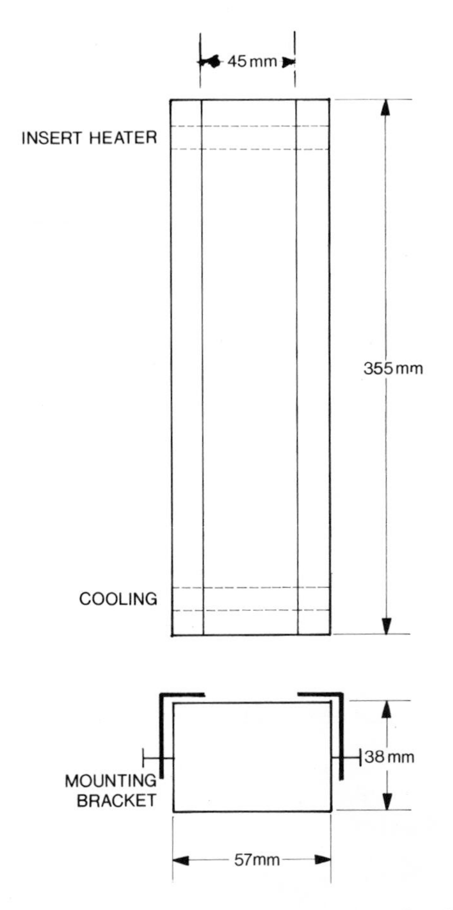


Fig. 4-10. Schematic diagram of temperature gradient bar.

ature curves for polymers with various degrees of ionic binding. The dotted line denotes cohesive failure.

## EFFECT OF ADHESIVE THICKNESS

It is generally accepted that peel force increases with increasing adhesive thickness up to a certain limit. Further increase of the adhesive coating thickness does not increase resistance to peel. The results by Chan and Howard[23] shown in Table 4-2 support this observation as well as the data given by Johnston.[24] Adhesive coatings rarely exceed 0.05 mm in thickness, except for tapes that are applied against rough surfaces. Adhesive coatings

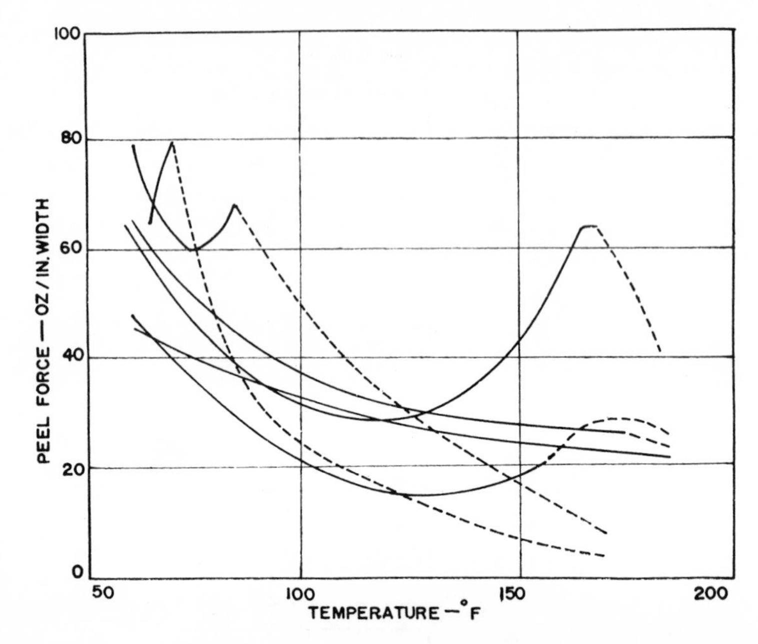


Fig. 4-11. Peel force of adhesives with various degree of ionic binding. Sodium content: curve top right 0.500%, 0.34%, 0.59%, 0.69%, 0.18%, curve bottom right 0%. (Reprinted from D. Satas and R. Mihalik, *J. Appl. Polym. Sci.* **12**, 2371–2379 (1968).

as thin as 0.025 mm are found satisfactory for thin film tapes which are applied over smooth surfaces.

Aubrey *et al.*[25] report that the adhesive thickness does not affect the peel force at low peel rates when the failure is cohesive. Increasing adhesive thickness causes the shift of cohesive-adhesive failure transition to higher peel rates. In the region of steady adhesive failure, the peel force increases with increasing adhesive thickness. His data were obtained on peel tests of

**Table 4-2. Variation of peel strength with adhesive thickness.**

| THICKNESS ($\mu$m) | PEEL STRENGTH (N/m) |
|---|---|
| 7.6 | 201–15 |
| 10.2 | 323–91 |
| 15.2 | 511–11 |
| 22.9 | 565–18 |
| 30.5 | 560–52 |

non-pressure-sensitive acrylic film of 1.2–258 $\mu$m thick. The increase in peel force is attributed to the increase of the area under stress. Similarly, Fukuzawa[26] has shown that the adhesive failure is more predominant for thick adhesive coatings. The peel force increased with increasing adhesive thickness in case of adhesive failure and also showed some increase in the case of cohesive failure. Gardon[27] also shows an increase of peel force with increasing adhesive thickness. Theoretical considerations by Gardon[28] show that

$$P = \alpha t_a^{0.25}, \tag{3}$$

where

$P$ = peel force,

$t_a$ = adhesive thickness,

$\alpha$ = parameter dependent on substrate thickness, adhesive and substrate moduli.

A similar relationship derived by Bikerman[29] is discussed by Rajeckas in Chapter 7.

## EFFECT OF THE BACKING

It is generally recognized that the backing thickness and modulus has an effect on the peel force. The effect of thickness is more pronounced in a 180° than in a 90° peel test.[30] The thicker the tape backing, the larger is the adhesive area under stress. This causes an increase of the peel force. The moment arm exerted on the adhesive bond increases with increasing thickness and this factor decreases the peel force. Data for 180° and 90° peel tests are shown in Figs. 4-12 and 4-13. The backings used were polyester film, aluminum foil, and polyurethane rubber film.

Aubrey *et al.*[25] have reported peel force data for various backing thicknesses at various peel rates. Polyester film bonded with poly(*n*-butyl acrylate) adhesive was used. At the low peel rates where the failure was cohesive, the peel strength was unaffected by the backing thickness in the range of 0.015–0.052 mm. The peel strength decreased with a higher backing thickness (up to 0.126 mm).

Johnston[24] presents some data on the effect of backing thickness on peel force. Polyester film 0.0064–0.035 mm and dead soft aluminum foil 0.0038–0.08 mm thick were used for tape backings. The increase of backing thickness changes the geometry of the peel test. Correlation between the backing thickness and the peel force becomes impossible because the tests

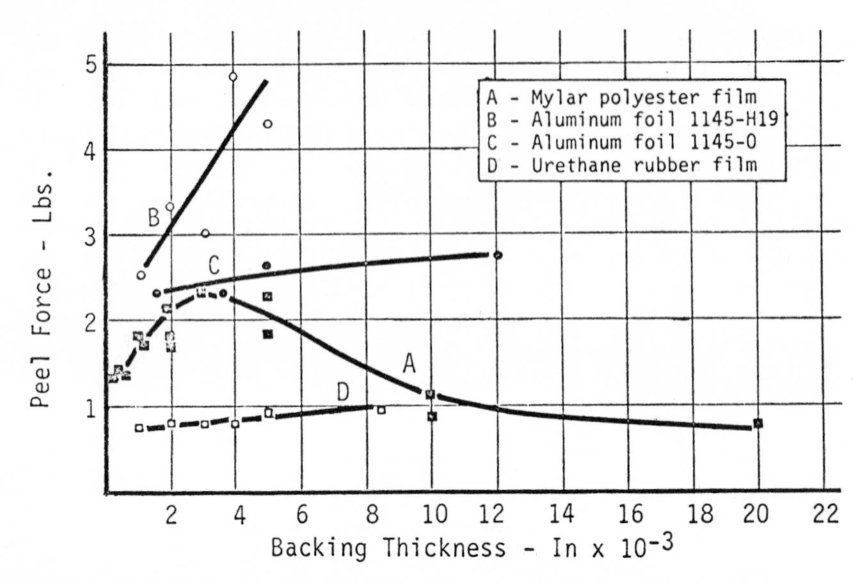


Fig. 4-12. The dependence of peel force on the backing thickness in a 180° peel test. (Reprinted from *Adhesives Age* **9** (8) (1966).)

conducted are not geometrically similar. Johnston suggests to modify the peel test so that the backing is peeled away from double-faced adhesive tape. The data shown in Fig. 4-14 indicate a linear relationship between the thickness of polyester film and peel force.

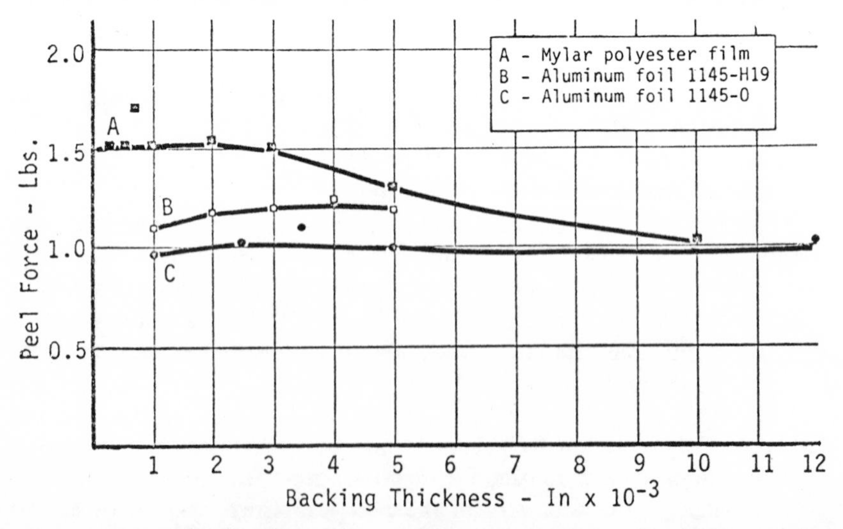


Fig. 4-13. The dependence of peel force on the backing thickness in a 90° peel test. (Reprinted from *Adhesives Age* **9** (8) (1966).)

Duke[17] suggests that four different relations exist between the peel force and adherend thickness. The peel force may be proportional to $T^0$, $T^{2/3}$, $T^{3/2}$, and the relation between $P$ and $T$ can be controlled by the increasing moment arm. At low adherend thickness with pronounced adhesive legging during peel, the peel force is independent of backing thickness. In the region following that, at higher backing thickness, legging is decreasing and $P \sim T^{2/3}$, where $P$ is the peel force and $T$ the thickness of the adherend. In the region of truly elastic failure, $P \sim T^{3/2}$. The increase of $P$ with $T$ is ultimately limited by either the inadequacy of adhesive adhesion or, more commonly, by the ability of joint configuration to impose increasing moment arms as peel or cleavage begins.

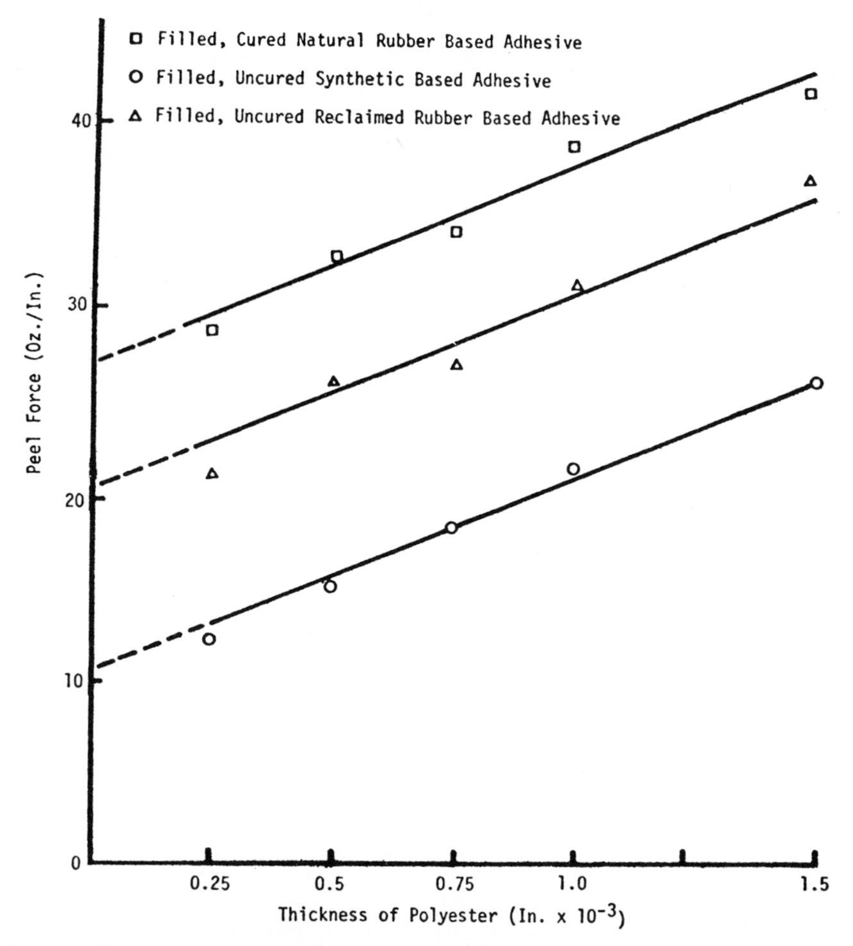


Fig. 4-14. The dependence of peel force on polyester film thickness using modified test method described by Johnston.[24] (Reprinted from *Adhesives Age* **11** (4) (1968).)

## EFFECT OF OTHER VARIABLES

The effect of peel angle has been reviewed by Gardon.[31] Kaelble[32] gives peel force data as a function of peel angle for various backings at different peel rates (0.05–50 cm/min). The peel force increased with increasing peel rate at all angles. At low peel rates, a maxima-minima is observed at 30–40° as shown in Fig. 4-15. This is attributed to the transition of the failure mechanism from cleavage to boundary shear.

Toyama *et al.*[33] have investigated the relationship between wetting and adhesion of pressure-sensitive poly(butyl acrylate) by applying the adhesive over adherends of varying surface tensions. Wetting of the surface by adhesive is incomplete and the wetted spots are scattered over the adherend surface. The density of these minute spots depends on the surface tension of the adherend. The unbonding process proceeds around the wetted spots. The failure in the area of the spot is cohesive in the sense that a minute amount of adhesive could be found on the adherend. It was not cohesive in the sense this term has been used in this book. Table 4-3 shows the data by Toyama,[33] illustrating the dependence of peel on the critical surface tension $\gamma_c$ of the adherend. For poly(butyl acrylate), $\alpha_c$ was found to be 31 dyne/cm.

It takes time for the adhesive to wet the surface properly and to devel-

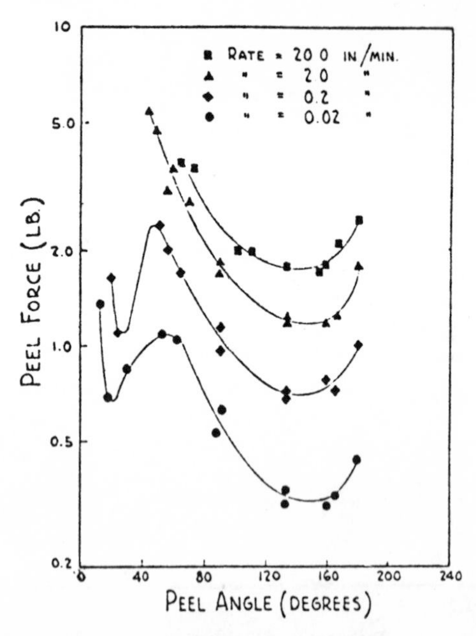


Fig. 4-15. Peel force as a function of peel angle for a glass cloth tape. (Reprinted from D. H. Kaelble, *Trans. Soc. Rheol.* **IV**, 45–73 (1960).)

**Table 4-3. Contact angle, tack values, and adhesive deposit of poly(butyl acrylate) adhesive on the surface of various adherends.**

| ADHEREND | CRITICAL SURFACE TENSION—$\gamma_c$ (dyne/cm) | COSINE OF CONTACT ANGLE[a] | ADHESIVE DEPOSIT (g/cm$^2$) | PEEL FORCE (g/cm width) | PROBE TACK[b] (g) |
|---|---|---|---|---|---|
| Poly(tetrafluoro ethylene) | 18.5 | 0.485 | 0.11 | 5 | 0 |
| Polypropylene | 29.0 | 0.883 | 0.03 | 96 | 96 |
| Low density polyethylene | 26.0 | 0.946 | 0.03 | 46 | 84 |
| High density polyethylene | 31.0 | 0.951 | 0.07 | 82 | Not measured |
| Polystyrene | 33.0 | 0.994 | 0.38 | 337 | 130 |
| Poly(methyl methacrylate) | 39.0 | 0.994 | 0.27 | 367 | 260 |
| Poly(hexamethylene capramide) | 42.0 | 0.917 | 0.07 | 330 | Not measured |
| Poly(ethylene terephthalate) | 43.0 | 0.968 | 0.08 | 266 | 160 |
| Stainless steel | — | 0.746 | 0.02 | 91 | 130 |

[a] Contact angle of adhesive on adherend was measured by placing a droplet of low molecular weight poly(butyl acrylate).
[b] Polyken probe tack, contact pressure 100 g/cm$^2$, contact time 1 sec, rate of separation 1 cm/sec, adhesive thickness 0.02 mm, 0.5 cm diameter probe.

op the complete bond. Dahlquist[34] has shown that the peel force increases significantly during the first 24 hr after the bond was made. The peel value levels off after 48 hr and there is little change afterwards.

The glass transition temperature $T_g$ of the adhesive has an effect on the mechanical properties of the polymer and is expected to have an effect on the peel force as well. Aubrey[35] suggests that pressure-sensitive adhesive properties could be correlated with $T_g$ of acrylic polymers. The peel force vs. peel rate curve as shown in Fig. 4-1 (the peel force vs. temperature curve) can be shifted along the x-axis by changing $T_g$ of the adhesive polymer. Raising $T_g$ values moves the curve to the left. A usable pressure-sensitive adhesive should have a high resistance to peel at a peeling rate of about 1 cm/sec and the failure mode should be interfacial. If adhesive fails cohesively at this rate, $T_g$ should be increased by choosing the proper comonomer. If the adhesive exhibits stick-slip failure at lower peel rates, $T_g$ should be lowered.

Shoraka[36] has reviewed the dependence of peel strength of pressure-sensitive adhesives on contact time, temperature, peel rate, and adhesive thickness. The heat of peeling was investigated by actual measurements of the temperature rise in the adhesive. The work by Good[37] was followed.

## STRESS DISTRIBUTION

The distribution of stresses in the adhesive bond during a peel test must be considered to gain a better understanding of the test and of the relation of peel force values to the properties of pressure-sensitive adhesive. It has been recognized by DeBruyne[38] and discussed by many other authors that compressional stresses are developed in the adhesive during peeling.

An apparatus for measuring the normal stresses during the peeling of an adhesive bond has been described by Johnson and Kaelble[39] and later discussed in greater detail.[40] Figure 4-16 shows a schematic diagram of such an apparatus, designed for mounting on an Instron Tester. The substrate to which the adhesive tape is applied is divided into two parts. The rightmost section is rigidly supported and its position is insensitive to the normal forces of peel. The substrate to the left of the division is sensitive to normal forces and deflects slightly in response to the normal bond stresses in peel test.

Figure 4-17 shows a schematic diagram of a profile of a pressure-sensitive adhesive tape undergoing peel at 90°. A typical stress distribution profile as recorded by the measuring instrument is shown. The stresses are highly localized at the boundary of the bond. Adjacent to the boundary is a region of tensional stress. A zone of compressional stress with a maximum compressive stress $\sigma_c = 5.7$ kg/cm$^2$ and a compressive zone length $\lambda/2 = 0.034$ cm is produced because of the leverage of the flexible member. A zone of low tensional stress appears next and then the normal stress drops to zero.

Table 4-4 shows the results of 40 repetitive 90° peel tests of a pressure-

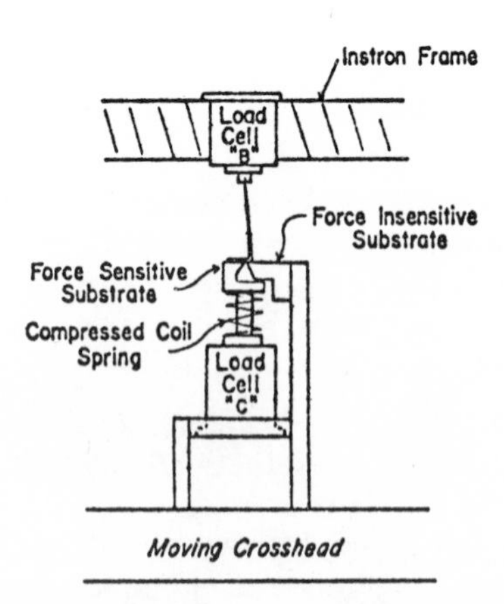


Fig. 4-16. Schematic of the bond stress analyzer. (Reprinted from D. H. Kaelble, *Trans. Soc. Rheol.* **9** (2) (1965).)

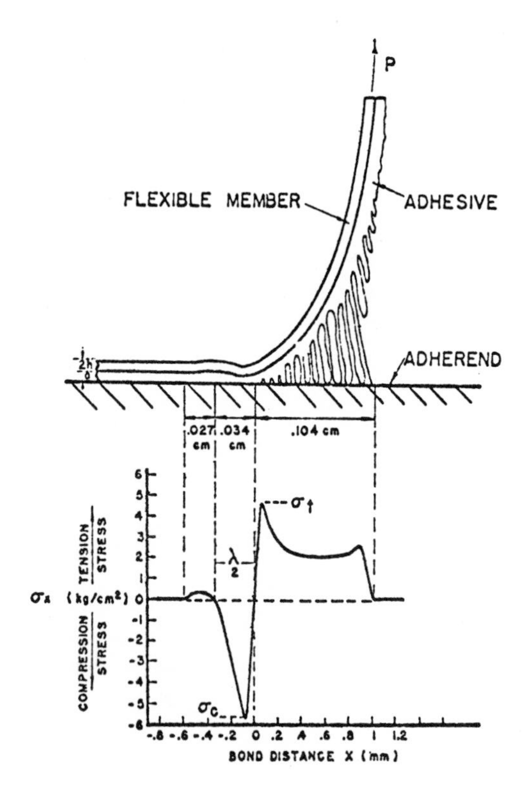


Fig. 4-17. Peel profile in a 90° peel test and the distribution of normal stresses. (Reprinted from D. H. Kaelble, *Trans. Soc. Rheol.* **9** (2) (1965).)

sensitive adhesive tape. The results are recorded in terms of peel force $P$, tensile stress $\sigma_t$, compressive stress $\sigma_c$, and compressive zone length $\lambda/2$ and are arranged in the order of the test number $n$. The peel force was essentially constant throughout the 40 tests, while internal stresses $\sigma_t$ and $\sigma_c$ and length factor values $\lambda/2$ varied considerably. Figure 4-18 shows the systematic variation of the magnitude of $\sigma_c$ in the plots of internal stress vs. bond length. The number $n$ is the test number in Table 4-4. The curves in Fig. 4-18 are drawn to a similar scale of stress (ordinate) and bond distance (abscissa). They are arranged according to the increasing value of $\sigma_c$. The top curve represents the lowest measured value of $\sigma_c$. The peel force was essentially constant, ranging from 230 to 240 g for the tests illustrated in Fig. 4-18. It may be also noted that the character of the stress distribution changes with a change of $\sigma_c$. The upper curves of low $\sigma_c$ values do not exhibit a sharp $\sigma_t$ maximum. The lower curves, characteristic of higher $\sigma_c$

**Table 4-4. Variability in peel and internal stress of a pressure-sensitive tape in a 90° peel test at a peel rate of 0.5 cm/min.** [a]

| $n$ | $P$ (g) | $\sigma_t$(g/cm$^2$) | $\sigma_c$ (g/cm$^2$) | $\lambda/2$ (cm) | $n$ | $P$ (g) | $\sigma_t$ (g/cm$^2$) | $\sigma_c$ (g/cm$^2$) | $\lambda/2$ (cm) |
|---|---|---|---|---|---|---|---|---|---|
| 1 | 240 | 3800 | 3800 | 0.038 | 21 | 240 | 3800 | 4300 | 0.039 |
| 2 | 250 | 3400 | 3400 | 0.040 | 22 | 230 | 3300 | 3400 | 0.039 |
| 3 | 220 | 4200 | 5100 | 0.038 | 23 | 230 | 3100 | 2700 | 0.044 |
| 4 | 230 | 3400 | 3400 | 0.044 | 24 | 240 | 3400 | 2200 | 0.050 |
| 5 | 240 | 3100 | 2100 | 0.057 | 25 | 220 | 2700 | 1100 | 0.069 |
| 6 | 220 | 4300 | 1000 | 0.075 | 26 | 230 | 2700 | 700 | 0.088 |
| 7 | 240 | 5700 | 4500 | 0.039 | 27 | 240 | 3100 | 2100 | 0.049 |
| 8 | 230 | 3800 | 3800 | 0.037 | 28 | 220 | 3500 | 3300 | 0.044 |
| 9 | 240 | 3100 | 2400 | 0.050 | 29 | 210 | 2900 | 2100 | 0.053 |
| 10 | 230 | 3000 | 3200 | 0.063 | 30 | 230 | 3900 | 4500 | 0.041 |
| 11 | 230 | 3900 | 5000 | 0.035 | 31 | 220 | 5300 | 5000 | 0.031 |
| 12 | 230 | 3300 | 3600 | 0.041 | 32 | 240 | 3100 | 3100 | 0.037 |
| 13 | 230 | 2900 | 2400 | 0.047 | 33 | 220 | 3900 | 4800 | 0.038 |
| 14 | 240 | 2600 | 1700 | 0.053 | 34 | 210 | 3400 | 4500 | 0.041 |
| 15 | 230 | 3400 | 3600 | 0.041 | 35 | 230 | 4300 | 4600 | 0.039 |
| 16 | 240 | 2600 | 2100 | 0.053 | 36 | 230 | 2600 | 1700 | 0.059 |
| 17 | 230 | 3100 | 2000 | 0.050 | 37 | 230 | 4600 | 5700 | 0.034 |
| 18 | 240 | 4100 | 5300 | 0.034 | 38 | 220 | 3600 | 4500 | 0.037 |
| 19 | 230 | 2700 | 1400 | 0.066 | 39 | 210 | 4600 | 3800 | 0.044 |
| 20 | 230 | 3800 | 4400 | 0.036 | 40 | 220 | 2800 | 2100 | 0.059 |

| Averages: | Standard deviations ($\pm$): |
|---|---|
| for $P = 229.8$ | for $P = 9.5$ |
| $\sigma_t = 3520$ | $\sigma_t = 702$ |
| $\sigma_c = 3260$ | $\sigma_c = 1285$ |
| $\lambda/2 = 0.0468$ | $\lambda/2 = 0.0122$ |

[a] Reprinted from D. H. Kaelble, *Trans. Soc. Rheol.* **9**(2)(1965).

values, exhibit a sharp $\sigma_t$ maximum and a definite boundary to the region of tensional stress.

Kaelble[40] also conducted further tests, including peel tests at 180°. The results of these tests are shown in Table 4-5. The variability of the peel force $P$ is slightly higher than in the test shown in Table 4-4 and the stress characteristics $\sigma_c$, $\sigma_t$, and $\lambda/2$ show less variation. The comparison of the plots of the internal stress functions is shown in Fig. 4-19 arranged in the similar manner to Fig. 4-18.

Kaelble and Reylek[41] describe the instrument modified to measure the normal stresses over three decades of the peeling rates. A schematic diagram of a 180° peel test and the distribution of normal stresses is shown in Fig. 4-20. The 180° peel test stress diagram differs from the diagram of 90° peel test by the presence of two distinct tensile stress maxima.

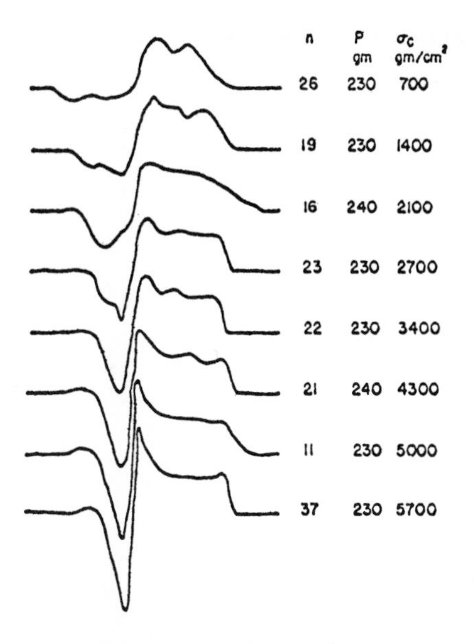


Fig. 4-18. Comparison of the forms of the internal stress function. (Reprinted from D. H. Kaelble, *Trans. Soc. Rheol.* **9** (2) (1965).)

## ANALYSIS OF PEEL

The analysis of the behavior of adhesive joints during peeling and the derivation of mathematical expressions, which would allow the prediction of the strength of such joints, has been attempted by many authors. Unfortunately, none of the analyses is entirely satisfactory. Many simplifying assumptions have to be made and they do not always represent the real mechanism of failure.

One of the early analyses was by Spies[42] where he calculated the internal bond stress distribution in peel for a peel angle of 90°. A solution for the peel force was offered, assuming adhesive failure of a Hookean interlayer.

In the simplest theory, assuming that the adhesive and the backing are Hookean solids and that shear stresses in adhesive are negligible, Bikerman[29,43] shows that the peel force $P$ is proportional to:

$$P \sim w t_a \sigma, \tag{4}$$

where

$w$ = width of the tape

$t_a$ = adhesive thickness

$\sigma$ = tensile strength of the adhesive.

**Table 4-5. Variability in peel and internal stress of a pressure-sensitive tape in a 180° peel test at a peel rate of 0.5 cm/min.[a]**

| $n$ | $P$ (g) | $\sigma_t$ (g/cm$^2$) | $\sigma_c$ (g/cm$^2$) | $\lambda/2$ (cm) | $n$ | $P$ (g) | $\sigma_t$ (g/cm$^2$) | $\sigma_c$ (g/cm$^2$) | $\lambda/2$ (cm) |
|---|---|---|---|---|---|---|---|---|---|
| 1 | — | — | — | — | 21 | 303 | 5710 | 7100 | 0.035 |
| 2 | 260 | 3100 | 5900 | 0.036 | 22 | 305 | 7450 | 6950 | 0.036 |
| 3 | 263 | 4880 | 4600 | 0.046 | 23 | 253 | 4600 | 6550 | 0.036 |
| 4 | 225 | 3850 | 3100 | 0.064 | 24 | 270 | 4720 | 5860 | 0.039 |
| 5 | 270 | 4800 | 5770 | 0.038 | 25 | 265 | 5300 | 5890 | 0.037 |
| 6 | 260 | 5700 | 6350 | 0.038 | 26 | 265 | 5850 | 5880 | 0.036 |
| 7 | 245 | 6000 | 6650 | 0.040 | 27 | 280 | 5300 | 5880 | 0.036 |
| 8 | 245 | 6210 | 6800 | 0.038 | 28 | 270 | 4990 | 5440 | 0.037 |
| 9 | 243 | 5550 | 5900 | 0.038 | 29 | 258 | 5800 | 6100 | 0.039 |
| 10 | 250 | 5700 | 6880 | 0.040 | 30 | 260 | 5720 | 6450 | 0.036 |
| 11 | 260 | 5700 | 6650 | 0.040 | 31 | 270 | 5750 | 5250 | 0.042 |
| 12 | 248 | 5100 | 6360 | 0.038 | 32 | 260 | 5120 | 5300 | 0.038 |
| 13 | 257 | 6560 | 7170 | 0.036 | 33 | 265 | 4600 | 5190 | 0.042 |
| 14 | 255 | 5720 | 6570 | 0.034 | 34 | 275 | 5350 | 5900 | 0.037 |
| 15 | 250 | 6130 | 7120 | 0.038 | 35 | 290 | 6190 | 5530 | 0.040 |
| 16 | 253 | 6420 | 6420 | 0.038 | 36 | 265 | 3320 | 4960 | 0.038 |
| 17 | 255 | 6130 | 5600 | 0.039 | 37 | 265 | 8700 | 6980 | 0.034 |
| 18 | 240 | 6420 | 6130 | 0.037 | 38 | 240 | 3260 | 4460 | 0.042 |
| 19 | 285 | 5440 | 5440 | 0.042 | 39 | 260 | 5600 | 6110 | 0.037 |
| 20 | 297 | 4360 | 6130 | 0.037 | 40 | 260 | 5910 | 6170 | 0.039 |

| Averages: | Standard deviations (±) |
|---|---|
| for $P = 262.5$ | for $P = 16.9$ |
| $\sigma_t = 5462$ | $\sigma_t = 1066$ |
| $\sigma_c = 5987$ | $\sigma_c = 828$ |
| $\lambda/2 = 0.0388$ | $\lambda/2 = 0.0048$ |

[a] Reprinted from D. H. Kaelble, *Trans. Soc. Rheol.* 9(2)(1965).

Jouwersma[44] elaborated further on Bikerman's work. Bikerman and Yap[45] have pointed out that peeling of the adhesive joint is a rheological phenomenon as long as the adhesive is liquid. Peeling depends on the viscosity of the adhesive and the rigidity of the backing. The peel tests cannot give any information on the adhesion between the adhesive and the substrate.

Chang[46] discussed the peel of pressure-sensitive tapes specifically using a mechanical model consisting of three elements. Peel of pressure-sensitive adhesives has been reviewed by Dahlquist.[50] Dahlquist[34] has also discussed the estimation of the work of adhesion between a pressure-sensitive adhesive and substrate. Peel has been also reviewed by Gardon.[31] Gardon[49] has also derived equations showing the change of peel force with adhesive thickness. Non-pressure-sensitive polymers were used for his study. Math-

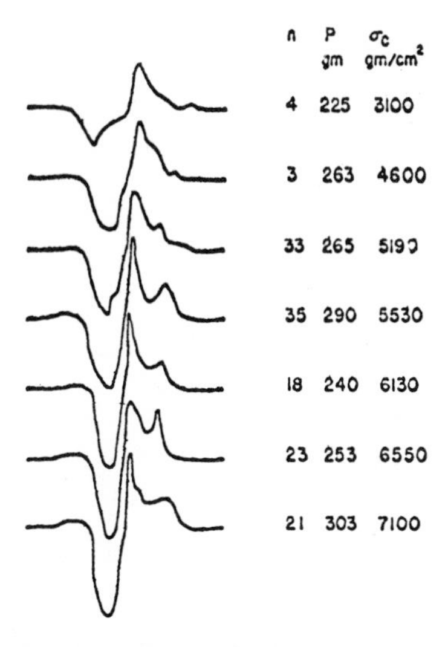


Fig. 4-19. Comparison of the forms of the internal stress function corresponding to Table 4-5. (Reprinted from D. H. Kaelble, *Trans. Soc. Rheol.* **9** (2) (1965).)

ematical expressions describing the tensional and compressional forces during peel have been given.[28]

The most prolific author on the subject of peel is Kaelble. Peel has been reviewed.[47] Steady state unbonding, assuming a mechanism of boundary cleavage (and in another case, of boundary shear), has been analyzed.[48] The

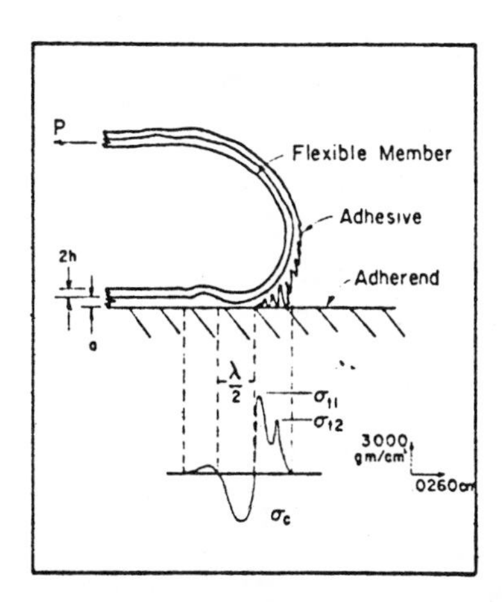


Fig. 4-20. Peel profile in a 180° test and the distribution of normal stresses. (Reprinted from D. H. Kaelble and R. S. Reylek, *J. Adhesion* **1**, 124–135 (1969).)

analysis was extended[32] to provide expressions for the boundary stresses and bond stress distributions.

Peeling has been reviewed by Duke.[17] He points out that the theories of peel adhesion (Spies, Jouwersma, Kaelble, Gardon) have assumed that the adhesive is elastic. The stresses applied to the adhesive bond during peeling are less that those required to cause an inelastic response. This assumption also implies that the adherend is stiff and does not undergo a plastic flexular yield. According to Duke, this latter assumption is incorrect and the elastic peel mechanism describes only a very limited number of cases. It is of practical interest mainly in case of joints between thick, rigid adherends. An assumption of plastic behavior of adhesive and adherend better describes most bond failures by peeling.

## REFERENCES

1. Hendricks, J. O., and Dahlquist, C. A. Pressure-Sensitive Adhesive Tapes. *Adhesion and Adhesives*, Vol. 2. Houwink, R., and Salomon, G., (eds.) Amsterdam: Elsevier Publishing Co., 1967.
2. "Standard Methods of Testing Pressure-Sensitive Adhesive Coated Tapes Used for Electrical Insulation. ANSI/ASTM D1000-78.
3. Tentative Method of Test for Peel Resistance of Adhesives (T-Peel Test). ASTM D1876-61T.
4. Satas, D. *Adhesives Age* **13** (6): 38–40 (1970).
5. Diefenbach, W. T. TAPPI **45** (11): 840–842 (1962).
6. Šalkauskas, M. J., and Paulavičius, R. B. *Zavodskaya Laboratoriya*. **1976**, (8): 879–880.
7. McLaren, D. D., and Seiler, C. J. *J. Polym. Sci.* **4**, 63–73 (1949).
8. Huntsberger, J. R. *J. Polym. Sci.* **A1**, 2241–2250 (1963).
9. Bright, W. M. *Adhesion and Adhesives, Fundamentals and Practice* Clark, F., Rutzler, J. E. and Savage, R. L., (eds.) Soc. of Chem. Ind., 130–138. New York: John Wiley & Sons, Inc., 1954.
10. Kaelble, D. H. *J. Adhesion* **1**: 102–123 (1969).
11. Gent, A. N., and Petrich, R. P. *Proc. Roy. Soc.* **A 310**: 433–448 (1969).
12. Greensmith, H. W., and Thomas, A. G. *J. Polym. Soc.* **18**: 189 (1955).
13. Greensmith, H. W. *J. Polym. Sci.* **21**: 175 (1956).
14. Dannenberg, H. J. *J. Appl. Polym. Sci.* **5**: 125 (1961).
15. Gardon, J. L. *J. Appl. Polym. Sci.* **7**: 625 (1963).
16. Lannus, A., and Zdanowski, R. *Resin Review* **16**: No. 4, 15 (1966).
17. Duke, A. J. *J. Appl. Polym. Sci.* **18**: 3019–3055 (1974).
18. Krotova, N. A., Kurulova, N. M., and Deryagin, B. V. *J. Phys. Chem. USSR* 1921–1931 (1956).
19. Deryagin, B. V., *Research* **8**: 70 (1955).
20. Salomon, G. *Adhesion and Adhesives* Houwink R. and Salomon G., (eds.). Vol. 1, p. 119, Amsterdam: Elsevier Publishing Co., 1967.
21. Voyutskii, S. S. *Autoadhesion and Adhesion of High Polymers*, p. 162. New York: Interscience Publishers, 1963.
22. Satas, D., and Mihalik, R. *J. Appl. Polym. Sci.* **12**: 2371–2379 (1968).
23. Chan, H-K., and Howard, G. J. *J. Adhesion* **9**: 279–304 (1978).

24. Johnston, J. *Adhesives Age* **11**(4): 20–26 (1968).
25. Aubrey, D. W., Welding, G. N., and Wong, T. *J. Appl. Polym. Sci.* **13**: 2193–2207 (1969).
26. Fukuzawa, K. *J. Adhesion Soc. Japan* **5**: 294 (1969).
27. Gardon, J. L. *J. Appl. Polym. Sci.* **7**: 625–641 (1963).
28. Gardon, J. L. *J. Appl. Polym. Sci.* **7**: 643–665 (1963).
29. Bikerman, J. J. *The Science of Adhesive Joints*, Second Ed. New York: Academic Press, 1968.
30. Satas, D., and Egan, F. *Adhesives Age* **9** (8) 22–25 (1966).
31. Gardon, J. L. *Treatise on Adhesion* Vol. 1 Patrick, R. L., (ed.). New York: Marcel Dekker, Inc. 1967.
32. Kaelble, D. H. *Trans. Soc. Rheol.*, **IV**, 45–73 (1960).
33. Toyama, M., Ito, T., Nakatsuka, H., and Ikeda, M. *J. Appl. Polym. Sci.* 17: 3495–3502 (1973).
34. Dahlquist, C. A. *ASTM Special Technical Publication No.* 360, 66th Annual Meeting Papers, Atlantic City, N. J., June 26, 1963.
35. Aubrey, D. W. *Development in Adhesives I* Wake, W. C., (ed.) London: Applied Science Publishers, Ltd., 1977, 127–156.
36. Shoraka, F. *Adhesion and Heat of Peeling of Pressure-Sensitive Tapes.* Ph.D. Thesis. State University of New York at Buffalo, May 1979.
37. Good, R. J. *Aspects of Adhesion* Alner, D. J. (ed.) Vol. 6, London: University of London Press, 1971.
38. DeBruyne, N. A. *Nature* **180**: 262 (1957).
39. Johnson, E. G. and Kaelble, D. H. "Apparatus for the Measurement of the Normal Stress Distribution in Adhesive Bonds During Peel." Presented at the Meeting of the American Physical Society, Cambridge, Mass., April 1959.
40. Kaelble, D. H. *Trans. Soc. Rheol.*, **9**(2): 135–163 (1965).
41. Kaelble, D. H., and Reylek, R. S. *J. Adhesion*, **1**: 124–135 (1969).
42. Spies, G. J. *Aircraft Eng.* **25** (289): 64 (1953).
43. Bikerman, J. J. *J. Appl. Phys.* **28**: 1484 (1957).
44. Jouwersma, C. *J. Polym. Sci.* **45**(145): 253–255 (1960).
45. Bikerman, J. J., and Yap, W. *Trans. Soc. Rheol.* **II**: 9–21 (1958).
46. Chang, F. S. C. *Trans. Soc. Rheol.* **IV**: 75–89 (1960).
47. Kaelble, D. H. *Chemistry of Adhesion.* New York: Wiley-Interscience, 1971.
48. Kaelble, D. H. *Trans. Soc. Rheol.* **III**: 161–180 (1959).
49. Gardon, J. L. *J. Appl. Polym. Chem.* **7**: 643–665 (1963).
50. Dahlquist, C. A. *Treatise on Adhesion* Vol. 2. Patrick, R. L. (ed). New York: Marcel Dekker, Inc.

# Chapter 5

# Creep

**Carl A. Dahlquist**
*St. Paul, Minnesota*

If a substance exhibits the property we call pressure-sensitive adhesiveness, i.e., it adheres when brought into contact with a surface under light pressure, and has sufficient cohesiveness that it can be peeled away from the surface without leaving a residue, the substance must have rather specialized and characteristic rheological properties. It must have sufficient compliance to conform to surface rugosity, and it must undergo relaxation so that the stored energy due to elastic forces will be dissipated before they can overcome the forces of adhesion. Yet the pressure-sensitive adhesive must exhibit an elastic cohesiveness and a resistance to flow under stress. It is this resistance to flow that is commonly associated with creep or resistance to creep, but in a broader sense, the entire stress-strain behavior of a pressure-sensitive adhesive can be treated as a creep phenomenon.

The creep of a soft polymer is conventionally regarded as the increase in strain with time after stress is suddenly applied and maintained constant. It can be shear deformation under constant shearing stress, compressive strain under constant compressive stress, or tensile strain under constant tensile stress. The creep response can be expressed either as the creep modulus vs. time or as creep compliance vs. time. Modulus is the ratio of stress to strain, and compliance is its reciprocal. The engineer usually has been trained to think of modulus as the property describing purely elastic stress-strain behavior. The creep of pressure-sensitive adhesives almost invariably involves some viscous flow in addition to the elastic strain. It may be technically distasteful to the engineer to speak of a modulus which involves nonrecoverable strain. Though it may be technically no more acceptable, it seems less offensive to regard compliance as the ratio of total strain to stress, and, unless otherwise indicated, this interpretation of compliance will be used throughout this dissertation. It is fully recognized that, depending on the time scale, creep compliance has all of the counterparts of modulus, namely, storage, loss, elastic, and viscous components.

Although the term "creep" has a connotation of slow processes involving minutes, days, or months, one can, in principle, determine the short-term, or dynamic, stress-strain response of a viscoelastic solid as well as the long-term response by means of creep measurements and time-temperature superposition. Difficulties arise in the so-called "instantaneous" application of the load and in inertial effects associated with rapid response, so that it is more expedient to use oscillatory methods when the response time of interest is less than 1 sec, but in principle any deformation which has an associated retardation time is a part of the creep spectrum. It is also possible by well-known mathematical treatments to obtain the elastic and viscous components of the creep spectrum. In the discussion to follow, attention will be focused mainly on the long-term creep response of pressure-sensitive adhesives, but compliance and the retardation spectrum play such significant roles in the quick stick or tack of pressure-sensitive adhesives that some attention must be given to the short-term response as well.

## SHORT-TERM CREEP AND TACK

The creep compliance-time spectra of a gum rubber and a pressure-sensitive adhesive are very similar, but there are subtle differences that distinguish the nontacky rubber from the tacky pressure-sensitive adhesive. Used in this context, the word "tack" does not mean self-adhesion, as in the tire makers' building tack, but in the ability to stick under light pressure to foreign surfaces. Figure 5-1 shows the creep compliance vs. time of a lightly milled Hevea rubber and of the pressure-sensitive adhesive prepared by blending 59 parts of weight (pbw) of the milled rubber with 41 pbw of a polymerized $\beta$-pinene resin. It is found that the pressure-sensitive adhesive is less compliant than the rubber when the stress interval is 0.01 sec or less, and increasingly more compliant at stress intervals exceeding 0.01 sec.

The shift to a less compliant response at time intervals of very short duration, say 0.001 sec or less, can be accounted for quite well by the shift in the glass transition temperature $T_g$ which occurs when the resin is added. The glass transition temperature of the rubber is 200°K, and that of the adhesive 212°K. The time shift, log $a_T$, that would correspond to the $T_g$ shift from 200°K to 212°K, can be estimated from the WLF equation

$$\log a_T = \frac{-17.4(T - T_g)}{51.6 + T - T_g}, \quad \text{amounting to 0.53 units} \tag{1}$$

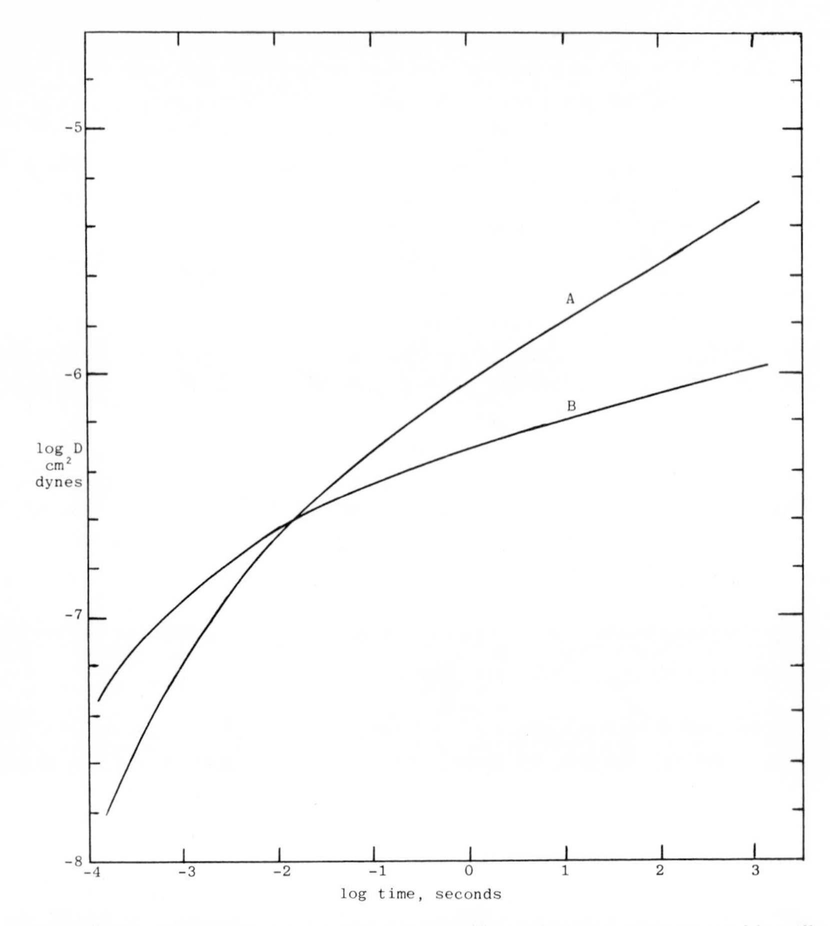


Fig. 5-1. Creep compliance in tension of a natural rubber-polyterpene pressure-sensitive adhesive (A) and the natural rubber base of the pressure-sensitive adhesive (B).

of logarithmic time. If the compliance-time curve of the rubber is shifted toward longer time by this amount (Fig. 5-2), it shifts into near coincidence with the compliance-time curve of the pressure-sensitive adhesive in the region where the stress duration is about 0.001 sec.

As the stress interval increases beyond 0.001 sec, the pressure-sensitive adhesive shows increasingly greater compliance than the rubber base. Thus the tackifying resin acts as a stiffening agent when stress duration is in the millisecond range, and as a plasticizer when the stress interval is greater by about an order of magnitude.

It is of interest that the time scale in which the crossover in the compliance-time curves takes place is precisely the time scale of viscoelastic

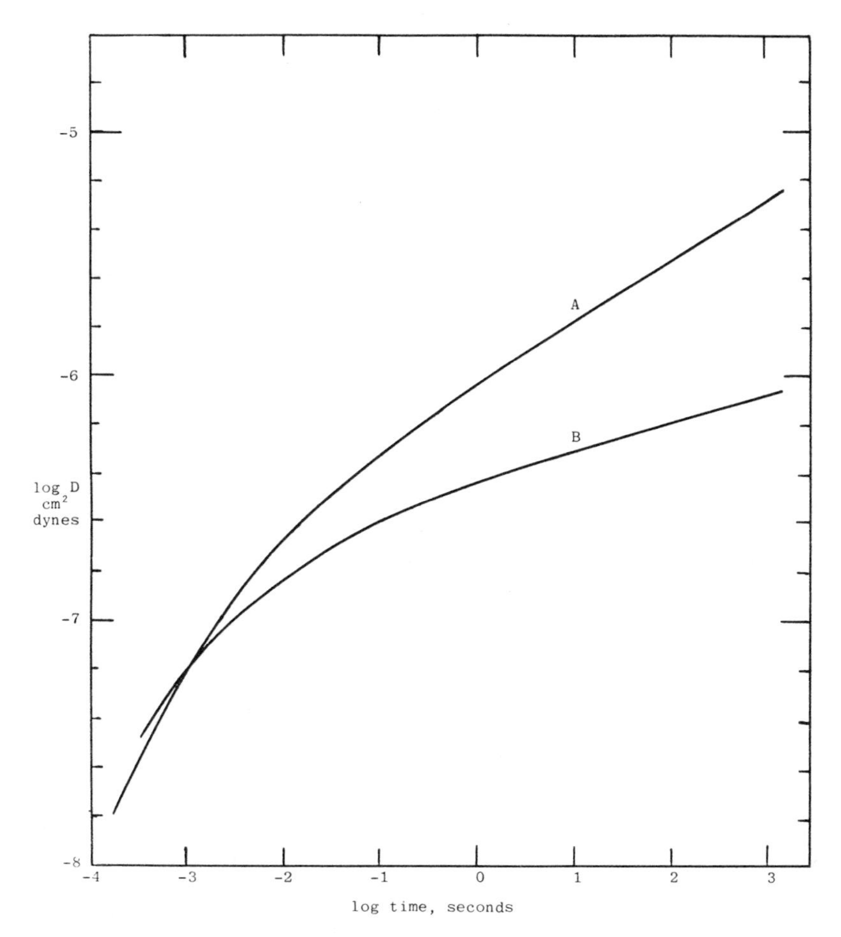


Fig. 5-2. Creep compliance curve A of Fig. 5-1 shifted to place A and B on an equivalent $T_g$ basis.

response in many of the conventional tests which measure the tack and peel adhesion of pressure-sensitive adhesives and pressure-sensitive adhesive tapes. In a simple probe tack test, it is quite common to allow a contact time of about 1 sec, and the separation, which generates the force taken to be a measure of tack, may occur within a time interval of 0.01–0.001 sec. The pressure-sensitive adhesive exhibits tack, and the rubber exhibits little or no tack. The pressure-sensitive adhesive is more compliant than the rubber in the time scale of bond making, but less compliant in the time scale of bond breaking.

Similarly, in peel adhesion tests running at very slow rates, say

1 cm/min. or less, the adhesive may undergo very high deformation, and even cohesive failure, before releasing from the adherend. At high rates, say 100 m/min., the peel process may become stick-slip and show all of the characteristics of brittle fracture.

The author has attempted, with some success, to evaluate the role of the modulus (reciprocal of compliance) in tack in a more quantitative way.[1] An observation well supported by experiment is that the compression modulus in the time frame of the bonding process cannot exceed $10^7$ dyne/cm$^2$, and preferably should approach $10^6$ dyne/cm$^2$. When the compression modulus is greater than $10^7$ dyne/cm$^2$, the degree of conformability to the surface contours of even so-called "smooth" surfaces is not sufficient to effect the desired adhesive bonding. Figure 5-3 illustrates this relationship between modulus and tack when both are measured over a

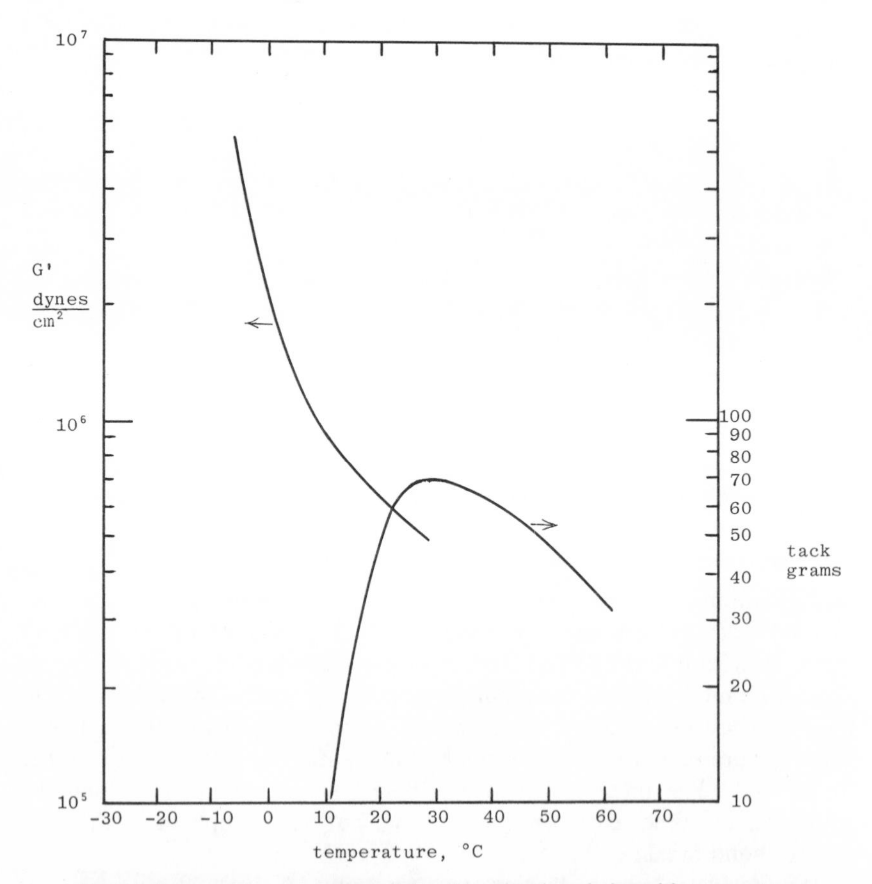


Fig. 5-3. Dynamic shear storage modulus, $G'$, and tack variation with temperature.

range of temperature. A specific instrument is used for the tack measurements.[1,2]

High compliance alone, however, is not a sufficient criterion to insure pressure-sensitive tack. For example, one can break down natural rubber by milling so that it will have more than adequate compliance for bonding, yet it will display little or no tack at room temperature. If, however, it retains sufficient compliance when chilled, it may exhibit appreciable tack at a lower temperature. The author has attributed this development of tack at low temperature to the increase in strength associated with lower temperature and high rate of extension in the bond breaking process. Recently Kraus *et al.*[3] have proposed that the loss modulus plays an important role in tack. As yet, this proposition seems to be somewhat speculative, though the "lossiness" of the adhesive would have a bearing on the rate of decay of compressive stress, and hence the rate of diminution of restoring force, during the bonding process.

## LONG-TERM CREEP

Resistance to long-term creep is a highly desirable property in pressure-sensitive adhesives. Generally, the requirements are dictated by end-use, where the stress conditions depend upon the application. For example, tape used for book mending or envelope sealing may never be subjected to severe stress. Yet, even here, when a transparent tape is used to mend a torn page in a book, the pressure-sensitive adhesive must resist the capillary forces, or else it will wick through the page and render it translucent. This process may take months to years. In other applications, pressure-sensitive adhesive tapes are required to support loads, sometimes for relatively short times at elevated temperatures, e.g., masking tapes supporting protective paper aprons in automotive painting during baking cycles, and sometimes for very long periods of time at room temperature, as when objects are attached to walls with double-faced tapes.

A dramatic example of undesirable creep is the "telescoping" of rolls of pressure-sensitive tape (Fig. 5-4), where the compressive stresses in the rolls are not directed uniformly toward the core, but have unidirectional components of shear stress which cause each successive layer of tape to slide sideways on the one beneath, and the accumulated movement causes the roll distortion known as "telescoping." This problem was especially prevalent and annoying in the early days of the plasticized vinyl film tapes when plasticizer migration caused the pressure-sensitive adhesives to become very soft and the techniques for winding tapes under controlled and uniform tension had not been perfected.

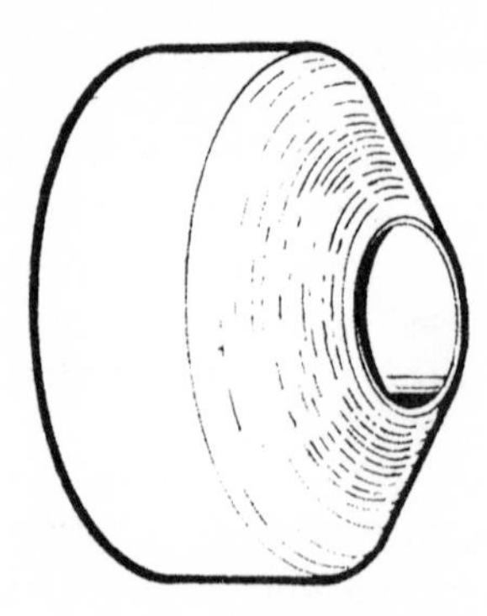

Fig. 5-4. Illustration of the telescoping of pressure-sensitive adhesive tape.

## LONG-TERM CREEP AND STEADY STATE VISCOSITY

When shearing stresses are small and the rate of creep is very slow, the creep rate depends primarily upon the steady-flow viscosity of the pressure-sensitive adhesive. If shearing stresses are high and the rate of shear correspondingly high, the viscosity may vary as the shear rate changes (non-Newtonian viscosity).

The factors that determine the steady flow viscosity are generally the same in pressure-sensitive adhesives as in all soft polymers. The steady-flow viscosity is strongly dependent on the molecular weight of a polymer. Measurements on series of monodisperse homopolymers have shown that the steady-flow viscosity varies as the 3.4 power of the molecular weight. In our own laboratories, Mr. Guy Fortry* investigated the rheological properties of *cis*-polyisoprene prepared in fairly narrow molecular weight distributions by *n*-butyl lithium polymerization (Table 5-1). Although the polymers were not ideally monodisperse, the proportionality between the steady-flow viscosity $\eta_0$ and the molecular weight raised to the 3.4 power is well supported by the grouping of data points about the line of slope 3.4 in the log $\eta_0$ vs. log $M_w$ plot (Fig. 5-5).

Polymers used in commercial pressure-sensitive adhesives are rarely of narrow molecular weight distribution, and deviation from $\eta_0 \propto M_w^{3.4}$ must be expected. Nevertheless, it is useful to know that the strong dependence of steady-flow viscosity on molecular weight exists, because a small increase in molecular weight can produce a significant improvement in capacity to support a load.

For example, two laboratory batches of a tacky copolymer of *n*-butyl

* Unpublished work.

**Table 5-1. Steady state viscosity vs. molecular weight of cis-polyisoprene.**

| $M_w$ | $M_w/M_n$ | $\eta_0$ (poises) |
|---|---|---|
| 125,000 | 1.23 | $1.2 \times 10^6$ |
| 350,000 | 1.27 | $4.0 \times 10^7$ |
| 590,000 | 1.28 | $2.0 \times 10^8$ |
| 830,000 | 2.50 | $4.0 \times 10^8$ |

vinyl ether and *i*-butyl vinyl ether were prepared under similar conditions, but one batch had an intrinsic viscosity of 4.4 deciliters (dl)/g and the other 3.5 dl/g. The intrinsic viscosity-weight average molecular weight relationship was taken to be

$$[\eta] = aM^{2/3} \tag{2}$$

The $M_w$s then stand in the ratio $[4.4/3.5]^{1.5}$, or 1.4, from which steady-flow viscosities would be expected to stand in the ratio $(1.4/1.0)^{3.4}$, or 3.1/1.0. The steady-flow viscosities were measured by creep and recovery and were found to be $2.1 \times 10^9$ poises and $5.9 \times 10^8$ poises, yielding a measured ratio of 3.6/1.0. In consideration of the uncertainties involved in the estimation of the molecular weights and lacking knowledge of molecular weight distributions, the agreement between the measured and the estimated ratios of steady-flow viscosities is not bad. Since the ability of a pressure-sensitive adhesive to sustain loads for long times is directly proportional to its steady-flow viscosity, the significance of molecular weight is readily apparent.

A small fraction of high molecular weight material can make a substantial contribution to creep resistance. The first elastomer used as a base for pressure-sensitive adhesives was natural rubber. High quality natural rubber gum has a bimodal distribution of molecular weights, and carries a fraction that is very high in molecular weight. For a practical reason, namely, to produce solutions of suitable coating viscosities and adequate solids content, the rubber was milled. It soon became apparent that the milling conditions had a great effect on the creep resistance, and therefore the shear holding ability, of the pressure-sensitive adhesives compounded from natural rubber. The high molecular weight fraction breaks down faster than lower molecular weight fractions, and therefore the steady-flow viscosity of natural rubber is rapidly reduced by milling. SBR gum, which does not have the bimodal distribution, breaks down much less rapidly than

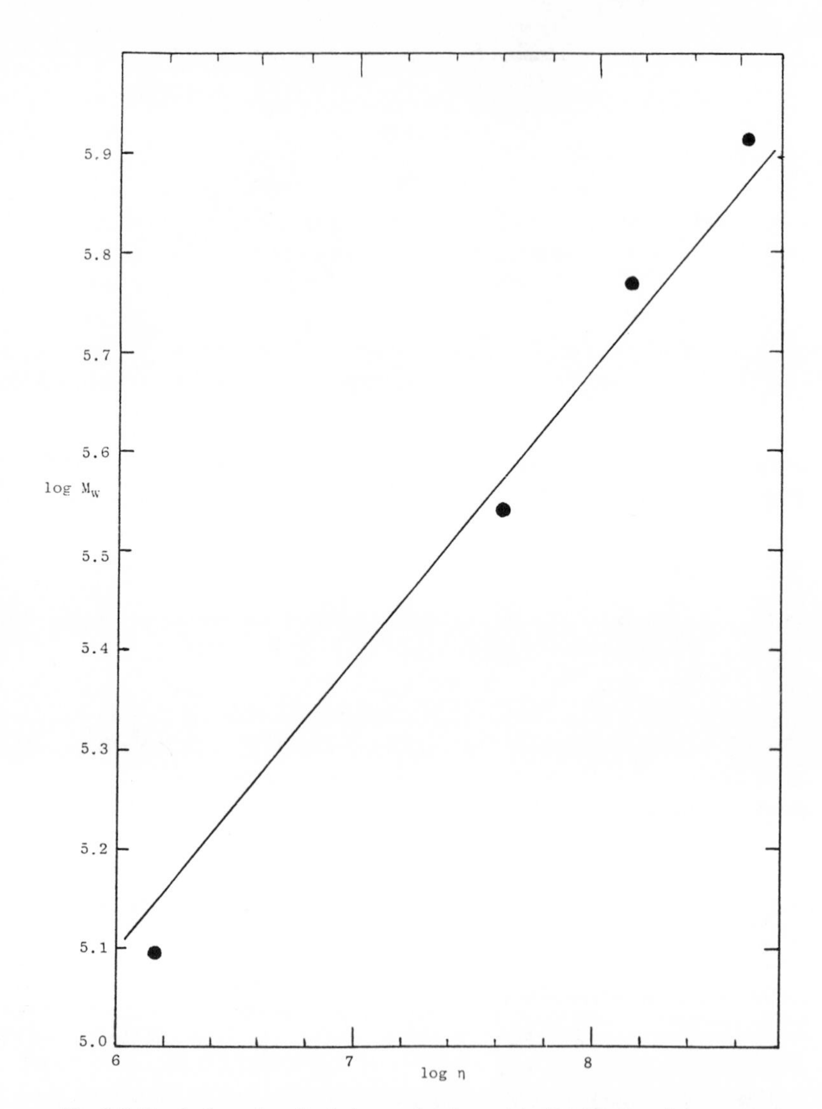


Fig. 5-5. Steady-flow viscosity ($\eta_0$) vs molecular weight ($M_w$) in *cis*-polyisoprene.

natural rubber gum, but it does not have the creep resistance provided by a high molecular weight fraction.

The contribution of high molecular weight to creep resistance is illustrated by an experiment in which small percentages of a synthetic pressure-sensitive adhesive copolymer of high molecular weight were added to a lower molecular weight copolymer of similar composition. Again, these were copolymers of fairly high polydispersity, and the applicability of viscosity-molecular weight relationships that hold for monodisperse systems is open to criticism. Nevertheless, useful guidelines become apparent in analysis of the observed results.

The weight average molecular weight of copolymer A was $7.2 \times 10^6$; that of polymer B, $4.5 \times 10^5$. Polymer A was added to polymer B in the amounts of 0, 4.14, and 6.80 percent by weight. After these compositions were coated on polyester film to give a dry coating thickness of 0.005 cm, overlap shear tests were made using polished steel as a test surface, an initial area of shear $1.27 \times 1.27$ cm, and a load of 1000 g. When times to failure in such a test are long enough so that the initial rate of shear is low and the slip is dominated by the steady-flow viscosity, the time to failure is given by

$$T = \frac{L^2 W \eta_0}{2Mtg} \tag{3}$$

where

$T$ = time to failure (sec)

$L$ = overlap (cm)

$W$ = width (cm)

$\eta_o$ = viscosity (poises)

$t$ = thickness of adhesive interlayer (cm)

$M$ = load ($g$)

$g$ = acceleration due to gravity (981 cm/sec$^2$)

Since the load and all of the geometrical parameters are fixed in each set of static shear tests, the time to failure becomes directly proportional to the viscosity of the adhesive.

The measured shear times, the steady-flow viscosities calculated from them, and the steady-flow viscosities that one would predict from the $M_w$ increase resulting from addition of the high molecular weight copolymer are given in Table 5-2. The excellent correlation between the observed and calculated steady state viscosities is possibly fortuitous, but it is obvious that small amounts of high molecular weight polymer can markedly increase the steady-flow viscosity and the creep resistance.

**Table 5-2. Effect of high molecular weight polymer fraction on shear time and steady–flow viscosity ($\eta_0$)**

| COMPOSITION OF BLEND (pbw) | | OBSERVED $M_w$ | CALCULATED $M_w$ | OBSERVED SHEAR TIME (sec) | $\eta_0$ (poises × $10^{-6}$) | |
|---|---|---|---|---|---|---|
| A | B | | | | Obs. | Calc. |
| 0 | 100 | $4.50 \times 10^5$ | $4.50 \times 10^5$ | 100 | 0.48 | 0.48 |
| 4.14 | 95.86 | $7.49 \times 10^5$ | $7.29 \times 10^5$ | 408 | 2.58 | 2.48 |
| 6.80 | 93.20 | $9.25 \times 10^5$ | $9.10 \times 10^5$ | 834 | 5.28 | 5.26 |

## CREEP AND NON-NEWTONIAN VISCOSITY

When stresses are such that the rate of creep, and hence the rate of shear, exceed minimum values that are not easily defined, the viscosity becomes non-Newtonian. It becomes dependent on the rate of shear and decreases as the rate of shear increases. Some years back, W.A. Vievering and the author investigated the shear rate dependence of the viscosity of several pressure-sensitive adhesives by subjecting them to various rates of shear in a shearing jig designed for an Instron tensile tester. At the time, we were interested in the holding power of automotive masking tapes.

The measurements were made at 121°C. The jig had a rectangular parallel plate configuration. The layers of pressure-sensitive adhesive were sandwiched between three plates and the restraining force was measured as the middle plate was pulled out of the sandwich. An elastic strain followed by degeneration into viscous flow was observed. Typical load vs. displacement curves are shown in Fig. 5-6. At the onset of viscous flow, the resisting force diminishes in direct proportion to the decrease in area under stress.

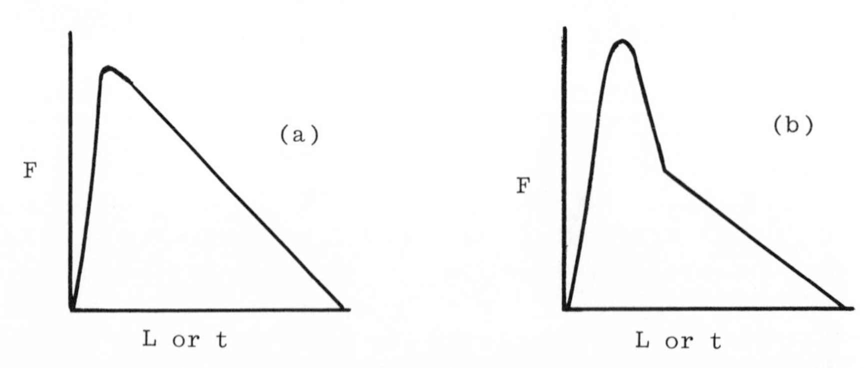


Fig. 5-6. Force vs length of overlap (L) or time ((t) in steady rate pull of a pressure-sensitive adhesive overlap shear bond. a. Elastic strain followed by viscous slip. b. Elastic strain followed by shear breakdown, then viscous slip.

A log-log plot of viscosity vs. rate of shear for one of the pressure-sensitive adhesives is shown in Fig. 5-7. Although a tendency toward steady-flow viscosity is indicated when the shear rate is less than 0.01/sec, the viscosity is non-Newtonian over the entire range. Overlap shear bonds were prepared and the times to failure under various loads were measured. Times to failure were calculated using an equation which the author has derived for situations where the viscosity-shear rate dependence can be expressed by a straight line on the log-log plot, such that

$$\eta = \eta_1 - a \log \gamma \tag{4}$$

$\eta_1$ is the viscosity when $\gamma = 1/\text{sec}$, $a$ is the slope, and $\gamma$ is the rate of shear.[4]

In the shear tests, the load was varied from 100 to 800 grams per 1.27 × 2.54 cm. The initial overlap was 1.27 cm and the thickness of the pressure-sensitive adhesive layer 0.0127 cm. The times to failure (Table 5-3) were such that the shear rates varied too much to fit a single straight line on the log-log plot. By fitting two lines to the data, as shown in Table 5-3,

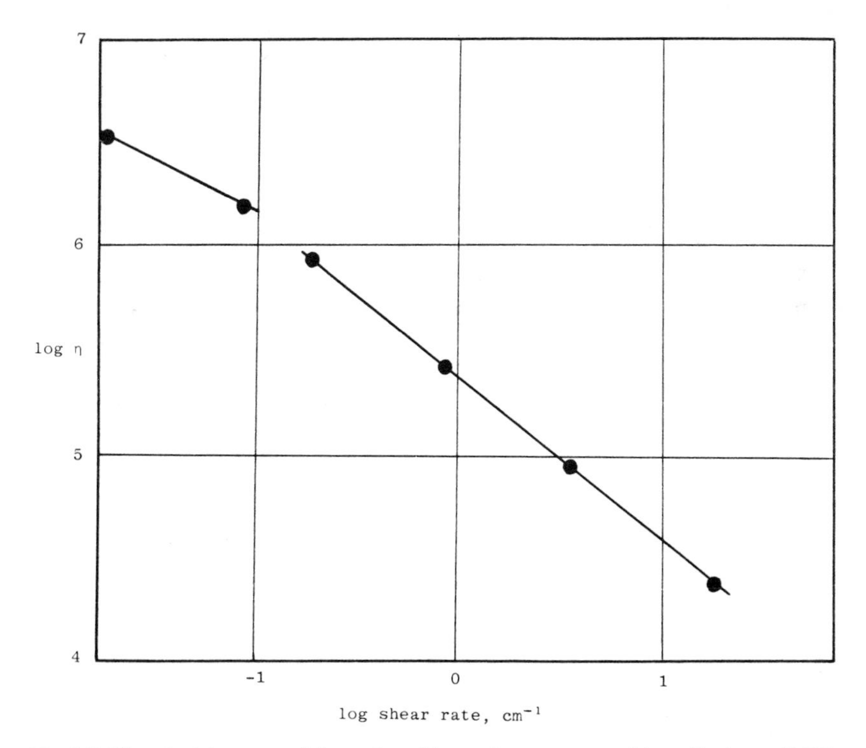


Fig. 5-7. Viscosity ($\eta$) vs. rate of shear of a rubber-resin pressure-sensitive adhesive at 121°C.

**Table 5-3. Non-Newtonian flow of pressure-sensitive adhesive in shear.**

| LOAD (g) | TIME TO FAILURE (min) | |
|---|---|---|
| | OBSERVED | CALCULATED |
| 100 | 128.5 | 127[a] |
| 200 | 27.3 | 31.8[a] |
| 400 | 2.9 | 2.9[b] |
| 800 | 0.25 | 0.18[b] |

[a] $\eta = 4.6 \times 10^5 - 0.5\gamma$
[b] $\eta = 2.1 \times 10^5 - 0.75\gamma$

excellent agreement between calculated and observed times to failure was obtained, justifying the observation that the creep in shear was dependent on the viscosity. Obviously, the fitting of curves to viscosity vs. shear rate data and the calculation of times to failure in shear when non-Newtonian flow prevails can be done much better with present-day computer technology.

Cognizance of the shear rate dependence of the steady-flow viscosity must be taken both in testing and in use of pressure-sensitive adhesives. Thus, one cannot assume that doubling the stress will halve the time to failure in a shear test, and when pressure-sensitive adhesives are loaded in shear, overloading can reduce holding times far out of proportion to the performance anticipated from a high and constant steady-flow viscosity.

## CHEMICAL CROSSLINKING AND CREEP RESISTANCE

An obvious way to raise the steady-flow viscosity and increase the creep resistance of a pressure-sensitive adhesive is to take advantage of crosslinking. If one can join each pair of molecules by a crosslink, thereby doubling the molecular weight, the steady-flow viscosity should increase by a factor of $2^{3.4}$, or 10.5. Crosslinking must be done judiciously, however, because "overcure" can have deleterious effects on the tack and adhesion.

Figure 5-8 illustrates the effect of crosslinking on the creep and recovery of a synthetic pressure-sensitive adhesive polymer, which, in the uncrosslinked state, has a molecular weight $M_w$ of about 500,000, a Gaussian distribution of molecular weights, and an $M_w/M_n$ of about 10. The steady-flow viscosities before and after crosslinking indicate a 1.8 fold increase in molecular weight or about 0.9 crosslink per molecule.

The patent literature reveals that "curing" of pressure-sensitive adhesives to improve creep resistance was practiced early in the development of

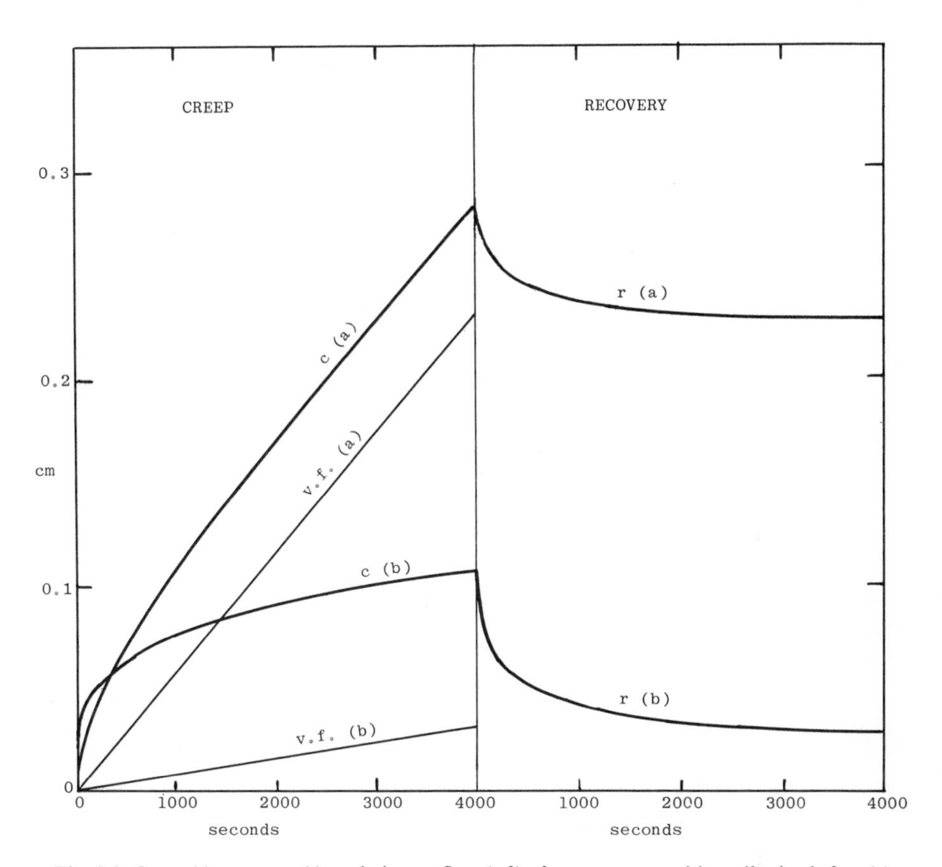


Fig. 5-8. Creep (c), recovery (r), and viscous flow (v.f.) of a pressure-sensitive adhesive before (a) and after (b) crosslinking.

pressure-sensitive adhesive tape technology. It was found to be beneficial to cure rubber-resin pressure-sensitive adhesives with oil-soluble phenolic resins.[5] Peroxide curing of synthetic polymer pressure-sensitive adhesives has been practiced for a long time,[6] as has crosslinking by ultraviolet and high-energy radiation.

In the early days, control of the degree of crosslinking was very much an art. Modern methods for measurements of molecular weights, molecular weight distribution, and crosslink density have placed the technology on a much more scientific basis. Currently, when attention has been turned to solventless processing, which generally requires low molecular weight polymers for practical coating speeds, web curing is essential and the advances in the characterization of crosslinked systems have taken on added importance.

One might ask what level of steady-flow viscosity is a reasonable objective for adequate performance in a pressure-sensitive adhesive. If we take as a requirement that an overlap shear test in which a dead load of 500 g is applied to a $1.27 \times 1.27 \times 0.0025$ cm specimen should survive for at least 100 minutes, it can be calculated from Equation (4) that a steady-flow viscosity of $7.2 \times 10^6$ poises is required. This could be a typical objective for so-called general-purpose use. If, on the other hand, a double-faced tape having on each side a 0.005-cm layer of pressure-sensitive adhesive is required to hold a 1.0-g load per $1.27 \times 1.27$ cm for a year with an allowable slippage of only 0.127 cm (10%), the steady-flow viscosity must be $5.4 \times 10^8$ poises. This approximately 100 fold spread in steady-flow viscosity, and therefore in creep resistance, is quite typical of the spread found in commercial pressure-sensitive adhesives, and even a 1000 fold spread is quite possible. In synthetic polymer pressure-sensitive adhesives, this would indicate a 4 to 8 fold spread in molecular weights.

If the molecular weight increase is to be realized simply by crosslinking a polymer of low molecular weight, say $2 \times 10^5$ to $5 \times 10^5$ daltons, the augmentation of molecular weight cannot be effected without some gel formation. Now, ideally, a gel should reach an equilibrium compliance and therefore exhibit infinite steady-flow viscosity, but due to imperfections in networks produced in commercial polymers of broad molecular weight distributions, pressure-sensitive adhesives can have a substantial gel content and still exhibit nonrecoverable creep. For example, the crosslinked copolymer of Fig. 5-8 had a gel content of about 50% as measured by 48-hour immersion in tetrahydrofuran.

## PHYSICAL CROSSLINKS

The concept of entanglement molecular weight has been introduced into the theory of long-term compliance and steady-flow viscosity of polymers. Long, linear, flexible polymers are believed to become entangled in interlinked loops which behave as pseudo-crosslinks, which, though they will eventually disentangle under stress, contribute long retardation times and high steady-flow viscosities. The contribution to high steady-flow viscosity and creep resistance will depend upon the number of entanglements per molecule. The lower the entanglement molecular weight, the higher the number of entanglements for a given molecular weight. Polymers with large side groups such as the acrylic and vinyl ether polymers commonly used in pressure-sensitive adhesives have high entanglement molecular weight, in the range of 40,000 to 50,000. On the other hand, the entanglement molecular weight of common synthetic rubbers is on the order of $10^4$ or less. Thus the number of entanglements per molecule of a typical synthetic polymer pressure-sensitive adhesive of molecular weight $10^6$ is 20 to 25; that of natural rubber of the same molecular weight, about 100.

The addition of low molecular weight diluents reduces the number of entanglements per unit volume and increases long-term compliance and flow. Natural rubber, with its large number of entanglements per unit volume, can tolerate a considerable dilution and yet retain adequate creep resistance.

Copolymerization of monomers having bulky aliphatic ether or ester groups with monomers which provide hydrogen bonding, ionic bonding, or acid-base exchange has an effect on compliance similar to that of increasing the prevalence of chain entanglement, thereby raising the holding power of pressure-sensitive adhesives. The groups that enhance chain interaction generally raise the glass transition temperature. This will shift the whole mechanical spectrum such that the compliance is reduced at all time intervals of stress application. This can result in insufficient tack at room temperature, and shocky or stick-slip peel, if the introduction of the strongly interacting groups is not done with impunity.

The introduction of the ABA type of block copolymers[7,8,9] confronted the pressure-sensitive adhesive industry with technology for physical crosslinking and control of copolymer structure and molecular weight not previously available. The prospect of copolymers that could be melt processed with ease, yet could retain the properties of cured rubbers over an acceptable range of use temperatures, was, to say the least, exciting to adhesive formulators.

For those concerned with pressure-sensitive adhesive technology, the perfection of the network structure and the high level of crosslinking provided by the uniform size of the glassy A segments and the rubbery B segments, and the complete incorporation of the A blocks into separate glassy domains, posed a difficult tackification problem. This is obvious from the high level of plasticizing oils and soft tackifying resins used in the early formulations. While a high level of creep resistance is retained at room temperature, the migration of low molecular weight compounding agents may cause problems in long-term adhesion and in creep resistance at temperatures much lower than the $T_g$s of the glassy blocks, this last problem resulting from plasticization of the glassy blocks by the tackifying agents.

A compromise is possible by formulating these pressure-sensitive adhesives primarily on AB block copolymers, which associate through the A blocks to have an effective molecular weight at room temperature several times the molecular weight of the individual AB molecule. The steady-flow viscosity at room temperature then stands in relation to the melt viscosity as $n^{3.4}$, where $n$ is the degree of association of the ABs. (Appropriate account of viscosity dependence on temperature must also be taken.) The AB block copolymers can be tackified with smaller amounts of tackifying agents than the ABAs, and if further augmentation of creep resistance is desired, this can be done by addition of ABA at low levels, or by low level chemical crosslinking.[10]

## PRESSURE-SENSITIVE ADHESIVE DEGRADATION AND CREEP

Because of the high sensitivity of long-term compliance to molecular weight, compliance measurements are very useful in detecting polymer degradation in pressure-sensitive adhesives. This came to the author's attention very forecefully when a problem arose due to iron salt contamination in a rosin tackifier used in a natural rubber-based pressure-sensitive adhesive. The iron catalyzed oxidation initially caused chain scission of the rubber, with a dramatic increase in long-term compliance. Short-term compliance was not greatly affected.

As time went on and oxidation continued, the pressure-sensitive adhesive turned leathery and eventually resinified. Creep compliance then decreased accordingly (Fig. 5-9). The degradation produced first a loss of holding power, and eventually a complete loss of tack, but the creep compliance measurements were more sensitive to the early changes than these functional tests.

## MECHANICAL PROPERTY MEASUREMENTS

In early investigations of the creep of pressure-sensitive adhesives, the author measured creep in tension.[11] Bubble-free films of pressure-sensitive adhesive were cast from solution, nominally to 0.025–0.050 cm in thickness. Strips of predetermined length were cut and were loaded by a weight which had the shape of a hyperboloid of revolution. The weight sank into a cylinder filled with water, and the buoyant force reduced the load to the extent required to maintain constant stress per unit of cross-sectional area as the specimen underwent extension in creep.

The method of maintaining constant stress by means of a hyperboloid weight and buoyancy was first published by Andrade.[12] Another method for maintaining constant stress in tensile creep involves a cam and a wheel on a common shaft set in low friction bearings. The periphery of the cam is grooved and a string which suspends a weight lies in the groove. Another string goes around the periphery of the wheel and is attached to the specimen. The shape of the cam is such that the torque diminishes as the cam and wheel turn due to the stretching of the specimen.

The chief advantage of tensile creep over shear creep in the early days

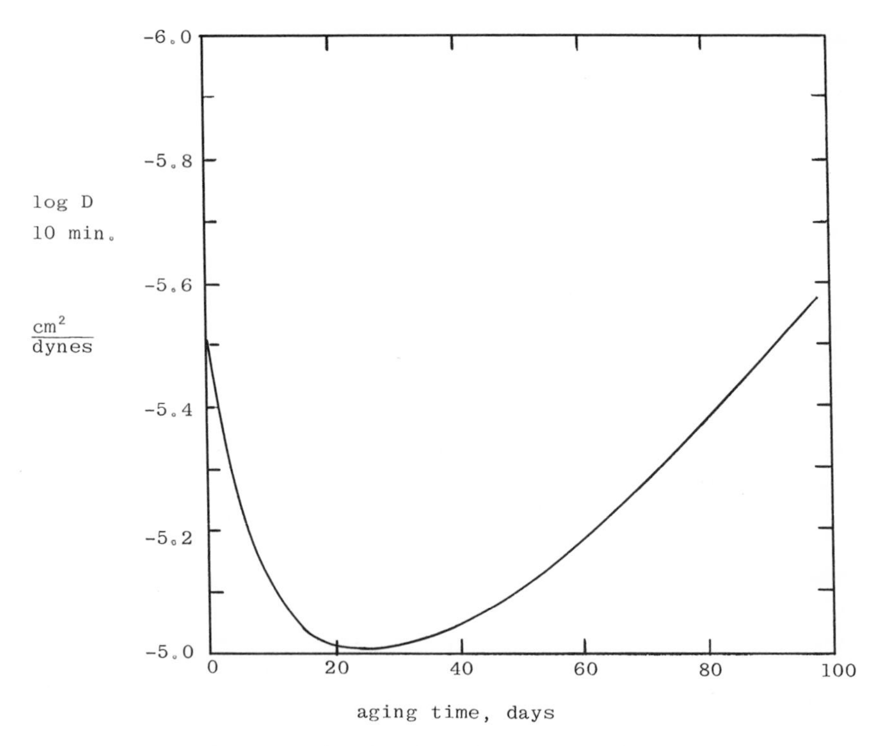


Fig. 5-9. Aging of an iron contaminated rubber-resin pressure-sensitive adhesive. Change in 10-minute tensile compliance with aging time.

was that the strain in tensile creep involved displacements that were large enough to be recorded and measured accurately with crude equipment. This advantage has long since disappeared with advances in technology for measuring and recording small displacements. Creep in shear is now the method of choice.

Several shear creep methods are discussed by Ferry[13] in general terms. The one chosen by this author and his associates is the one we call "parallel plate double sandwich," where two layers of pressure-sensitive adhesive are sandwiched between three plates and load is applied to the middle plate. Contrary to Ferry's advice, we have found it expedient to arrange the plates vertically and counterbalance the middle plate so that recovery proceeds unrestrained when the load is removed. The two layers of adhesive are typically 1 mm thick and $2 \times 3.5$ cm in area, giving a total area of 14 $cm^2$. The typical load is 500 g. The creep with time is readily measured by a suitable device for measuring small strains, such as a linear variable differential transformer (LVDT) and the strain versus time is strip chart recorded. Upon removal of the load, the recovery is recorded over a time span equal to the creep time.

The measurement of the steady-flow viscosity from creep measurements is subject to some uncertainty. In principle, the permanent set that remains after complete recovery of the elastic deformation will be the consequence of viscous flow, and from it one can calculate the steady state viscosity. Some authors have proposed that the steady-flow viscosity can be obtained by subtracting the recovered strain from the creep strain at equivalent times, and the difference then generates the time vs. displacement curve due to steady state flow. This is valid only when the creep time is sufficiently long so that little or no elastic creep remains. Viscoelastic processes with long retardation times are difficult to distinguish from viscous flow. The problem is generally encountered in pressure-sensitive adhesives containing very high molecular weight polymers, so that there is a high degree of entanglement, and in partially crosslinked pressure-sensitive adhesives.

If one wishes to measure the steady-flow viscosity directly, it can be done, in principle, by capillary rheometry. This is not a creep experiment. It has its own difficulties upon which the author does not wish to elaborate.

As mentioned earlier, if one wishes to measure mechanical properties directly in time regimes of less than 1 sec, one has to resort to oscillatory methods. There are many instruments which can be used and the author does not wish to endorse any specific one, except to say that from personal experience it has been found advantageous to have an instrument that will span the frequency range of about 10–2000 Hz.

In the mid-1940s, before commercial instruments were available, Dr. W.W. Wetzel and the author measured the dynamic tensile moduli of pressure-sensitive adhesives in the range of about 100–2000 Hz by measuring the velocity of longitudinal vibrations in strips of pressure-sensitive adhesive films. The method has been described by Nolle.[14] The method yields approximate complex tensile moduli which can be resolved into storage and loss components, but instrumentation now available offers precision and much better temperature control.

Both of these features were offered by the Fitzgerald instrument[13] which the author used with good results for several years, but the excessive time needed to reach temperature equilibrium renders this instrument very costly to operate. It measures the storage and loss shear moduli over a frequency range of 20–5000 Hz, although the author encountered problems with system resonances at frequencies above 2000 Hz.

The Weissenberg Rheogoniometer and the Rheometrics Mechanical Spectrometer are elegant instruments that utilize cone-and-plate geometry and are versatile in that they can be used for fluid viscosity, creep compliance, and dynamic shear modulus measurements. However, they are limited to frequencies less than 100 Hz because of inertial effects.

The Rheovibron is now widely used, and though originally designed for obtaining mechanical properties of relatively rigid materials, it now comes equipped with accessories for measuring the dynamic shear moduli of soft polymers.

The Melabs Rheometer, an instrument which the author and his associates found particularly suitable for obtaining dynamic shear moduli of pressure-sensitive adhesives over the 20–2000 Hz frequency range and a temperature range from near $T_g$ to 130°C, offered advantages in ease of operation and temperature control. Unfortunately, it is no longer manufactured commercially and it required drastic modification of the stress sensor. Its main feature is that the driver is a stack of piezoelectric wafers.

Devices which utilize the vibrating reed, one of which is manufactured by Bruel and Kjaer, or the torsion pendulum, which is a low frequency device, when used on pressure-sensitive adhesives require supplementary stiffness such as a steel spring support for the pressure-sensitive adhesive to counter the damping of the adhesive. This tends to reduce sensitivity, since the contribution to the stiffness provided by the support must be subtracted from the total stiffness. However, the composite vibrating reed has been used with some success.[15] As a general rule, the dynamic mechanical methods subject the specimen to very low shear amplitudes. This has to be taken into consideration when one uses the data to predict stress-strain response when strains are high. However, in one instance, the author subjected a synthetic copolymer pressure-sensitive adhesive to dynamic shear strains as high as 5 and found good agreement between the dynamic moduli at the high strain with those obtained at extremely low strains. The main discrepancy could be attributed to a temperature rise.

## SUMMARY

The creep compliance of a pressure-sensitive adhesive is significant in tack, adhesion, peel and resistance to failure in loading. Both retarded elastic processes and steady-flow viscosity play important roles, the former in quick-stick, bonding, tack, and peel, the latter in long-term creep. In some circumstances, the non-Newtonian viscosity of pressure-sensitive adhesives must be taken into account.

The steady-flow viscosity is highly dependent upon molecular weight. Long-term creep can be greatly reduced by crosslinking, both chemical and physical. Long-term creep is also very sensitive to polymer degradation.

## REFERENCES


1. Dahlquist, C. A. Tack, *Adhesion, Fundamentals and Practice.* London: McLaren and Sons, Ltd., 1966.
2. Hammond, F. H., Jr. Polyken Probe Tack Tester, *Adhesion.* Am. Soc. for Testing and Mtls. Tech. Publ. No. 360, 1964.
3. Kraus, G., Rollmann, K. W., and Gray, R. A. Tack and Viscoelasticity of Block Copolymer Based Adhesives. *J. of Adhesion* **10**(3): 221 (1979).
4. Dahlquist, C. A. Pressure-Sensitive Adhesives, *Treatise on Adhesion and Adhesives.* New York: Marcel Dekker, Inc., 1967.
5. Drew, R. G. U.S. Patent 2,410,053 (October 29, 1945).
6. Ulrich, E. W. U.S. Patent 2,973,286 (February 28, 1955).
7. Holden, G. U.S. Patent 3,231,635 (January 25, 1966).
8. Zelinski, R. P. U.S. Patent 3,251,905 (May 17, 1966).
9. Harlan, J. T. Jr., U.S. Patent 3,239,478 (March 8, 1966).
10. Dahlquist, C. A., and Kolpe, V. V. U.S. Patent 3,787,531 (January 22, 1974).
11. Dahlquist, C. A. Elongation of Soft Polymers—A Constant Stress Method. *Ind. Eng. Chem.* **43**: 1404 (1951).
12. Andrade, E. N. daC. On the Viscous Flow in Metals, and Allied Phenomena. *Proc. Roy. Soc. (London)* **84**: 1 (1910).
13. Ferry, J. D. *Viscoelastic Properties of Polymers.* New York: John Wiley and Sons, Inc., 1970.
14. Nolle, A. W. Methods for Measuring Dynamic Mechanical Properties of Rubber-Like Materials. *J. Appl. Phys.* **19**: 753 (1948).
15. Test Method for Measuring the Vibration Damping Properties of Materials. ASTM Committee E33 Proposed Standard Method Draft No. 5, revised November 15, 1979.

# Chapter 6

# Electrical Properties

**Erhard Braeunling**
*Beiersdorf AG*
*Hamburg, Germany*

Pressure-sensitive adhesive tapes are used for various applications in the electrical industry. They offer advantages in some production processes and in most cases, the pressure-sensitive tapes are used to assist in production operations. In many cases the tapes have become an important element in the construction of electrical products. Certain production processes became possible only through the introduction of pressure-sensitive tapes.

Important areas of application for these tapes are the construction of transformers and electrical motors. A large amount of tapes is consumed in electrical installation work and in cable production. Pressure-sensitive adhesive tapes are also used in relays and switches. For example, in the production of transformers, the use of tapes allows for an improved operation. The tapes are used in the following production steps: core insulation, interlayer insulation, securing of winding start and winding end, securing of windings in multiple winding coils, insulating soldered joints at lead wires, stress relief of lead wires, and protective covering of the outer winding layer.

The manufacture of electrical harnesses is another example of a purposeful use of pressure-sensitive tapes. A bundle of cables is wrapped with a conformable adhesive tape in a spiral with a certain overlap. Difficult constructions with several branches can be easily covered with flexible tape manually or by a machine.

These few examples show that the proper choice of the adhesive tape is important for such specialized applications. Besides the mechanical and adhesive properties, certain chemical and thermal resistance is also required. Finally, quite expectedly for electrical applications, the electrical properties must be acceptable.

It is expected that the electrical pressure-sensitive tapes have a definite value of the dielectric strength and other desirable dielectric properties. A common requirement is the noncorrosiveness of the adhesive. Some users require that the quality of the adhesive tapes is not affected much by aging at 155°C or even at 180°C. Another application requirement is the stability to the resin and lacquer impregnants.

The tapes must undergo a rigid quality control in order to ensure that the products of constant properties are delivered to the user. The test methods must be closely related to the actual application. Naturally, these tests must give reproducible results to all testers. These tests have been nationally and internationally standardized after a careful selection of testing procedures. Only in exceptional cases are so-called house methods used after an agreement between the manufacturer and its customer.

The most important test methods will be discussed. These are related to tests designed for other electrical insulation materials.

## DIELECTRIC STRENGTH

The value of dielectric strength gives the user an indication of the stability of pressure-sensitive adhesive tape in strong electrical fields. The dielectric strength is defined as a quotient of breakdown voltage and the thickness of tape being tested. Breakdown voltage is defined as the effective value of a sinusoidal alternating voltage at which the insulation capability of the adhesive tape is suddenly lost because of its structural destruction.

The test apparatus for the breakdown voltage determination consists of transformers with a high rated voltage equipped with controls that allow an increase in voltage at a desired and constant rate from zero to the breakdown value. The voltage is imposed on the test specimen with metallic electrodes in the form of plates, spheres, or rods of different dimensions. For accuracy of measurement, it is important to locate the electrode centers in a homogeneous electric field which decreases in strength near the electrode edges.

The dielectric strength is not a material property. It depends on many outside factors. For example, a spherical pair of electrodes will give a different breakdown voltage than the one obtained by employing plate electrodes. It has been also observed that a thicker tape will give a lower dielectric strength value as compared to a thinner tape of identical chemical composition. Other factors which affect the dielectric strength are the frequency and curve form of the test voltage, as well as the rate of voltage increase.

Generally, the short time breakdown voltages are determined. This means that the voltage in the test apparatus is increased at a constant rate over a short time period until the breakdown occurs. In practical applications, it is often more important to know the stability of the tape against a constant voltage of a longer duration. This is obtained by measuring the so-called withstand voltage and *n* minute test voltage. The withstand voltage is the voltage which can be withstood by the adhesive tape for a longer period of time under specific test conditions without a dielectric breakdown. A *n* minute test voltage is a withstand voltage that lasts *n* minutes. These continuous voltages usually are considerably lower then the measured short time voltage of the same specimen.

In most cases the tests are conducted at standard conditions of 23/50 (i.e., 23 ± 2°C and at 50 ± 5% relative humidity). Since some of the adhesive tapes might be exposed to moisture during use, it is important to check the dielectric strength after an exposure to water.

An example of the dielectric strength and its dependence on time of exposure to voltage and on humidity is given in Table 6-1 for plasticized PVC tape and for a polyester (polyethylene terephthalate) film tape. Both adhesive tapes were tested after aging at standard conditions and once by the 15-minute withstand voltage method. In another experiment, the tapes were soaked in water for 24 hr. and tested using the short time voltage method. The results show that the exposure to withstand voltage and to soaking in water decreases the dielectric strength. The effect of humidity is greatest in the case of plasticized PVC film tape and of withstand voltage in the case of polyester film tape.

This short discussion should show the importance of method of determination on the value of breakdown resistance. It also shows that simply a statement of a measured value is meaningless, unless all testing conditions and details are stated as well. The test methods and conditions for breakdown resistance determination are described in several specifications.[1,2,3,4]

**Table 6-1. Dielectric strength and its dependence on the voltage duration and degree of humidity.**

| TEST METHOD | BREAKDOWN RESISTANCE (kV/cm) | |
|---|---|---|
| | PLASTICIZED PVC TAPE 100 μm thick | POLYESTER TAPE 25 μm thick |
| Short time test after standard aging at 23/50 | 560 | 980 |
| Short time test after soaking in water for 24 hr. | 150 | 910 |
| 15-minute withstand voltage after standard aging at 23/50 | 400 | 420 |

## ELECTROLYTIC CORROSION

Pressure-sensitive adhesive tapes, in contact with conductors through which electrical current is flowing, may cause corrosion under high humidity conditions. Such corrosion can increase the resistance of thin wires by decreasing their cross-sectional area and even cause a break in the current flow. This phenomenon is the result of the ionic conductivity of the adhesive tape. It is especially accelerated by direct current.

The corrosive action of the adhesive tapes cannot be definitely predicted by a chemical analysis. Dependable results can be obtained only through the physical measurement methods. The most common test methods are the indirect measurement of electrolytic corrosion, wire tensile strength method and the visual inspection method.

The most often conducted test is the indirect measurement of electrolytic corrosion. By this method, the conductive constituents of the adhesive tape are determined by a quantitative measurement of the electrical resistance. A layer of an adhesive tape is placed between two electrode clamps. The electrodes are made from brass. Their size and the distance between electrodes is fixed. The sample is aged for 18 hr in a chamber at 23°C and 96% R.H. The resistance at the electrodes to a direct voltage of 100 volts is determined immediately after aging. The results are reported in M ohms.

A direct method to determine the corrosive action of an adhesive tape is the so-called wire tensile strength method. Thin, bare copper wires are arranged as anode and cathode on the adhesive tape by being placed parallel with fixed distances between them. A potential of 250 volts is imposed on the wires for 24 hr in an atmosphere at 50°C and 100% R.H. The effect of electrolytic corrosion is determined by the measurement of tensile strength of the copper wires. If the adhesive tape has ionic ingredients, the cross-sectional area is changed, causing a decrease in tensile strength. The ratio of the tensile strength after to before the effect of the adhesive tape is called the corrosion factor. It has a value of 1.0 for noncorrosive adhesive tapes and a value of $< 1.0$ for corrosive tapes.

The visual inspection method allows a direct observation of the effect of electrolytic constituents in the tape. The tape to be tested is pressed against two juxtaposed brass foil electrodes, which are connected to a direct current at 100 volts. The test is conducted for four days in a chamber at 40°C and 93% R.H. If the adhesive tape contains ionic materials, color changes are observed on the brass foils. At the anode, zinc is removed and the excess of copper causes a red coloration. At the cathode, oxidation causes a black coloration. The appearance of the brass foils is compared to the reference color tables and a value is assigned. Unobjectionable adhesive tape might have a value of $A = 1.0$, while very bad tape might have a value of $B = 3.0$.

Which of the above methods for corrosion determination is chosen depends on the technical requirements of the application. All methods have their advantages and disadvantages. It should be mentioned that the experience shows that the visual method is the most sensitive, although it does not give an exact, objective value that is read from an instrument.[5,6,7,8]

## INSULATION RESISTANCE, VOLUME RESISTIVITY

Electrical resistance is defined as a quotient of electrical potential at the ends of a conductor and electrical current (Ohm's law). Adhesive tapes that have many unbonded and, under the influence of electrical potential, able-to-move electrons and ions, will have a low insulation resistance.

Generally, any resistance of an insulator, measured between two electrodes, is denoted as the insulation resistance. In addition, a difference should be made between the insulation resistance internally and at the surface. In case of pressure-sensitive adhesive tapes intended for electrical applications, the volume resistivity is especially important.

The volume resistivity is a physically defined value. It is obtained by multiplying the volume resistance of the adhesive tape by electrode surface and dividing the product by the adhesive tape thickness. The volume resistance itself is obtained in accordance to the Ohm's law by suitable measuring devices employing voltmeter-ammeter measurements or with help of a bridge circuit. The values of resistance are comparable to one another, regardless of the method of determination. Small differences might be caused by different measuring potentials. These differences can be disregarded within potential range of 100–1000 volts.

German Industrial Specification DIN 40 633, p. 1, section 4.6.1 describes a good test which is close to the practical use of the tape. An 8 mm-thick copper rod is wrapped with an adhesive tape over a length of 400 mm with an overlap of 70%. The rod is bent in a U-shape and is immersed in a container with conductive water so that 250 mm of the wrapped rod is under water. After soaking for 2 hr in water, a direct voltage of 1000 volts is imposed between the copper rod and water. The volume resistance is measured after waiting for 1 min. For calculation of the volume resistivity, the surface of the wrapping immersed in water is chosen as the electrode surface and three times the tape thickness is used for the sample thickness.[9,10,11,12]

## DIELECTRIC CONSTANT, DIELECTRIC DISSIPATION FACTOR, LOSS INDEX

The dielectric constant and the dielectric dissipation factor are basic material constants which characterize an insulator. Both values may depend on the frequency, temperature and voltage. They give the user useful indication about the suitability of the insulator, in our case the pressure-sensitive adhesive tape. For example, the use of pressure-sensitive adhesive tapes in insulation technology should give low dielectric losses. Their maximum should be shifted to the highest possible temperature range in order to avoid heat penetration. If the pressure-sensitive adhesive tape is used for condenser construction, a high dielectric constant would be advantageous.

The dielectric constant is the factor by which the capacity of a condenser increases, when another insulator such as a dielectric is placed between the condenser plates instead of a vacuum. This factor is brought forward by polarization effects which are related to the structure of the insulator.

Energy losses in the electric alternating field are caused by the rhythmic polarization changes. A measure of these losses is the dielectric dissipation factor. It is expressed as a tangent of the angle of effective and reactive currents. It should be remembered that the effective current, which causes the energy loss in the insulator, always has the same direction as the corresponding voltage. The capacitative reactive current precedes the voltage by a one-quarter period.

In practice, the use of loss index is preferred. It is a product of the relative dielectric constant and the dielectric dissipation factor.

A general qualitative rule is important that, in case of nonpolar or slightly polar materials, the dielectric values are nearly constant in a broad frequency range. In case of polar substances, the dielectric constant varies depending on the frequency and temperature. In case of the dissipation factor, several maxima are observed as the result of dependence on frequency and temperature. The dielectric properties are determined by bridge, resonance, or other methods. The values obtained by different methods are comparable.[13,14,15,16]

## TEMPERATURE–TIME LIMITS

The size of electrical equipment is decreasing for weight and cost reasons, such as in case of electric motors, and the materials used must be capable of meeting stricter thermal requirements. The determination of temperature-time limit is an important method gaining in importance for stating the thermal expectations of electrical pressure-sensitive adhesive tapes.

Temperature-time limit is the highest temperature at which the adhesive tape can be used continuously for a period of 20,000 hr. During this time, the electrical and mechanical properties must not fall below a minimum value. For testing purposes, the dielectric breakdown voltage is considered as representative of all electrical properties and the weight loss of all mechanical properties. The dielectric breakdown voltage was chosen as a representative of all electrical properties, because during thermal aging tests, informative curves are obtained. This, for example, is not the case with the dielectric dissipation factor. A similar statement can be made about the weight loss. In this case, the aging data curves appear very uniform and intersect the level of the lowest requirement by a sharply dropping branch of the curve.

The conduction of temperature-time limit determination will be described by the method of breakdown voltage only. A larger number of 8 mm in diameter copper rods are covered for a length of 200 mm with the tape. The ends of the copper rods are left free. These copper rods are stored in three ovens at different high temperatures. The oven temperatures are 10–50°C above the expected temperature limit. After 1, 2, 4, 8, 16, and sometimes 32 weeks, five rods are removed from each of the ovens and cooled down to room temperature. The rod wrapping is covered for a length of 100 mm with a layer of silver. Immediately, the breakdown voltage of the adhesive tape is checked by imposing a sinusoidal alternating voltage of 50 Hz between the copper rod and silver coating. The evaluation of the obtained test data is displayed in Figs. 6-1 and 6-2. The determined breakdown voltage is plotted against the logarithm of aging time for each of the three test temperatures $T_1$, $T_2$, and $T_3$ (Fig. 6-1). Three resulting

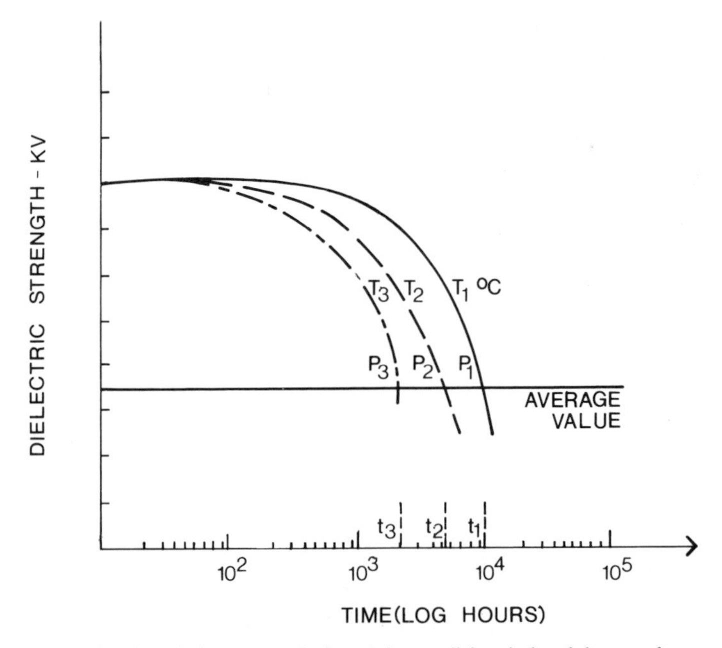


Fig. 6-1. Determination of time to reach the minimum dielectric breakdown voltage value at various temperatures.

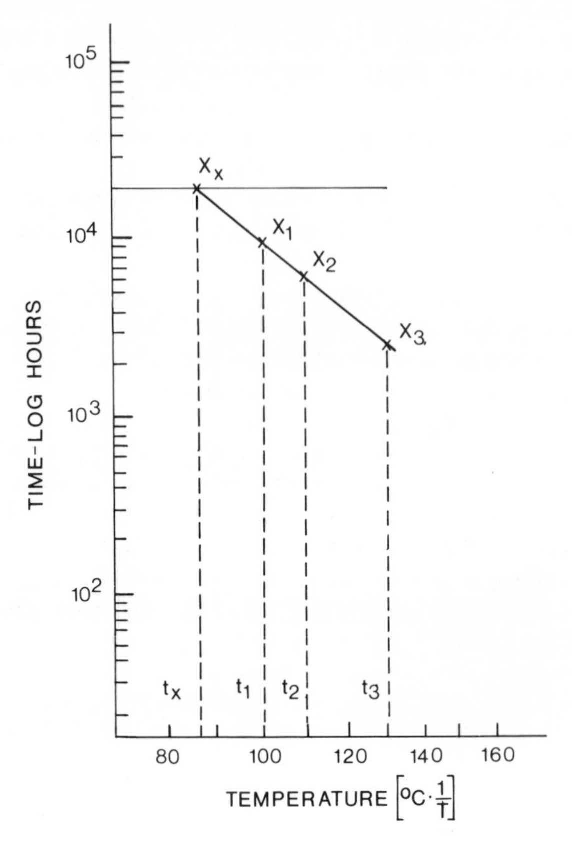


Fig. 6-2. Thermal aging curve for determination of temperature limit.

curves intersect the level of minimum breakdown voltage at points $P_1$, $P_2$, and $P_3$. Using the times $t_1$, $t_2$, and $t_3$ which are obtained at these intersections and the corresponding temperatures, a second diagram is constructed (Fig. 6-2) by plotting the time logarithm vs. reciprocal temperature. The resulting points $X_1$, $X_2$, and $X_3$ lay approximately on a straight line which is extended to the time of 20,000 hr. The temperature at the intersection $X_x$ is the sought temperature limit.

The determination of temperature limit based on the weight loss data is carried out in analogous manner.[17,18,19,20]

## OTHER IMPORTANT TEST METHODS

Several test methods have been discussed in some detail because they are considered important for the electrical pressure-sensitive tapes. For the sake of completeness, other methods which are equally important for some segments of the electrical industry, should be mentioned. In many specialty tapes, the temperature curing properties and saturant stability must be checked out in order to meet the requirements of the user during production.

Lately, the flammability of electrical tapes is queried and the answers are provided by the tests conducted according to the accepted specifications. Other requirements of the user are related to the adhesion of tapes at temperatures below 0°C. The appropriate specifications describe the testing of these properties.[21,22,23,24,25]

## INTERNATIONAL SPECIFICATIONS

The importance of the pressure-sensitive adhesive tapes for the electrical industry is recognized worldwide. Therefore, the International Electrotechnical Commission (IEC), an international umbrella organization of all national electrical specification associations, has carried out a wide-reaching network to assemble the specifications in the electrical adhesive tape sector into a recognized body. This work, released as Publication 454, consists of 3 parts:

Part 1: General Requirement
Part 2: Methods of Test
Part 3: Specification for Individual Materials

Part 3 so far published consists of:

Page 1: Requirements for plasticized PVC with nonthermosetting adhesive.

Page 2: Requirements for polyester film tapes (PET) with thermosetting adhesive.

Page 3: Requirements for polyester film tapes (PET) with nonthermosetting adhesive.

Page 4: Requirements for cellulosic paper, creped, with thermosetting adhesive.

Page 5: Requirements for cellulosic paper with thermosetting adhesive.

Page 6: Requirements for polycarbonate tapes with nonthermosetting adhesive.

Page 7: Requirements for polyimide tapes with thermosetting adhesive.

The national branches work actively not only within IEC, but are also prompted to include the results into their own specifications. In the near future, the specifications for the electrical pressure-sensitive adhesive tapes will be uniform worldwide, only written in different languages.

## REFERENCES

1. IEC Publication 243.
2. ASTM D149.
3. DIN 53 481.
4. VDE 0303, Teil 2.
5. IEC Publication 426.
6. ASTM D1000.
7. DIN 53 489.
8. VDE 0303, Teil 6.
9. IEC Publication 167.
10. DIN 53 482.
11. VDE 0303, Teil 3.
12. ASTM D257.
13. ASTM D150.
14. ASTM D1531.
15. DIN 53 483.
16. VDE 0303, Teil 4.
17. IEC Publication 454-2A.
18. IEC Publication 216.
19. DIN 40 633, Blatt 1.
20. VDE 0340, Teil 1.
21. VDE 0304, Teil 2.
22. IEC Publication 254-2.
23. VDE 0340, Teil 3.
24. SEMKO 26-1964.
25. NEMKO 1963.

# Chapter 7

# Bond Strength and its Prognosis

**Valentinas Rajeckas**
*Kaunas Polytechnic Institute,*
*Lithuania, USSR*

## STRENGTH CONCEPT AND FAILURE MECHANISM

The resistance of a body against failure when it is subjected to the action of mechanical forces is referred to as the mechanical strength or simply as strength.

The kinetic concept of the strength, which has been lately developed by the Zhurkov school,[1,2] considers failure as the final result of a gradual formation of cracks, their storage and growth. This process is due to thermal fluctuations, and the mechanical stress encourages and directs it. In the light of this interpretation, the basic measure of strength is long-term strength of a stressed body, i.e., the time from loading to failure. The fact that strength is often defined not by the time until the failure, but by the force (or stress) at which the failure occurs, does not change the concept of strength radically, if the force (or stress) acting on the body is considered to be a function of time. The force during the failure corresponds to the long-term strength which is achieved by a body subjected to a time-varying load.

It follows from the above considerations that the general discussion of the strength is pointless if the failure conditions are not described, i.e., temperature and the way of loading. The dependence of strength on these factors does not show up for the polymers below brittleness temperature. If the test temperature is above brittleness, where the effect of temperature and loading time (or rate) is important, the description of testing conditions is absolutely necessary. There are many experimental facts proving that the strength of a body is dependent not only on chemical bonds, but also on the energy of intermolecular (van der Waals) bonds as well. The strength depends on the molecular and the supermolecular structure and on the bond energy in those microvolumes where the concentration of the stresses is created, and the conditions are made for crack growth.[1,2,3,4] There are also other strength concepts worthy of mention.[5,6,7,8,9]

The destruction of polymers in the rubbery state at higher temperatures, when their deformation capability is increased, may occur due to cavitation and the formation of strings. The focuses of failure may become the points of chain fracture and separation of major structural elements. Eventually, when the flow temperature is approached, the deformation of polymers proceeds in accordance with the mechanics of plastic or quasi-plastic flow, i.e., the elements of supermolecular structure are displaced nonreversibly and the chains slide past one another. In such cases, the destruction of chemical bonds proceeds on a smaller scale than the process at low temperatures, and the strength may be determined by intermolecular bonds. It is known that rigid polymers subjected to a constant load show a long-term strength dependence on stresses $\sigma$ and temperature $T$. The temperature-time relationship of strength may be described by an exponential expression:

$$\tau = \tau_0 \exp(U_0 - \gamma\sigma/KT), \tag{1}$$

where

$\tau$, $U_0$, and $\gamma$ = certain parameters,

$K$ = Boltzmann's constant, for which the universal gas constant $R$ may be substituted

The empirical formula of this type, expressing temperature-time relationship of strength when the polymer destruction is treated as the kinetic process of crack growth, was suggested by W. Busse,[10] S. Zhurkov,[1] and by F. Bueche.[11]

If the failure occurs through the chemical bonds, then the parameters of this relationship in terms of Zhurkov's interpretation may acquire the following physical meaning: $\tau_0$ is constant, equal to $10^{-12}$–$10^{-13}$ sec, and is numerically close to the oscillation period of an atom in a solid body; $U_0$ has the energy dimension kcal/mole, which readily correlates with activation energy of breaking chemical bonds (for polymers, it correlates with the activation energy of thermal decomposition); $\gamma$ is the structure sensitive coefficient (kcal-mm$^2$/mole-kg) that depends on the dimensions and structure of the body. Thus $\gamma$ may be dependent on the degree of orientation or other technological factors. It quantitatively expresses the nonuniform distribution of stress and its concentration in the area where the focus of failure occurs.

Underexponential numerator in the expression $U = U_0 - \gamma\sigma$ in Equation (1) may be treated as the energy of activation of thermofluctuation process.

However Equation (2) may be better suited for elastomers tested in the creep mode within a moderate load range:

$$\tau = C\sigma^{-b} \exp(U/RT), \tag{2}$$

where

$C$, $b$ and $U$ are certain parameters of the equation.

The value of activation energy of destruction process may be attributed to the parameter $U$. For nonfilled rubbers, the value is 12–15 kcal/mole, and it shows that the prevalent mechanism of failure is through the intermolecular bonds. For some types of filled rubbers, however, the relationship of Equation (1) holds better. $C$ and $b$ are the parameters, the values of which depend on the type and structure of the material.

The relationship of Equation (2) can be transformed into Equation (1), assuming that the parameter $\gamma$ is dependent on the temperature, stresses, and on the ratio of parameters $\tau_0/C$. Such transformation may be useful for the prediction of strength.

Equations (1) and (2) are versatile, and can be applied for the materials at low or high deformations, respectively. The parameters of these relationships, such as $\gamma$, $U_0$ in Equation (1) and $C$, $b$, $U$ in Equation (2) may be dependent of the variables (temperature and stress) and their physical meaning may be interpreted in different ways.

For very small stresses, the relationships in question have no practical value, because they cannot be tested experimentally due to very great long-term strength values (when $\sigma \to 0$, $\tau \to \infty$).

Practically, the determination of relations between long-term strength, temperature and stress (or load) is of great importance in structural calculations as well as in strength prognosis. Sometimes the problem of the physical meaning of the parameters of long-term strength equations may not even arise.[1,12]

## BOND STRENGTH

Adhesive bond failures are most commonly divided into adhesive (through the interphase surface between the adhesive and the adherend), cohesive (through the adhesive or adherend in the area of the bond) and mixed.

According to Bikerman,[13] the adhesive failure is impossible, if the bonding process is carried out properly and a molecular contact between the adhesive and the ahderend surfaces is achieved. Theoretically, this assumption can be justified, but in practical situations, full contact in all actual (not geometrical) surface areas being adhered can hardly be achieved, because of the incomplete removal of absorbed gas layer from the surface, especially from surface fractures, pores, impurities, and weakened surface layers (e.g. oxidized surface).

If the adhesive is used in a solid form, i.e., the bonding is performed using pressure-sensitive adhesive tape, then the contact is achieved due to the plastic properties of the adhesive layer, and apparently not over the whole surface of the adherend. Therefore, adhesive failure is possible and, sometimes as in the case of temporary adhesive bonds where great strength is not required, it may even be desirable.

For strength evaluation and prognosis, it is essential to state what kind of bonds are established between the adhesive and the adherend, because the total energy value of bond failure and the destruction kinetics of the bond are greatly dependent on it.

Numerous experimental data are available which prove the existence of various bonds between the adhesive and the adherend. Bonds of molecular interaction as well as those of a chemical nature are possible. For both types of bonds the mechanism of failure, in terms of kinetic conceptions, is thermofluctual and the failure above brittleness temperature does not occur instantaneously, but during a time interval. These time intervals are considerable when glass transition (or crystal flow) temperature is approached, especially after a transition to the rubbery state.

For the prognosis of bond strength, it is important that the temperature-time relationship of polymers is qualitatively analogous, irrespective of the kinds of bonds responsible for failure, and that it can be described by Equations (1) and (2). The parameters of these formulas, however, may acquire different meanings in different cases.

Thus, Bikerman's view, [13,14] to treat the adhesive bond strength in the same way as the strength of homogeneous materials, is correct. The data[2,3] about the possibilities of polymer failure through molecular interaction show that there is no principal difference between the failure of materials or their adhesive bonds, and consequently, between their physical strength evaluations.

The numerical values of strength characteristics depend not only on energy required for bond failure, but also on that required for the deformation of the body under a load. Therefore, the mechanical deformation properties of the bond elements—adhesive and adherend—may have a great effect on the strength, especially for soft materials.

The stress concentration on the macro* level is also of great importance to a numerical evaluation of strength.

Generally, in terms of phenomenology, adhesive bond strength is dependent on the overall bond strength between adhesive and adherend, on the contact surface, and on the strength of the adhesive. Sometimes the failure may take place in the adherend. This occurs if both the adhesive bond and the adhesive itself are stronger than the adherend. It can also occur due to the concentration of stresses in the bond which in turn depends on the bond construction and the mode of force application.

Bond strength between the adhesive and the adherend, as well as the strength of the adhesive in the bond, may depend on many other factors alongside the ones already discussed. In the case of pressure-sensitive adhesives, the bonding technique is rather simple and its influence is minor. The significance of deforming relaxational properties of the adherend and the adhesive, as well as the influence of other parameters, especially temperature and deformation rate, may be considerable because the pressure-sensitive adhesives are in a rubbery state with pronounced plastic properties.

## STRESS CONCENTRATION IN ADHESIVE BONDS

Bond stresses may be divided into internal (occurring without the effect of outside forces) and external (occurring under the effect of outside forces).

If the adhesive layer in the bond does not harden after bond formation, but remains elastic and exhibits good relaxational qualities, the internal stresses do not affect the strength considerably. The stresses depending on outside forces are not uniformly distributed and thus their concentrations occur. In the bonds of soft materials, the stress concentrations cannot be avoided.

The quantitative index of stress concentration is of great importance in evaluating the bond strength. It is expressed by a concentration coefficient, which is defined as a ratio of maximum stress in its concentration zone to average stress that is equal to the force that is acting on the unit area of the bond.

It is possible to evaluate the stress concentration coefficient of the bonds of rigid materials for small elastic deformation within the limits of elasticity. For example, Volkersen's or Goland-Reissner's formulas may be used for this purpose in the case of lap joints. The latter is more accurate because it takes into account the bend of the bond in deformation.[16]

* The stress on the micro level is always distributed unevenly, not only in adhesive bonds but in homogeneous materials as well.

An experimental calculating method which enables one to determine the stress concentration coefficient is suggested for the bonds of soft materials for limited (not small) deformations, making use of the polar-optical method of tension measurements. The measurement method and mathematical apparatus of calculation is described.[15]

Here we shall mention some of the more interesting results achieved by testing the properties of stress distribution and its concentration in the bonds of soft materials. Stress tests were carried out for lap joints. Under the action of forces, the shear stresses at the adhesive and interphase surface are most hazardous for such bonds. Figure 7-1 shows the shear stress distribution of rigid bonds for small deformations within elastic range (Fig. 7-1a) and of soft bonds for rubbery deformations (Fig. 7-1b). In the first case, the tension is distributed unevenly only in the length direction.

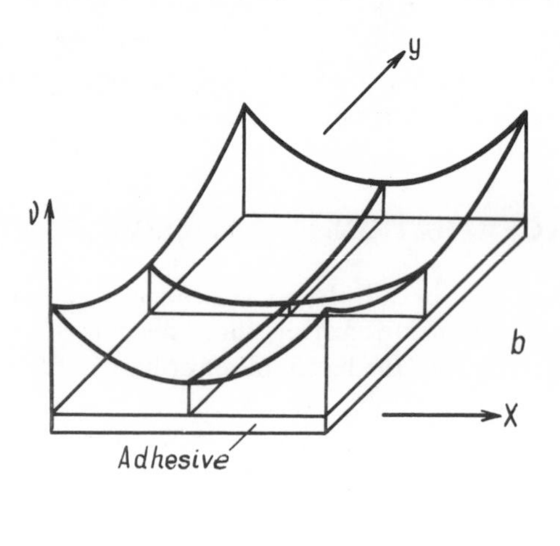


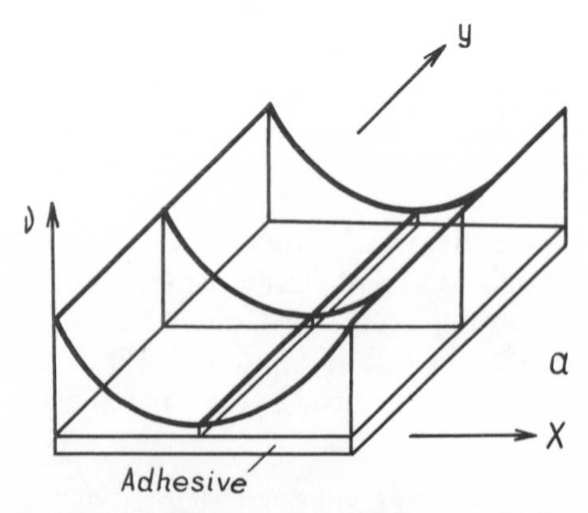


Fig 7-1. Stress distribution in adhesive bonds: (a) for solids; (b) for soft materials.

In the bond loaded by a single-axis load, a linear stress state is formed and stress concentration uniform over the width of the sample occurs at its ends. In the second case, the stresses are distributed unevenly, not only in the length direction, but transversely as well. A flat stress state is formed, although the load is a single-axial one. The maximum stress concentration occurs at the corners of the bond.

Transverse deformation is the reason for this kind of tension distribution. Relative transverse deformations of soft materials may be of the same order of magnitude as the longitudinal ones. At the ends of a lap joint, where there is a transition from a double thickness of the adherend to a single thickness, causing a decrease of rigidity, additional stresses in the transverse direction occur. For joints of hard material, when the adherend deformations are small until the failure, this effect does not show up. Thus, Fig. 7-1 shows two possible stress distributions, depending upon the magnitude of deformations of the adherend and of the bond.

The nature of tension distribution for soft material bonds depends on their width. Reduction of the width causes the flat state of stress to approach the longitudinal one as shown in Fig. 7-1a.

In double-axis loading of the bonds of soft materials the nature of stress distribution remains the same as in the case of single-axis loading, except the numerical values of the concentration coefficient are greater.

The value of stress concentration in the case of bonds of rigid materials[16] depends on the mechanical deformation properties and the thickness of the adherend and the adhesive. In lap joints, it depends on the length of bond. Such relationships hold for bonds of soft materials as well, except that the stress value depends on the bond width, unlike in the case of rigid materials (Fig. 7-2). This relationship is due to considerable transverse

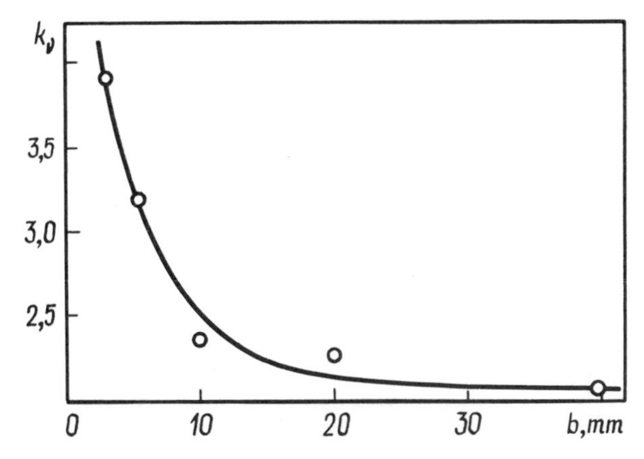


Fig. 7-2. Dependence of stress concentration coefficient $k_v$ on bond width (b).

deformations that depend on the bond width, occurring along with longitudinal ones. In bonded soft materials, Poisson's coefficient decreases when the sample width increases.[15] Thus, the narrower the sample, the greater is the relative transverse deformation (compared with the longitudinal one), and the greater are the stresses occurring at the concentration points.

With the decrease of the sample width, the longitudinal stress state is approached. The stress concentration is greater for narrower samples than for wider ones under the same deformations and other conditions, but the stresses at the end of the bond will be distributed more uniformly in the width direction. The wider the bonds of soft materials, the greater the possibility of failure at the corners of the bond.

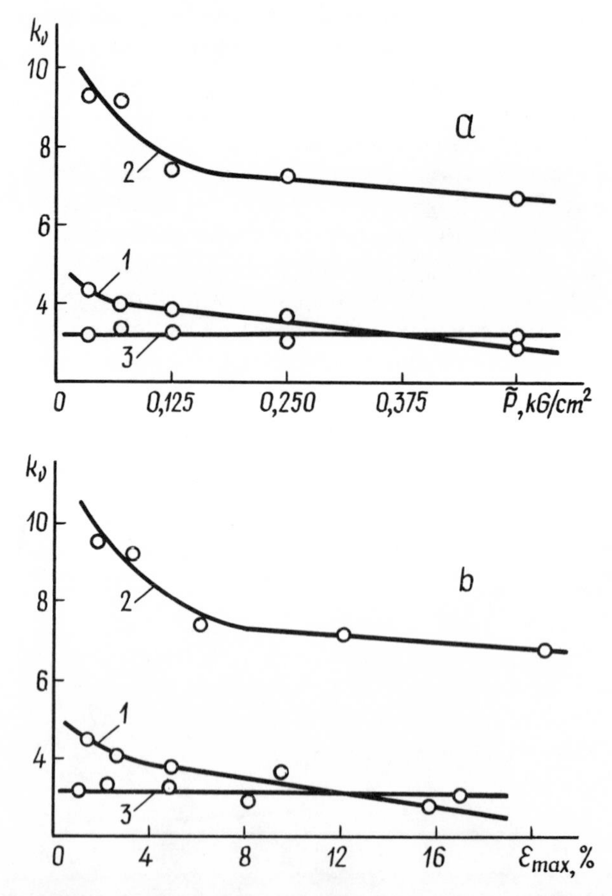


Fig. 7-3. Dependence of stress concentration coefficient on load (a) and deformation (b).

**Table 7-1. Data on polyurethane rubber joints bonded with various adhesives.**

| ADHESIVE | RELAXATION TIME INDEX (sec) | RELATIVE DECREASE OF STRESS CONCENTRATION INDEX $\Delta k_v$ (percent) |
|---|---|---|
| 1. Nitrile rubber | 2.5 | 34 |
| 2. Polychloroprene (100), butylphenolformaldehyde resin (40), zinc oxide (10, thiuram (5) | 15.5 | 28 |
| 3. Silicone rubber adhesive | 510 | 0 |

*Note*: $\Delta k_v = [(k_{v\bar{p}(\min)} - k_{v\bar{p}(\max)})/k_{v\bar{p}(\max)}] \cdot 100$, where $k_{v\bar{p}(\min)}$ and $k_{v\bar{p}(\max)}$ are the concentration coefficients for minimum and maximum loads used experimentally.

The stresses in the bond increase from the time of force application until bond failure. In the case of bonds of soft materials, this process is followed by elastic deformation and the time interval until the failure may be of the same order as the stress relaxation time characteristic for the particular adhesive and adherend. This may influence the stress concentration coefficient. Figure 7-3 shows the dependence of the stress concentration coefficient for lap joints on the force (Fig. 7-3a) and the deformation. The data on joints formed by bonding polyurethane rubber with different adhesives are shown in Table 7-1. The rubber was deformed only reversibly (the actual elastic deformation). The stresses were measured by the polar-optical method.[15] The samples were loaded by constant loads of various values until the time of failure. The deformations were measured in the area of stress concentration; in the adherend, at the end of the joint. Alongside with this, the relaxation time index was determined for the samples cut out of films cast on Teflon. The relaxation time index is defined as the time range in which the stresses decreases *e* times, *e* being the base of the natural logarithm.

It is apparent from Fig. 7-3 and Table 7-1 that the stress concentration coefficient is decreased considerably, if adhesives with good relaxation capabilities are used, such as adhesives 1 and 2. In this case, the stress relaxation is caused by small creep deformations which appear long before the time of bond failure. In the case of the silicone rubber-based adhesive, characteristic in reversible deformations, the decrease of stress concentration coefficient was not detected.

The stress relaxation due to conformative rearrangement of the chains was of no essential value, but stress distribution due to rearrangement of structural elements (creep kinetics) produced some effect during the short time of the load duration. This does not necessarily mean that at longer time intervals the first relaxation mechanism cannot have any influence on stress concentration coefficient.

The characteristic of stress distribution in lap joints is stress concentration at the ends of the bond, the dependence of concentration coefficient on the sample width and instability of stress concentration in those cases when the adhesive exhibits good properties of creep and elastoplasticity. Generally, these properties appear in other types of bonds of soft materials as well. The instability of stress concentrations may appear also in bonds of rigid materials, depending on the relaxational properties of the adhesive.

The above mentioned facts should be taken into account when evaluating the bond strength. The stress concentration at the ends is significant for evaluation of failure points in the structure. The effect of width shows that in comparing the relative strength of different bonds, their former width should be considered. Instability of stress concentration points to the possible relative strength increase when the adhesive has a high relaxational capacity. This often causes deviation from expected behavior.

## THE EFFECT OF MECHANICAL PROPERTIES OF THE ADHESIVE AND THE SUBSTRATE

The fact that the cohesive strength of the adhesive influences the bond strength is supported by various references in the literature.[13,18] The cohesive strength of the adhesive may be varied in different ways: by increasing its molecular weight, increasing the number of functional groups, etc. This increases the bond strength until the cohesive strength becomes greater than the strength of the adhesive bond, or until the stresses within the oriented adhesive layer concentrate and achieve critical value exceeding that capable of destroying the interaction bonds which are responsible for the adhesive strength in the nonoriented mass portion. In such cases, the direct link between bond strength and the cohesive strength cannot be expected. In evaluating cohesive strength, the stress state which is formed when the adhesive is in the bond is not usually considered.

Correlation of the mechanical deformation characteristics of free adhesive films and bond strength may be interesting.[15]

Various adhesives, those based on nitrile and polychloroprene rubbers and on chlorinated poly(vinyl chloride) resin, were studied. Seventy different bond variations were obtained, different in the composition of the adhesive or the adherend. The bond strength was experimentally determined by shear tests, evaluating the mode of the failure. The correlation between the bond strength and mechanical deformation characteristics of the adhesive films was analyzed.

Irrespective of the type of adhesive and adherend, the following relationships were observed.

1. In the case of cohesive failure, there exists a direct linear relationship between bond strength and the tensile strength of adhesive films, as well as their moduli of elasticity
2. In the case of interfacial failure (determined visually), there exists a direct linear relationship between bond strength and deforming capacity of the adhesive films, which is characterized by elongation at stresses of about 75% from the fracture stress. In both cases, the correlation coefficients are on the order of 0.85–0.95

Thus, in the first case, the bond strength can be increased by increasing the strength of the adhesive. In the second case, the bond strength may be increased by developing the deformation qualities of the adhesive. The latter may be understood if the formation of the contact between the adhesive and adherend is regarded as a kinetic process: better deformation qualities of the adhesive correspond to greater mobility of segment molecules. The increase of the adhesive bond strength may, in this case, result in the decrease of the cohesive strength of the adhesive and the required bond strength may not be achieved. The practical approach, for example, in developing adhesive formulations is the solution of an optimization problem. Here the application of the methods of mathematical experiment planning may be effective.

It is observed that there exists a close correlation link ($\Gamma \geq 0.9$) between the strength indices obtained by shear and peel tests, if the mode of the failure is the same in both cases. Such a link is absent if the nature of the failure is different. Thus it is possible to evaluate the strength of one case, if the data obtained in the other tests are known.

The bond strength is affected by mechanical deformation properties of the adherend. This fact was noticed long ago[16] and is explained as the effect of mechanical properties of the adherend on the value of the stress concentration coefficient. It decreases with the increase of adherend stiffness.

Numerous equations for peel strength evaluation[13,19,20,21,22] derived in accordance with some geometrical models suggest that the relationship of the form

$$P = aE_s^n \tag{3}$$

exists between peel strength $P$ and adherend modulus $E_s$ (or stiffness), where

$a$ is the coefficient depending on the mechanical properties and geometrical parameters of the adhesive and $n < 1$. In the equations suggested in References 13, 19, and 20, $n = 0.25$.

This relationship for lap joints was investigated experimentally by the author.[15] The stiffness of the adherend was varied in two ways: a thin layer of the adherent was backed up by another reinforcing layer, or the adherend was backed up by nitrile rubber strips of various degrees of vulcanization.

In case of an adhesive of a very high relaxational capability when the bond failure is cohesive, the relationship between the adherend stiffness $D$ * and bond strength $P$ is of an extremal nature (Fig. 7-4, 1). The strength maximum may occur at different values of $D$ and is dependent on the time range until failure, which in its turn depends on the deformation capability of the adherend and adhesive and of the bond strength. This generally determines the actual values of the deformations and the time to failure. When the adhesive does not display the above mentioned qualities and when the failure is interfacial proceeding at high speed, the relationship $P(D)$ does not exhibit extremal nature (Fig. 7-4, 2) and is readily approximated in semilogarithmic coordinates as a linear function. The part of extremal relationship $P(D)$ up to $D$ values close to the strength maximum is approximated rather well by this function.

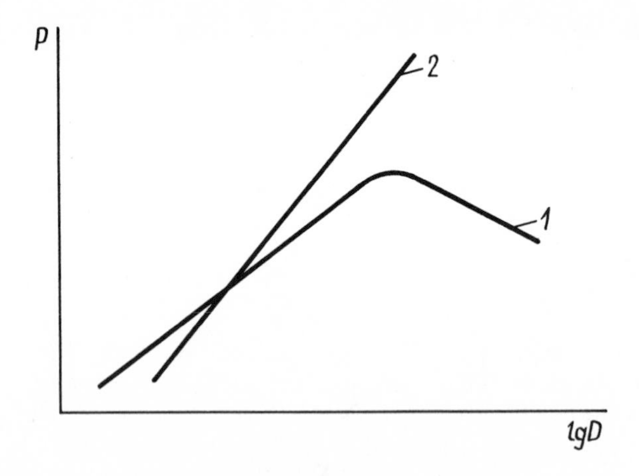


Fig. 7-4. Stress dependence on the stiffness of the adherend: (1) extremal; (2) nonextremal.

* Stiffness for tensile stress $D = E_s \cdot F$, where $E_s$ is a relative modulus of adherend elasticity (in this case, at deformations $\varepsilon \sim 75\%$) and $F$ is the cross-sectional area.

The nature of extremal relationship is relaxational. It is caused by the decrease of stress concentration under the load, before the beginning of the bond failure, as shown in Fig. 7-3.

## THE INFLUENCE OF SCALE FACTOR

The influence of scale factor on bond strength is very diverse. The dependence of strength on the adhesive thickness is most widely studied. The relationship depends on the test method.

For lap and butt joints, this dependence is hyperbolic (Fig. 7-5, 1) as shown by numerous experiments[13,15,23] to be:

$$P = ah^{-m} \tag{4}$$

where $P$ = strength
$h$ = thickness of the adhesive layer
$a, m$ = parameters

For most adhesives exhibiting elastoplastic properties, the value of $m$ is 0.1–0.3.

In T-peel tests, the strength usually increases with increasing thickness of the adhesive layer up to a certain limit. This follows from the theoretical equations[13,19,20,22] for the evaluation of the peel strength. At higher adhesive thickness, this no longer holds and sometimes the relationship $P(h)$ is of extremal character (Fig. 7-5.2). This could be understood in reference to the general laws of tear deformation kinetics. For low adhesive thicknesses, the

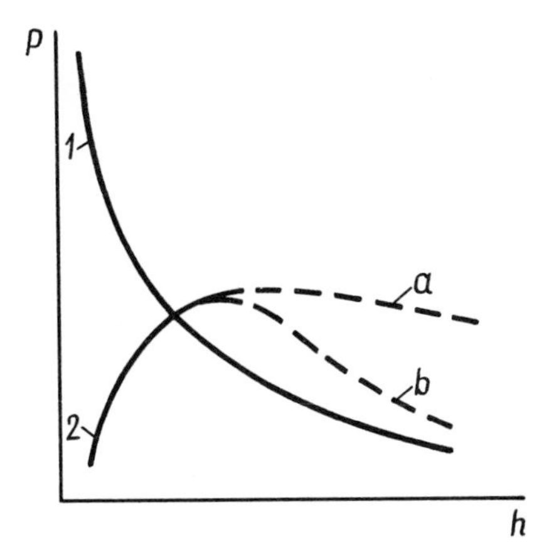


Fig. 7-5. Strength dependence on the thickness of the adhesive layer: (1) lap joints and butt joints; (2) T-peel joints.

deformation possibilities are limited and the process of failure proceeds in the form of small angle crack. The deformations in the adhesive increase with increasing thickness and the configuration of the crack changes: the angle increases, and at the same time, the stress concentration decreases corresponding to an increase in strength. After the adhesive layer thickness reaches a certain critical value, at which the local deformation in the crack area is sufficiently large, the effect of thickness may not be noticeable anymore (Fig. 7-5.2, a), or it may be similar as in the case of other tests (Fig. 7-5.2, b). The problem is discussed in greater detail in Reference 15.

The dependence of concentration coefficient on the bond width (Fig. 7-2) is also of some importance including its stability or instability under the load effect (Fig. 7-3), which is related to the relaxational capability of the adhesive.

In cases when the stress concentration coefficient does not decrease until the failure, relative strength usually increases with increasing bond width until a certain limit is reached. When the stress concentration coefficient decreases due to stress relaxation, the relationship $P(b)$ is nonmonotone in the limited range of bond width. Generally, the dependence of strength on the bond width is most noticeable in the samples of narrow width up to $b \sim 25$–30 mm.

## THE INFLUENCE OF TEMPERATURE AND LOADING REGIME*

It has been noted[15,22,24,25,26] that the nature of bond strength dependence on temperature is extremal (Fig. 7-6). This character with one or more maximum values does not depend on the test method. It has been shown that the strength maximum corresponds to the transition of the adhesive from a solid to a highly elastic state.[25] This shows that the strength maximum is of a relaxational nature, and the greater the stress concentration, the more pronounced the maximum. Sometimes the change of the mode of failure is observed from an interfacial to a cohesive failure. In cases when the failure mode does not change when passing through a maximum, a second local maximum related to the transition of the mode is observed.[24] For the adhesives of crystalline structure, local strength maximum is observed in the area of crystal flow.[25]

For the purposes of prognosis, the dependence of strength on temperature should be studied separately for the temperature range in which the adhesive is in a solid state (glassy or crystalline) and in a highly elastic state.

* The value of load and its variation in time is referred to as the loading regime.

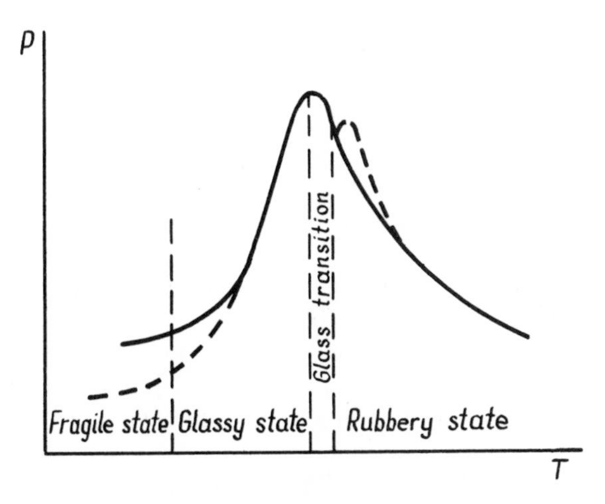


Fig. 7-6. Strength dependence on temperature.

Pressure-sensitive adhesives at normal temperatures are in the state of high elasticity with some plastic properties.

The most abundant information about the dependence of bond strength on temperature and loading mode is provided by the study of temperature-time dependence. The experimental data of bonds of soft materials, as well as bonds of thin layers of any stiffness, are most conveniently obtained by testing of lap joints in creep mode, i.e., making long-term strength experiments at constant loads.

The dependence of long-term strength $\tau$ of bonds on temperature $T$ and static load $\tilde{P}$ is described by the form of Equations (1) and (2) in which a load term is substituted for the stress:[15]

$$\tau = \tau_o^* \exp(U_0 - \gamma\tilde{P}/RT), \tag{5}$$

$$\tau = C\tilde{P}^{-b} \exp(U/RT) \tag{6}$$

The parameters $U_0$, $\gamma$, $C$, and $U$ could be given the same meaning as in Equations (1) and (2). As to the physical meaning of the parameter $\tau_o^*$, no uniform opinion has been achieved so far. Sometimes, when its values are $10^{-12}$–$10^{-13}$, it can be treated in the same way as in Equation (1). Often, however, for bonds $\tau_o^* > 10^{-12}$, it is possible to assume that $\tau_o^* = \tau_o \cdot C$ where $\tau_o \sim 10^{-13}$, and the values of $C$ may vary over a wide range and depend on the variables $T$ or $\tilde{P}$.

Equation (6) can, of course, be transformed to the form (5) analogous to the way Equation (2) can be transformed to (1). This is handy in theoretically studying the temperature-time dependence.

At constant temperature, Equation (5) acquires the form

$$\tau = A\ \exp(-\alpha\tilde{P}), \tag{7}$$

and at constant load

$$\tau = A_1\ \exp(U/RT). \tag{8}$$

Here, $A$ and $\alpha$ are the parameters dependent on temperature; $A_1$ and $U$, on the load ($U$ = constant is also possible).

Figure 7-7 schematically shows various possible and experimentally obtained relations between long-term bond strength and the load (at fixed temperatures, $T_i$ = constant) or the temperature (at fixed loading values, $\tilde{P}_j$ = constant).

It is obvious that long-term strength dependence on load and temperature in semilogarithmic coordinates may have various shapes. They may run to a single-point pole, with the ordinate value of $-13$ (i.e. $\tau_0^* \geq 10^{-13}$) or greater, and the abscissa value log $\tau$ and $1/T$ equal to zero or greater;

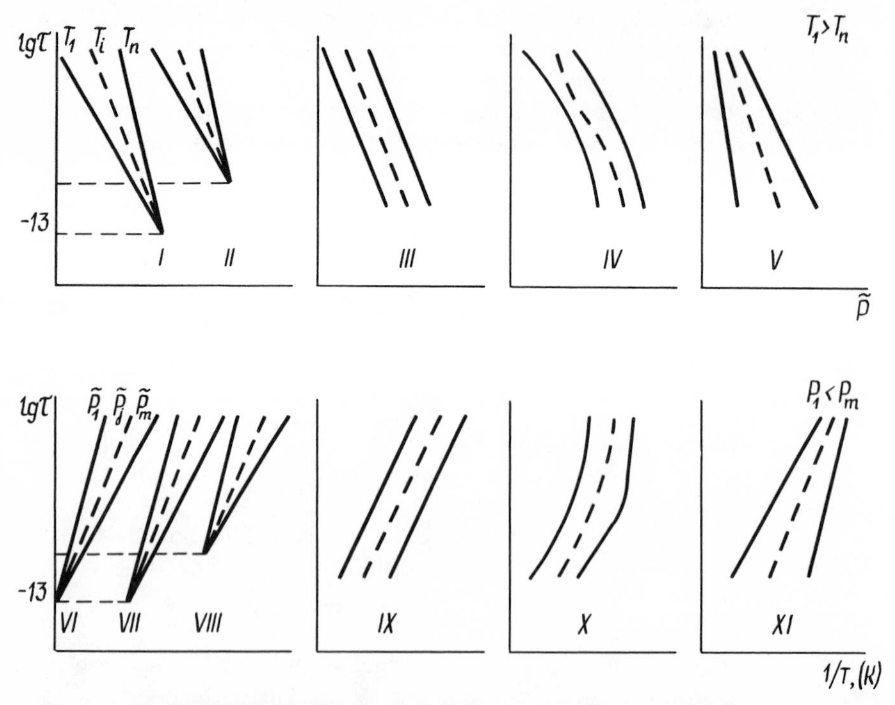


Fig. 7-7. Long-term strength ($\tau$) dependences on static load ($\tilde{P}$) and temperature ($T$) in semilogarithmic coordinates.

they may show parallel or converging linear relationships with the decreasing load and increasing temperature; they may be nonlinear and have different directions with regard to each other; they may be lines running from each other as the load increases and the temperature decreases.

The possibilities of defining the dependences $\tau(\tilde{P}, T)$ by Equations (5) and (6) were studied theoretically.[15] Referring to the abundant experimental data, the parameters $\tau_0^*$, $U_o$, and $\gamma$ of Equation (5) are shown to be constant only in some cases. In elastomeric adhesives with plastic properties, some of these parameters are dependent on the variables.

In seeking for qualitative or quantitative information on the long-term bond strength dependence on temperature and load, the first stage is an experiment. The values of long-term strength should be determined for 3 or 4 levels of temperature and 4 or 5 of constant load for each temperature value. The number of samples should be selected considering the high long-term strength test data dispersion. The variation coefficient, even for homogeneous materials, is 20–30%. It can be reduced to 10–12% only if the experiment is very thoroughly prepared, i.e. the differences in mechanical deformation characteristics and thickness of the adhesive layer are kept as small as possible.

The next stage is to construct a graph of the relationships $\tau(\tilde{P})_{T_i}$ and $\tau(T)_{\tilde{P}_j}$, as shown in Fig. 7-7. The relationship $\tau(\tilde{P}, T)$ is described by the form of Equation (6), if $\tau(\tilde{P})_{T_i}$ are not straight lines in the coordinates $\log \tau$ and $\tilde{P}$, but rectify in the coordinates $\log \tau$ and $\log \tilde{P}$. The plot of the relationships $\tau(\tilde{P})_{T_i}$ and $\tau(T)_{\tilde{P}_j}$ is sufficient for qualitative prognosis.

The quantitative prognosis and the evaluation of strength characteristics require the mathematical description of experimental data $\tau(\tilde{P}, T)$. The values of $\log \tau$ are distributed according to the normal distribution law of probabilities. Equation (5) parameter values are evaluated by regressive analysis. For this purpose, the parameters of long-term strength dependence on a single variable for various values of the second variable are evaluated. After that the form of the parameter dependence on another variable and its numerical values are determined and substituted into the equation of long-term strength dependence on the first variable. This gives the final form of the dependence on two variables.

It is methodically more correct to describe first the dependence on that variable which shows a simpler graphic form of the dependence with respect to its description by Equation (5). This can be verified visually or by linear approximation of the experimental data using the mean square method and evaluating approximation error. Usually the form of the dependences for the adhesive bonds is evaluated first according to Equation (7). Then the dependence of the parameters $A$ and $\alpha$ on temperature is evaluated and finally, substituting $A(T)$ and $\alpha(T)$ into equation (7) gives Equation (5).

In describing mathematically the dependences $\tau(\tilde{P}, T)$, whatever graphic form of the dependences $\log \tau(\tilde{P})_{T_i}$ and $\log \tau(1/T)_{\tilde{P}_j}$ may assume, the value of the parameter $\tau_0^*$ should be determined first. In the case of running these dependences into the pole, $\tau_0^*$ value could be evaluated from experimental data by approximation methods. The values of the parameters $U_0$ and $\gamma$ could be then evaluated accurately for the parameters as constants and as functions of variables. If the determination of accurate numerical values $\tau_0^*$ from the experimental data is not possible, then the accurate values of other parameters also remain unknown. Then, it is possible to consider that the description of the dependence $\tau(\tilde{P}, T)$ by Equation (5) is strictly empirical and the prediction possibilities of the equation qualitatively remain rather limited.

Another view to this question is also possible. Numerous experimental data show[1] that the pre-exponential multiplier $\tau_0$ in Equation (1) is a constant of the order of $10^{-12}$–$10^{-13}$ only for rigid materials, when the failure occurs through chemical bonds at small deformations. For elastomers, it has been noticed that this parameter may be equal to $10^{-5}$–$10^{-6}$, or even greater, depending on the material. Bartenev[4] explains this as a dependence of activation energy (parameter $U_0$) on temperature. In this case, as has already been mentioned, $\tau_0^* = \tau_0 \cdot C$. According to this interpretation in describing experimental data by Equation (5), it is possible for the pre-exponential multiplier to acquire the value of $10^{-13}$ and the multiplier $C$ may be placed under the exponent. Such an approach in describing experimental data mathematically could be applied in the case where the pole does not exist.

Here are some characteristic examples. In Figs. 7-8 and 7-9, the relationships $\tau(\tilde{P}, T)$ of cellulose diacetate films for the lap joints is represented. In the first case (Fig. 7-8), the films were bonded by a two-component adhesive, which consisted of polyurethane elastomer crosslinked by isocyanates. This polymer does not exhibit plastic properties due to its three-dimensional structure. In the second case (Fig. 7-9), the adhesive was nitrile rubber exhibiting plasticity, which is very clearly expressed at higher temperatures. The thickness of the layer of the adhesive in both cases is about 150 microns. This failure is cohesive. As seen from Fig. 7-8 in the first case, the pole exists at $\tau_0^* \sim 10^{-13}$. The relationship $\tau(\tilde{P}, T)$ is described by Equation (5), the parameters of which are $U_0$ = constant and $\gamma$ = constant. The numerical values are $\tau_0^* = 3.2 \times 10^{-13}$ sec, $U_0 = 28.6$ kcal/mole, $\gamma = 0.72$ kcal-mm$^2$/mole-kg.

The relationships obtained in the second case are described by Equation (6); here the relationships $\log \tau(\tilde{P})_{T_i}$ are straight lines in the coordinates $\log \tau$ and $\log \tilde{P}$. The parameter $U$ in Equation (6) is, in this case, a constant ($\log \tau(1/T)_{\tilde{P}_j}$, the lines are parallel). This means that the activation energy of

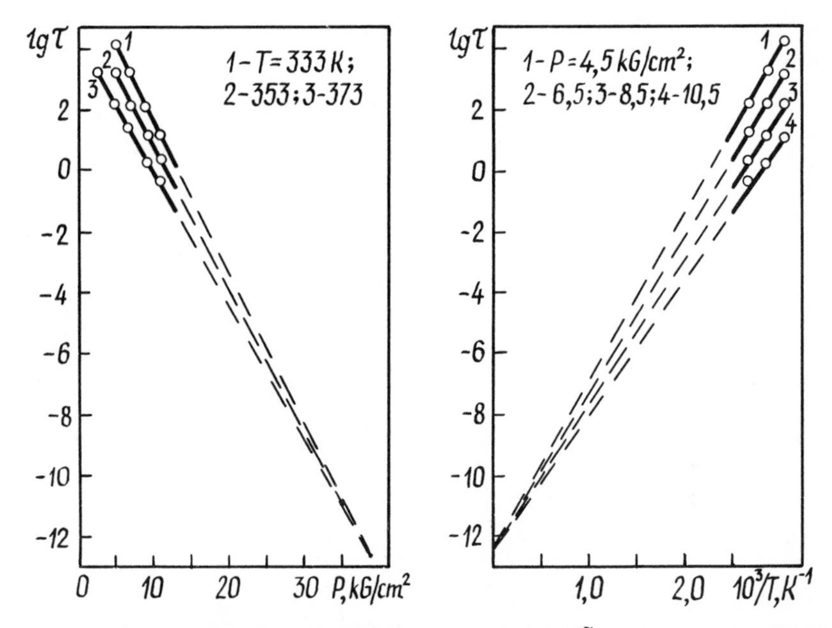


Fig. 7-8. Long-term strength ($\tau$) dependences on static load ($\tilde{P}$) and temperature ($T$) for cellulose diacetate film bonded with polyurethane adhesive.

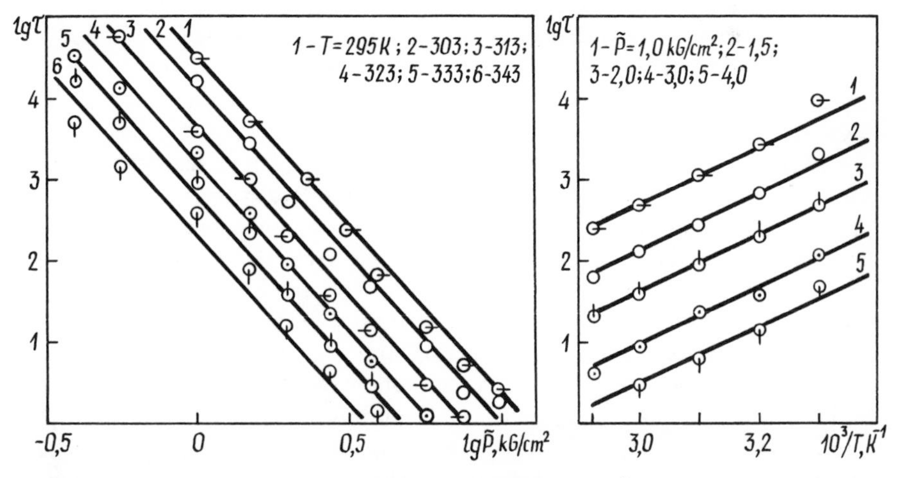


Fig. 7-9. Long-term strength ($\tau$) dependences on static load ($\tilde{P}$) and temperature ($T$) for cellulose diacetate film bonded with nitrile rubber adhesive.

the failure does not depend on the load. This is generally characteristic of elastomers.[4]

After transforming Equation (6) to (5), assuming that $\tau_0^* = 10^{-13}$, we obtain that the parameters of (5) are $U_0$ = constant and $\gamma = \gamma(\tilde{P}, T)$ of $\gamma = (a_1 \ln\tilde{P} - a_2)T/\tilde{P}$. The numerical values of these parameters for a given system are the following: $U_0 = 18$ kcal/mole, $\gamma = (0.031 \ln \tilde{P} - 0.075)T/\tilde{P}$ kcal-mm$^2$/mole-kg.

It is interesting to consider a case when the relationships $\log \tau(\tilde{P})_{T_i}$ and $\log \tau(1/T)_{\tilde{P}_j}$ are straight lines running away from each other with the increase of the load and decrease of temperature (Fig. 7-7). The relationship of this type is obtained experimentally for adhesive bonds where the failure was interfacial.[27] By changing the bonding technique, cohesive failure was obtained for the same system and the graphs of the relationship $\log \tau(\tilde{P})_{T_i}$ and $\log \tau(1/T)_{\tilde{P}_j}$ had another direction. They ran into the pole at $\tau_0^* \sim 10^{-10}$. This experimentally demonstrated fact can be interpreted in the following way: if the adhesive interaction is of adsorbing character and the bond failure occurs at the interfacial surface, the strength depends on intermolecular forces. As the temperature increases, the number of intermolecular bonds carrying the load decreases. The higher the temperature, the greater should be the intensity of long-term strength decrease with increasing load, because the stress per bond will increase more intensively, when the number of bonds on which the load is distributed is lowered. Similarly, the greater the load, the more significant will be the effect of temperature on long-term strength.

When the relationships $\log \tau(1/T)_{\tilde{P}_j}$ approach each other with decreasing temperature, the underexponential numerator in Equation (5), $U = U_0 - \gamma\tilde{P}$, becomes a growing function. This may occur when $U_0 \neq U_0(\tilde{P})$, if $\gamma = \gamma(\tilde{P})$ and $\gamma(\tilde{P})$ is a decreasing function, such that $|d\gamma(\tilde{P})| > d\tilde{P}$. If $\log \tau(1/T)_{\tilde{P}_j}$ straight lines approach each other with increasing temperature, then $U(\tilde{P})$ becomes a decreasing function.

All common types of bond dependence on load and temperature as shown in Fig. 7-7 will be generalized by selecting two extremal cases. One of them is when adhesive failure occurs only through chemical bonds. This corresponds to the cases of dependences $\log \tau(\tilde{P})_{T_i}$ and $\log \tau(1/T)_{\tilde{P}_j}$ as shown in Figs. 7-7, I, VI. The pole is present and for the relationship $\log \tau(1/T)_{\tilde{P}_j}$, it is on the ordinate axis, so that $\tau_0^* = \tau_0 \sim 10^{-12}$–$10^{-13}$. Another case concerns the failure occurring only through the intermolecular interaction, as shown in Figs. 7-7, V, XI. One would expect that the failure taking place through both chemical and intermolecular bonds will result in intermediate cases demonstrated by the displacement of the pole from the ordinate axis for the relationship $\log \tau(1/T)_{\tilde{P}_j}$, to the area of lower temperatures (VII, VIII), the parallelism of lines $\log \tau(\tilde{P})_{T_i}$ and/or $\log \tau(1/T)_{\tilde{P}_j}$ (III,

IX), and the nonlinearity of these relationships (IV, X). This has been experimentally verified. It is characteristic of the bonds when the adhesive is in a rubbery state, but not purely elastic (i.e., the effect of plasticity comes up under the load and the failure occurs according to the mechanical characteristic of elastomers when the processes of creep and cavitation develop). The stronger the elastoplastic properties of the adhesive, the more complicated is the form of the relationships log $\tau(\tilde{P})_{T_i}$ and, especially of log $\tau(1/T)_{\tilde{P}_j}$.

The practical significance of the above mentioned facts and the discussion of long-term strength dependence on load and temperature in Reference (15) is summarized below.

If the graphic picture of relationships log $\tau(\tilde{P})_{T_i}$ is the one shown in Fig. 7-7, I, and log $\tau(1/T)_{\tilde{P}_j}$ as in VI, it means that the failure occurs through the chemical bonds. Such bonds do not exhibit creep, at least in the range of loads and temperatures in which the experiment was carried out, and their failure (i.e., the growth of the main crack) occurs at relatively high speed. The magnitude of strength characteristics, is therefore, only slightly different at the beginning than at the end of failure. Cases II, VII, VIII are characteristic when intermolecular forces are also responsible for the strength. In this case, a lower thermal resistance and higher creep is expected. Cases III, IV, IX, X are characteristic of bonds where the adhesive, especially at higher temperatures, exhibits creep behavior and the failure through intermolecular forces prevails. Such adhesives have a lower thermal resistance, may be sticky and have a greater autoadhesive capability. In addition, cases IV, X, depending on the direction and slope of the curves, show that the structure of the adhesive is especially temperature-sensitive and at certain temperature intervals, nonmonotone variations of the structure take place. These variations are related to the changes of deformation properties and failure mechanism because of recrystallization, crosslinking, orientation, etc. Cases V, XI are characteristic of bonds where the adhesion is a result of intermolecular forces and the failure is interfacial.

The temperature and load sensitivity of different bonds can be seen from the slope of the curves.

The long-term strength can be quantitatively evaluated from Equation (5) for various $T$ and $\tilde{P}$ values.

Using the numerical values of the parameters of Equation (5), which were obtained experimentally at static conditions at constant loads and temperatures, quantitative information about the strength properties of bonds at variable loads and temperatures can be obtained.

As the data accumulated in the author's laboratory show, the value of errors resulting from long-term strength evaluation from Equation (5) is related to the extent of complication (Fig. 7-7) of the relationships $\tau(\tilde{P})_{T_i}$

and $\tau(T)_{\tilde{P}_j}$, and thus related to the form of the parameters of Equation (5). The error is least (10–15% or less) when these parameters are constant, and greatest (15–20%) when at least some of the parameters are dependent on both variables. Using Equation (5) for evaluation of other strength characteristics, as described below, the error may amount in some cases to 25–30%.

Thus, Equation (5) and occasionally (6) are sufficiently versatile and, in terms of information, sufficiently receptive. Certainly its derivation is more time-consuming than the process of standard dynamometric tests. But if the aim is to obtain detailed information on particular adhesives or adhesive bonds, and to accumulate a bank of information, then the tests described here are useful and effective.

## THE EVALUATION OF LONG-TERM STRENGTH AT VARIABLE LOADS

The bonds in structures and products are often subjected not to constant but to time-varying loads, which may be increasing, decreasing, cyclically applied, varying with rest periods. If the failure is assumed to be the overall result of individual nonreversible local failures, then the principle of superposition of failures, or Bailey's criterion can be applied for the determination of long-term strength. The mathematical expression of this principle at $n$ discrete loads is

$$\sum_{i=1}^{n} \Delta t_i/\tau(\tilde{P})_i = 1 \tag{9}$$

and at steady load

$$\int_0^{t_p} dt/\tau(P(t)) = 1 \tag{10}$$

where $t$ = time

$t_p$ = time corresponding to the failure of the load

$\tau(\tilde{P})$, $\tau(P(t))$ = the functional dependences of long-term strength on the load.

Using equations (9) and (10) to determine the long-term strength of the bonds, no limitations concerning the failure type need be applied *a priori*, whether the failure is cohesive or interfacial. It is known, however, that the principle of superposition does not always hold for the adhesive bonds[15] or for polymers in general.[1] Different structural changes may occur in the adhesive, when the time variable load is applied. Self-healing effect may be observed at higher frequency of cyclic loads. Alternating loading periods with rest periods can cause self-healing of the defects,[25] resulting in strengthening if the adhesive has good relaxational properties.

In such cases, Equations (9) or (10) cannot be used directly but the additional effects must be evaluated first. This can be accomplished by introducing additional terms into the equations. Thus, the possibility to evaluate long-term strength at varying loads and the validity of the superposition principle should be tested experimentally for every particular case. One of the ways to do this is the following. Long-term strength is determined experimentally at constant load $\tilde{P}_j$. Then the bonds are loaded by the same load $\tilde{P}_j$ and are kept for a shorter time than $\tau$, i.e., $\Delta t_1 < \tau$. Then, after the removal of the load, the rest period follows and the same load is applied again until the failure (time $\Delta t_2$). If $\Delta t_1 + \Delta t_2 = \tau$ is satisfied statistically, the principle of superposition holds. If $\Delta t_1 + \Delta t_2 > \tau$, then the rest period strengthens the adhesive bonds. If $\Delta t_1 + \Delta t_2 < \tau$, the rest period weakens it, compared to the long-term strength which would be attained at continuous load application.

The problems associated with the application of Bailey's criterion may be of a double nature: a so-called direct one, i.e. the evaluation of long-term strength at varying loads when the parameters of the equation at constant static load are known, and the inverse, i.e. the derivation of the long-term strength equation at constant static loads from the short-term tests at varying loads (using a conventional test with a dynamometer).

Attention should be directed to some interesting theoretical aspects of this question. Let us assume that the superposition principle holds and at a constant static load the relationship $\tau(\tilde{P}, T)$ is described by Equation (5). Bailey's criterion, then, has the form

$$\int_0^{t_p} \frac{dt}{\tau_0 \exp(U_0 - \gamma P(t)/RT)} = 1 \tag{11}$$

Two characteristic aspects of long-term strength may be solved in accordance with the scheme of the direct problem:

1. Long-term strength may be evaluated from equation (11) at varying load and at any constant temperature $T_i$, if the parameters $\tau_0^*$, $U_0$, and $\gamma$ of Equation (5) and the law $P(t)$ of the load variation with time are known
2. Long-term strength may be evaluated, in cases when the load variation $P_1(t)$ changes into $P_2(t)$

Assume that load $P_1(t)$ was in effect during time $t_1$ (where $t_1 < t_p$), and after that, its effect changed to $P_2(t)$. Then the remaining time until the failure after $t_1$ will be $\Delta t = t_p - t_1$. It may be calculated using the equation

$$\int_0^{t_1} dt/\tau(T_i P_1(t)) + \int_{t_1}^{t_p} dt/\tau(T_i P_2(t)) = 1. \tag{12}$$

This method, of course, is also applicable when the law of force changes $n$ times. Then the first member of Equation (12) becomes the sum of integrals.

In solving an opposite problem at constant temperature values ($T_i$ = constant, $i = 1, 2, \ldots$), with the help of Bailey's criterion

$$\int_0^{t_p} dt/A \exp(-\alpha P(t)) = 1,$$

the values of the parameters of Equation (7), $A$ and $\alpha$, are found. Then, assuming that $A = A(T)$ and $\alpha = \alpha(T)$, and having determined the form of these relationships, the values of the parameters $U_0$ and $\gamma$ of Equation (5) may be found.

While determining $A$ and $\alpha$ values at fixed temperatures, the relationship $P = P(t)$ of the load variation with respect to time is selected, and the values of time to failure $t_{P(e)}$ are experimentally determined. The load variation should be so selected that the integral of Bailey's criterion is of the simplest possible form. For this purpose, the linear function of the form $P = wt$ suits best, where $w$ is the intensity of load variation.* To avoid great errors in determining the values of parameters $A$ and $\alpha$, some short–time experiments should be carried out at different values of $w$ ($P = w_k t$, where $k = 1, 2, \ldots j \ldots, l \ldots n$.) From the data obtained, $C_n^2$ (number of combinations of $n$ in 2) equations of the following type are formed:

$$\left.\begin{aligned} \int_0^{t_{P,j}} dt/A \exp(-\alpha\, w_j t) = 1, \\ \int_0^{t_{P,l}} dt/A \exp(-\alpha w_l t) = 1, \end{aligned}\right\} \tag{13}$$

where $1 \neq j$. Parameters $A$ and $\alpha$ are determined as the mean values of the solutions of separate sets of equations. Since the parameters $\alpha$ and log $A$ are distributed normally (this follows from the normal distribution of the values of log $\tau$ for bonds), an $\alpha$ value is calculated as an arithmetic mean and an $A$ value as a geometric mean.

---

* The tension machine (dynamometer) must work according to this deformation pattern, rather than according to the more conventional method of constant deformation rate, $V = \varepsilon/t$.

If the relationship $\tau(\tilde{P}, T)$ is described by Equation (6) (at constant temperature, $\tau = BP^{-b}$) and the load varies in accordance with the linear relationship $P = wt$, then the set of equations in (13) acquires the following form:

$$\left.\begin{aligned}(w_j t_{P_j})^{b+1} &= Bw_j(b+1)\\(w_l t_{P_l})^{b+1} &= Bw_l(b+1)\end{aligned}\right\} \tag{14}$$

Eliminating the value of $B$ from Equation (14) we obtain:

$$b = \ln t_{P_l} - \ln t_{P_j} / \ln w_j t_{P_j} - \ln w_l t_{P_l} \tag{15}$$

Here the value of $b$ is calculated as an arithmetic mean and $B$ as a geometric mean.

The application of Bailey's criterion has been rather limited, even for the evaluation of long-term strength of polymer materials and especially for bond strength. The data accumulated by the author indicate that such a conservative attitude does not have a firm basis. A sufficient accuracy is obtained even for bonds based on viscoelastic adhesives, if the principle of superposition holds for the given load application.

Good agreement between experimental and calculated data is obtained when the load $P$ varies continuously in time $t$, for example, it increases accordingly with the linear relationship $P = wt$. If we have long-term strength in the form of Equation (7), obtained at constant loads $\tilde{P}_j$, using Equation (10) (subintegral function $1/\tau(P(t))$ of form (7) where $P \sim wt$), the time to failure and the force can be calculated. Controlled experiments, in which the bond failure strength was determined dynamometrically, show that calculated results usually differ from the experimental ones by 1–1.5% in the range at low rate of load variation. The higher rates compare quantitatively to the ones which occur in conventional dynamometric tests at deformation speed of 1–2 mm/sec.

Equally accurate results can be obtained by applying a nonlinear relationship of load variation. In this case, a mathematical expression of dynamometric curve of force vs. deformation must be obtained.

The practical value of such calculations is that if the long-term strength formula is known, dynamometric experiments are not necessary to predict the bond performance. Besides, using Equation (5) instead of (7), there are possibilities to evaluate the bond strength (break stress) at various temperatures or at least in the temperature range in which the parameter values of Equation (5) were experimentally determined. Naturally, all this is also possible in cases where the relationship $\tau(\tilde{P},T)$ is described by Equation (6).

The calculating possibilities of long-term strength at some other re-

**Table 7-2. The Results of a test of bonded cellulose diacetate film.**

| The Expression of Bailey's Criterion (I) | $T$(°C) | $\Delta t_1$ (sec) | $\dot{w}$ (kg/cm²sec) | $\Delta t_n$ Evaluated (sec) | $\Delta t_n$ From the Experiment (sec) | $/\delta/\%$ |
|---|---|---|---|---|---|---|
| $2m\int_0^{\Delta t_1} dt/\tau(wt) + m\Delta t_1/\tau(\tilde{P}_1)$ | 30 | 180 | | 385 | 386 | 0.3 |
| | | | $8 \times 10^{-3}$ | | | |
| $+\int_0^{\Delta t_n} dt/\tau(wt) = 1$ | 40 | 90 | | 300 | 332 | 9.6 |

| The Expression of Bailey's Criterion (II) | $T$(°C) | $\tilde{P}_1$ (kg/cm²) | $\tau(\tilde{P}_1)$ (sec) | $\Delta t_1$ (sec) | $\tilde{P}_2$ (kg/cm²) | $\tau(\tilde{P}_2)$ (sec) | $\Delta t_2$ Evaluated (sec) | $\Delta t_2$ from the Experiment (sec) | $/\delta/\%$ |
|---|---|---|---|---|---|---|---|---|---|
| | 30 | | $11.5 \times 10^3$ | $3.6 \times 10^3$ | | 329 | 606 | 599 | 1.2 |
| | | 1.00 | | | 2.0 | | | | |
| $\frac{\Delta t_1}{\tau(\tilde{P}_1)} + \frac{\Delta t_2}{\tau(\tilde{P}_2)} = 1$ | 40 | | $2.4 \times 10^3$ | $0.9 \times 10^3$ | | 165 | 105 | 106 | 0.95 |

| The Expression of Bailey's Criterion (III) | $T$(°C) | $\tilde{P}_1$ (kg/cm²) | $\tau(\tilde{P}_1)$(sec) Evaluated | $\Delta t_1$ (sec) | $\Sigma_{i=1}^{n} \Delta t_1$ (sec) from the Experiment | $/\delta/\%$ |
|---|---|---|---|---|---|---|
| | 30 | | 879 | 150 | 1256 | 42.9 |
| $\sum_{i=1}^{w} \Delta t_i/\tau(\tilde{P}_1) = 1$ | | 2.0 | | | | |
| | 40 | | 165 | 30 | 470 | 184.9 |

Note: $/\delta/$ is the relative error of experimental long-term strength values compared with the calculated ones.

lationship of load variation with time are discussed below. The experimental results obtained by testing bonds of cellulose diacetate films bonded by the nitrile rubber adhesive are used for illustration. Load variation program is shown in Fig. 7-10 and the calculated results in Table 7-2 along with the values of the parameters of Equation (5). As can be seen from Table 7-2 for load variation programs I and II (Fig. 7-10), the calculated long-term strength values correspond sufficiently well to the ones obtained experimentally. At the same time, experimentally obtained values for load variation program III considerably exceed the calculated ones. These should have corresponded to the long-term strength value $\sum_{i=1}^{n} \Delta t_i$, obtained under the continuous effect of load $\tilde{P}_1$, if the superposition principle would have held. This indicates that the rest between loading cycles was significant and resulted in the bond strengthening. Since the adhesive used was to some extent plastic, the strengthening was caused by self-healing of cracks. More examples are given,[15] showing that Bailey's criterion is applicable when the loading is continuous and its variation intensities are not great (such that internal heating does not result). It has been determined experimentally that when the rest period exists between load periods, long-term strength may either increase or decrease. For example, the decrease of long-term strength at normal temperatures is inherent for polychloroprene adhesives. So far the nature of this phenomenon is not sufficiently clear. With increasing temperature, this tendency becomes less obvious and eventually, even strengthening is possible.

Practical possibilities to solve the reverse problem will be illustrated by two examples. Figure 7-11 shows the relationship $\tau(\tilde{P})_{T_i}$ obtained experimentally at constant static loads (points), calculated by Equation (5) from the experimental data (continuous lines), and calculated from dyna-

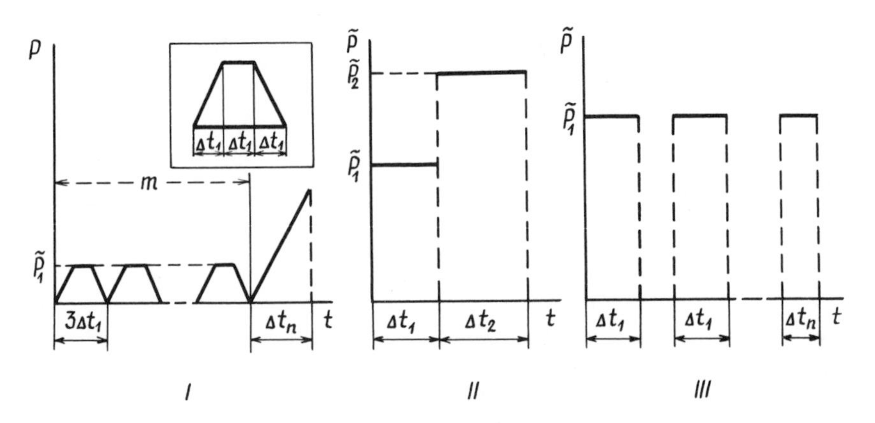


Fig. 7-10. Programs of load variation.

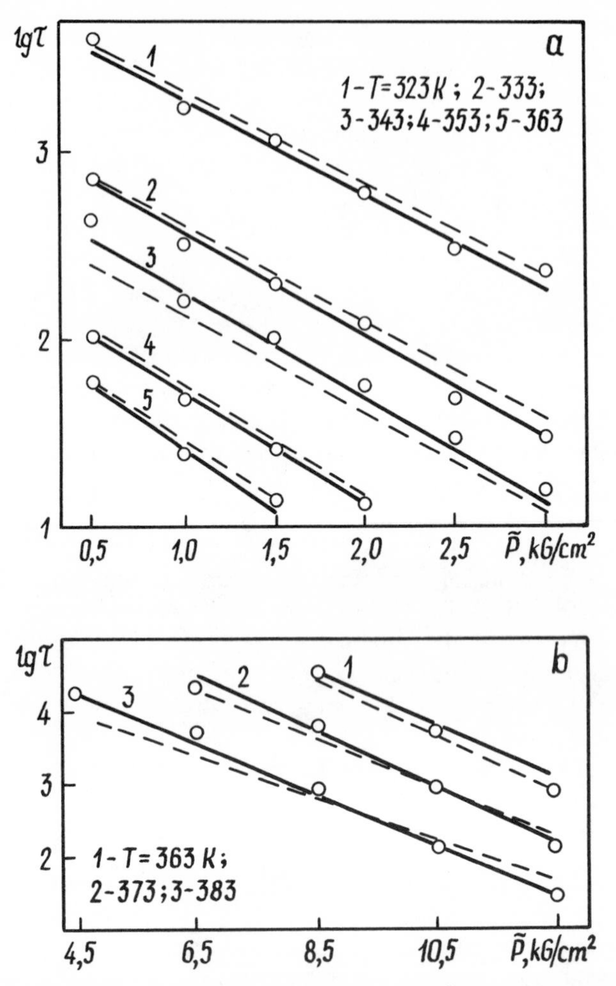


Fig. 7-11. Long-term strength ($\tau$) dependence on static load ($\tilde{P}$): (a) cellulose diacetate films bonded by polyurethane adhesive; (b) chrome leather bonded by polychloroprene adhesive. The points are experimental results; continuous lines are calculated according to Equation (5); dotted lines are the result of a solution of the reverse problem.

mometric tests, assuming a linear load variation (dotted lines). The fit is quite satisfactory for a complicated system, such as leather bonds. This shows that there are good possibilities to calculate parameter values of long-term strength Equation (5) from short-term dynamometric experiments. It should be noted that difficulties using this approach lie in the fact that parameters $A$ and $\alpha$ of Equation (7) are sensitive with respect to the experimental long-term strength values (time to failure $t_{P(e)}$) and relatively small variations of these values may provide great variation of values $A$ and $\alpha$. The variation of experimental long-term strength values in dynamometric tests may amount to 10–15% and higher. The methods of the determination of parameters $A$ and $\alpha$ suggested by the author and collaborators[15] decreases the inaccuracies, but on the whole, the methods to solve this problem are still to be developed. The reverse solution of the problem described here may be used as a fast method to approximate the long-term strength. The method is especially useful in comparative analysis of various bonds different from each other by the type of the adhesive and the adherends used.

## THE INFLUENCE OF TEMPERATURE AND THE DEFORMATION RATE

In dynamometric tests the strength is expressed by force $P$ or stress at the moment of sample failure which, in turn, depends on temperature and deformation rate. Such relationships have been widely discussed, including the behavior of rubbery adhesives.[13,15,20,21,23,25,29,30,31,32,33] The dependence on temperature $P(T)$ most often is of an exponential form[15,23]

$$P = A_0 \exp(U/RT), \tag{16}$$

where $A_0$, $U$ = parameters depending on the mechanical deformation properties of the elements of the bond.

Parameter $U$ may acquire the meaning of failure activation energy, but the quantitative correlation is not required within the wide range of temperatures including the glass transition area. The dependence $P(T)$ may be expressed by this equation but separately for different physical states, i.e., above and below the glass transition temperature.[15,23]

The deviations from this relationship in certain temperature area are notable, especially where the mechanical deformation properties vary more intensively because of crosslinking, recrystallization, etc., than expressed by Equation (16). In such cases, the relationship $P(T)$ may be treated as an overall result of two simultaneous processes: normal decrease of strength with increasing temperature and the local process associated with the changes of mechanical properties of the system elements within the temperature range. Then, it may be assumed that the parameter $U$, determining the strength intensity variation in Equation (16), is dependent on temperature:

$$U = U_1 + U_2(T). \tag{17}$$

The positive sign corresponds to the relative strengthening of the bond; a negative sign would be used in case of bond strength decrease.[15]

$U_1$ expresses the change of strength intensity corresponding to the relationship $P(T)$ of the form of Equation (16) and $U_2(T)$ expresses the deviation from it. Then

$$P = A_0 \exp(U_1 + U_2(T)/RT). \tag{18}$$

$U_2(T)$ can be expressed by the equation

$$U_2(T) = B_u \exp(-C_u(1/T - 1/T_0))^2, \tag{19}$$

where $B_u$ corresponds to the maximum deviation of $U$ from $U_1$ (at $T = T_0$), and $C_u$ characterizes the temperature range of the deviation of $P$ values from the ones corresponding to Equation (16).

The dependence on the rate $P(v)$ is weak when the adhesive is in a glassy state. For adhesives in a rubbery state, within a wide range of rates comprising 6–7 decimal orders, it is represented by a curve of the $S$ form. Therefore, the rate affects the strength values significantly only in a certain range of its mean values, about 0.015–15 mm/sec.

Within the range of rates, where the strength depends on the rate, relationship $P(v)$ is described by the function[15,23]

$$P = kv^m, \tag{20}$$

where $k$ and $m$ are parameters depending on the properties of bond elements and test conditions. In certain cases, relationship $P(v)$ deviates from the form of Equation (20) even in a narrow range of deformation rates. It is suggested that such deviations are associated with the change of failure mechanism.[31]

Numerous data are available on the reverse dependences of failure rate on constant load value.[24,30,34,etc.] Their form is analogous to that of the relationship $P(v)$.

The dependence of strength on two variables—temperature and deformation rate $P(T,v)$—may be described by the equation

$$P = kv^m \exp(U/RT), \tag{21}$$

similarly, as for polymers in the rubbery state.[2] The application of this equation for different bonds of soft and laminated materials was studied employing shear and peel tests.[15,23] The parameters $M$ and $U$ of Equation (21) are constants only if the relationships $\log P(1/T)_{v_i}$ and $\log P(\log v)_{T_j}$ are linear and the straight lines are parallel. Practically, this is not always the case. Often the relationship log $P(\log v)$ does not meet this requirement. Therefore, it is natural to suppose that the parameters of Equation (21) depend on the deformation rate. Such dependence can be approximated by the following functions:

$$U = U_1 v^{-b_1} \qquad \text{and} \qquad n = n_1(1 - n_2 \log v). \tag{22}$$

For example, for the cellulose diacetate film bonded with nitrile rubber adhesive in the interval $T \in (30;60)°C$ and $v \in (0.06;4)$mm/sec, the relationship $P(T,v)$ has the form:

$$P = 1.49 \times 10^{-5} \exp(7.6v^{-0.06}/RT)v^{0.85} \tag{23}$$

(the approximation error $\delta \sim 10\%$).

The dependence of parameters on variables (especially $v$) is stronger if the failure mechanism is more complicated. This is typical of the bonds of fibrous structures where the fibers reinforce the adhesive and must be broken.

In describing the experimental data of the relationship $P(T,v)$ by Equation (21), the error is usually less than 10–15%. On the whole, however, Equation (21) is not as suitable as (5) for the prognosis. The way to evaluate its parameters is the same as for Equation (5), but their physical meaning is not sufficiently defined.

The dependence $P(T,v)$ of the form of Equation (21) is very useful when the prognosis or calculations are based on the results of dynamometric tests. Sometimes, in solving problems of strength dependence, it is useful to hold one variable constant while the other is allowed to vary. Naturally, in describing strength dependences on a single variable $T$ or $v$ the approximation error is lower and usually does not exceed 10%.

## HEAT RESISTANCE

Heat resistance of adhesive bonds is understood as the property of maintaining the bond at higher temperatures. The critical limit at which the bond loses this property is its flow temperature. At this temperature, the failure of the bond may occur even without the effect of an outside force, for example, due to the weight of the adherends. At temperatures below the flow, the failure of adhesive bonds may also occur and the higher the temperature, the lower the creep resistance, because the molecular interaction weakens with increasing temperature. Therefore, heat resistance can be regarded as long-term strength dependency on load and temperature.

This may be evaluated experimentally or by using Equations (5), (6), (16), and (21).

Comparing different adhesives or structural properties of the bonds, heat resistance may be evaluated in terms of the temperature at which the bonds lose their stability and the bond fails. If the heat resistance is determined at nonisothermal conditions, with temperature increasing in some pattern, the temperature at which the bond failure occurs may be assumed to be a limiting (extremal) characteristic of heat resistance at a given load and temperature variation. It is named the stability loss temperature $T_{T_s}$.

The way to determine this characteristic is not complicated. The joint is placed into a heat chamber; it is loaded by a constant or periodically applied load and the temperature is raised gradually in accordance with some given pattern (e.g., linear, exponential, parabolic, etc., which may be accomplished by a simple programmed control). As a critical temperature for the particular bond is reached, the failure occurs. The temperature at the time of failure is recorded. For lap joints, it is the time when a nonreversible shift of the elements with respect to each other begins (nonreversibility may be checked by unloading). In the peel test, it is the beginning of peeling. Since this time may be difficult to fix, another characteristic may be used, i.e., by fixing the time when the crack depth will be, say, 2 mm.

The dependence of temperature $T_{T_s}$ on load $\tilde{P}$ and on the thickness of the adhesive $h$ is of a hyperbolic form:

$$T_{T_s} = a_1 \tilde{P}^{-k_1} \tag{24}$$

$$T_{T_s} = a_2 h^{-k_2} \tag{25}$$

where $a_1$, $k_1$ and $a_2$, $k_2$ are the parameters of these relationships. The parameter $a_1$ corresponds to the value of $T_{T_s}$ at $\tilde{P} = 1$ and may be regarded as a reference heat resistance characteristic of various bonds; $k_1$ shows the variation intensity of $T_{T_s}$ as the load varies, thus characterizing the creep resistance of the adhesive. Similar is the relationship of the parameters $a_2$, $k_2$ to the adhesive thickness.

It has been determined that for bonds with the adhesive in a rubbery state, the form of the relationship $T_{T_s}(\tilde{P})$ and $T_{T_s}(h)$ is not dependent on the chemical nature of the adhesive, type of bond, or on the method of temperature and force application. Thus, Equations (24) and (25) are quite versatile. It is interesting that the values of $T_{T_s}$ are not dependent on the form of the temperature variation. If a temperature $T_{T_s}$ is reached in time $\Delta t$, for example linearly, the $T_{T_s}$ value will remain the same after the change of the way the temperature increases (with an accuracy of 3–5°C). On the other hand, $T_{T_s}$ values depend greatly on the rate of temperature change, indicating that this index represents the heat resistance evaluation characteristic. Equation (2-4) defines the extremal operational limits of the bond in terms of temperatures and load.

Table 7-3 shows parameter values of Equations (24) and (25) for some lap joints.

In principle, the evaluation of $T_{T_s}$ values in terms of load is possible by applying Bailey's criterion

$$\int_0^{t_{T_s}} dt \Big/ \tau(\tilde{P}; T(t)) = 1, \tag{26}$$

without performing the experiments described here, if the long-term strength form, for example, of Equation (5) is known.

Here $\tau$ is assumed to be the time during which temperature $T_{T_S}$ is reached, i.e., $\tau = t_{T_{T_s}}$, as the temperature varies with time in accordance with $T(t)$ relationship. (The method of the solution of this problem is described in Reference 35.) It can provide results with the error up to 2–3%.

## THE APPLICATION OF ANALOGIES FOR STRENGTH PROGNOSIS

Different factors may have analogous effects on relaxational properties of polymers. For example, it is known that within certain intervals of temperature and deformation rate, there exists a similarity between the effect of temperature increase and rate decrease on the mechanical properties of polymers. This is one of the ways in which the temperature-time analogy reveals itself. The principle of this analogy is based on the experimental fact that the form of the dependence of the viscoelastic function of the polymer on the time logarithm $\log t$ does not change with temperature. The temperature effect is evaluated by a horizontal shift of the curves of the viscoelastic function along the $\log t$ axis by the value of $\log a_T$, which is dependent on temperature. This shift results in one curve referred to as a master curve (Fig. 7-12).

The dependence of $a_T$ on temperature is described by the WLF Equation[36] which has found a wide application:

$$\log a_T = -C_1^S(T - T_s)/C_2^s + T - T_s, \tag{27}$$

Here $a_T$ is the ratio of relaxation times at temperatures $T$ and $T_s$; $\log a_T$ is referred to as a reduction factor; $T_s$ is a standard (reduction) temperature; $C_1^s$ and $C_2^s$ are the coefficients. (Their physical meaning is explained in Reference (36).)

**Table 7-3. Parameter values of Equations (24) and (25) for some lap joints.**

| | Elements of the Bond | | Dependence $T_{T_s}(\tilde{P})$ | | | | Dependence $T_{T_s}(h)$ | | | |
|---|---|---|---|---|---|---|---|---|---|---|
| | | | Thickness of the Adhesive $h$(mm) | Parameters of Equation (24) | | Intervals of the Values $\tilde{P}$(kg/cm$^2$) and $T_{T_s}$(°K) | Load (kg/cm$^2$) | Parameters of Equation (25) | | Intervals of $h$(mm) Values |
| | Adherend | Adhesive | | $a_1$ | $k_1$ | | | $a_2$ | $k_2$ | |
| 1 | Cellulose Diacetate Film | Nitrile Rubber | 0.15 | 304 | 0.07 | 0.025–1.125<br>403–299 | 0.100 | 335 | 0.028 | 0.01–0.58 |
| 2 | Cellulose diacetate film | Polycholoprene Compound | 0.05 | 322 | 0.07 | 0.025–0.875<br>417–331 | 0.125 | 317 | 0.020 | 0.03–0.58 |
| 3 | Chrome Leather | Polyurethane Adhesive with 5% of polyisocyanate | 0.11 | 351 | 0.19 | 0.50–5.00<br>385–295 | 1.250 | 317 | 0.030 | 0.08–1.04 |

NOTE: Relative error of the experimental data approximation by Equations (24) and (25) did not exceed 1–1.5%, i.e., the experimental values of the dependences $T_{T_s}(\tilde{P})$ and $T_{T_s}(h)$ in the logarithmic coordinates fitted almost ideally on the straight lines. The experiment was performed at linear temperature variation law at a rate of 2°K/min.

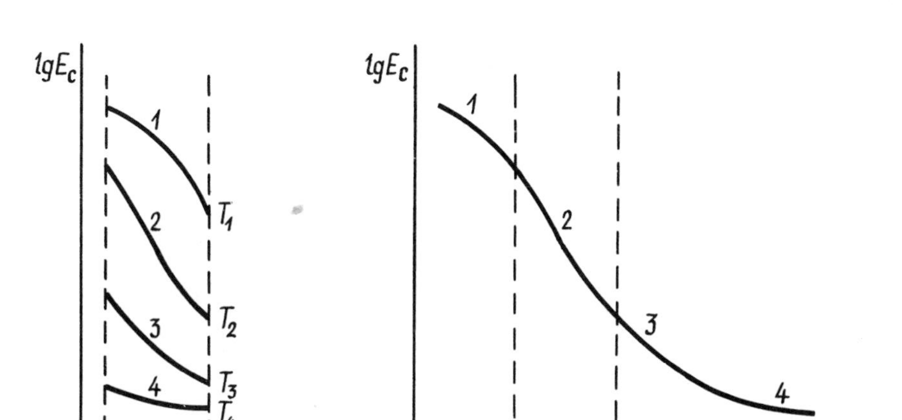


Fig. 7-12. Master curve for the dependence of creep modulus on temperature and time.

When the temperature-time analogy exists, the form of the WLF equation is not dependent on the selected value of $T_s$. Therefore, two ways of building the master curves are possible. The first is to assign to the coefficients $C_1^s$ and $C_2^s$ their universal values; $C_1^s = 8.86$ and $C_2^s = 101.6$. Then, most often $T_s = T_g + 50$ ($T_g$ is the glass transition temperature). If $T_s \neq T_g + 50$, the first stage in building the master curve is to determine the reduction temperature $T_s$. The essence of the second way is that the $T_s$ value is selected arbitrarily to be convenient in terms of building the master curve. Then, according to the empirical values of log $a_T$, optimal values of the coefficients $C_1^s$ and $C_2^s$ are to be determined.

The very idea of using analogies to obtain master curves proved very effective. It is used for the description of strength characteristics as well.[37] Theoretically, it is justified because the failure and deformation processes of polymers, especially above glass transition temperatures, are closely related and are followed by conformative rearrangement of macromolecules.[36] If relaxation times change by a factor $a_T$, due to a temperature change from $T_s$ to $T$, the amount of energy required for the failure must be accumulated during time $\tau/a_T$ ($\tau$ is long-term strength at temperature $T_s$ for deformation rate $va_T$). Therefore, in describing the dependence of failure force (stress) on temperature and deformation rate, a generalized strength curve is built on the coordinates log $PT_s/T$ and log $va_T$, and a long-term strength curve is built on coordinates log $\tilde{P}T_s/T$ and log $\tau/a_T$ (for polymers at constant stress log $\sigma T_s/T$ and log $\tau/d_T$). The ratio $T_s/T$ is referred to as a vertical shift factor. It evaluates the vertical shift of individual curves required for building a master curve. When the horizontal shift alone is sufficient, this

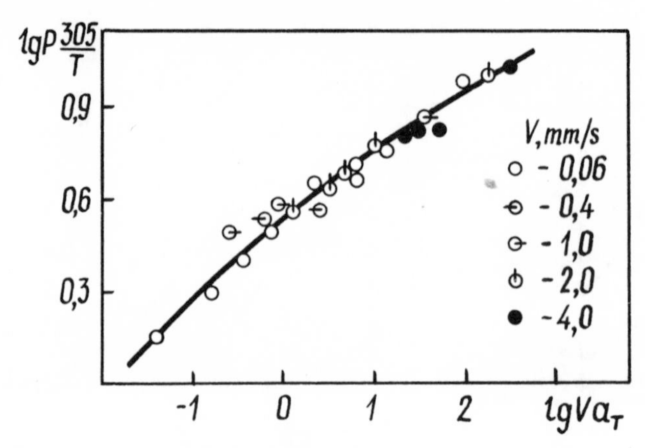


Fig. 7-13. The master curve of strength ($P$) dependence on temperature ($T$) and rate ($v$). Cellulose diacetate films bonded with nitrile rubber adhesive (Table 7-4).

factor can be neglected. Then master strength curves are built on the coordinates log $P$ and log $va_T$, and long-term strength curves in log $\tilde{P}$ and log $\tau/a_T$.

At present, a number of other analogies have been determined: vibro-time, concentration-time, stress-time.[38] The possibilities of the application of analogies for bonds have also been proven.[15,21,29,30,39,40] Temperature-time, concentration-time, temperature-concentration, and some other analogies were also determined.[15,41]

In all cases when one or another analogy exists, the reduction factor $a_x$

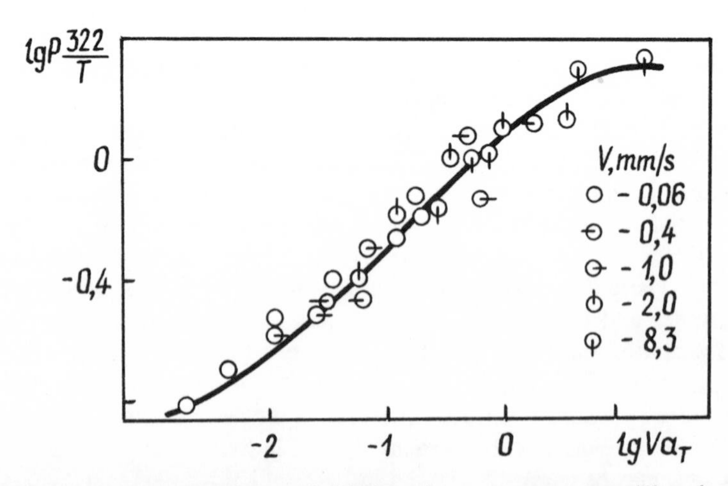


Fig. 7-14. The master curve of strength ($P$) dependence on temperature ($T$) and rate ($v$). Cellulose diacetate films bonded with polyurethane adhesive (Table 7-4).

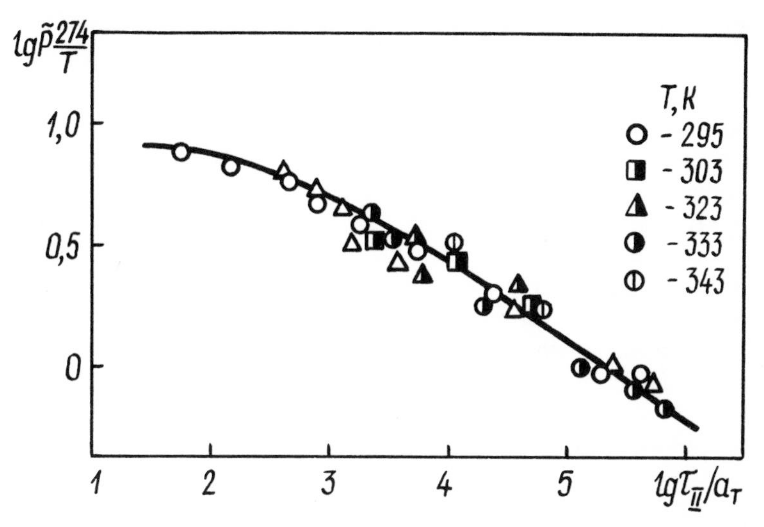


Fig. 7-15. The master curve of the long-term strength ($\tau$) dependence on static load ($\tilde{P}$) and temperature ($T$). Cellulose diacetate films bonded with polyurethane adhesive (Table 7-4).

is used for construction of master curves. It is calculated from forms analogous to Equation (27) for which other variables, instead of temperature, exist that effect the relaxation (e.g., vibration frequency, concentration of plasticizer in the polymer, stress intensity, etc.). The coefficients $C_1^s$ and $C_2^s$ also have other values. In such cases, the reduction factor $a_x$ represents the ratio of relaxation time at different values of factors influencing relaxation, similar to the temperature-time analogy.

$$\log a_x = \alpha(x - x_s)/\beta + x - x_s. \tag{28}$$

Equation (28) is the generalization of the WLF equation.

Below are some examples of the temperature-time analogy (Figs. 7-13, 7-14, and 7-15) obtained by testing cellulose diacetate film bonds. The data are represented in Table 7-4.

**Table 7-4. Tests of cellulose diacetate film bonds.**

| FIGURE NO. | ADHESIVE | TEST TYPE | $T_g$ (°K) | $T_s$ (°K) | $T_s - T_g$ |
|---|---|---|---|---|---|
| 7-13 | Nitrile rubber | Shear | 255 | 305 | 50 |
| 7-14 | Polyurethane and polyisocyanate (7%) | Peel | 282 | 322 | 40 |
| 7-15 | Polyurethane and polyisocyanate (7%) | Shear at $P$ = constant | 255 | 274 | 19 |

From the given illustrations, it is clear that experimental points lie sufficiently well on the master curve for strength characteristics obtained by dynamometric tests at different rates and temperatures (Figs. 7-13 and 7-14) and for long-term characteristics at different static loads and temperatures (Fig. 7-15). In building master curves for the adhesive bonds mentioned here and for many others,[15,40,41] certain regularities concerning the reduction temperature values have been observed. For lap joints by dynamometric tests, $T_s - T_g = 50$; for dynamometric peel tests, as well as in the case of long-term strength tests at constant loads, $T_s - T_g < 50$.

For pressure-sensitive adhesives, various temperature-concentration analogies are significant. For bonds with the same adhesive different in the degree of plasticization, the quantity of plasticizer in the adhesive $c$ may be considered as a factor affecting relaxation. Then the reduction factor is $\log a_c$, and Equation (28) has the form

$$\log a_c = -b_1(c - c_s)/b_2 + c - c_s. \tag{29}$$

The master curve is constructed in the coordinates $\log P\, c_s/c$ and $\log T/a_c$ or $\log P$ and $\log T/a_c$. (This analogy is described in References 15 and 40.)

The other modification of the temperature-concentration analogy is useful for bonding with the adhesives in solution form. After bonding, some solvent remains in the adhesive, which acts as a plasticizer. With time, the solvent gradually evaporates, often resulting in bond strengthening. The decrease of the quantity of solvent in terms of strength in such a case corresponds to the decrease of temperature. Therefore, from the experiments conducted at room temperature, at various times after bonding, the bond strength at various temperatures may be predicted, or a reverse problem may be solved: from the experimental data at various temperatures, bond strength corresponding to various times after bonding may be predicted.

Such an analogy was obtained experimentally (Fig. 7-16) for polychloroprene adhesive-bonded textile.[41] The main experiment was carried out by peel tests conducted 1 to 25 days (7 tests) after bonding at 9 fixed temperatures from 20 to 80°C. Each experimental point is the arithmetic mean of 6 to 8 tests from 60 to 80 strength values with a variation coefficient up to 13%). Control experiment was carried out at a single $c_s$ of residual solvent and at temperatures −10°C, −5°C, 0°C, and 55°C. We can see that the points of the control experiment fit well on the master curve (error 4–9%) and that it is within a reliable interval. The interval of prognosis information in terms of temperature is extended by 10 degrees.

Another modification of temperature-concentration analogy (Fig. 7-17) is where the quantity of resin in a polychloroprene adhesive stands for the

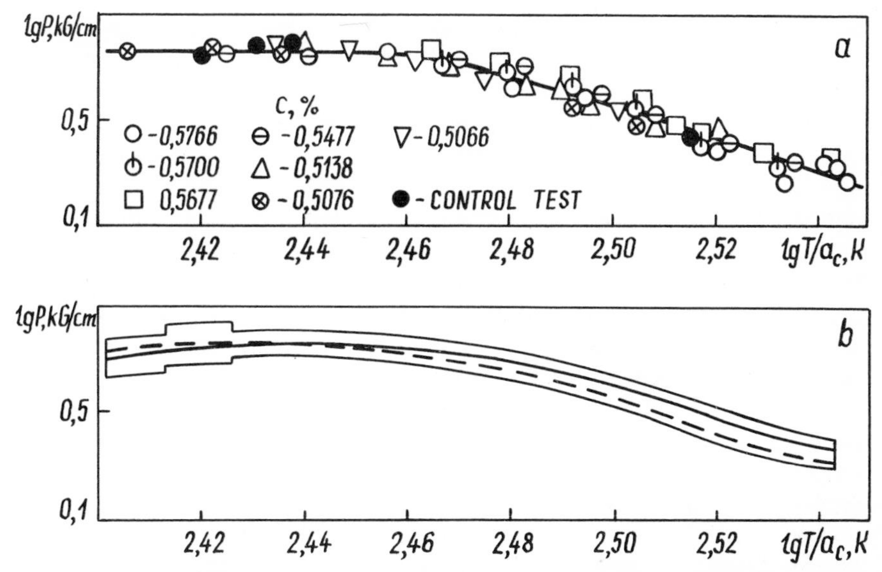


Fig. 7-16. Textile bonded with polychloroprene adhesive: (a) the master curve of strength ($P$) dependence on the remaining quantity of solvent in the adhesive ($c$) and temperature ($T$); (b) the reliable area of the master curve. The continuous line represents the control test; the dotted line represents the master curve.

factor affecting relaxation. In the experiment, butylphenol formaldehyde resin was used. Its quantity in the adhesive varied from 0 to 60%. The peel strength of textile strips was determined after 30 days in the temperature range of 20–80°C. Control experiment was carried out at temperatures

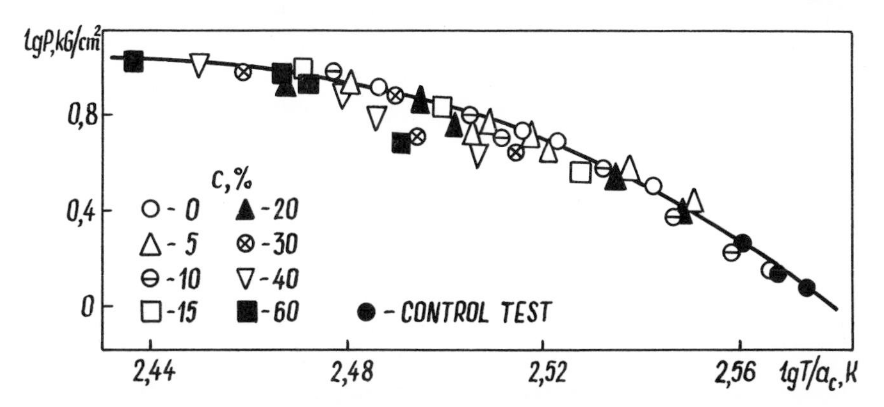


Fig. 7-17. The master curve of strength ($P$) dependence on the quantity of resin in the adhesive ($c$) and temperature ($T$). Textile bonded with polychloroprene adhesive.

90°C, 95°C and 100°C. Its points fit well on the master curve (error 2–4%). The prognosis interval in terms of temperature was extended up to 100°C.

The last example is temperature-time analogy obtained by making use of the same experimental data as for the analogy shown in Fig. 7-16. The time after bonding and the temperature were taken as variables (factors affecting relaxation). The control experiment at $T = 40°C$ was carried out after 25, 40, and 240 days. Its points fit well on the master curve (maximum error in 240 days is 6.7%, mean error is 2.5%). The prognosis interval here is extended from 7 minutes to 2.5 years in accordance with the master curve or at least until 240 days in terms of the control experiment. Thus, by making experiments shortly after bonding at various temperatures, the strength can be predicted at any temperature and at any time after bonding within the limits of prognosis interval, or at least within the limits of the control experiment.

From this discussion, one can get a picture of its applicability for prognosis. In some cases, the physical meaning or phenomenon may remain unclear. The existence of analogies should always be formally checked. This is proved by the alignment of the dependence of the reduction factor on the variable in the coordinates $(x - x_s) - (x - x_s/\log a_x)$ and at the same time, by the existence of the form of the WLF equation (28) or in some cases, Equation (27).

Generally, in applying the analogy method, the first step is the experiment. Then, if the fact of existence of the analogy for the factors in question is not certain, the second step is to check the existence of the analogy. Then the master curve is constructed and the control test is carried out.

To make use of the master curves for prognosis one can rely on various known analogies or look for new ones. It depends on the character of the expected information of prognosis and on which investigations or experiments are more convenient to carry out. To obtain the information more rapidly from the master curves, special nomograms may be used.[15,40] To draw master curves, a special program may be set up and computer techniques may be applied.[15]

## EVALUATION OF COMPLEX STRENGTH

Various methods have been used to evaluate bond strength characteristics. They are tests of strength at fixed conditions such as temperature, rate, environmental parameters, quantity of agressive agents, etc. This way, some information is obtained which often may be sufficient.

In some cases, more extended information on strength is desirable. As new types of adhesives are being developed, more information on their application technology is sought, such as information about strength at

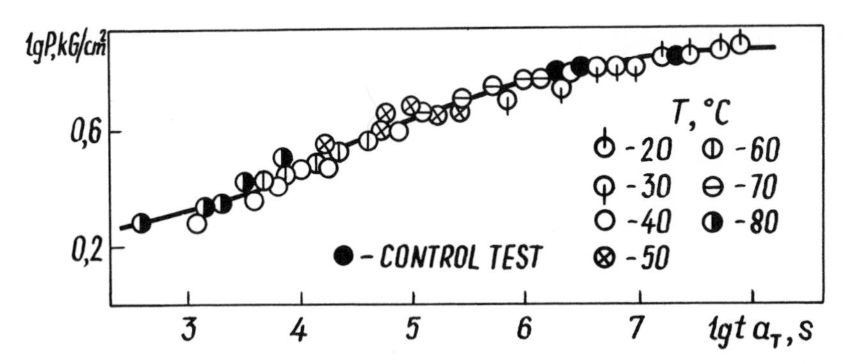


Fig. 7-18. The master curve of strength dependence of the time after bonding ($t$) and temperature. Textile bonded with polychloroprene adhesive.

various times after bonding, heat-resistance and fatigue at cyclic or static loads, etc. More extensive information requires more extensive experiments, which are not always effective enough in terms of information yield.

The way to increase the experimental efficiency is to make use of the complex evaluation approach. There is no method yet developed to evaluate the bond strength completely. The possibilities of developing such a method are sufficiently real as has been shown by the author's laboratory experiments. Besides, it should be stressed that there are strength evaluation methods which are quite informative and allow one to determine a series of strength values. They are the long-term strength equations described here, the use of stability-loss temperatures, Bailey's criterion, and the master curves.

In creating the complex strength evaluation methods, correlation links among the various experiment methods should be known. Then after certain formalization, the test complex can be expressed in the form of the informational matrix, which can be enriched with new results or new correlation links among the tests already known.

For the processing of the matrix, a special program for the computer should be developed. The computer informs us about the test combination which could supply the required information on the questions under consideration. With such information available, the laboratory could choose the desirable combination of tests or evaluation methods.

The same information may be obtained in a different way, if a sufficiently large quantity of correlation links among individual tests and methods is determined. The degree of reliability and the volume of tests may depend on the method chosen. Information of this kind can also be obtained from a computer.

## REFERENCES

1. Regel, V. R., Slutsker, A. I., and Tomashevskiĭ. E. *Kineticheskaya priroda prochnosti tverdykh tel.* Moskva, "Nauka" (1974).
2. Gul, J. E. *Struktura i prochnost polimerov*, Moskva, Khimiya (1971).
3. Slonimskiĭ L., Askadskiĭ A. A., and Kazanceva, V. V. *Polymer Mechanics*, No. **5** (1977).
4. Bartenev, G. L., Zuev, Yu. S., *Prochnost i razrushenie vysokoelasticheskikh materialov*, Moskva, Khimiya (1964).
5. *Fracture Processes in Polymeric Solids* B. Rosen, (ed.). New York: Wiley and Sons, 1964.
6. Andrews, E. H. *Fracture in Polymers* Edinburgh-L, (1968).
7. Bueche, F. *J. Appl. Phys.* **29**: 1231 (1958).
8. Tobolsky, H. Eyring *J. Chem. Phys.* **11**: 125 (1943).
9. Haplin, I. C. *J. Appl. Phys*, **35**: 3133 (1964).
10. Busse, W. F. *J. Appl. Phys.* **13**(11): 715 (1942).
11. Bueche, F. *Physical Properties of Polymers* New York: Wiley and Sons, 1962.
12. Regel, V. *Polymer Mechanics* No. 1, 1971.
13. Bikerman, J. J. *The Science of Adhesive Joints* New York: Academic Press, 1968.
14. Bikerman, J. J. *Recent Advances in Adhesion* Lee (ed.), Gordon and Breach, (1973).
15. Rajeckas, V. *Mekhanicheskaya prochnost kleevykh soedineniĭ kozhevenno-obuvnykh materialov* Moskva, Legkaya Industriya (1976).
16. De Bruine, N. A., and Houwink, R., (eds.). *Adhesion and Adhesives* Amsterdam: Elsevier Publishing Co., 1951.
17. Kaelble, D. H. *Physical Chemistry of Adhesion* New York: Wiley and Sons, 1971.
18. Houwink, R., and Salomon, G., (eds.). *Adhesion and Adhesives*, Amsterdam: Elsevier, (1965).
19. Jouwersma, C. *J. Polym. Sci* **45**: 253 (1960).
20. Gordon, J. I. *J. Appl. Polym. Sci.* **7**: 625 (1963).
21. Kaelble, D. H. *Adhesives Age* **3**(5): 37 (1960).
22. Hata, T. *Zairyo* (Japan) **13**: 341 (1964).
23. *Voprosy prochnosti kleevyhk soedineniĭ*, sb. statei Izd. Kaunasskiĭ politekhnicheskiĭ institut, 1971.
24. Fukuzava, K. *Setutiaku* (Japan) **11**: 250 (1967).
25. Malinskii, Yu. M., Prokopenko, V. V., and Kargin, V. A. *DAN SSSR*, **189**: 568 (1969); *VMS*, No. 10: 1832 (1964); *Polymer Mechanics*, No. **2**, **3**, **6** (1970).
26. Voyutskii, S. S. *Adhesives Age* **5**(4): 30 (1962).
27. Tikhonov, A. T., Tager, A. A., and Stepanov, Ye. *Polymer Mechanics*, No. 4 (1968).
28. Bailey, J. *Glass Ind.* **20**: 21 (1939).
29. Kaelble, D. H. *J. Coll. Sci* **19**: 413 (1964).
30. Hata, T., Gamo, M., Kozima, K., and Nakamura, T. *Chem. High. Polym.* (Japan) **22**: 160 (1965).
31. Aubrey, D. W., Wedling, G. N., and Wong, T. *J. Appl. Polym. Sci.* **13**: 2193 (1969).
32. Weidner, C. L. *Adhesives Age* **6**(7): 30 (1963).
33. Voyutskii, S. S. *Autohesion and Adhesion of High Polymers* New York: Wiley and Sons, 1963.
34. Imoto, T., *Sikizai Kyokaisi* (Japan) **40**(7): 304 (1967).
35. Rajeckas, V. *J. Adhesion* **9**: 51 (1977).
36. Ferry, J. D. *Viscoelastic Properties of Polymers* New York: Wiley and Sons, 1961.
37. Smith, T. L., and Stedry, P. J. *J. Appl. Phys* **31**: 1892 (1960).
38. Urzhumtsev, Yu. *Polymer Mechanics* No. 3 (1972).

39. Nonaka, J. *J. Soc. Mater. Sci.* (Japan) **16**: 460 (1967).
40. Rajeckas, V., and Pekarskas, V. *J. Appl. Polym. Sci.* **20**: 1941 (1976).
41. Zaletdinov, F. F., Pekarskas, V., and Rajeckas, V., *Izvestija vysshikh uchebnykh zavedeniĭ*, Tekhnologiya legkoĭ promyshlennosti, No. 5: 96 (1977).
42. Urzhumtsev, Yu. S., and Maksimov, R. D. *Prognostika deformativnosti polimernykh materialov*, Riga, Zinatne (1975).

# Chapter 8

# Aging Properties

**Keiji Fukuzawa**
*Consulting Chemical Engineer*
*Tokyo, Japan*

Many different elastomers are used for pressure-sensitive adhesives: natural rubber, SBR, polyisobutylene, butyl rubber, block copolymers, silicones, urethanes, and acrylics. Natural rubber and acrylic resins are the most important with block copolymers growing with increasing use of hot melt adhesives.

Unsaturated elastomers such as natural rubber, SBR and block copolymers (SIS and SBS) have double bonds in the molecule. Therefore, these polymers are inferior in aging properties to the saturated polymers. Since the pressure-sensitive application of these elastomers is usually of a short duration, it is more important to determine how well these adhesives retain their pressure-sensitive properties on aging in roll, rather than how they age after application. Pressure-sensitive adhesives based on acrylic resins and saturated rubbers such as polyisobutylene, butyl rubber, and others are employed for long-term uses. In this case, the properties after application also become important.

The rubber type pressure-sensitive adhesives are composed of rubber, tackifying resin, softener, and antioxidant. It is important to select stable components and a suitable antioxidant in order to maximize the retention of adhesive properties on aging.

## EFFECT OF TACKIFYING RESINS

Aging properties are heavily influenced by the tackifying resin used. Rosin has been widely used in pressure-sensitive adhesives and can give adhesives with good properties, especially at low temperature. However rosin, a natural resin, tends to oxidize easily because of its conjugated double bonds. The stability of rosin has been improved by disproportionation and by hydrogenation of the conjugated double bonds. Wetzel and Alexander[1] investigated the effect of various resins on aging properties of adhesives as shown in Fig. 8-1. Higuchi, Watanabe, and Toyama[2] have studied similar relations by the oxygen absorption method shown in Fig. 8-2. Figures 8-3 and 8-4 show data by Fukuzawa and Matsukawa[3] that compare various tackifying resins i.e., rosin, terpene, and $C_5$ petroleum resins and their behavior on aging. The peel adhesion of natural rubber-resin type adhesive has a tendency to increase in the region of adhesive failure and decrease in the region of cohesive failure resulting from the softening of the pressure-sensitive adhesive on aging. This is, of course, related to the decrease of cohesive strength of natural rubber on aging. Petroleum resins show better aging properties than natural resins.

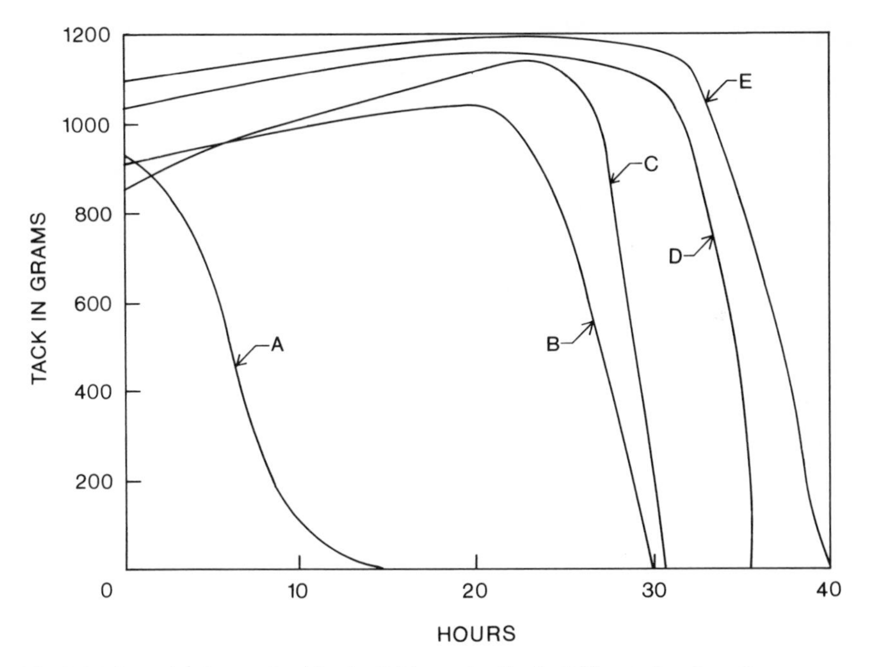


Fig. 8-1. The tack of natural rubber/tackifying resin blends (3/2) as a function of exposure at 93°C. According to Wetzel and Alexander, *Adhesives Age* **7** (1) 28 (1964): (A) Pentalyn H; (B) Resin H64–Pentalyn H; (C) Staybelite Ester 10; (D) Cellolyn 21; (E) ester of perhydrogenated rosin.

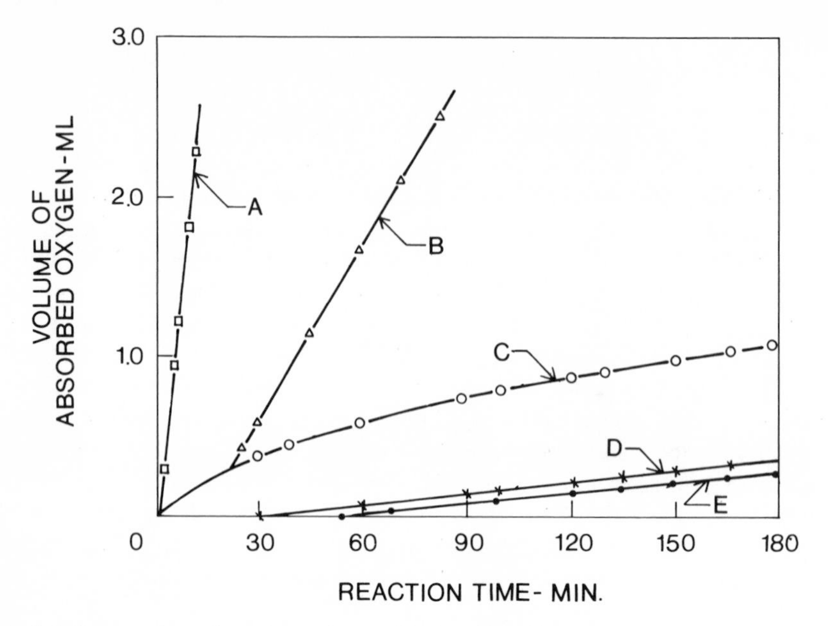


Fig. 8-2. Oxygen absorption curves of natural resins. Higuchi, Watanabe, and Toyama. Symposium on Adhesion and Adhesives (Japan), p. 17 (1965): (A) rosin; (B) glycerol ester of rosin; (C) pentaerythritol ester of partially hydrogenated rosin; (D) glycerol ester of partially hydrogenated rosin; (E) hydrogenated rosin. The sample consisted of 10 g resin solution in 10 ml o-dichlorobenzene. Reaction temperature 70°C, apparent oxygen pressure 760 mm Hg.

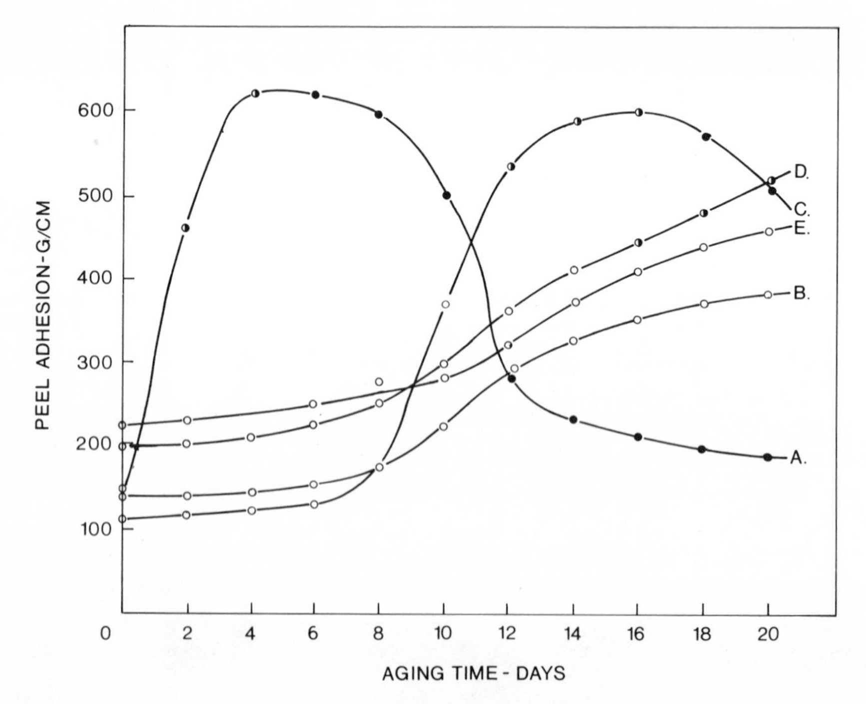


Fig. 8-3. Peel adhesion of natural rubber/resin blends (3/2) as a function of accelerated aging at 70°C.[3] ○ = adhesive failure; ◑ = transition area; ● = cohesive failure. (A) rosin; (B) Piccopale H2; (C) Ester Gum H; (D) Piccopale 100; (E) Piccolyte S-100.

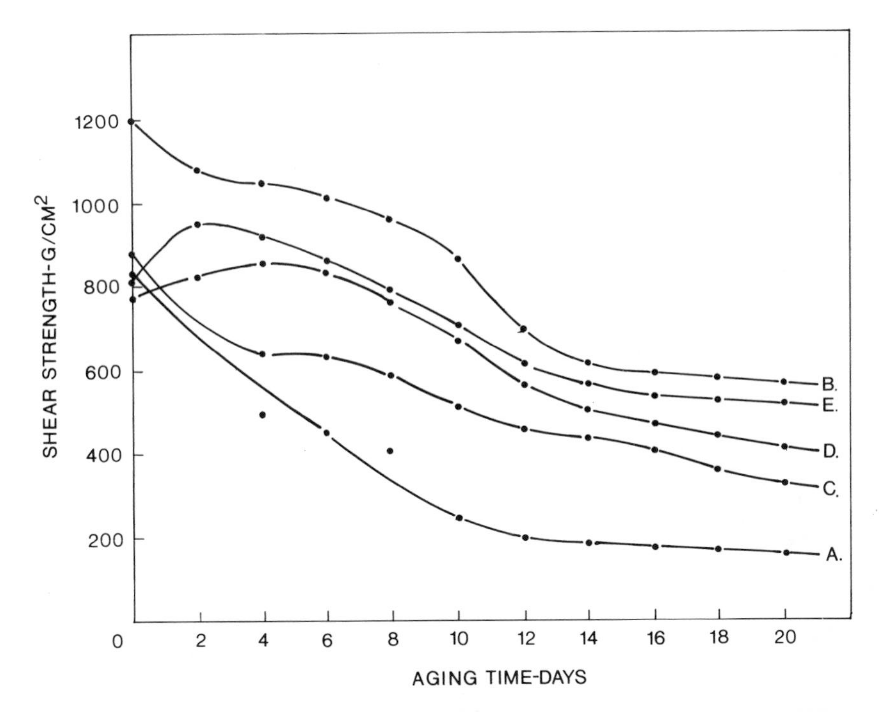


Fig. 8-4. Shear strength of natural rubber/resin blends (3/2) as a function of accelerated aging at 70°C.[3] (A) rosin; (B) Piccopale H2; (C) Ester Gum H; (D) Piccopale 100; (E) Piccolyte S-100.

## EFFECT OF ANTIOXIDANTS

Antioxidants provide an effective way to protect the unvulcanized rubber used in pressure-sensitive adhesive formulations from aging. Various chemicals have been found to be useful as antioxidants:

1. Secondary amines, diamines and their derivatives
2. Compounds of quinoline
3. Dithiocarbamates
4. Alkyl phenols
5. Phosphoric acid esters

Amine type antioxidants are rarely used for pressure-sensitive adhesives because of their staining. The most important antioxidants used for pressure-sensitive adhesives are listed in Table 8-1. The effectiveness of various antioxidants was evaluated by Fukuzawa, Nemoto, and Kosaka.[4]

**Table 8-1. Various antioxidants used for pressure-sensitive adhesives.**

| CHEMICAL NAME | TRADE NAME | M.P. (°C) | FDA STATUS | USES |
|---|---|---|---|---|
| 2,5-di(*tert-amyl*)hydroquinone | DAH,Santovar A | > 170 | Approved | PSA[a] |
| 2,5-di(*tert*-butyl)hydroquinone | DBH,NS-7 | > 200 | — | PSA |
| 2,6-di(*tert*-butyl)-4-methylphenol | BHT,Ionol | > 69 | Approved | PSA |
| Butyl hydroanisole | BHA | 57–65 | Approved | PSA |
| Styrenated phenol | S,SP, WS | Liquid | Approved | PSA |
| Hindered phenol | AgeRite,Geltrol | 217–225 | — | HMPSA[b] |
| 2,2′-methylenebis(4-methyl-6-*tert*-butylphenol) | NS-6,W-600,MBP LAO-5,Antioxidant 235 | > 120 | Approved | PSA |
| 2,2′-methylenebis(4-ethyl-6-*tert*-butylphenol) | Plastonox 425 NS-5,W-500,EBP Endox 22 | > 119 | Approved | PSA |
| 4,4′-butylidenebis(3-methyl-6-*tert*-butylphenol) | W-300,BBM,BB Santowhite powder | 208–210 | Approved | PSA |
| 1,1-bis(4-hydroxyphenyl)-*cyclo*-hexane | Antigen W | > 175 | — | HMPSA |
| 4,4′-thiobis(6-*tert*-butyl-3-methylphenol) | Yoshinox SR | > 150 | Approved | HMPSA |
| 4,4′-thiobis(6-*tert*-butyl-*m*-cresol) | Santowhite crystals | > 150 | Approved | HMPSA |
| Dilauryl thiodipropionate | DLTDP,LTDP, DLTP,TPL | > 38 | Approved | HMPSA |
| Tetrakis[methylene-3-(3′,5′-di-*tert*-butyl-4′-hydroxyphenyl)-propionate] methane | Irganox 1010 | 110–125 | Approved | HMPSA |
| Tris(mixed mono- and dinonylphenyl) phosphite | Polygard, Naugard P,TNP-S | Liquid | Approved | HMPSA |
| Zinc di-*n*-butyl dithiocarbamate | ZDBC,ZBUD,B8 | 95–104 | Approved | HMPSA |

[a] PSA: pressure-sensitive adhesive.
[b] HMPSA: hot melt pressure-sensitive adhesive.

The adhesive samples consisted of natural rubber (50 parts by weight), glycerine ester of hydrogenated rosin (50 parts) and antioxidant (2 parts). The adhesives were applied over fabric by calendering. The mass thickness was 0.07 mm. The samples were aged at 70°C for various time periods and the cohesive strength was tested in a T-peel test as shown in Fig. 8-5. The test temperature was 30°C, sample width 10 mm, jaw separation rate 20 mm/min. The test results are shown in Fig. 8-6. The effectiveness of antioxidants is demonstrated by comparing the properties of the sample without an antioxidant. 2,5-di-*tert*-butylhydroquinone was the most effective antioxidant for natural rubber in this test. The effectiveness of antioxidants depends on the elastomer used in the pressure-sensitive adhesive. The antioxi-

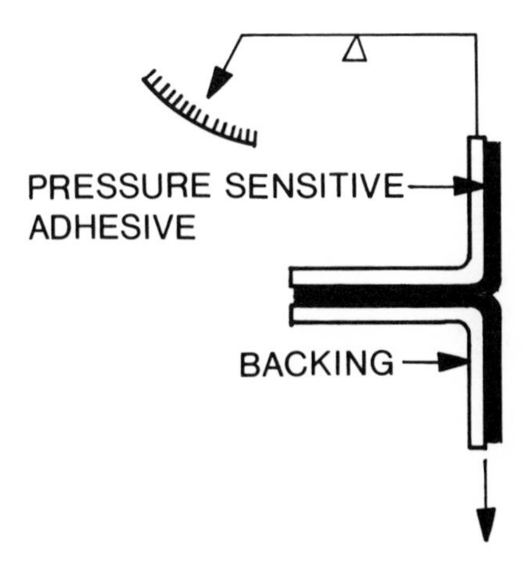


Fig. 8-5. T-peel test for measuring the cohesive strength.

dants used for hot melt pressure-sensitive adhesives are listed in Table 8-1 and zinc dibutyl dithiocarbamate and polyalkyl phosphite have been found especially useful.[5]

## THE CORRELATION BETWEEN ACCELERATED AGING TESTS AND THE SHELF-LIFE

The correlation between the accelerated aging tests and the actual shelf-life of the tape is difficult. Youman and Massen[6] reported the results of their investigation of the correlation between accelerated aging tests and shelf-life of vulcanized rubber using the physical property data collected over a period of 22 years. A correlation between the accelerated aging at 70°C and the shelf-life is shown. Fukuzawa and Matsukawa[3] have studied the correlation between 70°C oven aging tests and the shelf-life of surgical plaster. The formulation of the surgical plaster in parts by weight adhesive is given below.

| | |
|---|---|
| Natural rubber | 30 |
| Polyisobutylene | 15 |
| Glycerine ester of hydrogenated rosin | 23 |
| Zinc oxide | 10 |
| *l*-menthol | 10 |
| Methylis salicylas | 5 |
| Camphor | 2 |
| Antioxidants (2 types) | 4 |

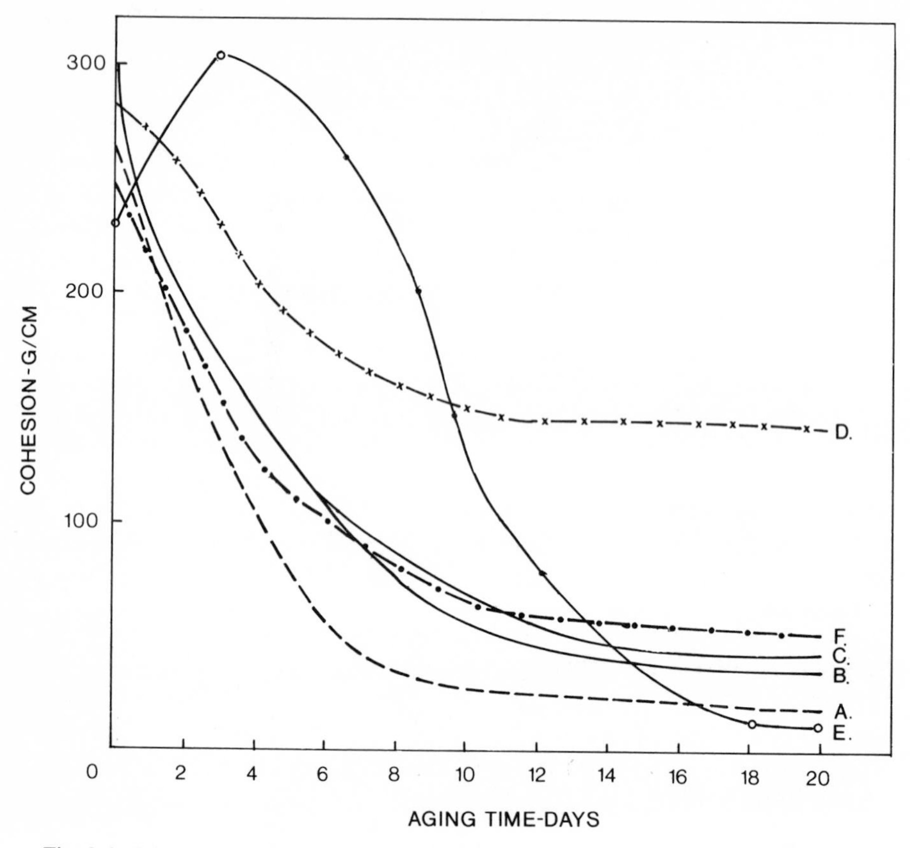


Fig. 8-6. Cohesive strength of adhesives protected by various antioxidants as a function of aging time.[4] (A) No antioxidant; (B) 2, 6-di-*tert*-butyl-4-methylphenol; (C) 4, 4′-butylidenebis-(3-methyl-6-*tert*-butylphenol); (D) 2, 5-di-*tert*-butylhydroquinone; (E) Mercaptobenzo-imidazole; (F) 2, 2′-methylene-bis-(4-methyl-6-*tert*-butylphenol).

| Formulation used: | | |
|---|---|---|
| | Natural rubber | 50 |
| | Ester Gum H | 50 |
| | Antioxidant | 2 |

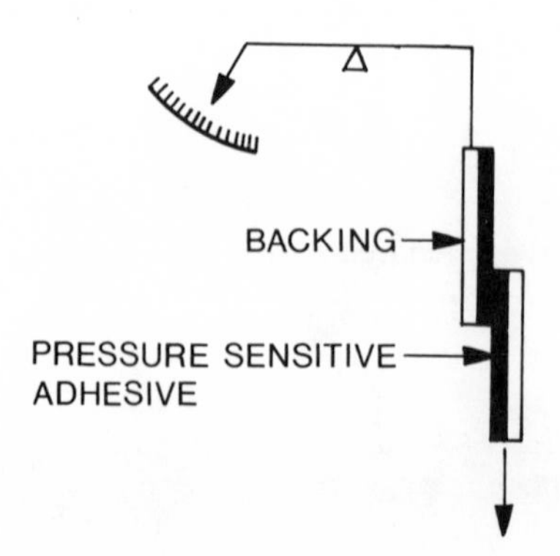




Fig. 8-7. Schematic diagram of the shear strength test method.

The 0.15 mm thick adhesive mass was applied by calendering. The adhesive surface was covered with a cellophane film liner and cut to 42 × 65 mm sheets. The samples were packed in moisture–proof cellophane and kept on shelf for 40 months. Part of the material was aged in a 70°C oven. An adhesive to adhesive lap joint 25 × 25 mm was made, pressed for 20 sec with a 200 g weight and the shear strength was measured as shown in Fig. 8-7. The tests were conducted at 30 ± 1°C and at a jaw separation rate of 20 mm/min. The failure was cohesive.

The test results are shown in Figs. 8-8 and 8-9. The cohesive strength has a tendency to decrease on aging. The cohesive strength was 1.6 kg after 25 months in the shelf-aged samples and the same value was reached after 15 days of accelerated aging. This would indicate that 15 days of accelerated aging is equal to two years of shelf aging for this product under the described conditions.

Tests of pressure-sensitive adhesive tapes were performed after exposure for 24 months in the Weathering Test Center of Choshi Exposure Laboratory (Japan) by a request of Japanese Ministry of International Trade and Industry.[7] Cellophane, kraft paper, and cloth tapes and acrylic adhesive labels were selected for testing. The exposure shelf was placed outdoors in a horizontal position. The shelf was covered by glass and by a light nontransmitting cover. The conditions were considered to be more severe than aging

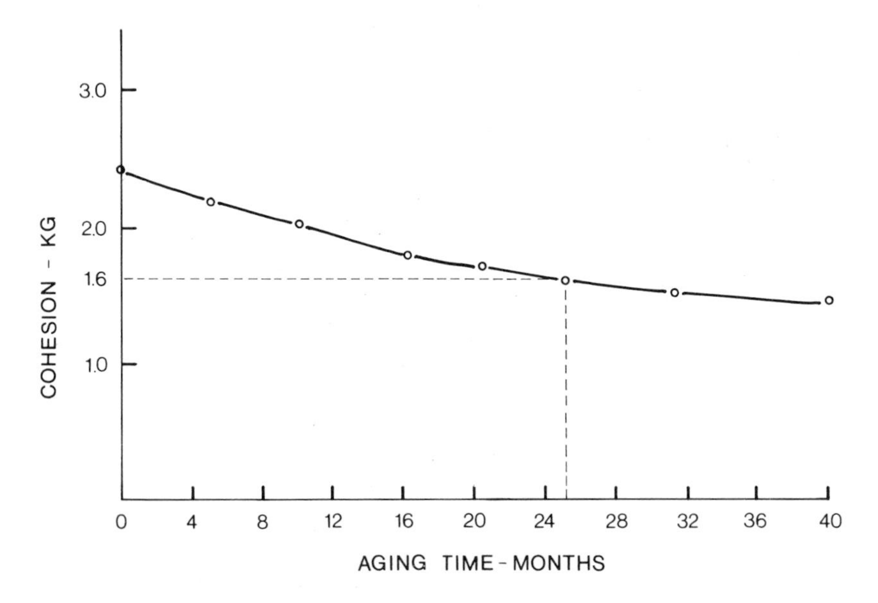


Fig. 8-8. The dependence of cohesion on aging time at room temperature.

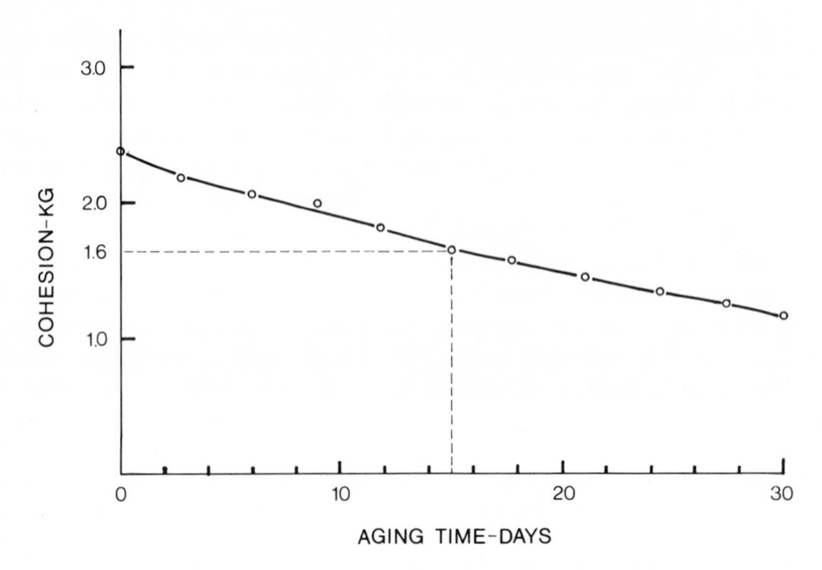


Fig. 8-9. The dependence of cohesion on aging time at 70°C.

indoors. The accelerated aging tests were conducted at 65°C and 80% relative humidity for comparison purposes.

The aging conditions might have been too severe for the cellophane tape that is generally used indoors. Telescoping was observed after 10 days in unpackaged rolls. Packaging is important to prevent moisture penetration. The tape retained its adhesive properties for nine months under no-light exposure conditions.

Kraft paper tape retained the adhesive properties in roll form for two years. The aging properties of this tape applied over a stainless steel plate were also investigated. This adhesive is based on natural rubber, is affected by light and softens on aging. An increase of peel adhesion is observed initially until the failure changes from an adhesive to a cohesive failure. After that, a decrease of peel adhesion is observed.

The cloth tape in a roll form showed little change after 24 months. Samples attached to stainless steel panel were also aged. In this case, the adhesive was greatly affected by light. This adhesive was based on reclaimed natural rubber and it appears that natural rubber adhesives are more sensitive to light than to heat. The pressure-sensitive tapes, however, are supplied in rolls and exposure to light is rarely a factor.

Polyvinyl chloride film labels were constructed from plasticized polyvinyl chloride, acrylic adhesive, and release paper. One of the samples was exposed directly and the other was under glass. The release paper could not be removed after 12 months in case of direct exposure and after 24 months in case of exposure under glass. Samples adhering to a glass plate were evaluated. The changes in peel adhesion were small, even after a long exposure, as compared to rubber-based adhesives. The peel adhesion changed greatly on accelerated aging, indicating that the resistance to light of acrylic adhesives is better than the resistance to heat.

Price and Nathan[8] reported the results of 40-week exposure of silicone and polyurethane pressure-sensitive adhesives. Four types of adhesives were evaluated: one uncatalyzed, two component chemically crosslinked, peroxide-cured silicones and an ultraviolet radiation-cured urethane adhesive. The adhesives were applied in 0.1–0.125 mm thickness over a 0.05-mm thick aluminum foil. The adhesives were faced with release papers after

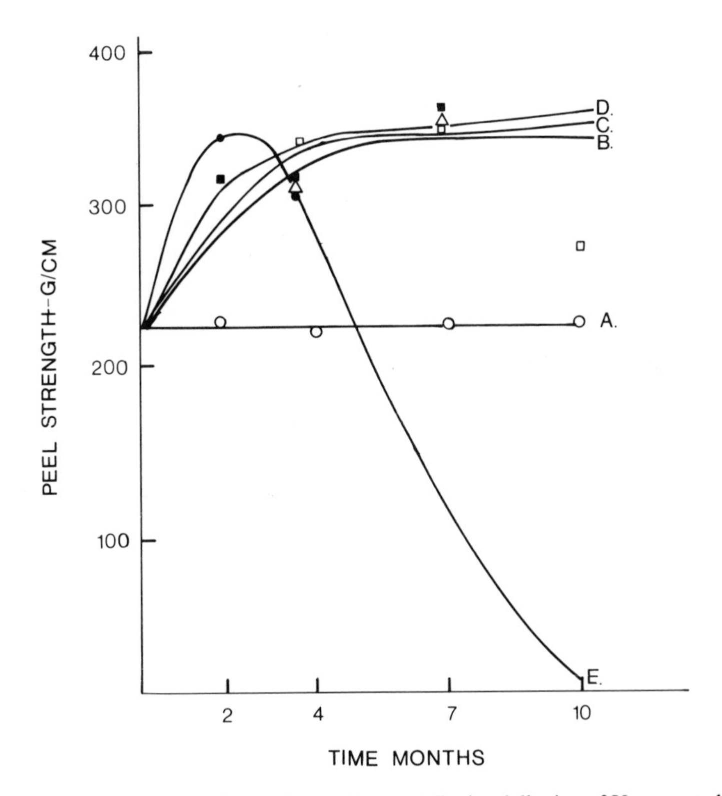


Fig. 8-10. Aging Properties of Uncatalyzed Silicone Adhesive-Adhesion of Unprotected Adhesive Film. (Reprinted from S. J. Price and J. B. Nathan, Jr. *Adhesives Age* **17** (8) (1974).) (A) –54°C; (B) 23°C; (C) 74°C; (D) 23°C, 95% R. H.; (E) temperature and humidity cycling.

proper curing. The foils were suspended vertically on a frame in the aging environment. The samples were aged at the following conditions.

At −53°C and 50% relative humidity
At 23°C and 50% relative humidity
At 66°C and 50% relative humidity
At 23°C and 95% relative humidity
At cycling temperature and humidity from −53 to 66°C and 95% relative humidity (according to MIL-STD-331)

The specimens were taken after 0, 8, 16, 28, and 40 weeks of exposure. The uncatalyzed silicone and peroxide-cured silicone systems still showed measurable pressure-sensitive adhesive properties at the end of 40 weeks exposure under all aging conditions. The two-component silicone continued

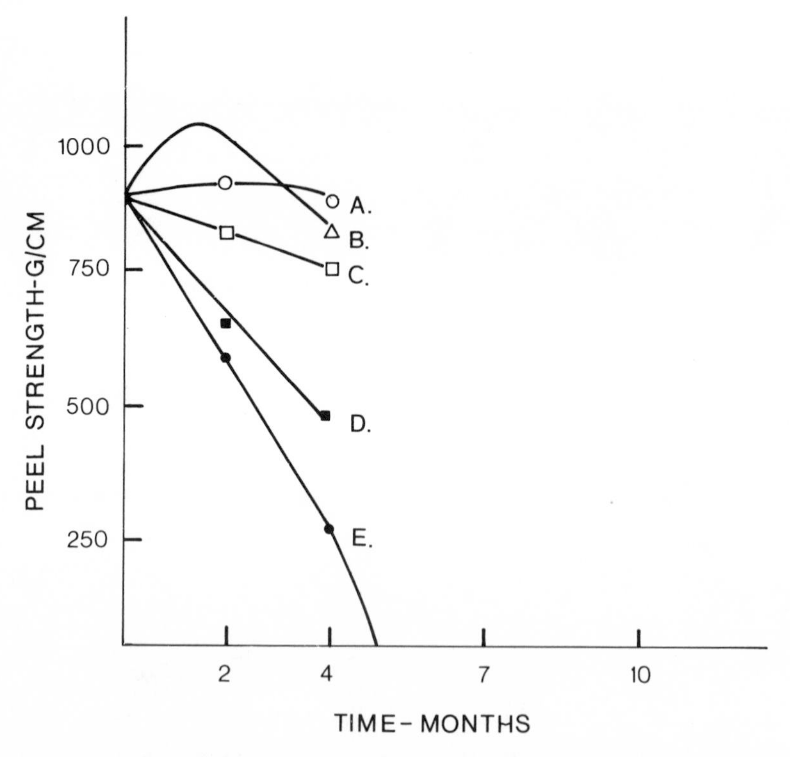


Fig. 8-11. Aging Properties of Two-Component Silicone Adhesive-Adhesion of Unprotected Adhesive Film. (Reprinted from S. J. Price and J. B. Nathan, Jr. *Adhesives Age* **17** (8) (1974).) (A) −54°C; (B) 23°C; (C) 74°C; (D) 23°C, 95% R. H.; (E) temperature and humidity cycling.

to crosslink during aging at high humidity. The adhesive lost tackiness after 16 weeks in the MIL-STD-331 environment. The polyurethane adhesive depolymerized to a liquid at elevated temperature and high humidity and it could not be tested at 16 week or longer exposures. Samples of two-component silicone and polyurethane adhesives were removed from testing after 16 weeks. The saturated release papers deteriorated and became brittle after 28 weeks exposure at high humidity and elevated temperature. The test data are shown in Figs. 8-10, 8-11, and 8-12.

Egan[9] has reported on aging of acrylic pressure-sensitive tapes for electrical tapes. The testing is complicated by the fact that the adhesive tapes are two-part systems consisting of backing and adhesive and both parts must function properly if the entire tape is to function. Therefore, the aging studies of such tapes included both backing and the adhesive. Sometimes there is an interaction between the adhesive and the backing and a difference in adhesive aging properties is observed depending on the backing

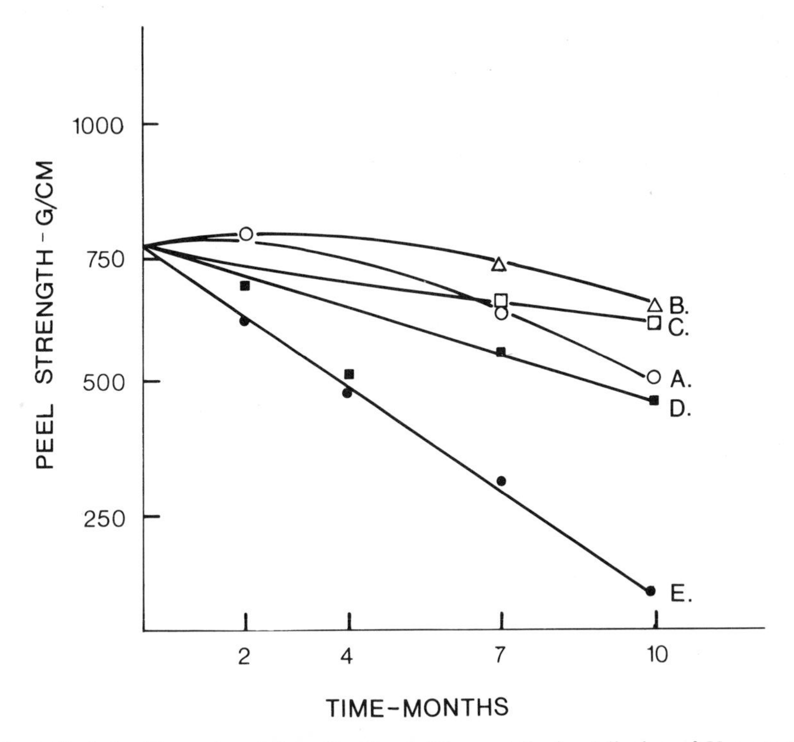


Fig. 8-12. Aging Properties of Peroxide Cured Silicone Adhesive-Adhesion of Unprotected Adhesive Film. (Reprinted from S. J. Price and J. B. Nathan, Jr. *Adhesives Age* **17** (8) (1974).) (A) −54°C; (B) 23°C; (C) 74°C; (D) 23°C, 95% R. H.; (E) temperature and humidity cycling.

used. For example, silicone pressure-sensitive adhesive ages faster on a woven glass fiber backing than on a Teflon film backing. If a metallic mandrel is wrapped with each of such tapes and placed in the oven at 250°C, the woven glass fiber tape with the silicone adhesive becomes embrittled after two months: the adhesive oxidizes to sandlike particles. The Teflon film-backed tape with the same adhesive exhibits no embrittlement after two years of exposure as reported by Egan. The difference is attributed to the greater air permeability of open-weave glass fiber cloth.

A 1.25 cm diameter metallic bar was wrapped with 2.5 cm wide tape using a 0.3 cm overwrap. The tape was wrapped under a specified constant tension of 900 g/cm width. The wrapped mandrels were then oven aged for various periods of time at three to four elevated temperatures. Suggested time and temperature sequence for thermal aging of this type can be found in ASTM D1304 (Thermal Aging of Rigid Electrical Insulating Materials). For example, if the material in question is considered for a continuous operation at 130°C, then a typical starting temperature for an accelerated life test would start at 25°C above the continuous operating temperature, or 155°C, and go on to 180°C and 205°C.

The dielectric breakdown was measured initially and after each aging

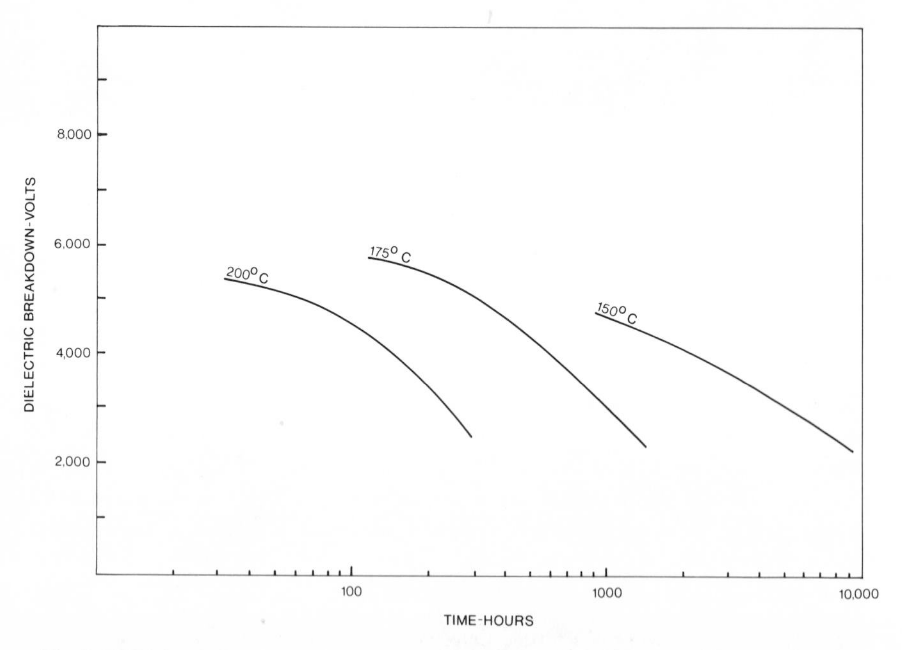


Fig. 8-13. Dielectric breakdown Voltage vs. Aging Time at Various Temperatures. (Reprinted from Egan.[9])

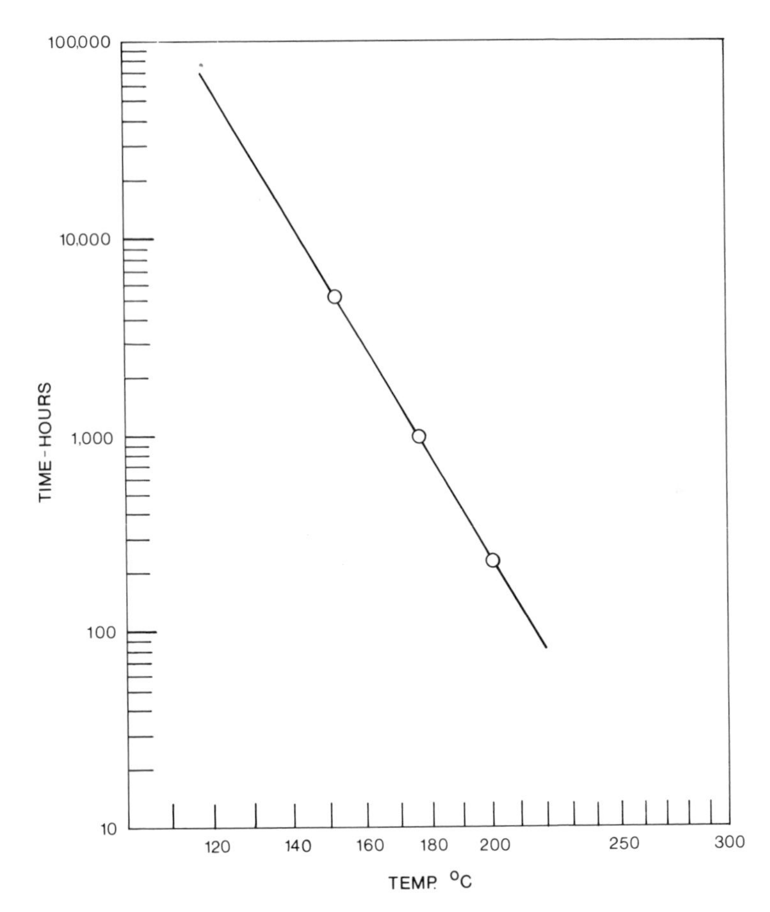


Fig. 8-14. The data shown in Fig. 8-13 replotted as half life time vs. temperature on logarithmic coordinates.

period, using a foil electrode system. The dielectric breakdown voltage then becomes the measured quantity indicative of the deterioration on aging. Some attempt was made to choose a sample geometric configuration so that both adhesive and backing properties are reflected. If the adhesive breaks down at the overlap, then the dielectric failure will occur at that point. If the backing deteriorates, then the dielectric failure will occur through the backing. An end point is chosen, usually a 50% drop in the dielectric breakdown value, and the samples are checked at various time intervals until the 50% drop is achieved. A plot of a typical curve is shown in Fig. 8-13. This type of data is then converted into an Arrhenius-type equation where the half-life log time in time units is plotted against the reciprocal of the absolute temperature. If the deterioration rate is controlled by a single chemical reaction, the plot will approximate a straight line as shown in Fig. 8-14.

The usual practice is to extrapolate the curve to some agreed time figure, typically 40,000 or 50,000 hr and report the corresponding temperature. In Fig. 8-15 of a polyester film tape with an acrylic adhesive, the extrapolated 40,000 hr temperature is 123°C. The neighboring curve shows the data obtained with rubber adhesive applied on the same polyester film backing. The extrapolated 40,000 hr temperature for this tape is 118°C or 5°C lower than for acrylic adhesive.

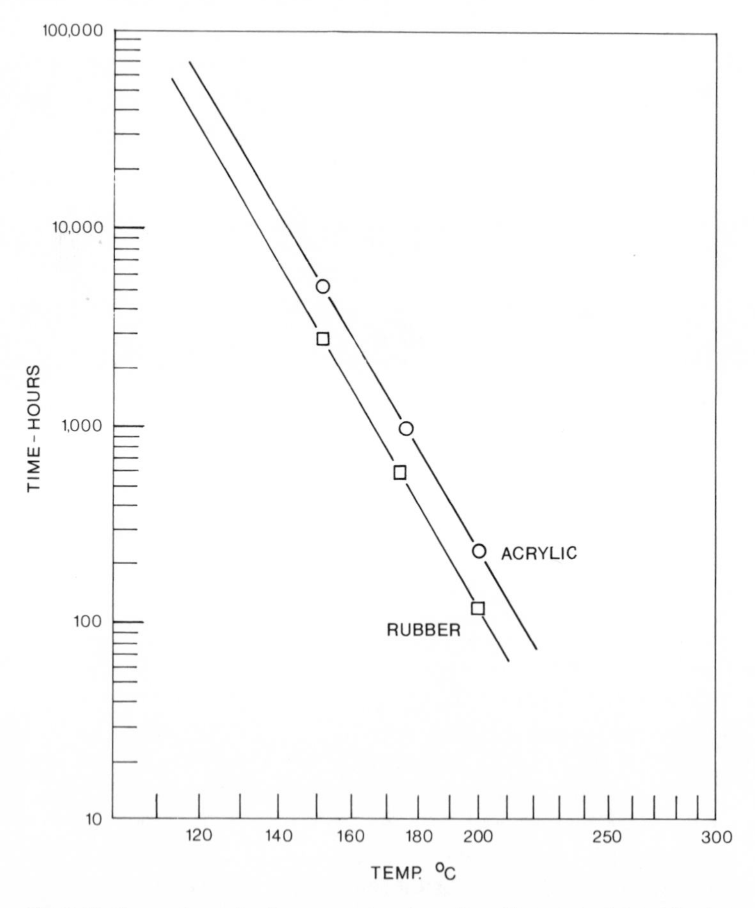


Fig. 8-15. Comparison of aging properties of acrylic and natural rubber adhesives.

This method appears to offer a means of evaluating the thermal aging properties quantitatively, not only of the adhesive alone, but of the complete tape.

## REFERENCES


1. Wetzel, F. H., and Alexander, B. B. *Adhesives Age* **7**(1): 28 (1964).
2. Higuchi, Watanabe, and Toyama. Symposium on Adhesion and Adhesives (Japan), p. 17 (1965).
3. Fukuzawa, K., and Matsukawa, H. *Engineering Materials* **16**(1) 17 (1968).
4. Fukuzawa, K., Nemoto, T., and Kosaka, T. The 7th Symposium on Adhesion and Adhesives (Japan), p. 81 (1969).
5. Formulating Kraton Thermoplastic Rubber Crumb. Shell Chemical Co. Bulletin SC:72–3.
6. Youman and Massen. *Ind. & Eng. Chem.* **47**: 1487 (1955).
7. Report on the aging properties of industrial materials and products. Japan Weathering Test Center (1976).
8. Price, S. J., and Nathan, J. B., Jr. *Adhesives Age* **17**(8) (1974).
9. Egan, F. *Physical Testing of Pressure-Sensitive Adhesives.* Paper given at ASTM seminar, March 22–23, 1973, Louisville, Kentucky.

# Chapter 9

# Analytical Techniques for Identification and Characterization

**David. N. Kendall**

*Kendall Infrared Laboratories*
*Plainfield, New Jersey*

The identification of pressure-sensitive adhesives can be aided by several analytical techniques. Infrared absorption spectroscopy (IR)[1,2] is probably the most useful because it can help identify all organic materials and most inorganics, except for those few having both monatomic cations and anions.

Ultraviolet absorption spectroscopy (UV)[3] is useful for helping to identify unsaturated (e.g., C = C) substances, such as determining whether a rosin component of an adhesive has been partially or totally hydrogenated. Ultraviolet has greater sensitivity than infrared for those materials to which it is applicable and therefore can be useful in quantitative work at very low concentration levels. Quantitation at a component level of 0.005% by weight is not unusual, while a level of 0.1% by weight is often encountered as the limit for practical quantitative IR methods, without resorting to specialized IR techniques.

Laser Raman spectroscopy can be useful in certain solvent analyses and for analysis of aqueous adhesive systems.

The power of chemical analysis often currently referred to as "wet chemical analysis" should not be overlooked in adhesive identification. Spot tests, as found in Feigl and Fieser, can give fast identification of critical functional groupings or elements present in an adhesive. Judicious use of sulfide formation at controlled pH conditions serves, for example, to identify Zn where the presence or absence of ZnO is sought as a filler and/or vulcanizing agent in a neoprene type pressure-sensitive adhesive.

Ashing an adhesive to "burn" off the organics leaves the inorganics in a more easily identified state by IR or chemical tests. Usually only a limited number of possibilities exist, so a tedious number of chemical tests is not required for identification.

Simple flame tests, using a Bunsen or Meker burner, can quickly establish the presence of sodium, potassium, calcium, lithium, and occasionally copper (especially when moistened with hydrochloric acid) when any are possibly present in an adhesive.

Differential thermal analysis[4] (DTA) examination can be useful in the identification of polymers present in a pressure-sensitive adhesive. Once the polymers have been separated, each can sometimes be identified by their glass transition temperatures ($T_g$'s). Moreover, the heats of transition or fusion determinable by DTA can measure the degree of crystallinity of a polymer, which can be used to identify polymers. Standard known polymers must already have been subjected to the technique, so that this information is available, just as a catalog of known spectra is normally necessary for a spectral identification.

Light microscopy can be useful in comparing the surface characteristics of pressure-sensitive adhesives. Where the surface microstructure is important to an adhesive's functioning, a study of that surface by scanning electron microscopy (SEM) techniques will reveal minute structure and surface differences.

In order to characterize pressure-sensitive adhesives, the concern is to a large extent with polymers and the determinants of polymer properties. Because the film-forming elastomeric material of a pressure-sensitive adhesive composition is generally a polymer, as is also the resin or other material used to impart the desired degree of tack, wetting power, and specific adhesion, polymer properties are critical.

Polymer properties are determined by molecular weight and its distribution, degree of crystallinity, chain branching, side-chain substitution, and crosslinking. The ability to crystallize, the flexibility of chain bonds and the nature, number, and spacing of polar groups come into consideration. For crystalline polymers, the type of crystal structure, degree of crystallinity, size and number of spherulites, and orientation are important to polymer properties. Addition of other materials to polymers in the pressure-sensitive adhesive formulation has an important effect on polymer properties. Plasticizers, fillers and other modifiers come under this category.

The most widely applicable technique for identifying and analyzing pressure-sensitive adhesive formulations is IR.

## INFRARED METHODS

The atoms of any molecule not at absolute zero are continuously undergoing oscillations of various kinds. Fortunately, these molecular motions are of the same order of magnitude ($10^{13}$–$10^{14}$ cps) as those of IR radiations. When the frequency of molecular motion is the same as that of IR radiation impinging on that molecule and when there is a change in dipole moment during the vibration, the molecule absorbs by resonance all, or a portion of, the radiation incident upon it. A permanent dipole in a molecule is not necessary for IR absorption. Carbon tetrachloride ($CCl_4$), for example, absorbs IR because there are vibrations which the C-Cl groupings undergo, when they are not in phase, during which a change in dipole moment occurs.

While the IR region extends from about 0.75 to 1000 $\mu$m (13,333 to 10 $cm^{-1}$), the most useful range for chemical analysis is from 2 to 25 $\mu$m (5000 to 400 $cm^{-1}$). Approximately 95% of all such work is carried out in the 2 to 16 $\mu$m (5000 to 625 $cm^{-1}$) region.

All organic chemicals and most inorganics can be handled by IR. Inorganics having monatomic cations and monatomic anions (NaCl, KBr, and CsBr, used as windows or prisms for dispersing IR) do not yield characteristic absorption spectra, except for their lattice vibrations. Neither do most metals. Diatomic molecules, such as $H_2$, $O_2$, $N_2$, $Cl_2$, etc., cannot be identified or determined by IR. They do not give useful IR absorption spectra, since no change in dipole moment occurs during their lone stretching vibrations which are electrically symmetrical.

No two substances which absorb IR radiation absorb it at the same frequencies to the same extent. Therefore, an IR absorption spectrum is a "fingerprint" of a substance as unique as the fingerprint of a human being and as useful in differentiating one molecule from another as human prints are in differentiating one individual from another. Moreover, the spectrum of a mixture, excepting certain hydrogen bonding situations or chemical reactions taking place, is simply the sum or superposition of the spectra of the individual components comprising the mixture. Accordingly, IR is the most widely used technique for identifying unknown substances and mixtures.

Infrared spectrophotometers, the instruments used for chemical analysis by IR, comprise three essentials: the source of IR radiation, a dispersing means to separate the radiation by frequency (wavelength), and a detector which changes the received radiation into an electrical signal. Other components are necessary, such as slits, collimators, mirrors, amplifiers, and recording mechanisms.

Sources can be a nichrome helix, a Nernst glower (a mixture of rare earth oxides), or a Globar (a silicon carbide rod which requires water or air cooling). Dispersing elements are commonly gratings, now having replaced prisms used in the earliest IR spectrophotometers. Filter-grating types of grating spectrometers can cover a range up to about 333 $\mu$m (30 $cm^{-1}$).

Detectors are radiation thermocouples, bolometers, pneumatic detectors, or triglycine sulfate (TGS).

One of the more recently developed instruments is the Fourier transform infrared spectrometer (FT-IR). This technique uses an interferometer of Michelson type[18] to produce an interferogram of the sample. A fast digital computer quickly transforms this to the familiar transmission or reflection type spectrum. The FT-IR began to be a practical proposition when Fellgett[19] pointed out the advantages of its multiplex operation. Today, modern FT-IR instruments incorporating sophisticated computers offer high resolution (0.08 $cm^{-1}$ is not uncommon), many rapid scans to average out "noise" and quick subtraction of the components of a mixture.

## MULTIPLE INTERNAL REFLECTION

When a pressure-sensitive adhesive sample has the adhesive surface exposed, or when it can be exposed by simply removing a release layer, the multiple internal reflection (MIR) technique can be used. The MIR technique was formerly known as attenuated total reflectance (ATR). It is a kind of internal reflection spectroscopy. The tacky pressure-sensitive adhesive surface is placed against the MIR crystal surface on one or both sides of the MIR crystal, the angle of incidence of the incoming IR beam adjusted to yield maximum energy at 5 $\mu m$ (2000 $cm^{-1}$), and the adhesive's spectrum scanned from 2.5 to 50 $\mu m$ (4000 to 200 $cm^{-1}$). Often, a scan from 2.5 to 15 $\mu m$ (667 $cm^{-1}$) is sufficient. Thallous bromide-iodide (KRS-5) is the most useful crystal with AgCl a distant second. Germanium and Irtran crystals also find use.

The MIR crystal is normally incorporated in a small accessory which slides into the sample holder and sampling compartment of commercial IR spectrometers, and which can be manually adjusted via knobs to vary the angle of incidence of the incoming IR beam. One can also be placed in the reference beam of double-beam instruments, if desired, to produce a greater spread between absorption band maxima and minima. A common type of MIR accessory provides about 25 internal reflections. The different MIR crystals of differing refractive indices allow varying angles of incident IR radiation to be used, from about 20° to 55°. Satisfactory MIR spectra are produced by judicious selection of the appropriate ratio between the refractive index of the crystal and that of the sample, of the angle of incidence which the radiation makes with the crystal-sample interface, and the contact between the sample and the crystal. Good contact is vital. Unless it can be made, a useful MIR spectrum is not attainable. This limits the technique to relatively smooth-surfaced samples.

While liquids can be scanned by smearing them onto an MIR crystal, normally no advantage is gained over sample transmission techniques. The MIR technique is more tedious. Moreover, the MIR spectrum is very similar but not identical to the transmission spectrum. The MIR technique should be viewed as a specialized one of limited application.

Illustrative of a specialized application is the use of MIR to determine if the plasticizer in a pressure-sensitive adhesive may have migrated from the backing layer of the substrate through to the surface of the adhesive layer. One wishes to examine the pressure-sensitive adhesive layer as is. Comparing the MIR spectrum of this surface to that of a freshly prepared specimen as a control could show whether or not plasticizer migration had taken place. Poor adhesion of a pressure-sensitive tape could suggest that the examination just described should be made, particularly if it is known that the backing layer did contain a plasticizer. In such circumstances, a barrier layer should be present to prevent migration, but barrier layers can be misapplied.

Occasionally, it may be easier to place the MIR crystal in contact with a very small liquid film sample, which will transfer to the crystal and can thus be scanned. This situation could arise where one wishes to avoid extracting the liquid film with solvent by eliminating the solvent, and then scanning in the usual transmission mode. For example, one might wish to avoid injuring or contaminating the substrate beneath the liquid film.

Since the impinging IR radiation beam penetrates the sample surfaces up against the MIR crystal roughly to the depth of the wavelength of the IR beam, MIR is essentially a technique for examining surfaces more deeply than the scanning electron microscope (SEM) or Auger spectroscopy.

## ANALYSIS OF BACKING MATERIAL

When identification or other analysis of the backing material is sought, the adhesive is first removed by scraping off with a knife wet with an appropriate solvent, one that will loosen the adhesive. Acetone, methyl ethyl ketone (MEK), carbon tetrachloride, or toluene may be appropriate.

If the backing is paper, an analysis of coating, saturants, fillers, etc., may be desired. A coating can be examined by MIR spectroscopy if the coating is smooth enough. If not, it can be extracted with a solvent. One normally tries solvents with low boiling points initially, since they are easier to remove from the extracted coating. Toluene, MEK, $CCl_4$, or ortho-dichlorobenzene (ODCB) may be required to remove the coating. After concentrating, the extract is cast onto NaCl plates, the solvent removed by heating, and the room temperature coating film on the NaCl plate scanned in the IR, usually from 2.5 to 16 $\mu$m.

Water-soluble additives in the paper can be extracted with boiling water and after concentration cast onto AgCl or Irtran plates, the water removed by heating in an oven, and the coating scanned in the IR.

Inorganic fillers in the paper backing can be identified by carefully ashing a portion of the sample over Bunsen or Meker flames or in a muffle furance—then scanning the ash, not only from 2.5 to 16 $\mu$m but out to 25 $\mu$m or even 50 $\mu$m, since useful identifying inorganic absorptions are often found in the region from 15 to 50 $\mu$m (200 $cm^{-1}$).

Once the coatings, saturants, and fillers have been identified, one may wish to check to determine if the backing is really paper. Mulls can be made either from the backing without the pressure-sensitive adhesive or from such a backing after removing the coatings. To about 15 mg of the paper a drop of mineral oil, such as Nujol, is added to the sample on a glass plate and thoroughly ground and dispersed using a glass paint muller. The objective is to obtain as fine a particle size as possible to minimize scattering of the IR beam impinging on the sample. Scattering is radiation that impinges on the sample. But rather than being transmitted through it, radiation is scattered aside without getting into the monochromator of the IR instrument. Scattering is effectively eliminated if all mulled particles are below 2 $\mu$m in size. This is not always possible to obtain.

It may be necessary to add more Nujol or occasionally more sample during the mulling process. A satisfactory mull will usually be translucent to visible light, after being placed between two NaCl plates and squeezed out. The adequacy of the mull can be judged by noting the slope of the spectrum between 2 and 4 $\mu$m. The worse the deviation from linearity, the greater the scattering and the poorer the mull and the recorded spectrum. Some slight upward slope from 2 to 4 $\mu$m can be tolerated. A comparison of the spectra against the spectra found in reliable literature will enable one to judge how much scattering can be tolerated without loss of accuracy.

After the mineral oil mull, a mull may be made using perfluorokerosene or hexachlorobutadiene (HCBD) and scanned over only those regions where mineral oil absorbs, namely, 2 to 4 $\mu$m, 6.5 to 7.8 $\mu$m, and 13.5 to 14.3 $\mu$m. Thereupon, one has the complete 2.5 to 15 $\mu$m spectrum of the solid sample free of mulling agent absorptions.

If the pressure-sensitive adhesive backing should be a fabric, one can identify the fibers by scanning Nujol and HCBD mulls. (Admittedly, the mulling process may be a bit difficult, but mulling such fabrics keeps the hand and arm in shape!) Nylon type backings can be identified by scanning IR spectra as evaporated films on NaCl plates, cast from solutions in cresols or formic acid.

A detailed examination of the fabric may be desirable. For silk and synthetic filaments, the term denier is employed. The denier is the weight in grams of 9000 m of filament or yarn. International scientific and standards organizations use the tex system. The tex in grams per kilometer is the internationally preferred unit to describe linear density. The heavier the filament or yarn, the greater is the denier or tex number. A vibrascope can be used to determine quickly and accurately the denier of individual fibers 2–4 cm in length, according to the equation

$$m = T/4\, l^2 f^2 \times 10^5,$$

where $m$ = linear density (g/cm)
$T$ = fiber tension (dyne)
$l$ = effective fiber length (cm)
$f$ = fundamental resonant frequency (cycles/sec).

For natural fibers such as wool, cotton, and flax, the denier is not commonly measured, since staple length and diameter are usually considered sufficient to define fineness. In cotton the longer the staple length, the better the quality. Staple length can be measured using a Fibrograph.[5] The diameter can be measured with the light microscope. The quality or grade of wool is determined on a statistical basis of distribution of fiber diameter. Both the mean value and distribution around the mean are measured.

The fineness of cotton can be determined using the Micronaire, which is manufactured by a Bendix Corporation subsidiary. It measures, under precise conditions, the air flow resistance of a plug of cotton of specified dimensions. Other things being equal, the air flow is a function of the fineness of the fiber. Also, the density of fibers can be determined by precise microscopic measurement of length, cross-sectional area, and weight.

Nonwoven fabrics are fibrous webs bonded by the addition of chemicals or fusion of some of the constituent fibers. Web bonding comprises production of a network structure in the web using adhesives and/or entanglements to bond the constituent fibers. Wet or dry bonding systems can be used. Since polymeric emulsions featuring acrylics, butadiene copolymers with styrene or carboxylated styrene, acrylonitrile-butadienes, acrylates and their copolymers, vinyl acetate-based polymers, or vinyl chloride polymers may be present, the choice of solvent to extract the binder is wide. Typically, MEK, toluene, and ODCB should be tried. Because some binders are self-crosslinking or formed from a reactive additive, some binders may be insoluble in most solvents. ODCB, however, will normally dissolve enough of the low molecular weight ends of an insoluble binder to allow it to be identified by an appropriate IR scan. Here, microsize plates and ordinate scale expansion may be necessary to obtain IR spectra adequate for identification of the binder.

It is sometimes profitable to scan in IR the nonwoven backing without the pressure-sensitive adhesive prior to solvent extraction and after solvent extraction, to learn whether or not the binder has been practically completely removed. This can be done as mineral oil and HCBD mulls. The IR spectra of the fabric itself, with binder, needs to be subtracted out where appropriate.

Foams can be urethanes or thermoplastics foamed by incorporating a blowing agent which decomposes to a gas at elevated temperature, or by incorporating an inert gas. Rubber latices are foamed by mechanical whipping to a froth. Some double-faced tapes have a backbone of foam, since foams provide a cushion between two surfaces and fill the gaps between two irregular surfaces. For identification of foams by IR, proceed as for organic polymers or films, as previously described: scan as evaporated films from solution or as mineral oil and HCBD mulls, if the foam is solvent-insoluble.

## IDENTIFICATION OF COMPONENT PARTS

A pressure-sensitive adhesive is usually composed of an elastomer, a tackifying resin, and an oil. More than one resin may be present to provide properties not available from a single resin. Fillers are often added to raise the viscosity, add color, increase the specific gravity, or reduce the cost. Antioxidants stabilize the adhesive against heat and light degradation, as well as against oxidation.

Adhesives for electrical applications or high temperature use sometimes incorporate a minor concentration of a heat-curing, oil-soluble phenolic resin. To promote reaction between the phenolic resin and the rubber component, a catalyst such as an alkaline filler or zinc resinate is used.

### Elastomer Identification

Natural or reclaimed rubber, SBR (styrene–butadiene rubber), polyisobutylene or butyl rubber, higher polyvinyl alkyl ethers, Buna N (butadiene-acrylonitrile copolymer), higher polyacrylate esters, styrene-butadiene-styrene (S-B-S), and styrene-isoprene-styrene (S-I-S) block copolymers are typical of the rubbery elastomers used in pressure-sensitive adhesives.

To determine whether an elastomer is or is not a thermoplastic rubber of the block copolymer type, a glass transition temperature $T_g$ measurement will suffice. The thermoplastic rubber will have two $T_g$'s, while a random SBR copolymer will have only one. An S-B-S block copolymer with the same styrene/butadiene ratio as a random SBR copolymer will show two $T_g$'s.

An IR identification of the elastomer in a pressure-sensitive adhesive can be made sometimes on the spectrum of the adhesive, scanned as is and squeezed out between NaCl plates. Elimination of fillers, when present, may be helpful. The elastomer can be extracted away from the fillers using an appropriate solvent, such as toluene or ODCB. A concentrated solution or adhesive without fillers is cast onto NaCl plates, the solvent evaporated off by heating in an oven, and the evaporated film scanned in the IR. A substantial catalog of IR spectra of known elastomers for pressure-sensitive adhesives is a must.

If both the fillers and tackifying resin can be removed to leave essentially only the elastomer, then identification can be made by scanning the pyrolyzate of the elastomer and comparing the resulting spectrum against a catalog of pyrolyzate spectra of known elastomers. Pyrolysis equipment is available from Foxboro Analytical, Harrick Associates, and Harshaw Chemical. These accessories enable control of time and temperature at which pyrolysis is carried out and the volume of the pyrolyzing chamber. When precisely reproducible conditions are unnecessary, one can prepare a satisfactory pyrolyzate using a borosilicate test tube of appropriate sample size.

A gram or less of the sample as is, or cut into pieces to conform to the test tube, is placed in the bottom of the test tube and a Bunsen flame applied to the tube held with its open end slightly higher than the action end. Should some moisture be produced by heating the sample initially, it is well to evaporate this out of the tube before continuing to heat and collect the liquid pyrolyzate produced in the cooler open end of the test tube. This pyrolyzate material is placed between appropriate salt plates, such as NaCl, KBr, etc., and scanned in the IR. Comparison against known pyrolyzate spectra serves to identify the elastomer.

If one has known pyrolyzate spectra of elastomer and tackifying resin, of course, then identification can proceed from pyrolysis of the unknown combination of elastomer and tackifier.

An identification of whether or not an acrylonitrile-butadiene (AB) copolymer is present in an elastomer component simply involves scanning the pressure-sensitive adhesive between 4 and 5 $\mu$m. The presence of an absorption at about 2250 $cm^{-1}$ indicates a nitrile, and an acrylonitrile copolymer should come under consideration. The lack of a band at 2250 $cm^{-1}$ indicates no nitrile and thus no possibility of an AB copolymer.

## Tackifying Resin Identification

Commonly used tackifying resins or tackifiers in pressure-sensitive adhesives are gum rosin, rosin esters and other rosin derivatives, polyterpene resins, coumarone-indene resins, oil-soluble phenolic resins, and petroleum hydrocarbon resins. While sometimes these can be identified by IR scanning of the as is adhesive and comparing against known tackifier spectra, such is not always possible. Critical portions of the tackifier spectra may be swallowed up by absorptions of the elastomer. Therefore, separating of the elastomer and tackifier may be necessary. A styrene-containing elastomer can often be separated from the tackifier and oil of a pressure-sensitive adhesive by extraction with an alcohol such as *n*-butanol. The tackifier and oil will be soluble in the *n*-butanol, while the elastomer will be practically insoluble. An IR scan of the spectrum of tackifier and oil, as evaporated films from *n*-butanol on NaCl plates, will lead to identification of the tackifying resin, since the oil will normally contribute few interfering absorptions. As a bonus to this separation, IR scanning of the elastomer above, which has been separated, may lead to more certain identification of the elastomer.

The separation of the three main pressure-sensitive adhesive ingredients may also be done by liquid chromatography.[6]

Pyrolyzates of separated tackifiers can be made, scanned in the IR, and identification made by comparison against pyrolyzate spectra of knowns produced under the same conditions. Keep in mind that more than one tackifier may be present in an adhesive formulation under investigation. A mixture of resins may provide a balance of properties not obtainable with a single resin.

### Plasticizer Identification

The plasticizers used in pressure-sensitive adhesives can be mineral oils, liquid polybutenes, liquid polyacrylates, certain phthalates, and lanolin, as typical examples. Some adhesives contain no plasticizers.

Perhaps mineral oils are the most commonly used. These often can be identified in the combination of tackifying resin and plasticizer (oil) that is separated from the elastomer using *n*-butanol as previously described. If separation of tackifier and plasticizer is necessary for identification of the latter by IR, this can be done by liquid chromatography. Sometimes column chromatography will accomplish the separation. Once the plasticizer is alone, infrared identification by comparison of its spectrum against the spectra of knowns can proceed. This can be a painstaking process, since a number of plasticizers, particularly of the petroleum oil type, have very similar IR spectra. When aromatic $C=C$ unsaturation is present, a fine distinction between otherwise very similar oils can be made by UV analysis.

### Filler Identification

Fillers in pressure-sensitive adhesives may be $CaCO_3$, $TiO_2$, ZnO, clays, pigments, and aluminum hydrates. Ashing the total composition may be the easiest way to bring about filler identification. The elastomer, tackifier, and plasticizer usually are organic and burn off, leaving the inorganics—the fillers. The latter can then be scanned from 2 to 50 μm in the IR as mineral oil and HCBD mulls between CsI plates. For inorganics in particular and most specifically for oxides, the region between 16 and 50 μm is very useful for identification purposes.

Quantitative ashes where calcium carbonate is involved must take into account the breakdown of $CaCO_3$ to CaO and $CO_2$ and the taking up of $CO_2$ from the air to form some $CaCO_3$ again, after cooling begins. Changes can also take place in certain clays, and degrees of hydration should be kept in mind as one makes comparisons of IR spectra among those of ashes vs. known spectra of inorganics.

Fillers can often be left in the solvent-insolubles from the separation of elastomer, tackifier, and oil. Infrared scanning of these, after heating off the solvent, should then lead to identification of the fillers.

## Antioxidant Identification

Antioxidants used in adhesives[7] include rubber antioxidants such as aromatic amines, substituted phenols and hydroquinones, metal dithiocarbamates such as zinc salt, and metal chelating agents such as *n*-propyl gallate and copper-8-quinolinolate.

Most of the antioxidants in adhesives can be separated in a very concentrated form by extraction of the adhesive with acetonitrile[8] (methyl cyanide) or methanol. The latter tends to extract more oil, which interferes with identification of the antioxidant.

Ten grams of the adhesive are shaken with 50 ml of acetonitrile for 30 min. The mixture is filtered and the filtrate evaporated to about 20 ml on a 60°C hotplate. The solution remaining is cooled to −20°C for 2 to 3 hr. Most of the small amount of oil initially extracted with $CH_3CN$ collects on the bottom of the vessel. The mixture is decanted and the filtrate evaporated to dryness on a 60°C hotplate. The remaining residue is very concentrated in antioxidant. This residue can be used for identification of the antioxidant by IR. Nuclear magnetic resonance (NMR) and mass spectroscopy (MS) can also be used for the antioxidant identification. Using all three, when possible, is best for a sure identification, since NMR gives the ratio of aromatic to aliphatic protons, IR supplies functional group analysis, and MS provides molecular weight data. As low as 0.2% by weight of an antioxidant in an adhesive formulation can be detected and identified by the technique described.

Scanning of the antioxidant already separated from the other adhesive components in the UV [9] from 200 to 400 μm, and comparison against UV spectra of knowns can also aid in identifying antioxidants.

## CHARACTERIZATION OF PRESSURE-SENSITIVE ADHESIVES

Once the identification of the components of a pressure-sensitive adhesive formulation has been completed, there may be a need to characterize further the total formulation or the polymers comprising the elastomer or tackifier. While IR can tell the elastomer present is an SBR, it cannot take molecular weight (MW) into account. An SBR with a MW of 1000 and one of 2000 would yield essentially identical infrared spectra in the 2 to 50 $\mu$m region. IR sees nearest neighbors for the most part.

### Molecular Weight Determination

Molecular weight determination of any polymers present in an adhesive formulation such as the elastomer or the tackifier can be helpful to improving adhesive properties. Knowing the molecular weight and solubility parameter $\delta$ for a series of tackifiers helps to optimize cohesive strength and other properties of the adhesive. The value of the solubility parameter is a function of molecular weight, because $\delta$ is a measure of the total intermolecular bond energy holding the molecules of a substance together.[7]

For two materials to be soluble in each other, the free energy of mixing must be negative. This occurs when the values of $\delta$ for the two materials are close. As a general rule, the higher the molecular weight of one or both materials is, the closer the values of $\delta$ must be for miscibility to occur.

For nonpolar substances, $\delta$ is a relatively simple number, but for polar materials and situations where intermolecular hydrogen bonding is possible, $\delta$ is a three-dimensional quantity. In the latter two cases, data are meager and therefore, currently, there is no way to handle accurately situations involving polar or hydrogen bonded components. The solubility parameter, however, may serve on an approximation basis for planning experiments and interpreting findings.

Molecular weight affects the solubility parameter and many other polymer properties. The molecular weight of a polymer is the product of the molecular weight of the repeat unit and the degree of polymerization. The repeat unit is usually equivalent or nearly equivalent to the monomer or starting material from which the polymer is formed. The length of the polymer chain is specified by the number of repeat units in the chain, i.e., the degree of polymerization (DP). Taking poly(ethyl acrylate) as an example, a polymer of DP of 1000 has a molecular weight of $100 \times 1000 = 100{,}000$. Most useful polymers have molecular weights between 10,000 and 1,000,000.

Control of molecular weight is of decided commercial importance, e.g., in the polymerization of butadienes to synthetic rubbers. Often, chain transfer agents[10] which help to depress molecular weight are essential in producing a synthetic rubber which can be readily processed.

Molecular weights of polymers can be determined by end group analysis, vapor pressure lowering, boiling point elevation, freezing point depression, and osmometry. These methods yield number average molecular weight ($\bar{M}_n$), where the number of molecules in a known mass of material is counted to yield the average value. Of the techniques described, osmometry is by far the one of greatest applicability.

Weight average molecular weight ($\bar{M}_w$), where the mass of molecules in a known mass of material gives the average value, can be determined by light scattering, ultracentrifugation, and solution viscosity. Strictly speaking, solution viscosity determinations are applicable only to linear polymers and yield a viscosity average molecular weight ($\bar{M}_v$) which is closer to $\bar{M}_w$ than to $\bar{M}_n$.

An osmometer holds a solution and the solvent in compartments separated by a semipermeable membrane, such as crosslinked polyvinyl alcohol or gel cellophane. Polymers with molecular weights between 15,000 and 1,000,000 can be accommodated by osmometry.[11] The method is limited at the low molecular weight end by diffusion of the solute through the membrane and at high molecular weight by the small magnitude of the osmotic pressure. The latter is developed when solvent molecules pass readily through the membrane, but solute molecules cannot.

Determination of the melt viscosity of a polymer, separated from an adhesive formulation, gives a strong handle on weight average molecular weight and crosslinking as well as an indication of chain branching.

## Molecular Weight Distribution Determination

Gel permeation chromatography (GPC) provides a convenient method for determining molecular weight distribution of a polymer up to molecular weights of about 20 million. The stationary phase in GPC is either a crosslinked polymer that swells when placed in contact with a solvent or a porous polymer of rigid structure. The pore size is critical, since molecules above a certain size will be unable to enter the pores and the pores will be only partially accessible to all but the smallest molecules.

The mobile phase solvent is selected for its ability to dissolve the sample and for low viscosity at operating temperature. Flowing through the column, the solvent takes the larger molecules through unhindered. Smaller molecules travel slower, the rate depending on their ability to penetrate the gel. This GPC technique separates materials of similar chemical composition in order of their molecular weight with the higher molecular weight components eluting first. Differential refractometer and UV detectors can be used.

## Chain Branching Determination

Chain branching in a polymer of interest to pressure-sensitive adhesives can be detected by IR spectroscopy. For long chain branches, an observable effect will be seen on the solution viscosity of the polymer. Comparison of the viscosity of a branched polymer against that of a linear polymer of the same molecular weight can be used to study chain branching. Short chain branches, such as ethyl or butyl groups, usually have little effect on solution viscosity.

Solubility measurements can be used to determine the controlled chain branching or crosslinking introduced into styrene copolymers by addition of small amounts of, say, divinylbenzene during styrene polymerization. About 0.0025% divinylbenzene yields soluble branched polymers. Approximately 0.1% divinylbenzene results in an insoluble, nonthermoplastic polymer which swells greatly in solvents. When 5–10% divinylbenzene is used, the product is a hard, brittle, crosslinked material.

## Degree of Crystallinity

The degree of crystallinity of a polymer can be measured by specific volume determination, X-ray diffraction and IR spectroscopy. One can determine the crystalline specific volume $V_c$ from X-ray unit cell dimensions, the amorphous specific volume $V_a$ from extrapolation of the specific volumes of polymer melts, and the specific volume of the sample concerned $V$ from curves relating specific volume as a function of time at temperatures below the crystalline melting point. Then the crystallinity ($S_c$) of the sample is given by

$$S_c = (V_a - V)/(V_a - V_c)$$

This equation assumes additivity of the specific volumes and a sample free of voids.

The X-ray technique permits calculation of the relative amounts of crystalline and amorphous material in a specimen, if it is possible to resolve the contributions of the two types of structure to the X-ray diffraction pattern. If the intensity of X-rays scattered by a sample is plotted vs. the diffraction angle, the amorphous peak is broad, while the crystalline peaks are narrower and usually more intense. Areas under peaks or peak heights can be used to estimate the degree of crystallinity.

The IR method can be used to determine the degree of crystallinity. In polyethylene, for example, an absorption at 7.67 $\mu$m can be identified with a methylene wagging motion in the amorphous regions.[12] In nylon 610,[13] the band at 10.6 $\mu$m was found to have absorbance vary linearly with specific volume. For polypropylene,[14] it is possible to use the ratio of the 995 to 974 $cm^{-1}$ bands as a measure of the crystalline-amorphous content. The mostly crystalline isotactic polypropylene has strong bands of nearly equal intensity at 995 and 974 $cm^{-1}$. The mostly amorphous atactic polypropylene has a strong, slightly broader band at 974 $cm^{-1}$ but practically none at 995 $cm^{-1}$. By measuring the absorbance ratios of these bands from suitably prepared standards and plotting a working curve, the amorphous-crystalline content can be measured with reasonable accuracy.

The absorbance ratio of the 720/730 $cm^{-1}$ doublet can be used for IR crystallinity determinations in polyethylene,[15] but films taken from melted samples should be used. This is because orientation in molded films causes variations in the absorbance ratio of the 720/730 $cm^{-1}$ doublet.

## Side-Chain Substitution and Crosslinking

Side-chain substitution in polymers can be evaluated by comparing the effects on $T_m$ (crystalline melting point) or $T_g$ (glass transition temperature) of substituting nonpolar groups for hydrogens of a polymer chain. Crosslinking can be ascertained by the effect on $T_g$ and the decreased solubility as compared to the uncrosslinked polymer.

Substitution of nonpolar groups for hydrogens of a polymer chain can lead to a reduction in $T_m$ (or $T_g$) or even a complete loss of crystallinity. Measurement of $T_m$ can sometimes distinguish between a side-chain substituted polymer and an unsubstituted one. For polyethylene, the crystalline melting point is lowered about 25°C going from the linear to the branched material. An increase in the length of the side-chain of a polymer often results in a looser crystal structure with an increasingly lower $T_m$. But an increase in the bulkiness of the side-chain increases $T_m$, because rotation in the side-chain is hindered in the liquid state.

Measurement of the glass transition temperature can throw light on side-chain substitution. In poly(*n*-alkyl methacrylates),[16] increasing chain length from one to 12 carbons reduced $T_g$ from 105 to $-65$°C.

Crosslinking of polymers can be assessed at least qualitatively by solubility determinations. An uncrosslinked polymer may be completely soluble in a given solvent, say, toulene, while its crosslinked counterpart may show varying degrees of insolubility from complete to partially soluble.

If crosslinking in a polymer system has been brought about by radiation, ions or radicals will normally be present which can be detected by conductivity, chemical tests, or by electron paramagnetic resonance (EPR) spectroscopy. Infrared is almost always productive in disclosing the effects of crosslinking. The formation of *trans*-vinylene unsaturation often accompanies radiation crosslinking. This can be followed by observation of the characteristic IR spectra of *trans*-unsaturated functional groups.

Irradiation of pressure-sensitive adhesive components, such as polystyrene, polyacrylates, and natural and synthetic rubbers, leads to crosslinking as the principal reaction. Gel formation accompanies crosslinking and ultimately leads to insolubilization of the entire sample.

Elastomers can be crosslinked by a great variety of substances, including many rubber accelerators. The latter can sometimes be detected by means of extraction with a 1:1 mixture of alcohol and chloroform, provided residual crosslinker remains in the composition.

Crosslinking in phenolic resins, such as those used in pressure-sensitive adhesives for high temperature use or electrical applications, can be followed and studied by IR spectroscopy.

## Orientation in Polymers

Orientation in polymers of pressure-sensitive adhesives can be determined by X-ray diffraction, birefringence changes, and IR dichroism studies. Orientation can occur in noncrystalline polymers. This can be measured by the force with which a specimen tends to retract on heating above $T_g$, as well as by birefringence.

X-ray diffraction patterns of oriented vs. unoriented crystalline polymers can follow the alignment of the polymer chains in the direction of the applied stress which produced the orientation. Measurements of tensile strength and stiffness of polymers with increasing orientation will show the increase in tensile strength and stiffness that normally occur.

## Effect of Additives

The effect of additives on the polymer properties of the polymers in an adhesive, such as plasticizers, can be measured by the changes in $T_g$, as well as by measurement of stiffness, hardness, and brittleness.

Plasticizers can often be extracted from a pressure-sensitive adhesive formulation with a selected alcohol or other solvent in which the elastomer and tackifier are insoluble, or substantially so.

Measurement of the melt viscosity of a plasticized polymer, determination of the dilution ratio, readings of the viscosity of dilute solutions of the polymer in the plasticizer, and measurement of electrical or mechanical properties of the plasticized polymer can all help in selection of the optimum plasticizer for a pressure-sensitive adhesive.

Plasticizers used in pressure-sensitive adhesives are mostly common oils such as mineral oil, liquid polybutenes and polyacrylates, and lanolin. Characterization of these is often made difficult by the small percentages used in formulations and by the necessity to separate them from other components present—often not a simple task. Once separated, however, they can be characterized by IR spectroscopy, especially from spectra scanned on thick films of these oils.

The effect of fillers on pressure-sensitive adhesives can be measured by tensile strength, elongation at break and 400% modulus (tensile stress at 400% elongation, a measure of stiffness). Inert fillers, such as clay and $CaCO_3$, have little effect on physical properties, although they make an elastomer mixture easier to handle. Reinforcing fillers do have an effect on physical properties such as increasing tensile strength and stiffness.

Fillers in adhesive formulations can be separated for property measurements by ashing off the other components, which are normally organic. Identification can be made by IR scans from 2.5 to 50 $\mu$m.

Stress-strain curves of filler-reinforced vs. unreinforced elastomers will show what stiffening effect a filler has introduced.

Measurement of tear resistance and abrasion resistance can indicate how a particular reinforcing filler has contributed to improvement in these properties of an elastomer.

Determination of the optimum loading with a given filler in an elastomer formulation can be ascertained by graphs of tensile strengths vs. volumes of filler per 100 parts of elastomer. Normally, increasing amounts of reinforcing filler cause continuing improvement in tensile strength until a maximum is reached, representing optimum loading. Beyond this, additional filler acts as a diluent and properties deteriorate.

The use of antioxidants to stabilize an adhesive against oxidation and heat and light degradation is common. Normally, they are the type used in rubber formulations. An antioxidant is itself readily oxidized or it combines with the oxidizing polymer to form a stable product. Rubber literature[17] on antioxidants should be consulted. Simulated, stepped-up conditions to produce oxidation quickly can be used to evaluate antioxidants. Actual trial of antioxidants is best.

The use of one's own judgment and common sense is often the best and simplest approach in selecting techniques for the identification or characterization of the components in a pressure-sensitive adhesive.

Figures 9-1 and 9-2 illustrate the use of infrared spectroscopy for identifying and duplicating of commercial adhesives.

Figure 9-1, the infrared spectrum of a commercial pressure-sensitive adhesive, shows absorptions (bands) at 1600, 1580, and 1493 $cm^{-1}$, indicat-

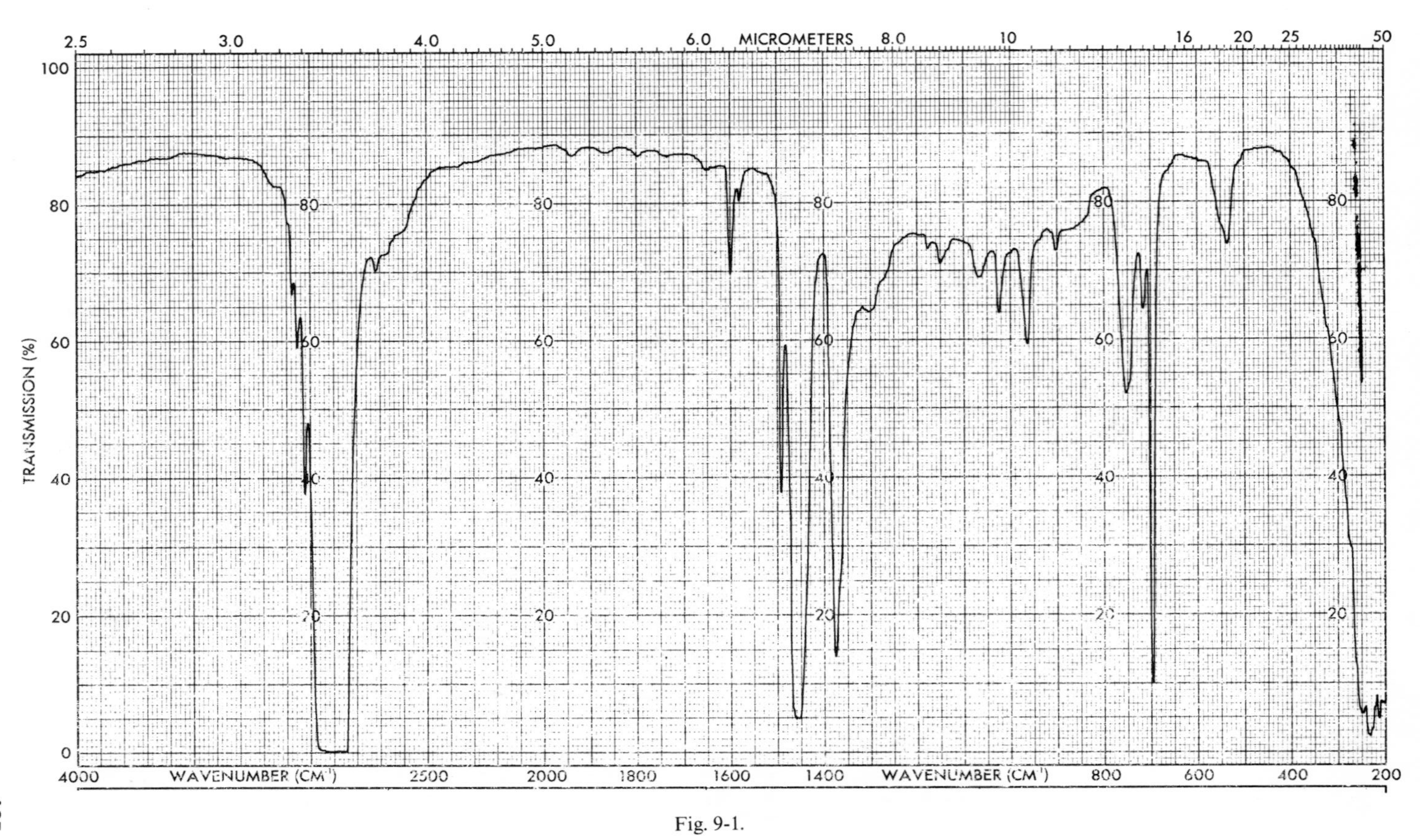


Fig. 9-1.

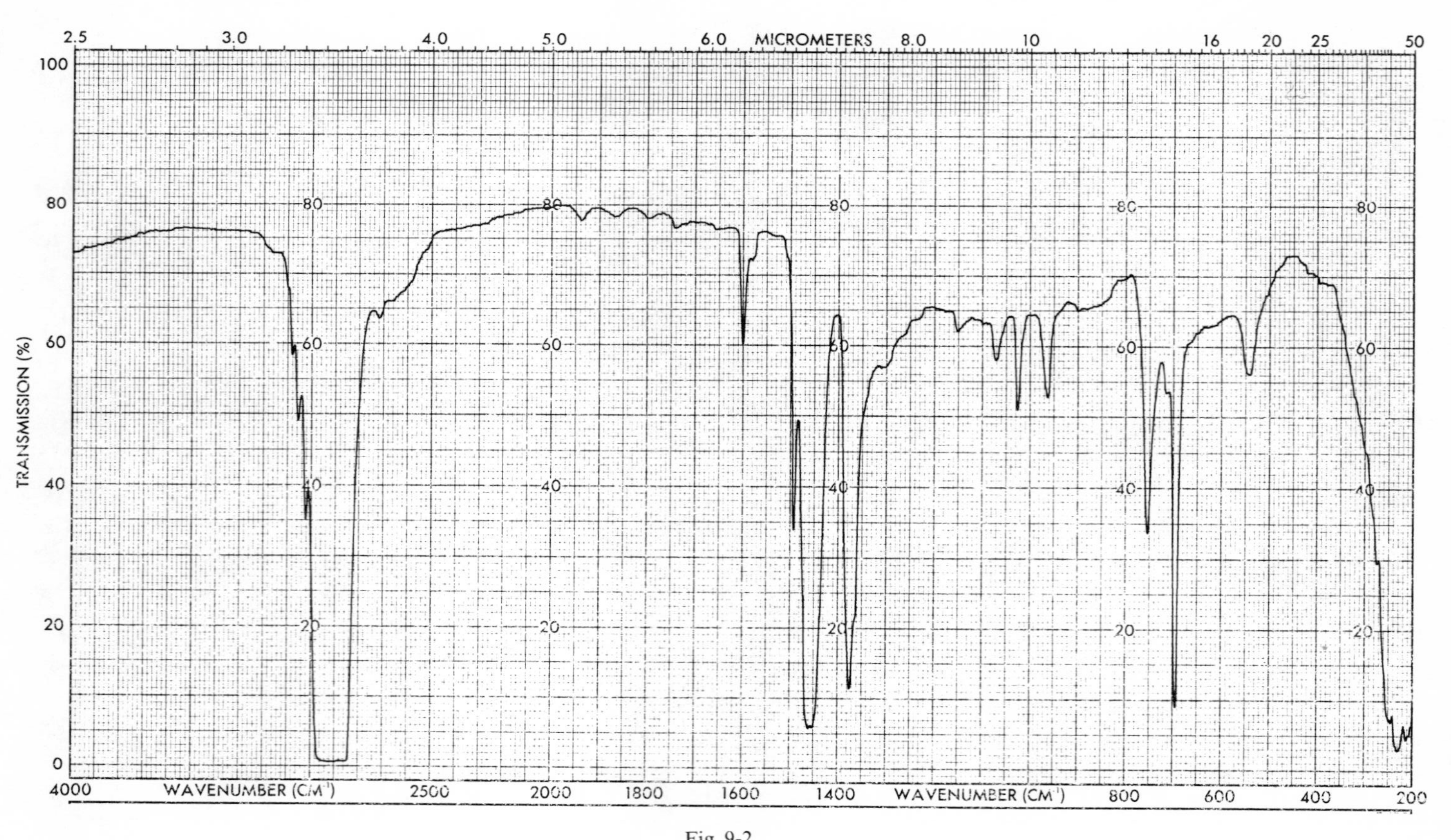
MICROMETERS
2.5
3.0
4.0
5.0
6.0
8.0
10
16
20
25
50
TRANSMISSION (%)
100
80
60
40
20
0
WAVENUMBER (CM⁻¹)
4000
2500
2000
1800
1600
1400
800
600
400
200

Fig. 9-2.

ing a benezene derivative. Bands at 696 and 756 $cm^{-1}$ suggest the aromatic is monosubstituted as do the four bands between about 1740 and about 1940 $cm^{-1}$. The sharp shoulder absorptions above 3000 $cm^{-1}$ indicate aromatic C—H. Bands at 964 and 900 $cm^{-1}$ together with the monosubstituted benzene derivative already indicated suggest a butadiene-styrene copolymer. Comparison against spectra of such knowns confirms this elastomeric copolymer is present.

The weak band at about 1658 $cm^{-1}$ could arise from the C=C of an unsaturated terpene. The 964 $cm^{-1}$ band (which had already suggested polybutadiene) perhaps indicates also the *trans* —CH=CH— of an unsaturated terpene. Consideration should be given to the presence of such a terpene resin.

Very little evidence can be seen in Fig. 9-1 showing the identity of the plasticizer. The very weak band at about 718 $cm^{-1}$ suggests that a long methylene chain type substance could be present. A hydrocarbon type oil may be present, but one cannot tell from this spectrum. Liquid chromatography or solvent separation techniques could be tried, in order to separate the oil, if present, from the elastomer and resin of this pressure-sensitive adhesive. One should scan the separated plasticizer and then meaningful comparisons can be made against the spectra of likely knowns.

Figure 9-2 shows the 2.5 to 25 $\mu$m (4000 to 400 $cm^{-1}$) infrared spectrum of the first attempt to match the composition of the pressure-sensitive adhesive of Fig. 9-1. Present in this known is a styrene-butadiene elastomer, a terpene type for resin, and a hydrocarbon type oil for the plasticizer.

While the spectral similarity between the unknown and the attempted match is encouraging, it is by no means a spectral match. Note the broader band near 753 $cm^{-1}$ in Fig. 9-2 as compared to the corresponding absorption in Fig. 9-1. The weak band at 718 $cm^{-1}$ in Fig. 9-2 is relatively stronger than its counterpart in Fig. 9-1. The doublet band near 540 $cm^{-1}$ in Fig. 9-2 contrasts with a singlet band in Fig. 9-1. Adjustments must now be made in the proportions and perhaps also the compositions of the elastomer, resin, and plasticizer selected for the second attempt to match the qualitative and quantitative composition of the pressure-sensitive adhesive of Fig. 9-1. Once a good spectral match has been obtained, one can be reasonably certain the composition of the unknown has been matched. Performance tests of the matching formulation vs. the commercial material will confirm or deny this spectral conclusion.

## REFERENCES

1. Kendall, D. N. (ed.) *Applied Infrared Spectroscopy*. New York: Van Nostrand Reinhold Co., 1966.

2. Smith, A. L. *Applied Infrared Spectroscopy*. New York: John Wiley and Sons, 1979.
3. Jaffe, H. H., and Orchin, M. *Theory and Applications of Ultraviolet Spectroscopy*. New York: John Wiley & Sons, 1962.
4. Wendlandt, W. W. *Thermal Methods of Analysis*, 2nd Ed. New York: Wiley-Interscience, 1974.
5. ASTM, Book of ASTM Standards, Parts 24–25, Philadelphia, 1979.
6. Snyder, L. R. and Kirkland, J. J. *Introduction to Modern Liquid Chromatography*. New York: John Wiley and Sons, 1974.
7. Skeist, I. (ed.) *Handbook of Adhesives*, 2nd Ed. New York: Van Nostrand Reinhold Co., 1977.
8. Carlson, D. W., Hayes, M. W., Ransaw, H. C., McFadden, R. S. and Altenau, A. G. *Anal. Chem.*, **43**: 1874 (1971).
9. Brock, M. J. and Louth, G. D. *Anal. Chem.* **27**: 1575 (1955).
10. Billmeyer, F. W., Jr. *Textbook of Polymer Science*. New York: Wiley-Interscience, 1962.
11. Kirk–Othmer (eds.) *Encyclopedia of Chemical Technology*, 2nd Ed. Vol. 16: New York: Wiley-Interscience, 1968.
12. Miller, R. G. J., and Willis, H. A. *J. Polymer Sci.* **19**: 485 (1956).
13. Sandeman, I. and Keller, A. *J. Polymer Sci.* **19**: 401 (1956).
14. Luongo, J. P., *Anal. Chem.* **33**: 1919 (1961).
15. Luongo, J. P. *Polymer Letters* **2**: 75 (1964).
16. Rogers, S. S., and Mandelkern, L. *J. Phys. Chem.* **61**: 985 (1957).
17. Morton, M. (ed.) *Rubber Technology* 2nd Ed. New York: Van Nostrand Reinhold Co. 1973.
18. Michelson, A. A. *Light Waves and Their Uses*. Chicago: Univ. of Chicago Press, 1902.
19. Fellgett, P. B. Ohio State Symposium on Molecular Spectroscopy, 1952.

# Chapter 10

# Natural Rubber Adhesives

**G. L. BUTLER**
*DRG Tapes and Adhesives*
*Borehamwood, England*

The history of pressure-sensitive adhesives is comparatively short; natural rubber was the elastomer used in the earliest examples of this type of adhesives, and it still remains today one of the major elastomers used in the manufacture of these adhesives, though the development of synthetic elastomers such as polyacrylates and S-B-S and S-I-S block copolymers is threatening the leading position that has been held by natural rubber.

The first use of natural rubber pressure-sensitive adhesives was in surgical plasters. An early U.S. patent[1] taken out by Shecut and Day in 1845 describes a mixture of India rubber, turpentine extracts, Peruvian balsam, and pine gum.

However, the real development of the pressure-sensitive adhesive industry began with the inventions of R. G. Drew in the late 1920's and early 1930's. Drew's inventions were primarily concerned with the development of crepe paper-based masking tapes and clear tapes based on regenerated cellulose film. He needed adhesives for these products which would adhere well under applicational pressure, and could be subsequently peeled off the surface without depositing the adhesive. A further necessary property of the adhesive was that the tape should be capable of being unwound without the adhesive offsetting to the reverse side of the tape. In these early days, primer coats to bond the adhesive to the web had not been developed, so the composition of the adhesive had to be such that allowed satisfactory unwinding characteristics.

One of Drew's earliest patents[2] describes an adhesive composition for masking tape comprising plasticized plantation natural rubber, ZnO filler tackified with either coumarone-indene resin or a natural resin such as Burgundy Pitch. His early patent[3] for cellulose tape describes an adhesive based on masticated pale crepe latex, coumarone-indene tackifier, and liquid paraffin, claimed to increase the tack of the adhesive without causing the adhesive to offset, which would have occurred with added resin content.

By the early 1930's, the need to impart some temperature resistance to masking tapes was recognized. This was needed to achieve clean removal of masking tape after it had been exposed to paint stoving cycles. This need was met by the earliest examples[4,5] of the uses of vulcanized natural rubber in adhesive tapes. Curing was achieved with the use of sulfur together with diphenyl guanidine or other accelerators.

In the mid-1930's, wood rosin was well-recognized as a tackifier for natural rubber, but the end of this decade saw the development of the use of the less acidic ester gum as the main tackifier resin for natural rubber.[6] This type of resin was found to give considerably improved aging resistance in comparison with the unesterified rosin. In particular, it was found to be free from the development of small crystals in the adhesive that was observed with gum rosin. Also the more acidic rosin used to impart a dark color to the adhesive, an undesirable feature on cellulose-based tapes. The use of ester gum (normally a glycerol ester of wood rosin with an acid number in the range 2–10) also led to a firmer adhesive mass, which overcame the problem of adhesive "oozing" at the edges of the roll of tape. An adhesive composition from Minnesota Mining Company's British Patent 514,402 (1938) was

| | |
|---|---|
| Milled latex type rubber | 250 parts |
| Less acidic ester gum | 175 parts |
| Antioxidant | 1.25 parts |

This type of composition was widely used in the rapid expansion of pressure-sensitive adhesives and tapes in the postwar period.

## COMPOSITION OF ADHESIVES

As indicated above, this type of adhesive comprises essentially natural rubber, tackifying resins, and antioxidant. Other materials can be added for special purposes, including plasticizers, pigments, and curing agents to vulcanize the adhesive partially. The adhesives are normally coated from organic solvent solutions; typical solvents include fast-drying, nonpolar ones such as toluene, heptane, etc.

## Natural Rubber Type

The type of natural rubber used in pressure-sensitive adhesive will depend to a large extent on the end application. In the case of clear filmic tapes such as those based on cellulose film, it is important that the color of the adhesive should be kept as light as possible. For this reason, it is normal to use a pale crepe grade of rubber. Where color is not important (e.g., dark colored tapes, certain masking tapes, cloth tapes, etc.), darker grades such as ribbed smoked sheet or SMR 5 are often used, as these grades are less expensive than pale crepe.

Natural rubber has to be masticated before it can be used in a pressure-sensitive adhesive. This is necessary to break down the gel content and reduce the molecular weight of the natural material. Without mastication, it would be very difficult to obtain the desired level of tack in the adhesive and also, as most adhesive systems are coated from organic solvent solutions, the unmasticated rubber does not give a free-flowing adhesive suitable for subsequent coating without the use of excessive quantities of solvent.

## Tackifying Resins

Pressure-sensitive adhesives require a balance of three main properties: peel adhesion, cohesive holding power, and surface tack. Natural rubber alone has a very low tack and adhesion to surfaces. Consequently, it is necessary to add tackifying resins to the elastomer to produce the required balance of tack, peel adhesion, and resistance to shear forces.

Three main classes of tackifying resins have been used with natural rubber. These are:

1. Wood rosin and its derivatives
2. Terpene resins
3. Petroleum-based resins

**Rosin and Derivatives.** Apart from the wood rosin and its glycerol ester (gum ester), a number of other derivatives of this resin type is used with natural rubber. To vary the softening point, different esters are used e.g., pentaerythritol ester. The abietic acid molecule in wood rosin contains unsaturated double bonds which are susceptible to oxidation. Therefore, hydrogenated rosin derivatives are frequently used in adhesives requiring long-term aging properties. These include glycerol ester of partially hydrogenated rosin and the pentaerythritol ester.

**Terpene Resins.** This type of resin, essentially consisting of polymerized β-pinene, started to be used in the late 1950's and early 1960's. It can be used in a variety of softening points. Terpene resins became the standard with natural rubber for many years. This resin shows a more improved balance of tack and shear resistance than the rosin derivations, and it also possesses excellent aging properties. Although still used in speciality adhesives and those requiring its unique balance of properties, this resin has declined in use in recent years due to its high cost and variable availability. The corresponding and more plentiful pinene resin, polymerized α-pinene, is not such an efficient tackifier for natural rubber.

**Petroleum Resins.** These resins generally comprise polymerized C5 or C9 or polycyclic streams derived from the naphtha cracking process. Although used as tackifiers for natural rubber for more than 20 years, the earlier resins of this type did not yield very good tack, although they gave good holding power, and were generally used because of their lost cost. This class of resins has seen many developments in recent years, and there is now available a wide range of resins of differing softening points yielding adhesives with good balances of tack and cohesion, when used in natural rubber formulations.

## Effects of Choice of Resin and Resin Concentration

As indicated earlier, pressure-sensitive adhesives require a balance of peel adhesion, cohesive strength and surface tack. Generally, the formulator is trying to optimize all three properties, and with natural rubber-based adhesives, this is largely done by selection and the quantity of resins used in the formulation. Within certain limits, the higher the resin content, the higher is the peel adhesion and surface tack, but the lower the cohesive strength. Consequently, many natural rubber-based adhesives tend to be compromises between adhesion and tack on the one hand and internal cohesive strength on the other.

Much work has been carried out to determine the optimum resin type and concentration to achieve the best balance of properties. The studies carried out by Wetzel and Alexander[7,8] illustrate the effect that the concentration and softening point of the tackifying resin has on tack properties of natural rubber-based adhesives in the case of the resin ester family of resins (Figs. 10-1 and 10-2).

These figures show that as the tackifier concentration is increased, little or no increase in tack is observed until a resin concentration of 20–40% is reached. Thereafter, tack increases quite rapidly to a maximum; the concentration of tackifier in natural rubber at which maximum tack is obtained

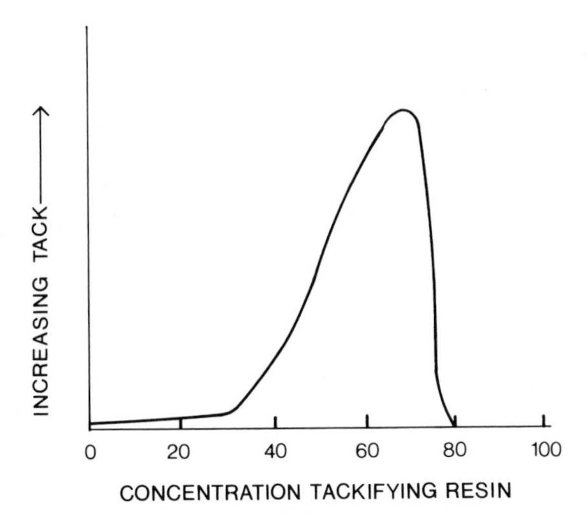


Fig. 10-1. Tack vs. concentration of tackifying resin in natural rubber for high and low softening point resins (after Wetzel).

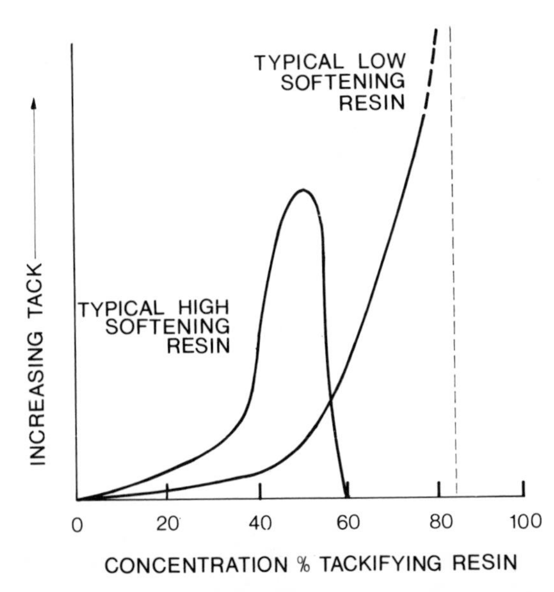


Fig. 10-2. Tack vs. concentration of tackifying resin in natural rubber for intermediate softening point resins (after Wetzel).

varies from 50–75%. Maximum tack with these resins can be correlated inversely with the viscosity of the resin, as indicated by its softening point. Additionally, resins of lower softening points (i.e., less than 95°C) provide higher maximum tack than resins of higher softening point.

Tack, however, is only one of the three main parameters that determine the properties of an adhesive. As we have seen, higher tack is generally achieved with resins of lower softening points, and with these resins generally, the higher the resin concentration, the higher is the tack. However, resins with the lowest softening points have the lowest viscosities and hence produce adhesives with poor internal strength and shear resistance. Consequently, choice of resins has to be a compromise and is frequently achieved by blends of one or more resins to give the desired balance of properties.

The contrasting effects of resin concentration on the three main adhesive parameters are illustrated in Fig. 10-3. The resin used in this example is

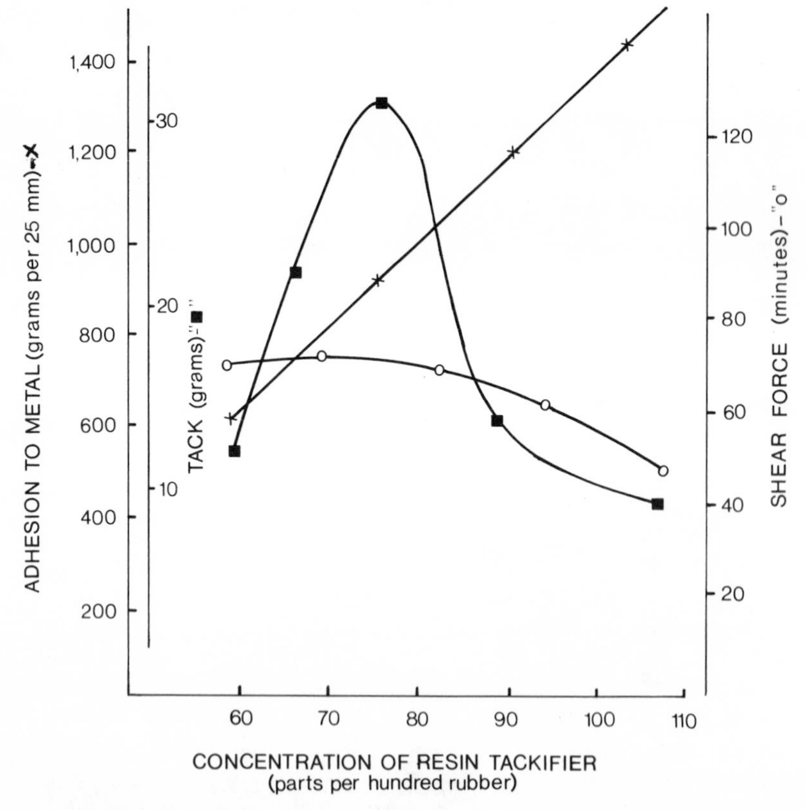


Fig. 10-3. Effect of resin concentration on peel adhesion (x----x), tack (■----■), and shear resistance (O----O) for a hydrocarbon resin in natural rubber.

a hydrocarbon resin. Shear resistance was measured on 1.25 × 2.5 cm contact area with a 1.5 kg load at 40°C. Tack was measured by the rotating wheel method described by R. F. Bull.[9] As can be seen from the graph, the best balance of properties obtained with this resin occurs in the region 80 parts per hundred rubber (phr). Different resins will give different balances of these three main properties, but the performance of this resin is typical of those used in natural rubber pressure-sensitive adhesives.

Apart from the three main parameters discussed above, there are other special factors which affect the choice of tackifying resins. If the adhesive is destined for application for a clear tape, then color is an important factor. In this respect, terpene resins are particularly useful as are the hydrogenated resins, both rosin derived and also the hydrogenated polycyclic petroleum hydrocarbon resins. Not only do these resins show good initial color in a rubber-based adhesive, but also give subsequent aging resistance due to the saturated nature of the resin molecule.

C. A. Dahlquist[10] has described a natural rubber adhesive tackified with alkylated polystyrene resins for a particularly good water-white color and good UV resistance. A typical resin is poly *tert*-butyl styrene with a molecular weight in the range 1000–2000, and can be used at a concentration of about 77 phr to give useful adhesives with milled natural rubber.

Most resins used to tackify natural rubber are compatible with the polymers. But it has been claimed[11] that the addition of small quantities of resins which either have a borderline compatibility or are incompatible with natural rubber, do lead to adhesives with considerably increased peel adhesion values. Examples of such resins include maleic modified ester of rosins, oil-soluble phenol aldehyde resins such as *p*-tertiary amyl phenol formaldehyde, etc. These resins are added in addition to the normal tackifiers in amounts of less than 15% by weight of the rubber. The theory of this process is that the incompatible resins tend to separate from the body of the adhesive during application of the adhesive to the backing, and provide a resin-rich area at the surface, hence increasing adhesion, without affecting the cohesiveness of the adhesive layer as a whole.

## Antioxidants

The *cis*-polyisoprene molecule of natural rubber contains unsaturated double bonds along the chain, and is subject to oxidation attack. This action on natural rubber has a degrading effect of chain scission with consequent loss of cohesive strength. This oxidation attack is normally increased by exposure to elevated temperature or to UV light. Consequently, it is necessary to include antioxidants in the formulation of natural rubber-based adhesives to ensure both adequate shelf-life of the adhesive after it has been converted to tapes or labels, and also to afford a degree of protection to the adhesive in use.

In general, three main types of antioxidants have been utilized with natural rubber adhesives. Firstly, there are the amine antioxidants, such as *N–N′* di-$\beta$-naphthyl-1,4-phenylenediamine (AgeRite D). This type of material, while acting as an efficient antioxidant, is a staining material, and can impart a slight stain to surfaces with which the adhesive comes into contact. The most widely used antioxidants have been the phenolic types; these include 2,5-di-(*tert* amyl) hydroquinone (Santovar A), 2-2′-methylenebis(4-methyl-6-*tert* butyl phenol) (Antioxidant 2246) and many others of similar types. Phenolic antioxidants are nonstaining and are particularly effective in protecting adhesives against the effects of UV and sunlight. The third class of antioxidants found useful in natural rubber adhesive is the dithiocarbamates. The most commonly used of them is zinc dithiodibutyl carbamate.[14] This type of antioxidant is effective against heat aging as well as imparting some resistance against UV light. These types are also largely nonstaining.

One of the problems with antioxidants is that particular ones tend to be effective against specific types of aging, whereas the adhesive formulator is looking for the best protection against all forms of aging (heat, light, and the effects of certain metal ions which catalyze rubber oxidation). Consequently, he is often forced to use a blend of antioxidants to achieve optimum protection. One such blend that is claimed[12] to be particularly effective is a mixture of di-organo hydrogen phosphonate together with either a phenolic or a secondary amine antioxidant. Among the phosphonates that can be used in this invention are diethyl hydrogen phosphonate. Antioxidants are normally added in the range of 0.5–2.0 parts per hundred of the rubber to achieve optimum protection.

## Plasticizer Addition

A variety of plasticizers can be used in natural rubber adhesives. Addition of plasticizer to the adhesive will always lower the cohesive strength, generally reduce the peel adhesion, and will have a variable effect on the tack depending on the type of plasticizer used. Plasticizers, such as mineral oil or lanolin, reduce the cost of the adhesive mass, and have a depressing effect on the peel adhesion. This can be useful for certain specific applications, such as peelable adhesive labels or printable tapes. It is claimed that lecithin is a particularly effective agent in reducing the peel adhesion to cellulose film. This can be useful in reducing the force required to unwind cellulose-based tapes.

### Pigments

Pigments are frequently added to adhesives for several purposes. First, for coloring reasons, many film-based packaging tapes have colored adhesives, the most common of these being the brown colored case sealing tapes. The most commonly used pigment is $TiO_2$, used for its opacity, together with colored toners to produce the desired color.

Second, pigments are added to reduce the cost of the adhesive mass. Clay and whiting fillers can be used for this purpose and are found particularly on tapes with heavy masses of adhesive, such as cloth-based tapes or masking tapes.

Third, pigments can be added to improve the performance of the adhesive. The addition of ZnO pigment tends to make the adhesive firmer. It has been claimed[13] that the addition of fine particle size silica improves the aging of natural rubber-based adhesives on fabric-based tapes. The addition of silica with particle sizes between 0.01 and 0.03 $\mu$m at levels between 2 and 8% of the adhesive is claimed to reduce the tendency of the roll of cloth tape to distort on aging and also to improve tack retention after aging.

### Blends of Other Elastomers with Natural Rubber

Other elastomers can be added to natural rubber-based adhesives for either cost or performance reasons.

Synthetic *cis*-polyisoprene polymers can be added to natural rubber-based adhesive. The main problem with commercially available *cis*-polyisoprene polymers is that they do not possess the same green strength as natural rubber and consequently, in uncured adhesives, this leads to loss of cohesive strength. Quantities of the synthetic elastomers that can be added are limited.

Styrene-butadiene copolymers can be usefully blended with natural rubber systems. Such blends generally result in improved aging of the adhesive.[15] In the oxidation aging process, natural rubber undergoes chain scission and hence becomes softer, whereas SBR elastomers crosslink and become harder. Consequently, the two materials tend to balance each other. As well as improving aging resistance, blending with SBR can also increase the creep resistance.

## MANUFACTURE

The vast majority of natural rubber pressure-sensitive adhesives are applied to their substrates from solvent systems. The initial process in the manufacture is milling or breakdown of the rubber to achieve the required viscoelastic properties. This can be carried out on either a two-roll rubber mill or in an internal mixer. The degree of milling required depends both on the state of the raw rubber, which being a natural product, can vary from batch to batch, and on the viscoelastic properties required in the end adhesive. For most purposes, the rubber would be milled to a Mooney viscosity of between 60–75. The milling process can be accelerated by the addition on the mill of peptizers, such as pentachlorothiophenol, in order to reduce times and milling energy. The dissolution of the rubber in the solvent can be a lengthy process. In order to speed this up as much as possible, it is necessary to granulate the rubber in order to create the maximum surface area for the solvent to attack the rubber.

Natural rubber is a nonpolar material. Consequently, nonpolar hydrocarbon solvents are used. The choice of the solvent is normally determined by the evaporation rate that is required in the coating and drying process. The most commonly used solvents are aliphatic hydrocarbons in the hexane–heptane range; additionally, aromatic solvents such as toluene are also used.

When the natural rubber adhesives are coated to film backings, such as cellulose, PVC, polyester, or polypropylene, it is necessary to apply a primer coat between the natural rubber adhesives and the film in order to bond the adhesive to the film. Release coats are often applied to the other side of the film to facilitate unwinding of the roll. These release coats are usually long, chained, waxy hydrocarbon materials or silicones. Adhesives, primer and release coats are normally applied in-line on modern adhesive coating machines.

## MECHANISMS OF TACK AND ADHESION

Much has been written on the mechanisms of tack and peel adhesion of natural rubber-based adhesives.

The concept of tack is difficult to define accurately and has different meanings in different industries. In the case of pressure-sensitive adhesives, tack can be described as the property whereby the adhesive will adhere tenaciously to any surface it comes into contact under light pressure. The strength of the bond will be greater under increasing pressure, hence the term pressure-sensitive. Tack has also been defined[16] as the force required to separate an adherend and an adhesive at the interface shortly after they have been brought rapidly into contact under a light load of short duration.

Basically, there have been two schools of thought regarding the theory of tack in rubber-resin blends. The first states that the tack of the system is essentially due to the morphology of a two-phase system. The second view is that tack is due to the deformation characteristics of the viscoelastic mass.

The two-phase theory was first proposed in 1957 by Wetzel[7] and further propounded by his co-workers at Hercules Powder Co., Hoch and Abbott.[17] In their work, Hoch and Abbott examined the adhesive film topography by electron micrography to demonstrate how changes in tack were accompanied by changes in topography. They showed with blends of natural rubber and various rosin ester tackifiers, that when high tack was obtained, a two-phase structure in the adhesive was observed. The continuous phase consisted of rubber saturated with resin, whereas the disperse phase comprised resin and low molecular weight rubber molecules. The number of disperse particles and also their size (from hundreds to a few thousand angstrom) vary in different systems, the number being greatest and size largest in systems having maximum tack. Furthermore, when these adhesives are aged, and the tack reduced or eliminated, the adhesive structure reverted to a single-phase system. This theory was further developed by Hoch[18] in 1963 when he states that tack is largely dependent on the disperse phase which is effectively a viscous fluid which can wet and flow to contact a surface intimately. This viscous layer is so thin that failure cannot occur in it. Consequently, ultimate adhesive properties are dependent on the continuous elastomeric phase. Hence the disperse phase contacts and wets the irregularities of the substrate, and the continuous phase carries the load.

The effect of solvent evaporation on the heterogeneity of rubber-resin films has been examined by Whitehouse, Counsell, and Lewis.[19] They investigated the chemical composition of natural rubber/rosin ester blends by ATR infrared spectroscopy, and found an excess concentration of resin at both surfaces of the adhesive layer, compared with the resin concentration in the bulk of the adhesive layer (Fig. 10-4).

Prolonged aging of the adhesive layer for about 500 days showed that the surface compositions of the films became identical to the bulk composition, thus indicating that the original heterogeneity of the adhesive films was due to the solvent evaporation stage. Various solvents were examined, and it was shown that the magnitude of the fractional surface excess resin concentration depends on the solvent used, the overall bulk resin concentration, the thickness of the adhesive film and the time for equilibration after the film has dried. The resin excess appears to be related to the mechanism of the film forming process in which the solubility characteristics and the rate of evaporation of the solvent are important parameters.

The two-phase theory cannot adequately explain the variation of tack with stress rate nor the tack generated by a resin which is fully compatible with natural rubber, such as a $\beta$-pinene resin. It is the view of several workers that pressure-sensitive adhesive tack is due to the viscoelastic properties of the adhesive layer and its reaction to deformation. Dahlquist[20] has shown that tackifiers have an effect on the adhesive

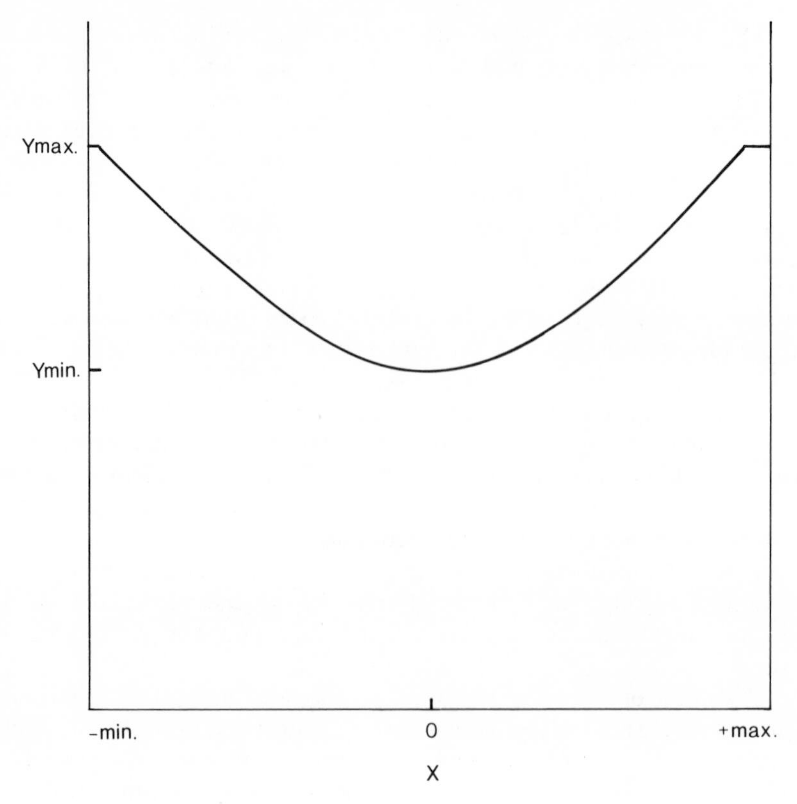


Fig. 10-4. Variation of volume fraction of resin 'Y' with distance 'X' from central plane of adhesive layer (after Whitehouse and Counsell).

moduli, and has also shown[21] that the apparent maximum tack, as a function of the resin concentration in a rubber-resin blend, could be altered by changing the speed of probe withdrawal in the probe tack test method.

Sherriff, Knibbs, and Langley[22] have investigated the effect of various tackifier resins on the viscoelastic properties of natural rubber adhesives and its relationship with tack. The resins they examined were Pentalyn H (Hercules Powder Co.), Piccolyte S 115 and Piccolyte S 70, both $\beta$-pinene resins supplied by Hercules Powder Co., and Akron P125 (Arakawa Forest Industries Ltd.) reportedly a polymerized dicyclopentadiene. Variation of the glass transition temperature $T_g$ for varying resin concentrations in natural rubber was determined by dilatometry, and these showed that three of the resins studied Piccolyte S 115, Piccolyte S 70 and Arkon P125 were compatible with natural rubber in all resin concentrations (Figs. 10-5, 10-6, and 10-7).

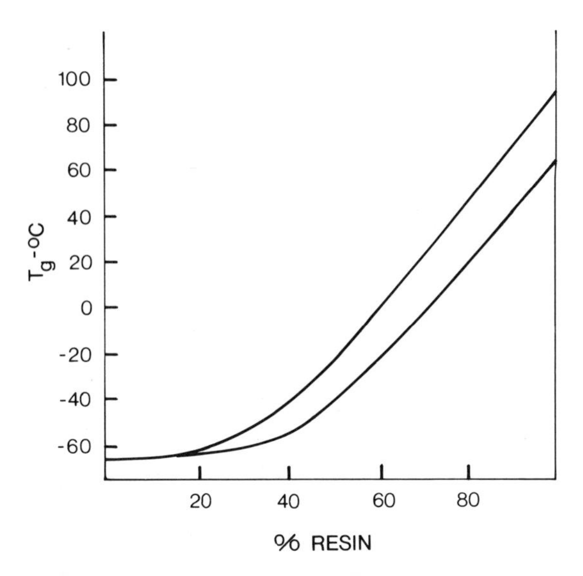


Fig. 10-5. Variation of transition temperatures with resin concentration with Piccolyte S70 (bottom curve) and S115 (top curve) in natural rubber (after Sherriff, Knibbs, and Langley).

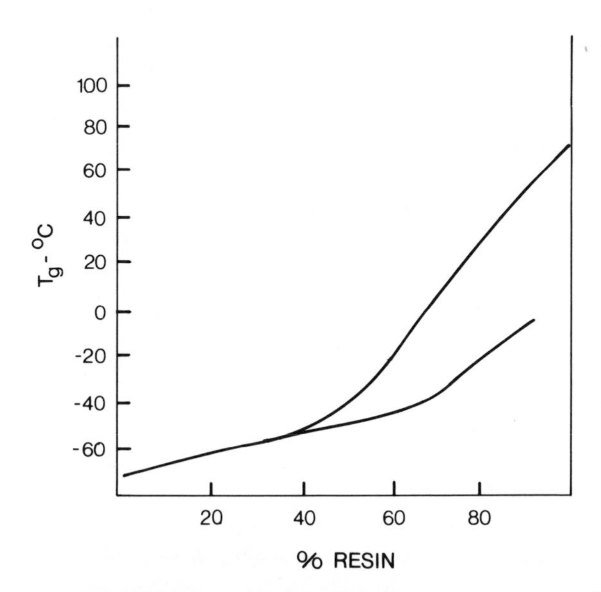


Fig. 10-6. Variation of transition temperatures with resin concentration with Pentalyn H in natural rubber (after Sherriff, Knibbs, and Langley).

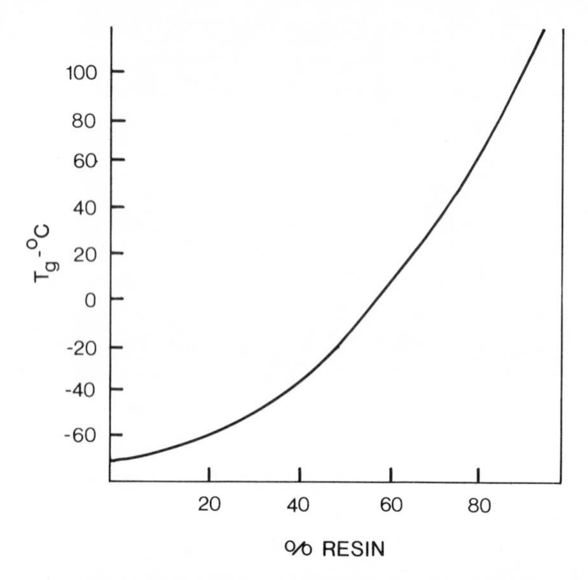


Fig. 10-7. Variation of transition temperatures with resin concentration with Arkon P125 in natural rubber (after Sherriff, Knibbs, and Langley).

Figure 10-6 tends to confirm Wetzel's theory for tackification of rosin esters, two phases are formed as is shown by the development of two $T_g$'s for concentrations in excess of 60% resin. It is interesting to note that $T_g$'s of both phases increase with increasing resin concentration, which does not suggest that one phase has become saturated as Wetzel indicates. Sherriff, Knibbs, and Langley show that variation to tack with resin concentration can be explained by the effect of the resin of the viscoelastic properties of the adhesive mass. The position of maximum tack in a rubber-resin blend is controlled by the $T_g$ of the blend. At resin concentrations at which the $T_g$ of the mixture approached room temperature, the adhesive will have difficulty in making contact with the adherend at very short contact times. These effects have been shown to be independent of the compatibility of the resin with natural rubber, so long as it is not fully incompatible. The softening point of the resin determines the resin content for maximum tack by its effect on the adhesive $T_g$.

As the tack of the natural rubber is determined by the response of a viscoelastic material to deformation under the effect of pressure, the effect of adhesive thickness becomes an important parameter in practical terms. The minimum coating weight of an adhesive is determied by that thickness which will allow sufficient viscoelastic response to form adequate tack properties.

D. W. Aubrey and his colleagues[23,24] at the National College of Rubber Technology in London have recently carried out work to study the peel adhesion behavior of natural rubber-resin blends and the degree of correspondence between peel adhesion and viscoelastic behavior of the adhesive.

## SPECIALLY MODIFIED ADHESIVES

The natural rubber adhesives that have been described, comprising essentially natural rubber and tackifying resins are perfectly adequate for many purposes, such as film-based packaging tapes, fabric tapes, and paper labels. However, many other applications of pressure-sensitive adhesives require modification. These modifications include imparting heat resistance, modifying electrical properties, giving a degree of solvent resistance, and increasing the creep resistance of the adhesive.

### Heat Resistance

The most important application for pressure-sensitive adhesives where heat resistance is required is in the field of masking tapes. These tapes are used in paint spraying operations where the tape has to undergo the paint stoving process, with a typical baking cycle of 40 minutes up to 130°C. After undergoing this exposure, the masking tape has to be capable of being peeled off without leaving any adhesive residue, and thus peeling can take place while the tape is still warm.

The usual construction of a masking tape comprises an impregnated creped paper, coated with a pressure-sensitive adhesive on one side and a paint-resistant back-coat on the other surface.

The conventional natural rubber adhesives previously described are quite obviously thermoplastic and do not have the required temperature resistance for a masking tape. Not only would the tape fall off in the oven, which could result in marking the uncured paint, but also it could not be removed without leaving adhesive residue behind. To impart the required degree of heat resistance to the adhesive, it is usual to effect a degree of crosslinking to the natural rubber. We have seen earlier that the tack and adhesion of pressure-sensitive adhesives depend upon the viscoelastic properties of the rubber-resin blend. Consequently, the degree of crosslinking has to be carefully controlled; if the degree of curing is too high, then the adhesive will lose its ability to deform readily and hence wet the surface to which it is being applied, but it has to be sufficiently cured to have resistance to the paint solvent, and also not to be softened too much by the stoving cycle.

It is important, therefore, to be able to study the effectiveness of crosslinking on the thermal stability of the adhesive. One way of doing this is to measure the peel adhesion of the tape at elevated temperatures. This can be done by applying the tape to a polished steel panel, heating the panel to a required temperature (120°C), and then peeling the tape at an angle of 180° at fixed speed at the elevated temperature and measuring the pulling force required. Figure 10-8 illustrates the effect of crosslinking on peel adhesion measured at 120°C.

In this case, a curable adhesive has been coated on to paper backing, the solvent removed, and the adhesive cured for 1 min at various temperatures.

The adhesive studied in this case is based on natural rubber/rosin ester tackifier, ZnO pigment together with a sulfur plus accelerator curing system. The vertical axis shows the peel adhesion in grams per 25 mm, using a peeling speed of 15 cm/min. The effect of crosslinking can easily be seen on this graph. On the left, at low curing temperature, there is relatively little crosslinking, the adhesive is thermoplastic, peel adhesion is low, and failure is largely cohesive.

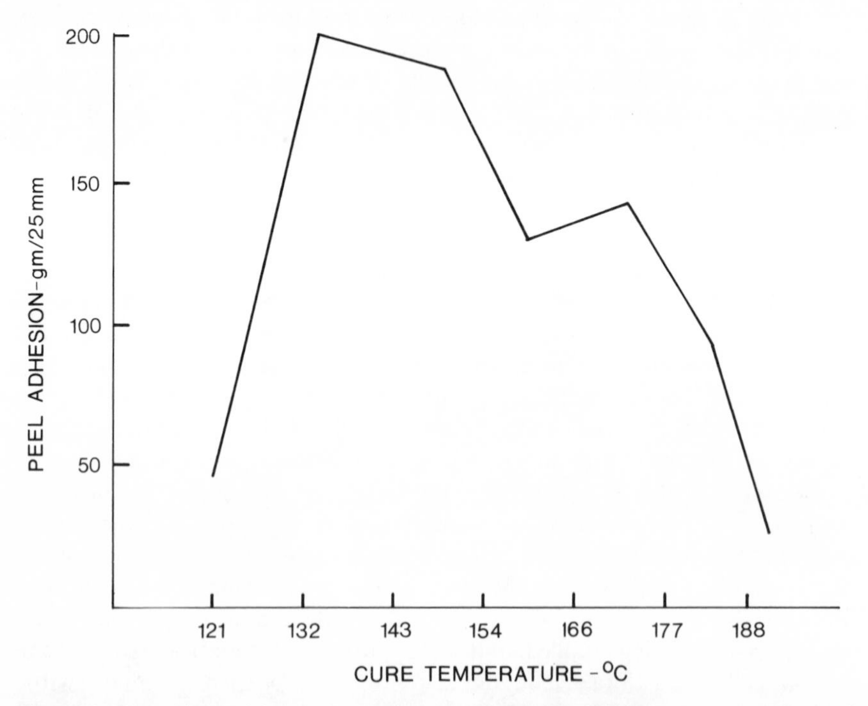


Fig. 10-8. Comparison of hot peel adhesion at 120°C with differing cure temperatures of adhesive.

Increasing the curing temperature gives an increase in peel adhesion up to a maximum. Increasing the curing temperature further causes a drop in adhesion. In this region of the graph, too much crosslinking has occurred, and the failure is adhesive, the adhesive having lost some of its ability to deform and wet the surface to which it is applied.

The curing or crosslinking of the adhesive is normally carried out in line with the adhesive coating operation, and consequently, the curing time needs to be kept as short as possible to enable the machine to run at economic speeds. Therefore, it is useful to the adhesive formulator to be able to study the comparative rate of curing of different adhesive systems. This can be effectively done by using the Wallace Shawbury Curometer. This instrument in effect comprises a small paddle which is oscillated between two cylindrically shaped specimens of adhesive. The specimens are placed between two electrically heated platens to effect curing of the adhesive. The oscillation of the paddle varies with the stiffness of the adhesive; the paddle is connected to a swinging stylus, which executes a magnified copy of the oscillation of the paddle. A chopper bar causes the stylus to mark a moving paper chart at the extremities of the oscillation. The chart then shows an envelope, which is illustrated in Fig. 10-9.

The widths of the envelope at any given line is in a measure if the deformation of the sample, which enables the stiffness of the sample and hence the state of cure to be examined. At the left side of the trace, it can be seen that the sample becomes less stiff for a short time. This is the period where the sample is heating up before curing starts, and on this trace,

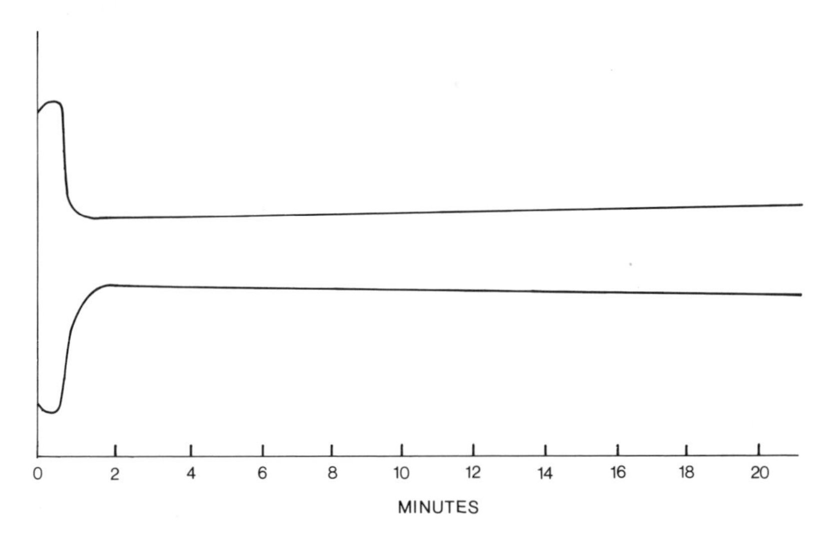


Fig. 10-9. Shawbury curometer trace, curing temperature 200°C.

curing is taking place almost immediately after this induction period as is shown by the rapidly increasing stiffness of the sample. As can be seen the sample is fully cured within 1 min at 200°C.

There are several curing systems for natural rubber adhesives of this type. Perhaps the most commonly used one is the phenol-formaldehyde resin cure systems. These have the advantage of imparting the required degree of cure at reasonable speeds, and they do not cause staining of any light-colored paints to which the tape may be applied. Phenolic resins of these kind must be compatible with natural rubber and heat advancing. Bemmels[25] has described such a resin based on the alkaline condensation product of nonyl phenol with *para*-formaldehyde. Bemmels states that it is necessary that the alkyl group of the alkyl phenol should contain at least eight carbon atoms to maintain good compatibility with the natural rubber. A number of commercially available phenol-aldehyde resins can be used in this application, including Amberol SF 137 (Rohm and Haas), Schenectady 1055 (Schenectady Chemicals, Inc.), or Catalin 9750 (Catalin Corporation).

As said earlier, the curing step is generally carried out in line with adhesive coating, so it is important that the curing reaction is carried out quickly to ensure economic production. Several methods have been proposed to accelerate the phenol-formaldehyde resin cure of pressure-sensitive adhesive systems. One accelerator system[26] claimed is the use of organometallic derivatives of elements of Group IV in the periodic table, for example zirconium tetraisopropoxide, or titanium tetrabutoxide.

Another accelerator is a compatible acid catalyst.[27] Acids are known to accelerate the phenol-formaldehyde resin cure reaction, but most acids are not compatible with rubber/resin adhesive systems and cause loss of tack. A compatible acid which can be used is octyl phenyl acid phosphate.

A more elegant solution to the problem of acceleration is provided in Johnson and Johnson's British Patent 1,172,670.[28] This invention claims the use of a halogenated phenol-formaldehyde resin, accelerated by zinc resinate. A typical formulation for an adhesive of this type is

| | |
|---|---|
| Natural rubber pale crepe | 100 |
| Piccolyte S 115 tackifier | 65 |
| Schenectady 1056[a] | 10 |
| Zinc resinate accelerator | 10 |
| Santovar A[b] antioxidant | 1 |
| Ionol[c] antioxidant | 1 |
| Zinc dibutyldithiocarbamate | 1 |

[a] Brominated phenol-formaldehyde resin.
[b] Registered trademark, Monsanto Co.
[c] Registered trademark.

The other method of curing to provide heat resistance to natural rubber adhesives is to cure by sulfur-donating compounds. A broad adhesive composition is quoted by Kisbany.[29]

| | |
|---|---|
| Natural rubber | 100 |
| Resin tackifiers | 50–120 |
| Inorganic activator (e.g., ZnO) | 25–50 |
| Inert filler ($CaCO_3$, $TiO_2$) | 35– 60 |
| Carbon black | 6–15 |
| Antioxidant | up to 1.5 |
| Sulfur or equivalent sulfur-donating compound | 0.5–2.25 |
| Zinc butyl xanthate | 0.5–3 |

Sulfur has the disadvantage of staining light colored paints. Consequently, sulfur is often replaced by a sulfur-donating compound such as alkyl phenol sulfide (e.g., Vultac No. 2, Sharples Chemicals) or thiuramtetrasulfide (Tetrone A, DuPont).

## Improved Solvent Resistance

Solvent resistance of pressure-sensitive adhesives is required in certain electrical applications, where the tape has been applied to electrical components which are subsequently dipped into varnish solutions. Typical varnish solvents are aromatic hydrocarbons (toluene) and alcohols (butanol).

Clearly, conventional uncured natural rubber-based adhesives have little resistance to hydrocarbon solvents. Degrees of curing described in the previous section, where the pressure-sensitive properties are still maintained, have slightly improved solvent resistance in comparison with uncured adhesives, but the elastomers can still be readily swollen by a solvent such as toluene. This problem is overcome by providing a tape with a pressure-sensitive adhesive which can be cured after the tape has been applied. After application, it is no longer necessary for the adhesive to remain pressure-sensitive, but it still must remain well-adhered to the adherend. The adhesive is in essence a thermosetting one.

The curing systems are generally similar to the phenol-formaldehyde cure formulations described in the section on heat resistance. In this case, larger quantities of phenol-formaldehyde resin are used in the formulation to obtain a higher degree of crosslinking. In a typical heat-resistant masking tape, the amount of phenol-formaldehyde resin would be on the order of 10 parts per hundred rubber, whereas in a thermosetting adhesive, this would be increased to 20–30 parts per hundred rubber. The degree of solvent resistance will depend on the time and temperature of the curing cycle, but a typical cure cycle to obtain maximum solvent resistance would be about 3 hr at 150°C. Phenol-formaldehyde curing resins are preferred for these thermosetting electrical applications, as they exhibit no curing of the rubber at room temperature, hence giving the tape a good shelf-life. Also they cause no corrosion of copper electrical conductors unlike sulfur curing agents.

Because of the high temperatures demanded in the curing cycle, rubber thermosetting pressure-sensitive adhesives have to be coated on to heat-resistant backings. The most widely used backing is biaxially-oriented polyester (e.g., DuPont Mylar). Others include glass cloth or specially impregnated papers. Methods of adhesive and tape manufacture are similar to those described in the section on manufacture of natural rubber pressure-sensitive adhesives.

## Improved Creep Resistance

As we have seen, uncured rubber-resin adhesives are perfectly adequate for many packaging and label applications. However, there are applications where it is necessary to improve the cohesive strength of the rubber-based adhesive. This is particularly the case with heavy-duty strapping tapes, where the adhesive can be subjected to considerable shear stress. Such tapes usually have a high tensile strength backing to provide the required strapping strength, and are based on films or paper with yarn or scrim reinforcement. More recently, these reinforced backings are being replaced by monoaxially-oriented polypropylene films, which give the required high tensile strength at greatly reduced cost.

Various techniques have been used to increase the cohesive strength of the adhesive. The phenol formaldehyde cure systems described in earlier sections cannot be used in these cases as the film backings are thermosensitive, and temperatures high enough to activate the phenol-formaldehyde resin cure cannot be attained.

One of the earliest examples of methods of increasing cohesive strength of natural rubber pressure-sensitive adhesives was the addition of di- or polyamines or hydrazines.[30] The addition of small quantities of compounds such as ethylene diamine, hexamethylene diamine, or hydrazine hydrate in quantities of 0.5–1% of natural rubber/$\beta$-pinene resin adhesives were claimed to give significant increases to shear resistance.

Another method that has been employed to improve creep resistance is the addition of titanate esters to the rubber resin adhesive.[31] Preferred additives include tetraisopropyl titanate and tetrabutyl titanate. The quantities of these materials added are generally in the range of 5 parts per hundred of natural rubber; increasing quantities tend to decrease the tack of the adhesive. One of the disadvantages of the titanate ester system is they tend to cause a very noticeable yellow discoloration to the adhesive layer.

Another system that can be used to increase the holding power of natural rubber adhesive is by crosslinking with poly or diisocyanates. Klepetar and Kucera[32] have shown that adhesives with good tack and improved shear resistance can be formulated with natural rubber in combination with a major amount of tackifying resin with the melting point above 105°C, a minor amount of tackifying resin of lower softening point and from 0.5 to 5% of a polyisocyanate. They claim that it is necessary to use the minor amount of low softening point resin in order to retain good tack in the shear resistant adhesive. Preferred tackifiers for the application are $\beta$-pinene resins of softening points of 115°C and 70°C. In this type of formulation, it is important to avoid the use of resins which will readily react with isocyanates, and in this respect the neutral terpene resins are ideal. The two isocyanates that are normally used in this type of system are toluene diisocyanate (TDI) and methylenebis-4 phenyl isocyanate (MDI). The former is generally more reactive, but suffers from a higher vapor pressure than MDI, which can cause handling problems with the very low concentrations of TDI. It is also important to remember that materials in the adhesive formulation should be free from water, as polyisocyanates react very readily with water.

Theoretically, it should be possible to increase the cohesive strength of rubber-resin adhesive by exposure to doses of high–energy electron beam radiation. Although a considerable amount of development has occurred in recent years in electron beam radiation equipment, little work appears to have been reported in its application to natural rubber adhesives. However, J.O. Hendricks[33] has described improvements that can be obtained in cohesiveness of rubber-resin adhesives by irradiation. By exposing a natural rubber Pentalyn A (Hercules Powder Co.) blend to varying doses of irradiation, the shear resistance measured at a temperature of 49°C was significantly increased. Using irradiation doses of 1–3 megarads at a distance of 10 cm from the adhesive surface, shear resistance of the adhesive at 49°C was increased from failure after 6 min to no failure after 22–35 min. Not only did the adhesive show an increase of shear resistance, but other properties of the adhesive, peel adhesion and tack, were not affected.

## Improved Electrical Conductivity

For many purposes, pressure-sensitive adhesives form a part of electrical insulating tapes. In this case, the conventional natural rubber adhesives that have been described show excellent electrical resistance. However, there are applications where it is desirable to have electrically conductive adhesive tapes; these tapes are generally based on a conductive metal foil backing (e.g., aluminum foil) coated on at least one side with a conductive adhesive layer. Such tapes can be used as solder replacements, pliable low voltage conductors, etc.

Electrical conductivity is imparted by incorporating conductive particles into the adhesive. One example[34] is to add milled aluminum spheres to a natural rubber/$\beta$-pinene blend. At a loading of 10.8 parts by weight per 100 of adhesive, the electrical resistance was lowered to 1.5 ohms/6.45 $cm^2$. However, the use of metallic particles does not always give very homogeneous results. An improved method[35] is to load the adhesive with conductive carbon black. Such a carbon black is Vulcan XC (Cabot Corporation) or acetylene black. In order to obtain reasonable electrical conductivity, it is necessary to use heavy loadings of carbon black—up to 70 parts of black for 100 parts of base adhesive material. To offset the effects of such heavy loadings on adhesive properties, especially tack, it is necessary to use a very soft base adhesive, which can be obtained by excessive milling of the natural rubber. This system gives homogeneous electrical properties, with electrical resistance which can be below 0.05 ohms/6.45 $cm^2$.

## EMULSION ADHESIVES

All the adhesives that have been discussed in this chapter have been made from an organic solvent solution. With increases in solvent costs and restrictive legislation on solvent emissions, interest in water emulsion natural rubber adhesives is being revived. Such adhesive systems have been known for some time. Eustis and Orrill[36] in 1940 described a pressure-sensitive adhesive comprising natural rubber latex, an emulsion of rosin ester together with a water absorptive agent such as casein.

Although the idea of using natural rubber emulsions is theoretically attractive, there are many practical difficulties. Whereas natural rubber is readily available as an aqueous latex, the dispersion of tackifier resins into the form of stable aqueous emulsions of consistent quality has presented problems. The more desirable tackifiers often have melting points above 100°C, and cannot therefore be readily emulsified by agitation with emulsifying agents at atmospheric pressure. Tackifiers first dissolved by solvents can easily be emulsified, but then of course, a totally solvent-free system will not be produced.

The next major problem is to obtain a tacky homogeneous film from the blended rubber tackifier emulsions. It will be appreciated that when resin and rubber emulsions are mixed, they remain as separate particles in the emulsion. In order to develop tack, the rubber and resin particles must dissolve each other, which is difficult to achieve consistently. Preferably, the drying operation needs to be carried out at a temperature above the softening point of the resins. But this cannot always be done, for example when the adhesive is coated on to a thermosensitive web, such as unplasticized PVC. The presence of a small quantity of a solvent in the emulsion will assist the development of tack. The consistent development of tack is also hindered by the fact that the molecular weight of natural rubber latex is high and variable; the rubber cannot undergo the milling step to break down the rubber molecules as is normally done in the solution coating technique.

One further disadvantage of the emulsion method is that the presence of the necessary emulsifying agents tend to create a haze in the finished adhesive film, an undesirable property in an adhesive that is applied to clear film such as cellophane or polypropylene.

Further work still needs to be done to achieve consistent pressure-sensitive properties from natural rubber latex systems. The goal of lower material costs, solvent-free emissions and lower total energy requirements will make this worthwhile.

## MAJOR APPLICATIONS

### Packaging Tapes

Perhaps the largest single use of natural rubber adhesives is in the field of pressure-sensitive packaging tapes. These products range from the well-known clear tapes used in the home to heavy-duty strapping tapes used in industry. For many years now, natural rubber has been the workhorse adhesive polymer on these tapes because of its unique balance of tack and cohesive strength combined with low cost.

The largest single use of packaging tapes today is the sealing of fiberboard cases. The penetration of the pressure-sensitive tape with this market has been more extensive in Europe than in the U.S., although market penetration in the United States is increasing. This market has been dominated in Europe by tapes based on unplasticized PVC film, coated with natural rubber adhesives at coating weights of 20–25 g/m$^2$. As this is a very cost-conscious market, production costs have to be kept to a minimum, and to date natural rubber tackified with synthetic hydrocarbon resins has represented the cheapest and most suitable adhesive systems. They meet the requirements of good tack, adhere immediately to the carton surface combined with good shear resistance to prevent the flaps of the carton opening up. Initially, these tapes were sold in comparatively short lengths of 50 or 60 meters and either applied by hand or by handheld tape applying devices.

Today, more and more cartons are sealed automatically on a packing line, applying a U seal of tape to top and bottom of the carton, using rolls of tape up to 1000 meters long to avoid frequent roll changes.

Figure 10-10 illustrates a carton sealer applying U seal to top and bottom of cartons, using 1000 m UPVC based tape coated with natural rubber-based adhesive.

Most of these tapes are brown, partly to match the cartons. This color is usually achieved with pigmentation of the adhesive layer. Within the last 1–2 years, polypropylene film has been increasingly used in this application due to its low cost. In the United States, the polypropylene film tape has occupied a dominant position among packaging tapes. Packaging tapes are used both in the home and industry for hundreds of applications, applied to many different kinds of surfaces, and the natural rubber-based adhesive has the advantage of displaying good tack to a wide variety of substrates, which make it a preferred polymer in these tapes.

## Masking Tapes

The most common and largest volume of masking tapes are those based on impregnated creped papers. These are mostly used in the holding of paper

Fig. 10-10. Packed cases being automatically sealed top and bottom with clear unplasticized poly (vinyl chloride) tape.

masks on vehicle bodies during spraying operations, both during the manufacture of vehicles, and perhaps more importantly, in vehicle repair shops. In the first case, paint stoving cycles of up to 40 min at 130°C can be encountered; in repair shops, lower temperatures of only 80°C are more common, and of course many paint spraying applications are air dried. For these types of tapes, natural rubber adhesives are still the most commonly used. Although it is common to blend the natural rubber with a compatible SBR polymer to improve aging, and the adhesives are nearly always cured, usually with the phenol-formaldehyde resin systems, discussed earlier. Figure 10-11 shows a masking tape being used to mask off the windows and stainless steel fittings of a fire tender.

However, there are other types of masking tapes and applications where natural rubber adhesives are used. These include a lithographic masking tape, where a semitransparent tape is required for a removable application to a photographic plate to mask out an area of the plate. Such

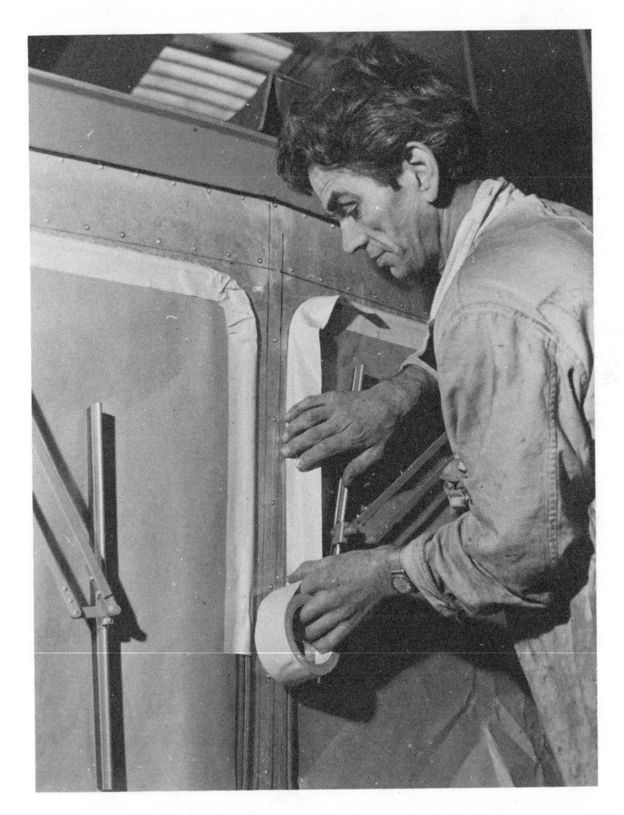

Fig. 10-11. Masking tape being applied to a fire tender prior to paint spraying.

tapes are made from semitransparent red films, cellulose acetate, or unplasticized poly(vinyl chloride), coated with a low tack natural rubber pressure-sensitive adhesive.

## Double-sided Tapes

One of the most rapidly expanding applications for pressure-sensitive adhesives is in the field of double-sided tapes. These products are finding an increasingly diverse range of uses in temporary or permanent mounting or splicing applications. These tapes usually comprise a carrier web, which can be either a film, paper, or fabric, coated on both sides with the pressure-sensitive adhesive, with an interleave to keep the two adhesive layers apart. The interleave is commonly paper siliconized on both sides to facilitate easy removal in use.

Double-sided tapes range from heavy-duty structures, using thick aggressive adhesive layers on fabric or paper webs to very thin products based on thin films. For many of the applications, natural rubber-based adhesives are still used because of their good all-around balance of tack and cohesive combined with low cost. Heavier duty products are used for applications which include mounting of flexographic printing stereos to cylinders, where it is essential to be able to remove the rubber stereo from the tape easily in order to reposition it properly.

Another typical application of this kind of heavy-duty product is for a carpet tape. Here, the double-sided tape is used to hold the carpet to the floor; a high adhesive mass with excellent tack is essential in order to be able to adhere to the very rough surface of the carpet backing.

Such tapes with heavy coatings of natural rubber adhesives ($>100$ g/m$^2$) are normally applied by a hot calendering operation, as it would be uneconomical to apply such coating weights from solvent.

Lighter weight products are usually based on thin films such as unplasticized poly(vinyl chloride), polypropylene, polyester, or cellulose, for many bonding, splicing, securing, and holding jobs in all industries as well as in the home. Again, the all-around properties of natural rubber adhesive, which shows good adhesion to many surfaces, make it a preferred system for such a wide range of applications. Figure 10-12 shows a double-sided thin unplasticized poly(vinyl chloride) film tape being used to hold the insole to the cast in the manufacture of ladies' shoes.

Speciality double-sided tapes include those with thermosetting adhesives to give improved heat and solvent resistance. These have many uses in electrical component construction, both as a temporary or permanent holding medium, particularly in stick-wound coil production.

Fig. 10-12. Double-sided unplasticized poly (vinyl chloride) film tape holding insoles to the last in shoe manufacture.

## Electrical Tapes

Two types of natural rubber adhesives are used on electrical tapes. First, they are uncured, noncurable rubber-resin systems on conventional insulating tapes. In these cases, the adhesives are coated on either plasticized PVC or polyethylene films; these substrates are very flexible and conformable, have good electrical insulation resistance and are ideal for general-purpose insulation tapes where operating temperatures do not exceed 80°C. Both films require a primer coat between film and adhesive; care must be taken with the plasticized PVC film due to the potential migration of the plasticizer from the film into the adhesive layers, causing it to become excessively soft and lose adhesion.

The second type of rubber-resin adhesive used in electrical applications is a thermosetting adhesive, the compositions of which have been previously described. This system is used in tapes used in electrical component manufacture, where the tape has to withstand a degree of solvent immersion (from varnish dipping) and subsequent high working temperatures. These adhesives are coated on temperature-resistant backings such as polyester film, or specially impregnated papers, flat or creped. The adhesive is cured after the tape has been applied to the component at temperatures between 100 and 150°C for up to 3 hr. After this thermosetting cycle, tapes based on the polyester films are capable of withstanding temperatures of 180°C for short periods and continuous operating temperatures up to 130–155°C. This is the highest continuous working temperature that can be achieved from a natural rubber-based adhesive. If higher temperature resistance is required, it is necessary to use either an acrylic or silicone adhesive system. Figure 10-13 shows a polyester thermosetting tape securely holding lead outs in position on a wire-wound component.

## Protection Masks

This application has also been a rapidly growing one for pressure-sensitive adhesives in recent years. These materials comprise either a filmic or paper

Fig. 10-13. Polyester thermosetting tape being used to hold lead-outs in a wire-wound component.

base web coated with a low tack adhesive. The protection masks are applied to a variety of substrates such as sheet metal, plastic, etc., to protect the metal or plastic surface against damage during fabrication, storage, transit, etc. The tape must have sufficient adhesion to adhere to the surface, but subsequently be capable of being easily removed, without leaving an adhesive residue on the metal or plastic surface. Protection mask materials are made in widths up to 2 meters, so it is essential for the mask to have a low tack adhesive for easy removal.

Rubber-resin adhesives are used in some of these protection masks, generally coated on thin polyethylene or unplasticized poly(vinyl chloride) films. Low tack is achieved by the use of lower quantities of tackifier resin (15–30 parts per hundred rubber) and lower coating weights (10–15 $g/m^2$), compared with 20–25 $g/m^2$ for a packaging tape).

The use of natural rubber adhesives is restricted in this application by the comparatively poor aging properties of natural rubber in comparison with more saturated polymers such as the acrylates, polyisobutylene, etc. Consequently, protection masks with natural rubber adhesives are generally restricted to those applications where only short-term indoor protection is required, for example, protection of stainless steel during drawing fabrication into vegetable dishes, kitchen sinks, etc. Aging can be improved by the use of an opaque pigmented film which will give protection to the adhesive against light degradation.

Protection masks also offer the opportunity for the use of aqueous natural rubber systems, as only a low degree of tack is required on these products. This can be readily achieved with natural rubber latex, usually together with a minor amount of tacky polymers in emulsion form. These adhesives are usually coated on flat papers to make the protection mask.

## Pressure-Sensitive Labels

Natural rubber adhesives have been widely used in the manufacture of pressure-sensitive label stocks; the largest volume of these have been labels based on unimpregnated paper, coated on one side with the pressure-sensitive adhesive and the adhesive layer covered with a siliconized paper. For certain special applications, films of various kinds can replace the label paper. Although there are many variations in kind of adhesive labels, they mostly fall into two categories, removable and permanent. The former has a less aggressive adhesive which allows the label to be peeled off a surface after it has been applied; the latter has an adhesive with high peel adhesion values, so that it is not possible to remove the label without delaminating it.

Natural rubber adhesives have been widely used for both types of adhesive. Rubber-resin adhesives are amenable to the two kinds of adhesive required; label adhesives do not require to have particularly good shear resistance, but they do need to be capable of being cut cleanly when the shaped label is being die stamped. Because the adhesives do not require good cohesive strength, this particular sector of the natural rubber pressure-sensitive adhesive industry has been more easily penetrated by hot melt applied adhesives based on EVA or SBR block copolymers or by acrylic emulsion systems.

## SUMMARY AND FUTURE TRENDS

Natural rubber has been the main workhorse polymer of the pressure-sensitive adhesive industry for many years. The wide range of properties obtainable from rubber-resin adhesives, their low cost, and good availability have meant that these adhesives have been, and are being used in tapes and labels of nearly all kinds. The main limitations is imposed by the unsaturated *cis*-polyisoprene molecule which makes it unsuitable for use at very high temperatures or for long aging, nonyellowing applications.

Today, the adhesive industry is undergoing more changes and is faced with more challenges than ever before. Natural rubber adhesives have historically been coated from solvent systems, which today suffer from rapidly rising costs, environmental difficulties and the fact that it is the highest energy-consuming process of all adhesive coating methods. Consequently, natural rubber adhesives are under threat and in some cases being replaced by acrylic emulsion or hot melt applied thermoplastic copolymer adhesives. With the current economic and environmental situations, this replacement process is likely to continue and accelerate. However, natural rubber adhesive systems are likely to be retained for many special applications, for which the newer polymer systems are not yet adequate. If natural rubber is to retain its dominant share of the pressure-sensitive adhesive market in the future, current solvent and energy costs and environmental considerations indicate that this will only happen with the successful development of emulsion-based rubber-resin adhesives.

## REFERENCES

1. Shecut, W. H., and Day, H. H. U. S. Patent 3,965 (1845).
2. Drew, R. G. British Patent 312,610 (1928).
3. Drew, R. G. British Patent 405,263 (1931).
4. British Patent 425,343 (1933) (assigned to Minnesota Mining and Manufacturing Co.).
5. British Patent 425,159 (1933) (assigned to Minnesota Mining and Manufacturing Co.).
6. British Patent 514,402 (1938) (assigned to Minnesota Mining and Manufacturing Co.).
7. Wetzel, F. H. *Rubber Age* **82**: 291–295 (1957).
8. Wetzel, F. H., and Alexander, B. B. *Adhesives Age* **7** (1): 28–38 (1964).

9. Bull, R. F., *et al. Adhesives Age* **11**; 20 (1965).
10. Dahlquist, C. A. U. S. Patent 3,681,190 (1970) (assigned to Minnesota Mining and Manufacturing Co.).
11. British Patent 650,001 (1947) (assigned to Johnson and Johnson).
12. Samour, C. M. U. S. Patent 3,162,610 (1961) (assigned to Kendall Co.).
13. Blackford, B. B. U. S. Patent 2,909,278 (1957) (assigned to Johnson and Johnson).
14. Bemmels, B. W. U. S. Patent 2,615,059 (assigned to Johnson and Johnson).
15. British Patent 694,190 (1949) (assigned to Minnesota Mining and Manufacturing Co.).
16. Wetzel, F. H. *ASTM Bulletin* **221**: 64 (1957).
17. Hoch, C. W., and Abbott, A. N. *Rubber Age* **82**: 471–475 (1957).
18. Wittock, C. *J. Polymer Sci.* **C** No. (3) 139–149 (1963).
19. Whitehouse, R. S., Counsell, P. J. C., and Lewis, C. *Polymer* **17**: 699–702 (1976).
20. Dahlquist, C. A. *Adhesives Age* **2**: 25 (1959).
21. Dahlquist, C. A. *Adhesives Fundamental & Practice.* Ministry of Technology, Maclaren, London (1969).
22. Sherriff, M., Knibbs, R. W., and Langley, P. G. *J. Appl. Polym. Sci.* **17**: 3423–3438 (1973).
23. Aubrey, D. W., and Sherriff, M. *J. Polymer Sci.* **C(16)**: 2631 (1978).
24. Aubrey, D. W., and Sherriff, M. Paper accepted for publication in *J. Polymer Sci.* (1980).
25. Bemmels, B. W. U. S. Patent 2,987,420 (1964) (assigned to Johnson and Johnson).
26. British Patent 960,509 (1963) (assigned to Johnson and Johnson).
27. British Patent 975,971 (1964) (assigned to Johnson and Johnson).
28. British Patent 1,172,670 (1969) (assigned to Johnson and Johnson).
29. Kisbany, F. N. U. S. Patent 2,881,096 (1959) (assigned to American Tape Co.)
30. British Patent 779,256 (1954) (assigned to Permacel Tape Corp.).
31. Crocker, G. J. British Patent 848,455 (1955) (assigned to Johnson and Johnson).
32. Klepetar, M., and Kucera, C. R. British Patent 1,234,860 (1969) (assigned to Johns-Manville Corp.).
33. Hendricks, J. O. U. S. Patent 2,956,904 (1960) (assigned to Minnesota Mining and Manufacturing Co.).
34. British Patent 1,169,946 (1966) (assigned to Minnesota Mining and Manufacturing Co.).
35. Stowe, R. H. U. S. Patent 3,778,306 (1971) (assigned to Minnesota Mining and Manufacturing Co.).
36. Eustis, W., and Orrill, G. R. U. S. Patent 2,429,223 (1940) (assigned to Kendall Co.).

# Chapter 11

# Block Copolymers

**William H. Korcz, David J. St. Clair, Earle E. Ewins, Jr.**
*Shell Development Co., Houston, Texas*

**and Dirk de Jager**
*Koninklijke/Shell-Laboratorium, Amsterdam, Holland*

Thermoplastic rubbers of the ABA block copolymer type, where A represents a thermoplastic polystyrene endblock and B represents a rubber midblock of polyisoprene, polybutadiene, or poly(ethylene/butylene), form a useful and versatile group of polymers for pressure-sensitive adhesives.

Because of their unique structure, thermoplastic rubbers offer processing advantages to the adhesive manufacturer as well as providing required performance in a wide range of formulated pressure-sensitive adhesive tape and label products. In preparing solvent-based adhesives, thermoplastic rubbers require no premastication, as they dissolve rapidly in common solvents. Usually finished thermoplastic rubber adhesives can be coated at much higher solids contents than adhesives based on conventional rubbers, as solution viscosities are lower. Also, because these rubbers are thermoplastic, they allow preparation of adhesives which can be mixed and coated as solvent-free, hot melt pressure-sensitive adhesives.

## NATURE OF THE BASIC MOLECULE

### Architecture

The simplest thermoplastic rubber consists of a rubbery midblock with two plastic polystyrene endblocks. This is pictured schematically in Fig. 11-1 where the diamonds represent monomer units in the polystyrene endblocks

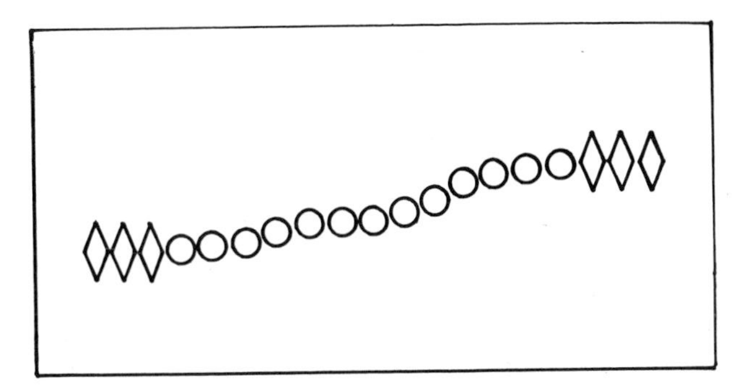

Fig. 11-1. Simplified representation of a thermoplastic rubber molecule.

and the circles represent monomer units in the rubbery midblock. There are two basic classes of thermoplastic rubber. One class consists of block polymers in which the rubbery midblock of the molecule is an unsaturated rubber. The two types of polymers in this class are the polystyrene-polybutadiene-polystyrene (S-B-S) polymers and the polystyrene-polyisoprene-polystyrene (S-I-S) polymers. This class of thermoplastic rubber was first marketed commercially in 1965 by Shell Chemical Company. The second class of thermoplastic rubber consists of block polymers in which the elastomeric midblock is a saturated olefin rubber. The polymers in this class are the polystyrene-poly(ethylene/butylene)-polystyrene (S-EB-S) polymers and the polystyrene-poly(ethylene/propylene)-polystyrene (S-EP-S) polymers. This class of thermoplastic rubber was first marketed commercially in 1972 by Shell Chemical Company.

The usual method of preparation of thermoplastic rubbers is by anionic polymerization in solvent using an alkyl lithium catalyst such as butyl lithium. S-B-S and S-I-S type block polymers may be synthesized by two basic routes: sequential polymerization of all three blocks or sequential polymerization of two blocks, followed by coupling. These two processes are illustrated below, the sequential process in Equation (1) and the coupling process in Equation (2), for the preparation of an S-I-S type polymer.

$$\text{BuLi} \xrightarrow{\text{Styrene}} \text{S-Li} \xrightarrow{\text{Isoprene}} \text{S-I-Li} \xrightarrow{\text{Styrene}} \text{S-I-S} \tag{1}$$

$$\text{S-I-Li} \xrightarrow{\text{Coupling Agent}} \text{S-I-I-S} \tag{2}$$

The sequential process results in what is called a linear S-I-S polymer. With the coupling process, use of a difunctional coupling agent gives a linear $(\text{S-I})_2$ polymer, while use of a multifunctional coupling agent gives a multi-armed $(\text{S-I})_n$ polymer. Usual coupling agents are esters or halogen containing molecules. Literature reports catalytic hydrogenation of S-B-S and S-I-S polymers to produce saturated analogues of the S-EB-S and S-EP-S types, respectively.[1]

Many variations can be made in the structure of a thermoplastic rubber. Among these are variations in molecular weight, styrene content, monomers used in the polymerization and the number of polymer arms coupled in the coupling reaction. The key requirement for a thermoplastic rubber is that the rubber midblock, having a glass transition temperature $T_g$ well below room temperature, must have terminal endblocks of a hard, glasslike plastic, also having a glass transition temperature well above room temperature. An additional requirement is that the plastic endblocks must be thermodynamically incompatible with the rubber midblock. When these requirements are fulfilled, the polymer can consist of two phases: a continuous rubber phase and a basically discontinuous plastic phase. A highly idealized representation of this phase separated structure is given in Fig. 11-2.

Domains, the plastic endblock phase, are shown in Fig. 11-2 in spherical form. These domains act as crosslinks between the ends of many rubber chains, thereby locking the rubber chains and their inherent entanglements in place. Thus, the thermoplastic rubber behaves like a conventionally vulcanized rubber that contains dispersed reactive filler particles. However, the thermoplastic rubber is physically crosslinked by the plastic endblock domains rather than being chemically crosslinked like a conventionally vulcanized rubber. As such, these physical crosslink sites (domains) can be reversibly unlocked and reformed by various means, i.e., solvation followed by solvent evaporation or through sufficient heating and shearing, then cooling.

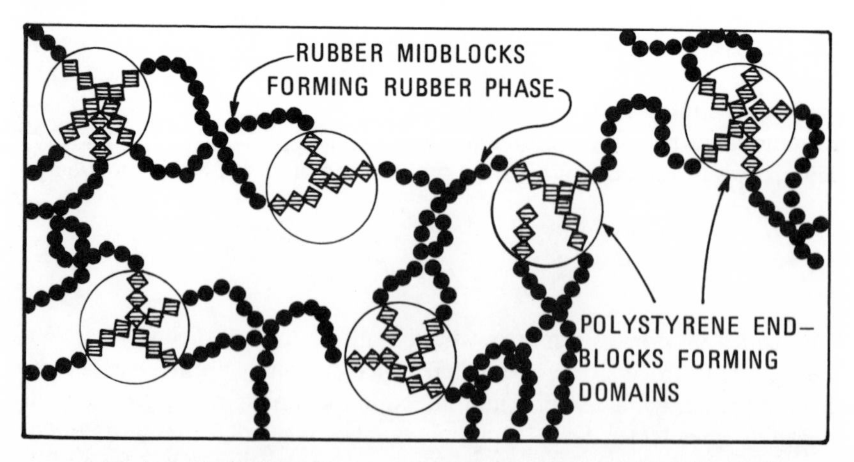


Fig. 11-2. Idealized two-phase network of thermoplastic rubber molecules.

The behavior of thermoplastic rubber is highly dependent on the morphology or geometry of the endblock domains as they are dispersed in the rubber matrix. In commercial thermoplastic rubbers useful for pressure-sensitive adhesives, the endblock phase is present in the smaller proportion and is dispersed in a continuous rubber matrix as suggested by Fig. 11-3. The uniform dispersion of spherical endblock domains shown in this figure (A, spheroids), however, is approached only in carefully prepared laboratory samples with low endblock phase concentration. Depending on the endblock phase concentration and on actual processing conditions used to prepare a given sample, the geometry or morphology of the dispersed phase may be spheroidal, rodlike, or lamellar as depicted in Fig. 11-3. The existence of these different morphologies has been confirmed by electron micrographs.[2,3]

In the latter two cases, the endblock phase may extend as a continuous plastic network throughout the rubber matrix. This tends to be the case when the endblock phase concentration is above about 20%w for neat S-B-S thermoplastic rubbers. In this situation, as the sample is stretched, the initial stress is borne by the plastic network, and the stress-strain properties are greatly affected as discussed below.

Under all conditions, the dimensions of the dispersed phase are restricted. Since the junctures between the endblocks and midblocks are located at the phase boundary, the domain thickness is limited to the distance which can be reached by the endblock segments extending into the domains from opposite sides. For commercial polymers, domain thicknesses have been both calculated and measured to be a few hundred angstrom or only a small fraction of the wavelength of visible light. If the domains are spheroidal, they do not scatter light. Thus, these thermoplastic rubbers are generally transparent, in spite of the large differences in refractive index between the two phases. With rodlike or lamellar morphologies, some light scattering and turbidity may exist.

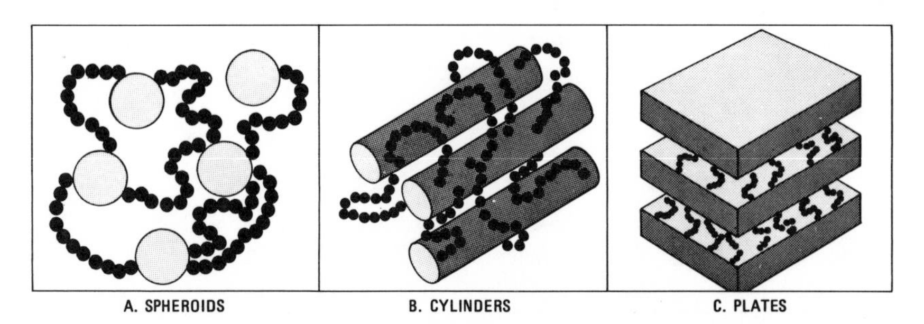


Fig. 11-3. Polystyrene domain configurations or morphologies.

### Properties of Neat Thermoplastic Rubbers

Thermoplastic rubbers are sold commercially in the form of powder, crumb, porous pellets or dense pellets. Bulk densities range from about 240 $kg/m^3$ (15 $lb/ft^3$) for crumb to about 480 $kg/m^3$ (30 $lb/ft^3$) for pellets. Some thermoplastic rubbers are dusted with about 1% of a dusting agent, such as talc, to minimize the tendency to block on long-term storage under warm conditions. The specific gravity of a solvent cast or molded film of the thermoplastic rubber is about 0.92.

Hardness and tensile properties of the thermoplastic rubber depend on the styrene content and the type of rubber midblock in the polymer. The styrene content of thermoplastic rubbers used in pressure-sensitive adhesives is usually between about 15%w and 30%w. Hardness varies from about 35 Shore A for a 15%w styrene S-I-S polymer to about 75 Shore A for a 30%w styrene S-EB-S polymer. The stress/strain curves for these two polymers are shown in Fig. 11-4. Both polymers show stress/strain properties characteristic of vulcanized rubbers. However, the soft, 15%w styrene S-I-S polymer has a low 300% modulus, 0.7 MPa (100 psi), a moderate tensile strength, 20 MPa (3000 psi), and high elongation at break (1400%). The harder, 30%w styrene S-EB-S polymer has a much higher 300% modulus, 5.9 MPa (850 psi), a high tensile strength, 41 MPa (6000 psi), and much shorter elongation at break (600%). The hardness and tensile properties of other neat thermoplastic rubbers are generally intermediate between the extremes of the two polymers just described.

Solution viscosities of thermoplastic rubbers depend strongly on the solvent composition. This behavior will be discussed later in this chapter. However, in a solvent such as toluene, which dissolves both the polystyrene endblocks and the rubber midblock, solution viscosities are relatively low. The solution viscosity of a 25%w solution of most thermoplastic rubbers in toluene will be from about 1.0 to 10 Pa·s at 25°C.

Melt viscosities of neat thermoplastic rubbers can be extremely high at low shear rates. Typically, melt viscosities are $10^4$ to $10^5$ Pa·s at about 170°C and a 1 $sec^{-1}$ shear rate. As will be discussed later in this chapter, melt viscosities at higher shear rates can be several orders of magnitude lower.

## GENERAL FORMULATING PRINCIPLES

### Phase Association of Ingredients and Morphology

As is the case with many conventional rubbers used in pressure-sensitive adhesives, such as natural rubber and SBR, thermoplastic rubbers have

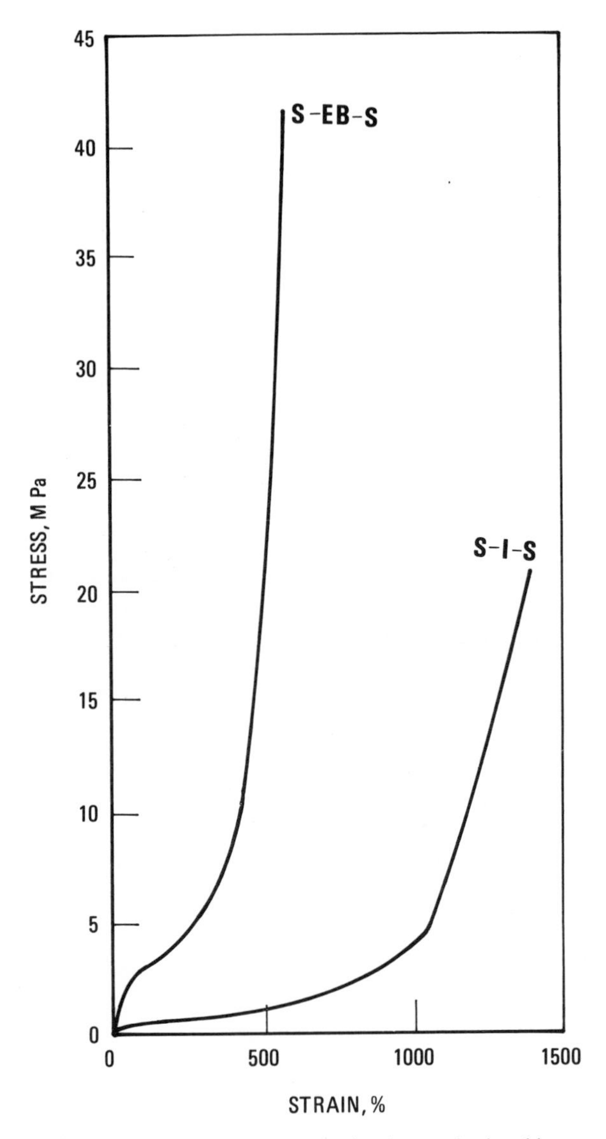


Fig. 11-4. Typical stress-strain curves for thermoplastic rubbers.

little inherent tack. Therefore, the rubber must be compounded with tackifying resin and, in some cases, with plasticizer to develop the required tack. However, since thermoplastic rubbers are two-phase systems, more care must be taken in formulation development with thermoplastic rubbers than with conventional, single-phase homopolymer or random copolymer rubbers.

In homopolymers or random copolymers, additives have only one phase with which to interact. In thermoplastic rubbers, additives have two phases with which to interact. Thus, an equilibrium will be established among the additives, the rubber network and the polystyrene domains. Possible interactions include an association of the additive with the rubber phase, with the polystyrene domains, with both phases, or with neither phase (resulting in formation of an independent third phase).

A simple but qualitative test which can be used to determine with which phase of a thermoplastic rubber a given resin is compatible is the following. Cast thin films from toluene solutions of a 1 : 1 mixture of resin and crystal grade polystyrene and of a 1 : 1 mixture of resin and polyisoprene or polybutadiene. Clarity of the dried film indicates mutual solubility; turbidity indicates incompatibility and phase separation.

The importance of the knowledge of the compatibility of a resin with each phase of the thermoplastic rubber lies in the effect of an endblock compatible resin and a midblock compatible resin on the stiffness and stress-strain properties of the rubber. The stress-strain properties of a thermoplastic rubber formulation are closely related to the morphology of the endblock phase and the midblock phase which is related to the volume ratio of the two phases.

**Effect of Phase Ratio.** As the endblock concentration in an unfilled thermoplastic rubber increases, the shape of the stress-strain curves change as shown in Fig. 11-5. These particular curves apply to solution-cast films prepared from a series of experimental S-B-S polymers. The total molecular weight for each polymer was held constant; only the styrene content was varied.

At polystyrene endblock concentrations of 20–30%w, the stress-strain curve resembles that of a vulcanized rubber. At concentrations above and about 33%w, the phenomenon of "drawing," commonly exhibited by thermoplastics, appears. This occurs when a continuous rodlike or lamellar endblock network exists as suggested in Fig. 11-3. When such a sample is stretched, an initial yield stress is observed. Then, as the relatively weak plastic structure is disrupted by further elongation, drawing occurs. When the stress is released, the plastic network will gradually reform. Higher temperatures will speed reforming of the plastic network. At higher endblock concentrations, the plastic phase is continuous and the midblock phase dispersed to give a system resembling high impact polystyrene.

**Effects of Additives.** The ratio of endblock phase to midblock phase can be varied by adding materials which associate preferentially with one

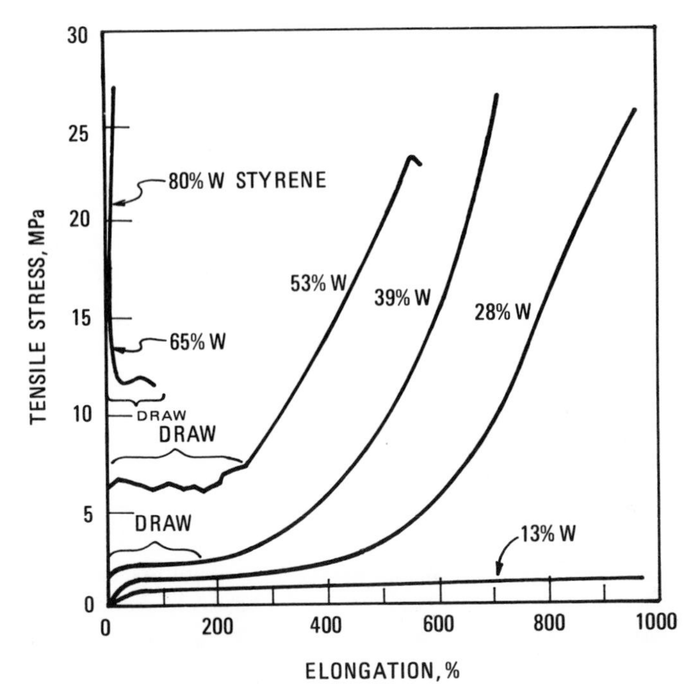


Fig. 11-5. Effects of polystyrene endblock concentration on linear S-B-S polymer tensile properties.

phase or the other. Coumarone-indene resins, for example, associate with the endblock phase in S-B-S and S-I-S polymers. Figure 11-6 shows how the initial portion of the stress-strain curve of the neat polymer (Curve B) is shifted (Curve A) when the endblock phase concentration is increased by adding an endblock resin. Curve C shows that the reverse occurs when a tackifying resin or a plasticizing oil which associates with the rubber phase is added to the neat polymer.

**Effect of Solvent Blends.** Because of the two-phase nature of thermoplastic rubbers, physical properties of films can be altered in a third manner. Films cast from mixed solvent systems best demonstrate the principle; in general, the properties will be determined by the component of the solvent system which evaporates last.

Figure 11-7 shows stress-strain curves for films of an S-EB-S rubber cast from different solvents. Curve A is for a film cast from a 60/40%v blend of *n*-hexane/MEK. In this case, the last component to evaporate is MEK, which dissolves the polystyrene phase only. Thus, the rubber midblock phase precipitates first and the polystyrene phase last, so that a continuous polystyrene phase tends to be formed rather than discrete polystyrene do-

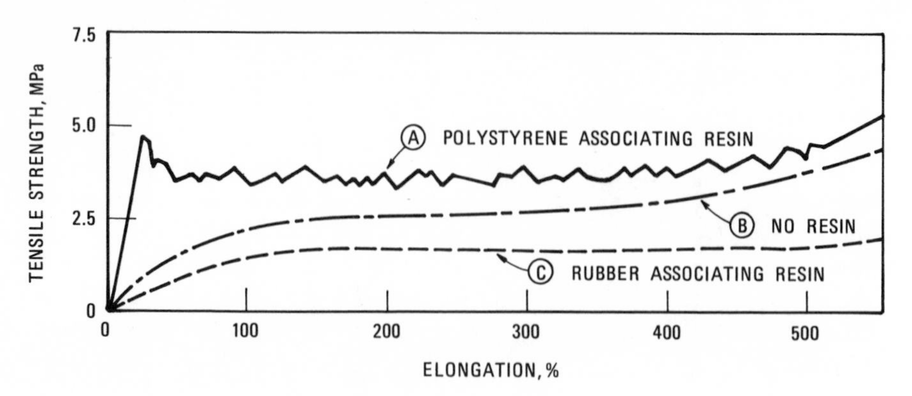


Fig. 11-6. Effect of polystyrene endblock and rubber midblock associating resins on tensile properties of thermoplastic rubber.

mains. The curve shows an initial yield, followed by cold drawing, and a final rise in stress to break point.

Curve B shows the stress-strain behavior for the film cast from solution in toluene, a relatively good solvent for both phases. This curve represents a structure close to an equilibrium state because the solvent is a good solvent for both phases.

Curve C is for a film cast from a 16/20/64%v blend of ethylbenzene/*n*-

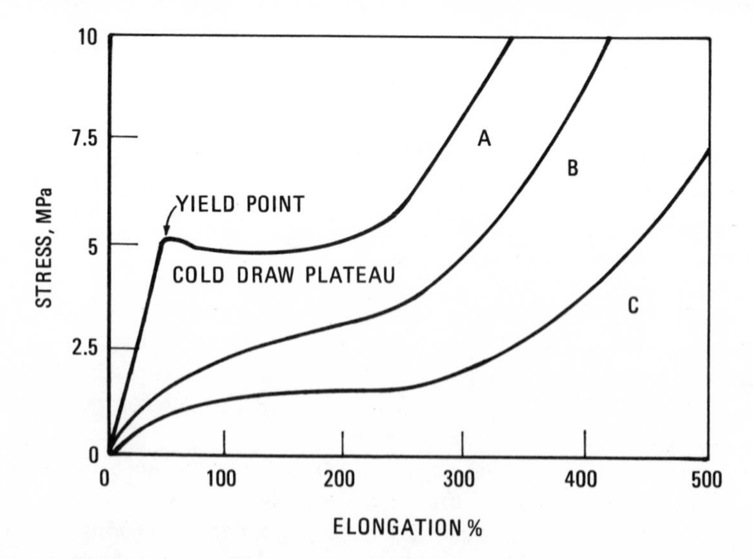


Fig. 11-7. Effect of solvent systems on physical properties of solution cast films of an S-EB-S polymer (Kraton® G 1650 rubber) at 23°C.

butyl acetate/Shell Sol 340EC. In this case, the last solvent to evaporate is Shell Sol 340EC, a very poor solvent for polystyrene, and the polystyrene is precipitated first, forming more discrete domains and fewer styrene continuous regions.

Properties intermediate to those shown by the three curves in Fig. 11-7 can be achieved by proper choice of solvents and solvent blends. Annealing the films represented by Curves A and C will tend to shift both curves toward the intermediate equilibrium position of Curve B. The rate at which the films anneal is temperature-dependent. Annealing effects may be encountered in processes where the film passes through a high-temperature drier.

## Choice of Compounding Ingredients

Resin selection is probably the most important aspect in the development of pressure-sensitive adhesive formulations based on thermoplastic rubbers. When considering the effects of resins on adhesive properties, the distinction must be made between the effects of midblock compatible resins and the effects of endblock compatible resins, since their effect on tack is widely different.

**Rubber Phase Associating Resins.** Aliphatic olefin derived resins, rosin esters, polyterpenes, and terpene phenolic resins derived from petroleum or turpentine sources and having relatively low solubility parameters tend to associate with the rubber midblocks and not with the polystyrene endblocks. As the molecular weights or softening points of these resins are lowered, their solubility in the endblocks increases. A partial list of commercial resins in this category is presented in Table 11-1.

The primary function of a midblock compatible resin is to develop tack in the thermoplastic rubber. This type of resin can also serve to improve the specific adhesion of the midblock phase toward polar substrates, as processing aids for the midblock phase, and to alter the modulus of the adhesive. Addition of predominantly midblock–associating resins tends to soften the compound by decreasing the modulus and reduces the tendency to draw as discussed earlier. These latter effects occur as the addition of midblock resin increases the fraction of the midblock phase in the system, causing the morphology of the dispersed endblock phase to become more spheroidal (less continuous).

Midblock compatible resins in this category with glass transition temperatures greater than that of the rubber midblock phase will increase the glass transition temperature of the rubber-resin blend. This will result in reduction of low temperature tack and flexibility. Partial compensation for this reduction can be obtained through the selection of midblock plasti-

**Table 11-1. Rubber phase associating resins for thermoplastic rubbers.**

| CHEMICAL TYPE | SOFTENING POINT (°C) | TRADE NAME | SUPPLIER |
|---|---|---|---|
| Polymerized mixed olefin | 100,115,130 | Eastman Resin | Eastman Chemical |
| Polymerized mixed olefin | 100,120 | Escorez 1300 series | Exxon Chemical |
| Polymerized mixed olefin | 100 | Hercotac 95 | Hercules |
| Polymerized mixed olefin | 85 | Imprez T85 | ICI America |
| Polymerized mixed olefin | 80,130 | Nevtac | Neville |
| Polymerized mixed olefin | 100 | Piccopale HM-200 | Hercules |
| Polymerized mixed olefin | 100 | Piccotac 95 BHT | Hercules |
| Polymerized mixed olefin | 85 | Quintone N180 | Nippon Zeon |
| Polymerized mixed olefin | 100 | Sta-Tac | Reichhold |
| Polymerized mixed olefin | 100 | Sta-Tac R | Reichhold |
| Polymerized mixed olefin | 100 | Super Nevtac 99 | Neville |
| Polymerized mixed olefin | 80,100 | Super Sta-Tac | Reichhold |
| Polymerized mixed olefin | Liquid[a], 75,115 | Wingtack series | Goodyear |
| Rosin Esters | | | |
| Glycerol ester of highly stabilized rosin | 82 | Foral 85 | Hercules |
| Pentaerythritol ester of highly stabilized rosin | 94 | Foral 105 | Hercules |
| Methyl ester of hydrogenated rosin | Liquid[a] | Hercolyn D | Hercules |
| Pentaerythritol ester of hydrogenated rosin | 104 | Pentalyn H | Hercules |
| Triethylene glycol ester of hydrogenated rosin | Liquid[a] | Staybelite Ester 3 | Hercules |
| Glycerol ester of hydrogenated rosin | 114 | Staybelite Ester 10 | Hercules |
| Polyterpenes and Derivatives | | | |
| Polyterpene | 85,135 | Piccolyte A | Hercules |
| Polyterpene | Liquid[a] | Piccolyte S10 | Hercules |
| Polyterpene | 85,115 | Zonarez 7000 series | Arizona Chemical |
| Polyterpene | 10,125 | Zonarez B series | Arizona Chemical |
| Terpene phenolic | 100 | Piccofyn A-100 | Hercules |
| Modified polyterpene | 110 | XPS 502 | Hercules |
| Low Softening Point Endblock Resins | | | |
| Alkylaryl | Liquid[a] | Piccovar AP10, AP25 | Hercules |
| Hydrogenated Resins | | | |
| Saturated alicyclic hydrocarbon | 85,125 | Arkon P series | Arakawa |
| Hydrogenated mixed olefin | 85,125 | Escorez 5000 series | Exxon |

[a] Because of their low molecular weight, these resins are soluble in both the endblock and midblock phases of thermoplastic rubbers.

cizers that have very low $T_g$'s. However, using low molecular weight plasticizer may also have an effect on the upper service temperature as the result of some unavoidable plasticizing of the endblocks. The importance and application of altering the $T_g$ of thermoplastic rubber compositions is discussed more fully under low temperature labels and tapes in the latter part of this chapter.

**Endblock Phase Associating Resins.** Polyaromatics, coumarone-indene resins and other high solubility parameter resins derived from coal tar or petroleum and having softening points above about 85°C tend to associate with the polystyrene endblocks and not with the rubber midblocks. As the molecular weights or softening points of these resins are lowered, their solubility in the midblocks increases. A list of typical and commercial endblock associating resins is presented in Table 11-2.

Addition of nonplasticizing endblock-associating resins tends to stiffen the compound by increasing the modulus and increasing the tendency to draw as discussed previously. This occurs as the fraction of endblock phase increases and its morphology becomes more rodlike or lamellar. In pressure-sensitive adhesives, only a limited amount of high softening point endblock compatible resin can be included in the formulation, before the stiffness increases and the aggressive tack is reduced to unacceptable levels. Somewhat higher levels of endblock resin can be included if the adhesive also contains a plasticizer to offset the reduction in tack.

Higher softening point resins in this category may also tend to increase the glass transition temperature $T_g$ of the endblock phase. This results in the

**Table 11-2. Endblock phase-associating resins for thermoplastic rubbers.**

| CHEMICAL TYPE | SOFTENING POINT (°C) | TRADE NAME | SUPPLIER |
|---|---|---|---|
| Polyalphamethylstyrene | 100,115,145 | Amoco 18 series | Amoco Chemical |
| Coumarone-indene | 155 | Cumar LX-509 | Neville |
| Coumarone-indene | Liquid[a] | Cumar P-25 | Neville |
| Coumarone-indene | 100 | Cumar R-16 | Neville |
| Alkylated aromatic hydrocarbon | 140 | Nevchem 140 | Neville |
| Heat-reactive hydrocarbon | 150 | Nevindene LX-685-180 | Neville |
| Polyalkylated aromatic polyindene | 70,140 | Picco 6000 series | Hercules |
| Polystyrene | 100 | Piccolastic D-150 | Hercules |
| Polyalphamethylstyrene/vinyl toluene copolymer | 100,120 | Piccotex | Hercules |
| Alkyl aromatic polyindene | Liquid[a] | Piccovar AP10, AP25 | Hercules |
| Alkyl aromatic polyindene | 130 | Piccovar 130 | Hercules |

[a] Because of their low molecular weight, these resins are soluble in both the endblock and midblock phases of thermoplastic rubbers.

formulation retaining its cohesive strength at higher temperatures. Low softening point endblock resins have the reverse effect, namely, to lower the tensile strength at elevated temperatures, to cause the tack in heat-activated adhesives to develop at lower temperatures and to reduce the temperature at which the formulation can be handled as a hot melt.

**Plasticizers.** Plasticizers perform the following functions in adhesives based on thermoplastic rubbers: decrease hardness and modulus at room temperature, eliminate drawing, enhance pressure-sensitive tack, reduce melt and solution viscosity, decrease cohesive strength or increase plasticity if desired, and substantially lower raw material costs. The properties of a plasticized formulation are highly dependent upon the plasticizer composition, its solubility parameter, and its molecular weight.

It is usually true that the best plasticizer for use with thermoplastic rubbers in pressure-sensitive adhesives is one which is completely insoluble in the endblock phase, completely miscible with the midblock phase, and low in cost. Low volatility, low viscosity, low density, and resistance to degradation are also desirable characteristics. Various hydrocarbon oils whose average solubility parameters are below those of the midblocks, but not too far below, satisfy these requirements reasonably well.

Hydrocarbon oils are usually mixtures of molecular species which can be classed as aromatic, naphthenic, and paraffinic. When these oils are added to thermoplastic rubber, fractionation may occur with the aromatics concentrating in the endblock domains. Reduced cohesive strength at ambient and elevated temperatures typically results from using oils containing as little as 2–3% aromatics.

Typical hydrocarbon plasticizing oils and oligomers are listed in Table 11-3. They are arranged in order of increasing solubility parameter. Oils with the lowest solubility parameter and highest molecular weight are the least soluble in the endblock phase and will have the least effect on high temperature strength of thermoplastic rubber formulations. These same oils will also have the lowest solubility in the rubber phase and thus have the most tendency to bleed out when present in high concentrations.

Selection of a plasticizer involves balancing the various plasticizer characteristics to fit best any specific application. Plasticizers that are readily compatible with the polystyrene endblocks can solvate the endblock network and prevent the formation of strong physical crosslinks between the thermoplastic rubber molecules. Endblock domains will form as a separate phase, but instead of being hard and rigid at room temperature, they will be soft and fluid. Stress applied to this plasticized network will cause permanent deformation and flow. However, there may be instances in which plasticizing of the polystyrene domains is desirable, e.g., low cohesive

## Table 11-3. Properties of plasticizing oils.

| TRADE NAME | SUPPLIER | SOLUBILITY PARAMETER $\delta_0$ (HILDEBRANDS)[a] | AVERAGE MOLE WEIGHT[b] | SPECIFIC GRAVITY (15.6°C) | VOLATILITY LOSS (22 HR @ 107°C, WEIGHT %) | OIL ABSORBED BY AN S-B-S POLYMER[c] (PHR) |
|---|---|---|---|---|---|---|
| Polypropene C-60 | Amoco Chem. Corp. | 6.55 | 800 | 0.86 | 0.1 | 25 |
| Polybutene-18 | Chevron Chem. Co. | 6.95 | 600 | 0.88 | 0.1 | 39 |
| Tufflo 6206 | Atlantic Richfield Co. | 7.06 | 660 | 0.88 | 0.05 | 31 |
| Polybutene-12 | Chevron Chem. Co. | (7.04) | 530 | 0.88 | — | (53) |
| Tufflo 6056 | Atlantic Richfield Co. | 7.18 | 550 | 0.87 | 0.3 | 56 |
| Polybutene-8 | Chevron Chem. Co. | (7.18) | 440 | 0.86 | — | (75) |
| Kaydol | Witco Chem. | 7.34 | 480 | 0.89 | — | 82 |
| Tufflo 6026 | Atlantic Richfield Co. | 7.29 | 410 | 0.86 | 1.0 | 76 |
| Polybutene-6 | Chevron Chem. Co. | 7.34 | 315 | 0.84 | 10 | 103 |
| Tufflo 6016 | Atlantic Richfield Co. | 7.51 | 390 | 0.85 | 2.0 | 106 |
| Tufflo 6204 | Atlantic Richfield Co. | 7.60 | 440 | 0.92 | 0.5 | 96 |
| Shellflex 371 | Shell Chem. Co. | 7.60 | 410 | 0.90 | 0.9 | 112 |
| Tufflo 6094 | Atlantic Richfield Co. | 7.60 | 410 | 0.92 | 0.8 | 95 |
| Tufflo 6054 | Atlantic Richfield Co. | (7.66) | 380 | 0.92 | 1.3 | (127) |
| Tufflo 6014 | Atlantic Richfield Co. | 7.73 | 320 | 0.89 | 12.0 | 214 |

[a] Calculated from experimentally determined surface tension, average mole weight, and specific gravity. Values in parentheses were interpolated.
[b] By ebullioscopic methods of Mechrolah osmometer.
[c] Grams of oil absorbed by 500 to 1000 μm films of Kraton D-1101 rubber at room temperature per 100 g original weight after soaking 100 hr. Films were prepared by casting from toluene solution on mercury and drying very slowly. Parenthetical values were interpolated or extrapolated from related data.

strength pressure-sensitive adhesives. This can be affected by choosing endblock plasticizing resins.

An example of an endblock plasticizer which destroys cohesive strength is dioctylphthalate (DOP), commonly used in plasticized polyvinyl chloride (PVC) compositions. Adhesives based on a thermoplastic rubber with polystyrene endblocks should be carefully tested when intended for direct contact with PVC highly plasticized with low molecular weight phthalate esters, which can diffuse into the endblock phase and destroy the cohesive strength. The most effective way to prevent the loss of cohesive strength in a situation of this type is to replace the DOP with a nonmigrating plasticizer, such as an intermediate or medium high molecular weight polymeric plasticizer (e.g., Paraplex G-50 or G-54 from Rohm and Haas Company). Another way is to use a primer or barrier coating which restricts the tendency for plasticizer migration into the adhesive mass.

**Other Polymers.** Low concentrations of thermoplastic rubber substituted into adhesives based on conventional unvulcanized rubber like natural rubber, polyisoprene, and SBR will upgrade cohesive strength, will lower solution viscosity, and may improve adhesive strength. Conversely, use of a limited amount of a conventional rubber in a thermoplastic rubber formulation may increase solution or melt viscosity and lower costs with limited loss in cohesive strength or other properties when these goals are desirable.

The degree of compatibility between other polymers must be considered in formulating adhesives. S-I-S polymers are essentially miscible with polyisoprene and natural rubber. S-B-S polymers mix well with SBR and polybutadiene rubbers. In other combinations, however, turbidity of cast films, slow phase separation in solution, and other evidence of molecular incompatibility may appear. Nevertheless, rubbers of the incompatible type, including neoprene and nitrile rubbers, may form commercially useful mixtures if mixing problems, tendency to phase separate, turbidity, etc., are appropriately handled. Methods of offsetting incompatibility include the use of high shear mixing, formulating for high solution viscosities, mixing immediately before use, and inclusion of compatibilizing resins in the formulation.

Blending thermoplastic rubber into ethylene-vinyl acetate (EVA) polymers with low to medium vinyl acetate content is roughly equivalent to increasing the vinyl acetate content of the EVA polymer. The mixtures become softer and exhibit better flexibility at ambient and low temperatures.

**Fillers.** Nonreinforcing fillers such as clay, talc, whiting, etc., can be used in pressure-sensitive adhesives to pigment the adhesive and to lower raw material costs. The amount of filler which can be used is limited because fillers increase the stiffness and reduce aggressive tack, although tack reduction may be useful in certain applications. Carbon black, which functions as a pigment and screening ultraviolet stabilizer, forms permanent gel structures with the unsaturated midblocks. This reduces solubility and thermoplasticity.

**Stabilizers.** Unsaturated thermoplastic rubbers are susceptible to attack by oxygen, ozone, and UV radiation, especially when stressed. Therefore, stabilizers should be incorporated in the formulation to provide resistance to attack during processing and to protect the finished adhesive during its service life.

Although the stability of the saturated thermoplastic polymers is substantially better, it is good practice to include stabilizer in formulated products containing these more stable polymers.

In choosing a stabilizer package, it should be noted that the rubber midblock is more susceptible to attack than the polystyrene domains. It is, therefore, more advantageous to use stabilizers which associate primarily with the rubber midblock and protect double bonds along the midblock chain.

The two types of rubber midblock—polybutadiene and polyisoprene—in the unsaturated thermoplastic rubbers behave differently when attacked by oxygen, ozone, or UV radiation. S-B-S polymers tend to crosslink, with films becoming hard and brittle. S-I-S polymers tend to undergo chain scission, whereby films become softer and tackier.

*Protection against oxygen attack.* Oxidative attack occurs in two ways. One is under normal end-use conditions; the other occurs at high temperature under shear. The latter instance would be typical during the mixing of hot melts, especially if the operation takes place in air over an extended period.

A list of some antioxidants found to be effective in thermoplastic rubber formulations is given in Table 11-4. Their relative physical effectiveness will vary with aging conditions (temperature, amount of contact with air) and with the criteria used to judge the amount of degradation. Combinations of antioxidants may be more effective than one alone.

*Protection against ozone attack.* Unsaturated thermoplastic rubbers are susceptible to degradation by ozone, particularly when under stress. Degradation is evidenced by surface crazing and hardening.

Table 11-5 lists some antiozonants for use in thermoplastic rubber formulations. Antiozonants such as nickel dibutyl dithiocarbamate (NBC) and Pennzone B improve the resistance of thermoplastic rubbers to ozone,

**Table 11-4. Antioxidants for thermoplastic rubbers.**

| CHEMICAL COMPOSITION | TRADE NAME | SUPPLIER | STARTING LEVEL (PHR) |
|---|---|---|---|
| Zinc dibutyl dithiocarbamate | Butyl Zimate | R. T. Vanderbilt Co. | |
| | Butazate | Uniroyal Chem. Co. | 1–5 |
| | Butyl Ziram | Pennwalt Co. | |
| Tetrabis methylene 3-(3, 5-di-*tert*-butyl-4-hydroxyphenyl)-propionate methane | Irganox 1010 | Ciba-Geigy Co. | 0.3–1 |
| 2,2-Methylenebis(4-methyl-6-*tert*–butyl phenol) | Plastanox 2246[a] | American Cyanamid Co. | 0.5–2 |
| 1,3,5-Trimethyl-2,4,6-tris(3,5-di–*tert*–butyl-4-hydroxybenzyl) benzene | Antioxidant 330 | Ethyl Corp. | 0.3–1 |
| 2-(4-Hydroxy-3,5-*tert*–butyl anilino) 4,6-bis (*n*-octyl thio)-1,3,5-triazine | Irganox 565 | Ciba-Geigy Co. | 0.5–1 |
| 2,2-Methylenebis(4-ethyl-6-*tert*-butyl phenol) | Plastanox 425[a] | American Cyanamid Co. | 0.5–1 |
| 4,4-Thiobis(6-*tert*-butyl-*m*-cresol) | Santowhite crystals | Monsanto Chemical Co. | 1–2 |
| Tri(nonylated phenyl) phosphite | Polygard[b] | Uniroyal Chem. Co. | 3–5 |

[a] May produce colored formulations.
[b] Polygard may be useful at high temperatures (150 to 200°C) at the high levels shown.

**Table 11-5. Antiozonants for thermoplastic rubbers.**

| CHEMICAL COMPOSITION | TRADE NAME | SUPPLIER |
|---|---|---|
| Nickel dibutyl dithiocarbamate | NBC | DuPont |
| Dibutyl thiourea | Pennzone B | Pennwalt |
| Undisclosed | Ozone Protector 80 | Reichhold |

but also produce discoloration and staining. Improvement is also provided by Ozone Protector 80, a nonstaining antiozonant, but considerably higher loadings are required than with NBC or Pennzone B. It should be noted that Pennzone B is not suitable for hot melt formulations, as it accelerates the crosslinking of the rubber segments.

*Protection against ultraviolet radiation.* Degradation by exposure to UV light is denoted by discoloration and embrittlement or strength loss of thermoplastic rubber compounds. In the vast majority of indoor applications, this type of degradation is not a consideration. If, however, direct exposure to sunlight is expected, unsaturated thermoplastic rubbers will be susceptible and must be protected, particularly with clear, nonpigmented formulations. In addition, care should be used to select stable resins and plasticizers in formulating UV resistant pressure-sensitive formulations.

Table 11-6 lists a variety of UV light inhibitors for use with thermoplastic rubbers. One or more of these stabilizers should be added during compounding, at the level of about 0.5 phr. With opaque products, even

**Table 11-6. Ultraviolet inhibitors for thermoplastic rubbers.**

| CHEMICAL COMPOSITION | TRADE NAME | SUPPLIER |
|---|---|---|
| 2,4-dihydroxybenzophenone | Uvinul 400 | GAF Corp. |
| Substituted hydroxyphenyl benzotriazole | Tinuvin 326 | Ciba-Geigy Co. |
| Substituted benzotriazole | Tinuvin P | Ciba-Geigy Co. |
| 2-Hydroxy-4-(2-hydroxy-3-methacrylyloxy) propiobenzophenone | Permasorb MA | National Starch and Chemical Corp. |
| Octylphenyl salicylate | Eastman OPS | Eastman Chemical |
| Resorcinol monobenzoate | Eastman RMB | Eastman Chemical |
| 1,3,5-Trimethyl-2,4,6-tris(3,5-di-*tert*-butyl-4-hydroxybenzyl) benzene | Antioxidant 330 | Ethyl Corp. |
| Octadecyl 3-(3,5-di-*tert*-butyl-4-hydroxyphenyl) propionate | Irganox 1076 | Ciba-Geigy Co. |
| Tetrabis methylene 3-(3,5-di-*tert*-butyl-4 hydroxyphenyl) propionate methane | Irganox 1010 | Ciba-Geigy Co. |

without UV stabilizers, the addition of up to five parts of a reflective filler like $TiO_2$ or a light-absorbing filler like carbon black will afford excellent protection.

## Basic Formulating Principles

The characteristics of formulations based on thermoplastic rubber depend on the various compounding ingredients chosen, the interaction of these ingredients with the two polymer phases present, and the concentration chosen. The following generalizations, although they may not be valid for all compositions, provide useful guidelines for formulation development studies.

1. Endblock resins will raise or lower the upper service temperature limit, depending on their softening point. Midblock resins will also increase or decrease service temperature limit depending on their softening point.
2. Midblock resins and plasticizers lower the room temperature modulus and soften adhesive formulations. Endblock resins tend to raise the modulus by increasing the percentage of hard endblock phase.
3. Pressure-sensitive tack tends to be favored by polyisoprene midblocks over polybutadiene midblocks; such tack is produced by some midblock resins and not by others. All solid endblock resins tend to lower pressure-sensitive tack by increasing modulus; rubber phase plasticizers tend to increase aggressive tack by lowering modulus.
4. Peel strengths tend to increase with increasing formulation modulus; specific adhesion to polar or metal substrates tends to be increased by polar, unsaturated, or aromatic resins and to be decreased by hydrocarbon plasticizers.
5. Melt and solution viscosities tend to be markedly decreased by both resins and plasticizers; inorganic fillers will tend to increase viscosity.

Other guiding principles can be established by experience and by considering a given application in the light of background information already presented in this chapter and specific examples of tape and label formulations discussed at its conclusion.

## Basic Formulation Development Technique

A final formulation developed for a specific end-use will contain some combination of the various ingredients just discussed. Since adhesive properties depend not only on the types of ingredients but also on their concentrations and their interaction with each other, finding the proper combination for a particular adhesive can be a large task. In this section, the use of adhesive property contour diagrams is suggested as a rapid and reliable approach to formulation development.

An adhesive property contour diagram can be thought of as a topographic map. The X and Y axes of the map are the concentrations of two of the compounding ingredients. The lines on this rectangular concentration grid are contours at which the particular adhesive property has a constant value. Thus, at any point on the grid representing a particular combination of the two ingredients, the value of the adhesive property is immediately known.

The construction and use of adhesive property contour diagrams can be illustrated by the following practical example on the development of a formulation for a hot melt pressure-sensitive adhesive for a general-purpose masking tape.

To begin, the adhesive formulator must select the critical property requirements which must be met. In this example, the adhesive must have a low value for rolling ball tack, moderate holding power to kraft paper, and low melt viscosity. Using the formulating principles presented, the formulator must then select the two ingredients most likely to affect these critical properties. In this example, the properties will be dominated by the concentration of midblock resin and midblock plasticizer. Adhesives are then prepared at compositions corresponding to selected points on the concentration grid covering the range of interest. In this example, sixteen adhesives were initially prepared containing Kraton D-1107 rubber, midblock resin (Wingtack 95) at 75, 100, 150, and 200 phr, each containing 0, 25, 50, and 100 phr of midblock plasticizer (Shellflex 371). The performance of these adhesives in the critical property tests is then determined and the results are written at the appropriate point on the concentration grid. From these data, contour lines are drawn through points at which it is estimated that the adhesive property will have a constant value. If necessary, regions of special interest may be further refined by preparing and testing more samples in the area.

There are many advantages of developing adhesive formulations using contour diagrams. One major advantage is that they clearly show how properties change with changes in the concentration of ingredients. The contour diagram of rolling ball tack in Figure 11-8 clearly shows that with increasing plasticizer content, higher loadings of tackifying resins can be tolerated without loss of tack. The contour diagram of holding power clearly shows the very detrimental effect of increasing plasticizer concentration on holding power.

Another advantage of contour diagrams is that by superimposing contour diagrams of the critical properties, the allowable concentration range

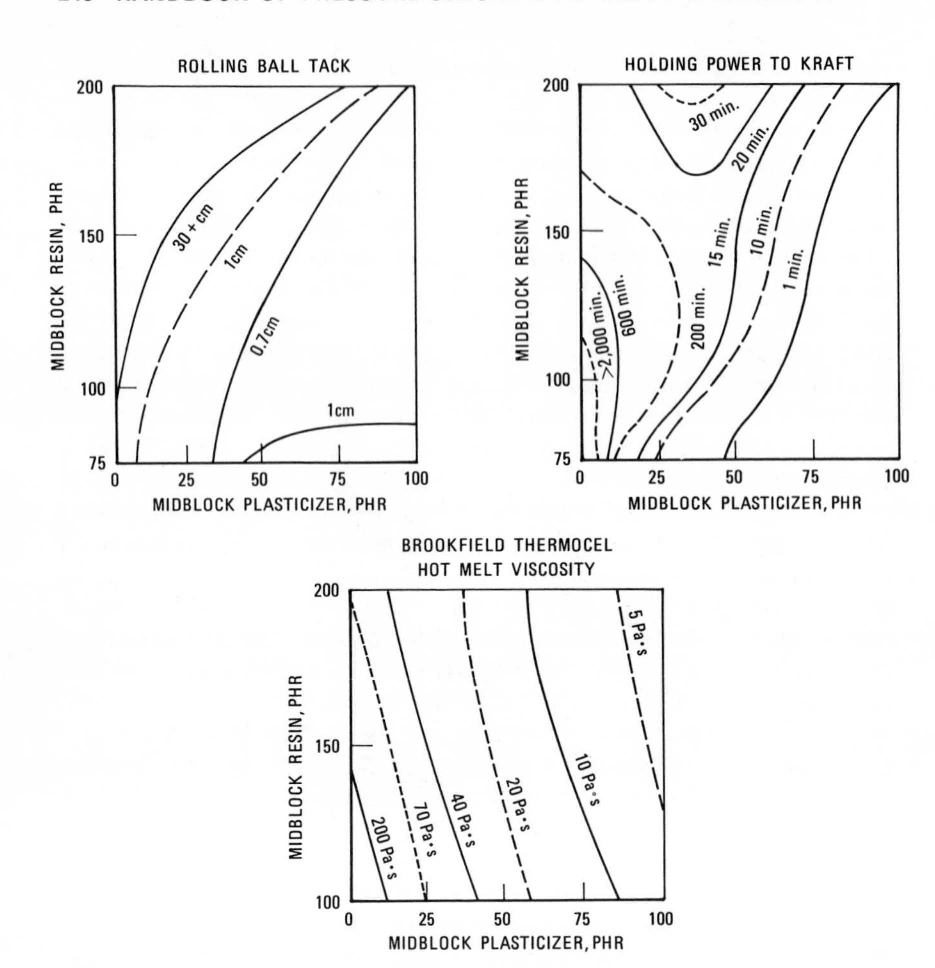


Fig. 11-8. Effect of midblock resin and midblock plasticizer concentration on adhesive properties: contour isopleths for tack, holding power, and hot melt viscosity.

in which satisfactory performance can be obtained is immediately apparent. If there is no region of overlap in which all critical properties are met, it is immediately apparent that it is impossible to reach the required performance with these particular ingredients. If the requirements of the general-purpose masking tape adhesive used in this example were that rolling ball tack must be 1 cm or less, holding power must be at 200 min and melt viscosity must be less than 40 Pa · s, superposition of the contour diagrams shows that the only practical formulation will contain about 125 phr of midblock resin and 25 phr of midblock plasticizer. The contour diagrams also show how sensitive the properties are to changes in concentration of the ingredients and thus how closely ingredient concentrations must be controlled during manufacturing to ensure satisfactory product performance.

Contour diagrams can be constructed using data obtained on samples cast from solutions, even if the intended use of the adhesive is for hot melts. Adhesive properties will depend upon the characteristics of the solvent from which the adhesive was cast. However, if the solvent used in preparing the adhesive is a good solvent for all ingredients in the formulation, the measured properties will be very similar to properties of the same adhesive prepared as a hot melt, assuming the hot melt has not been degraded. Reagent-grade toluene is usually a satisfactory solvent. There are two advantages of doing the formulation development in solvent rather than in hot melt. First, the manpower required to mix the adhesive and prepare test samples is substantially less and thus the time required to develop contour diagrams will be less. Second, heat history or degradation of the adhesive during processing, a very important variable which can significantly affect adhesive properties, is eliminated. Data from the contour diagrams generated from the solution cast adhesives indicate the most appropriate combination of ingredients for use in the hot melt adhesive. However, if the hot melt adhesive does not have exactly the required properties, the contour diagrams also show trends which indicate how to adjust the concentration of the ingredients of the hot melt adhesive to reach the target performance properties.

## APPLICATION OF FORMULATING PRINCIPLES

Understanding of the thermoplastic rubber molecule architecture, neat polymer properties, and general formulating principles forms the basis for making useful and competitive pressure-sensitive adhesive finished products.

Three general areas for consideration are identified:

1. pressure-sensitive adhesive product performance
2. adhesive manufacture
3. adhesive product manufacture (finished tape or label)

Guidance for the use of thermoplastic rubbers in solvent or hot melt pressure-sensitive adhesive systems is discussed in these categories in the following sections.

### Adhesive Product Performance

How do we attack the finished adhesive product performance issue and select the right thermoplastic rubber polymer and formulation to do the job? The answer is by first defining the required performance, be it a new adhesive application with thermoplastic rubber, or an established adhesive product where thermoplastic rubbers are being considered.

Each approach to answering the performance question has its traps and pitfalls, centered on one basic assumption: test methods and results measure the desired performance because a correlation exists between methods/results and actual performance. Decisions are made in judging formulations and adhesive quality on the assumption that this correlation is true. With materials that have a proven performance history in the field, it is a good assumption. For a different polymer or adhesive, the assumption may be faulty.

For example, superposition of temperature-time relationships to compare aging of different materials via measuring change in some property is a common tool (an Arrhenius plot). However, to make valid plots and comparisons, changes of state, e.g., glass transitions, crystalline melting points, must be considered in making tests and extrapolating results. For this reason, aging tests above the upper glass transition temperature $T_g$ of block copolymers ($T_g$ of styrene endblock about 92°C) are not valid indicators of block copolymer adhesive performance at lower temperatures for extended periods of time. Other polymers, natural rubber and acrylics for example, will not exhibit an elevated temperature $T_g$, and elevated temperature aging tests may be reasonable indications of long-term properties retention at lower temperatures. Care must be taken in generating and comparing such data.

These concerns suggest that past experience with familiar and established materials and the specifications established and existing for the materials and expected results in testing via standard methods are simply indicators of performance and not absolutes.

The following questions are relevant in this regard:

1. Are the test methods and criteria currently used in the familiar product related to the end-use performance?
2. What is the basis of the correlation between the test method and criteria which is established and the end-use performance?
3. Are the performance qualification test methods and criteria and the boundary range established for acceptable performance based on end-use performance need, or a property which characterizes the adhesive polymer type or its variability rather than finished product performance?
4. What other real performance characteristics of the established product or adhesive polymer are not defined or measured in the specifications or its tests, yet are known to be important in this product?

5. What new test method, criteria and guidelines should be established for a thermoplastic rubber adhesive which are relevant to the end-use performance, and will allow assessment of its performance on a sound basis? Is there a correlation with performance?
6. What structural differences between the current adhesive polymer and the block copolymer candidate may significantly alter results via the test method employed? Where can the structural differences in block copolymers be used to advantage?
7. What technical data need to be developed in order to verify performance via new test methods which reflect actual performance?

With any new or alternative adhesive system product, answers to these questions form a sound technical basis for judging the finished product, assuring quality in manufacture and developing product information which allows its introduction and successful use in the marketplace.

## Adhesive Manufacture

Adhesive formulations are most frequently prepared from crumb or pelletized thermoplastic rubber by solvent solution or hot melt mixing techniques. Commercial thermoplastic rubbers are not offered as emulsions.

Thermoplastic rubbers have two distinct characteristics that differentiate them as a group from conventional rubbers, and that are important in considering adhesive manufacturing via solvent or hot melt mixing techniques.

Styrene endblock type thermoplastic rubbers produce low solution viscosities in a good polystyrene solvent, e.g., toluene. Thermoplastic rubbers without premastication have lower solution viscosities at comparable concentrations than do elastomers such as SBR, natural rubber and polychloroprene after milling. This is because thermoplastic rubbers have relatively low, precisely controlled molecular weights. Molecular weights are allowed to be relatively low because the ABA structure and polystyrene endblock domain network afford high cohesive strength at these comparatively lower molecular weights. Unvulcanized conventional rubbers rely primarily on chain entanglements for strength, thus molecular weights must be high, resulting in higher solution viscosities. Premastication (milling) of natural rubber is beneficial to producing lower solution viscosity, but this is energy-intensive. The adhesive system will require vulcanization or expensive accelerators to regain initial strength levels through crosslinking. Thermoplastic rubbers reform the polystyrene endblock network with simple solvent evaporation, and form the high strength structure without curing or accelerators. Thus, higher solids solutions at given viscosity, i.e., less solvent

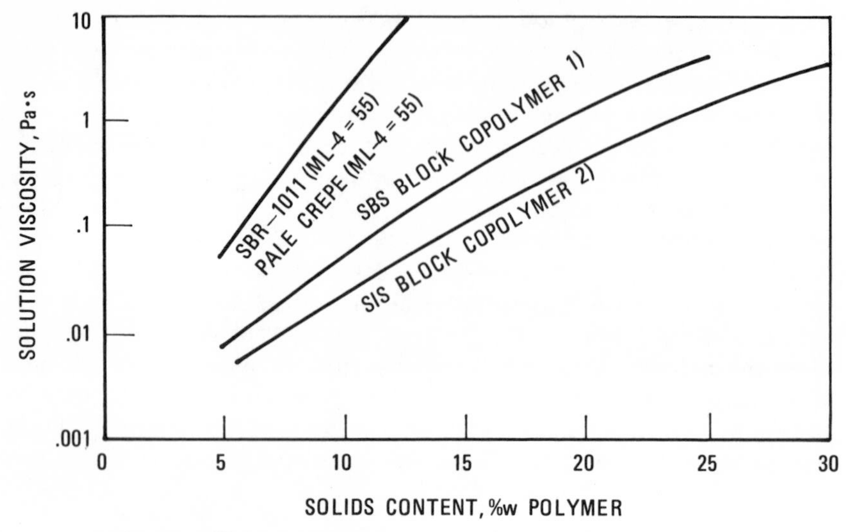


Fig. 11-9. Viscosity of toluene solutions at 25°C for various rubbers.

required initially or to be recovered, are possible with thermoplastic rubbers when compared with conventional rubbers.

Solution viscosity is shown in Fig. 11-9 for linear S-B-S and S-I-S thermoplastic rubbers at various concentrations in toluene, along with viscosity for natural rubber and SBR. These data illustrate the lower viscosity obtainable with thermoplastic rubber than with NR and SBR, even after those materials have been milled to reduce their molecular weight.

Thermoplastic rubber compositions are melt processible. For the same molecular structure reasons, thermoplastic rubber compositions are easily melt processible, while conventional rubber compositions are not. At hot melt mixing temperatures, the polystyrene endblock domain structure is made fluid with heat and shearing. The system viscosity is low because block copolymer molecular weights are relatively low and the system endblock structure is fluid. Conventional rubbers of much higher molecular weight produce very high hot melt viscosities.

The following sections serve as a guide to preparing solvent solution adhesives and hot melt adhesives with thermoplastic rubbers in greater detail.

**Solvent Behavior of Thermoplastic Rubbers.** The behavior of thermoplastic rubber toward solvents is unique because it is comprised of two different polymeric chains present in every molecule. Each midblock and endblock portion of the molecule has a characteristic solubility parameter.* Thus, the ability of the neat thermoplastic rubber molecule to be solvated is a strong function of the solvent or solvent blend chosen. Solution viscosities can be varied over a wide range as a function of solvent choice or blend at a given solids level concentration. This characteristic can be used advantageously in developing formulations where adhesive solids level and viscosity must be closely balanced, or optimized, for given solvent coating equipment, as well as a means to alter drying rate characteristics.

In addition, a judicious choice of resins and resin concentrations can lead to enhanced solution viscosities, and potentially higher solids systems with less costly or more environmentally acceptable solvent blends.

For ABA thermoplastic rubbers, two solubility parameters are involved, one for the polystyrene endblock and one for the rubber midblock. A good solvent for a thermoplastic rubber must therefore be one which dissolves both endblock domains and midblock segments.

To illustrate, Fig. 11-10A indicates the approximate range of solubility parameter of solvents for rubber midblocks. The boundaries are not sharp near the extremes because the polymer molecules tend to collapse while still solvated. Figure 11-10B indicates, in a similar way, the range of solubility parameters of solvents for polystyrene endblocks. Figure 11-10C shows what happens when both blocks are combined in a single block copolymer molecule. A central range of solvents readily dissolves both blocks and gives low viscosity solutions. With solvents at the lower solubility parameter end, the polystyrene endblocks tend to remain associated and form what might be termed solvated endblock domains. In this region, solution viscosity increases rapidly with decreasing solubility parameter until rigid crosslinked-like gels are formed. With solvents at the high solubility parameter end, the situation is somewhat different. First, the midblocks become less and less soluble and tend to associate together while the endblocks are highly solvated. Opalescent solutions with time-dependent vis-

* Hildebrand's solubility parameter (represented by $\delta$) is a measure of the total forces holding the molecules of a solid or a liquid together. It has the units of $(cal/cm^3)^{1/2}$. Every compound is characterized by a specific value of solubility parameter, although this value may not always be recorded in convenient literature references. Materials having the same solubility parameter tend to form homogeneous mixtures or to be miscible. Those with different solubility parameters tend to form separate layers or to be mutually insoluble. Other indexes of solvent compatibility (Kauri-Butanol value, aniline point, dilution ratio, etc.) can often be correlated with solubility parameter. For example, Kauri-Butanol values in the range of 20 to 100 for hydrocarbon solvents can be used to predict thermoplastic rubber solubilities. A solvent can be judged good or poor by comparison of its K-B value with those of known good or poor solvents. (Discussions of solubility parameter concepts are presented in References (5) and (6). K-B values and solubility parameters for a range of commercial solvents are listed in Reference (7).

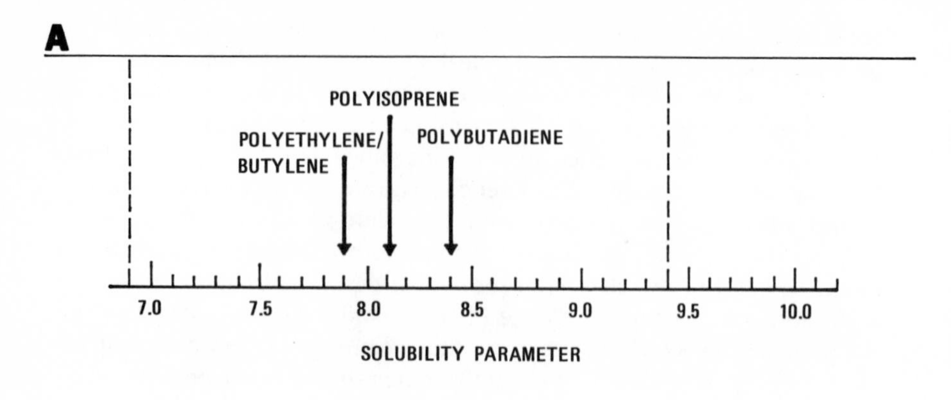


Fig. 11-10A. Solubility parameter range of solvents useful for rubber phase.

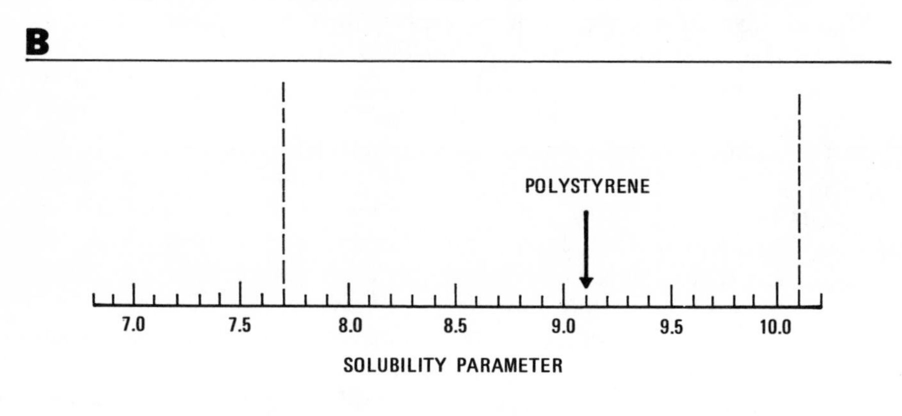


Fig. 11-10B. Solubility parameter range of solvents useful for polystyrene phase.

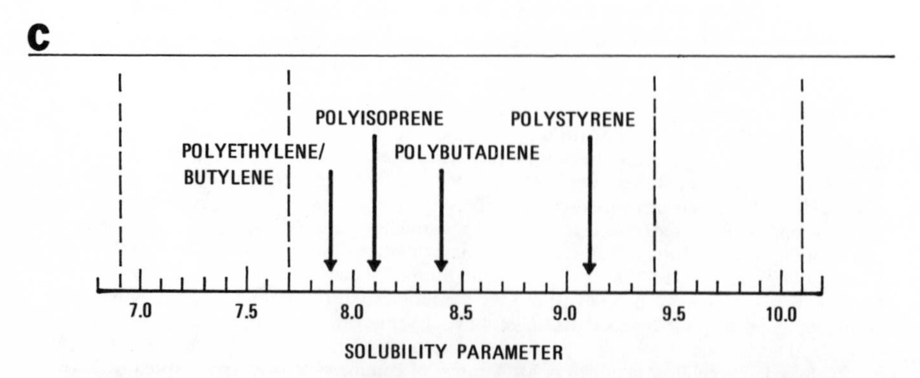


Fig. 11-10C. Solubility parameter range of solvents useful for thermoplastic rubbers.

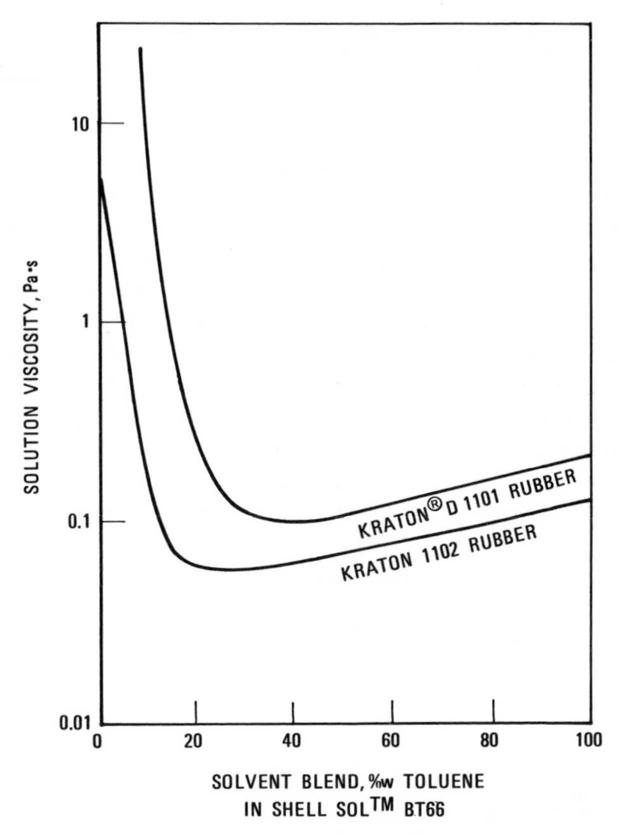


Fig. 11-11. Effect of toluene concentration on the solution viscosity of 15%W S-B-S thermoplastic rubbers in aliphatic hydrocarbon/toluene blends at 23°C.

cosities are formed. Thermoplastic rubber crumb can be dispersed in this sort of borderline solvent (for example, methyl ethyl ketone).

An example of these principles is shown in Fig. 11-11. These curves represent the viscosities of 15%w solutions of S-B-S thermoplastic rubbers in *n*-hexane/toluene solvent blends over a 0 to 100% toluene concentration solvent blend range. When an aliphatic solvent, such as *n*-hexane, or other pure solvents with solubility parameters below 7.4 are chosen, the S-B-S solution viscosity is much higher than that shown for a pure toluene solution of the same polymer. This is because *n*-hexane is a nonsolvent for the polystyrene endblock, while toluene dissolves both the midblock and endblock segments of the thermoplastic rubber molecule.

Increasing the toluene concentration in the blend or addition of other higher solubility parameter solvents, brings the polystyrene endblocks into solution, resulting in solvent blend solutions of lowered viscosity. Indeed, a minimum viscosity is shown in the 20-30%w toluene blend range for these polymers at a 15%w polymer concentration.

The solution viscosity is also dependent upon the molecular weight and a number of polystyrene arms in the thermoplastic rubber polymer, as well as actual weight percent polystyrene in the thermoplastic rubber polymer. These effects are shown for the two different linear S-B-S thermoplastic rubbers as shown in this same figure. While these curves are specific to these S-B-S polymers, similar behavior can be expected for S-I-S and S-EB-S thermoplastic rubbers in blended and pure solvents.

In the selection of a solvent system for a finished pressure-sensitive adhesive, one must consider all other components which are used in the thermoplastic rubber formulation. Resins and plasticizing oils are usually the most important in this regard.

Figure 11-12 demonstrates an interesting finding for the effect that addition of 100 phr of a tackifying resin (a rosin ester) has on solution

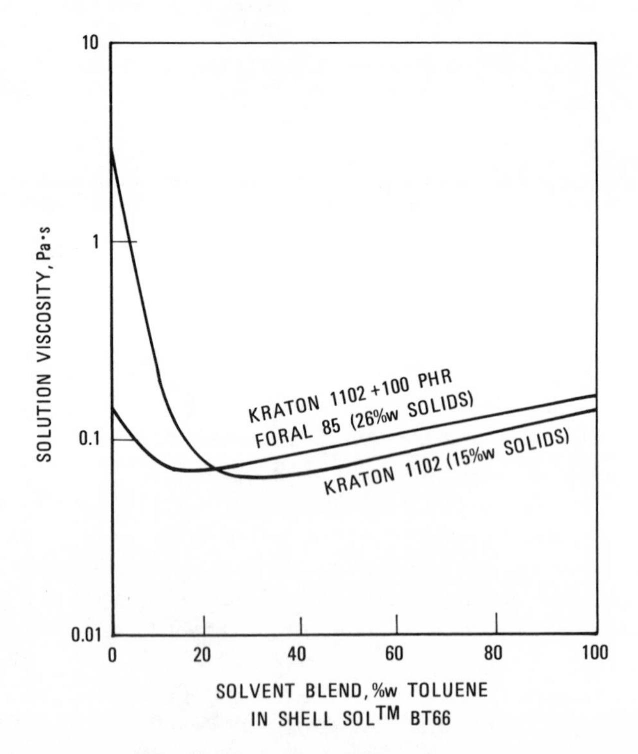


Fig. 11-12. Effect of a rosin-ester resin on the solution viscosity of an S-B-S thermoplastic rubber in aliphatic hydrocarbon/toluene blends at 23°C.

viscosity in solvent blends of *n*-hexane and toluene for 15%w solids S-B-S solution (plot similar to Fig. 11-11). The net effect is that the addition of this resin has shifted the minimum solution viscosity position of the rubber-resin blend to a point where the concentration of a good endblock solvent, toluene, is lowered when compared to the neat polymer thermoplastic rubber solution minimum viscosity point. This suggests that the solution viscosity as a function of solvent blend can be perturbed by the addition of resins which can act to alter the net solubility parameter of the solvent system and shift the minimum viscosity position. In this case, the resin is acting more like toluene solvent in the rubber-resin solution, allowing a lower concentration of this good endblock solvent to achieve a minimum viscosity.

Contours of such interactions could be plotted for various resin types, resin concentrations, solids levels, and polymer types in order to map and optimize solution viscosities and solvent blends.

Table 11-7A shows results for a series of solution viscosity measurements made on S-I-S linear thermoplastic rubbers when blended with midblock and endblock resins at a *constant solids level* (60%) in toluene solvent. Table 11-7B shows the composition of these blends.

Comparing samples 1 and 2 or 6 and 7, the expected change in viscosity

**Table 11-7A. Solution viscosity vs. resin level and type: constant 60%w compositions in toluene.**

| COMPOSITION NUMBER[a] | RESIN BLEND | RATIO OF S-I-S POLYMER[b,c] TO RESIN BLEND | SOLUTION VISCOSITY (Pa·s) |
|---|---|---|---|
| 1[b] | 100% midblock | 1 : 1 | 21.8 |
| 2[b] | 100% midblock | 5 : 7 | 11.4 |
| 3[b] | 6 : 1 (midblock: endblock) | 5 : 7 | 14.6 |
| 4[b] | 5 : 2 (midblock: endblock) | 5 : 7 | 20.0 |
| 5[b] | 5 : 2 (midblock: plasticizing endblock) | 5 : 7 | 9.2 |
| 6[c] | 100% midblock | 1 : 1 | 110.0 |
| 7[c] | 100% midblock | 5 : 7 | 33.4 |
| 8[c] | 6 : 1 (midblock: endblock) | 5 : 7 | 47.5 |
| 9[c] | 5 : 2 (midblock: endblock) | 5 : 7 | 88.4 |
| 10[c] | 5 : 2 (midblock: plasticizing endblock) | 5 : 7 | 10.9 |

[a] See Table 11-7B for detailed composition of solutions.
[b] Sample Nos. 1–5 based on Kraton D-1107 rubber.
[c] Sample Nos. 6–10 based on Kraton D-1111 Rubber.

**Table 11-7B. Compositions for solution viscosity measurements for two linear S-I-S thermoplastic rubber/resin blends at 60% solids in toluene (parts by weight)**

| INGREDIENTS SAMPLE NO. | KRATON D-1107 RUBBER | KRATON D-1111 RUBBER | ESCOREZ 1310 | CUMAR LX-509 | PICCOVAR AP-25 | SHELLFLEX 371 | IRGANOX 1010 | TOLUENE |
|---|---|---|---|---|---|---|---|---|
| 1 | 100 | — | 100 | — | — | 10 | 1 | 140.7 |
| 2 | 100 | — | 140 | — | — | 10 | 1 | 167.3 |
| 3 | 100 | — | 120 | 20 | — | 10 | 1 | 167.3 |
| 4 | 100 | — | 100 | 40 | — | 10 | 1 | 167.3 |
| 5 | 100 | — | 100 | — | 40 | 10 | 1 | 167.3 |
| 6 | — | 100 | 100 | — | — | 10 | 1 | 140.7 |
| 7 | — | 100 | 140 | — | — | 10 | 1 | 167.3 |
| 8 | — | 100 | 120 | 20 | — | 10 | 1 | 167.3 |
| 9 | — | 100 | 100 | 40 | — | 10 | 1 | 167.3 |
| 10 | — | 100 | 100 | — | 40 | 10 | 1 | 167.3 |

is noted when the polymer content of the 60% total solids polymer/resin blend is reduced by substitution with resin. In these cases, the viscosity at 60% solids decreases with this substitution of polymer by resin.

As endblock-associating resin is substituted for midblock resin (samples 3 and 4, 8 and 9), increase in solution viscosities is noted. In this case, the endblock resin substitution is increasing the integrity and extent of the endblock domain structure and concentration in solution, and the resultant effect of increasing viscosity is not expected. The trend is internally consistent for the two linear S-I-S polymers used in generating these data, Kraton D-1111 rubber, which is a 22% styrene polymer, and Kraton D-1107 rubber, which is a 15% styrene polymer. As might be expected, however, the higher styrene content polymer shows higher solution viscosities than the lower styrene content polymer as endblock resin is added and the solids level is held constant. Finally, the choice of a low softening, plasticizing endblock resin (samples 5 and 10) demonstrates its utility to act as a "solvent" and produce the lowest solution viscosities for the examples shown.

A method by which pressure-sensitive formulations can be compared is in examining the interaction of solvent blends and weight percent adhesive solids possible at a constant solution viscosity, where data much like that generated in Table 11-7A are used to develop the curves in Fig. 11-13. Higher solids content, i.e., less solvent, and the use of less costly and more available solvents in solvent blends, is the driving force for such analyses. While formulation variables are not controlled in the series of curves shown in Fig. 11-13, some general comments can be made on the method and its utility. The formulations shown in Figure 11-13 are based upon a linear 22% styrene S-I-S thermoplastic rubber, Kraton D-1111 rubber. The curves represent constant viscosity plots (50 Pa·s) for these formulations.

Focusing on the curves for formulations A and B at 40% toluene in the solvent blend (vertical line), the effect of 40 phr additional resin (Piccovar AP-25, a lower viscosity component and an endblock plasticizer) is apparent. A higher solids level is achievable at a constant 50 Pa·s viscosity for the formulation B. The additional resin provides a lower viscosity component to the solution, dilutes the rubber content, and also acts to lower solution viscosity as an endblock-plasticizing resin.

Alternately, constructing a horizontal line at 50 weight percent solids, we find that a greater ratio of toluene to Shellsol B is necessary to achieve this solids level at 50 Pa·s viscosity with the formulation containing no additional endblock plasticizing resin (formulation A). In cases where aliphatic solvents are less costly than aromatics or more environmentally acceptable, utilization of formulation B over formulation A may be advantageous.

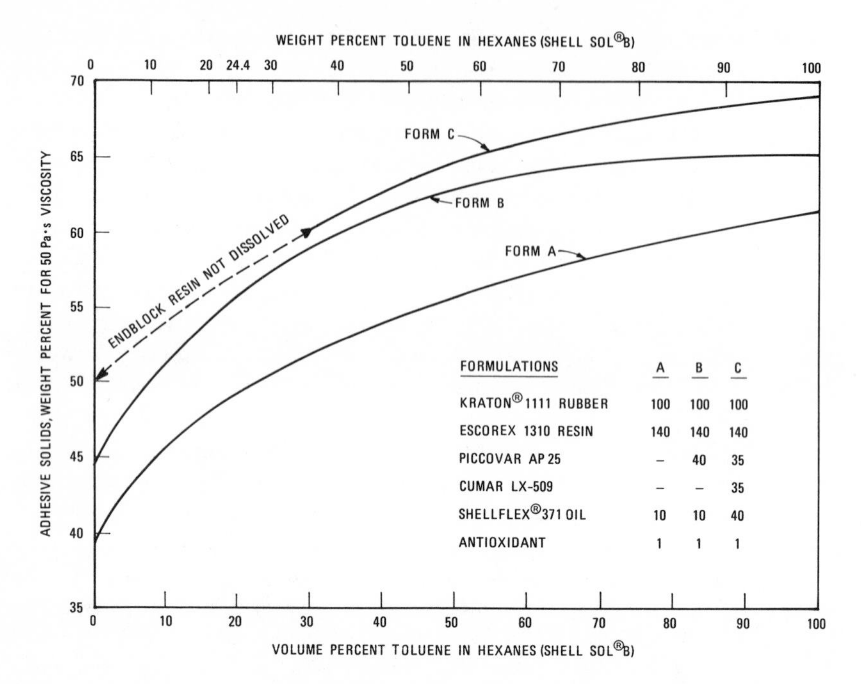


Fig. 11-13. Effect of toluene concentration on pressure-sensitive adhesive solids and composition in solvent blends to produce constant 50 Pa · s solution viscosities.

The plot for formulation C is simply shown to indicate that the high softening point, highly associative endblock resin used (Cumar LX-509) requires a given level of aromatic solvent in the solvent blend before it will dissolve in this system, about 30%. Further, formulation C is the most dilute polymer of the three, and shows the highest achievable solids at any given aliphatic-aromatic solvent blend ratio at a 50 Pa·s viscosity.

Drying times are affected by variations in the solids level of a formulation and by the choice of a solvent, or solvent blend. In this aspect of solution processing, the advantage lies with the less ideal solvents. That is, solvent release is faster with poor solvents than with good solvents because the last traces of poor solvent are not as tightly held by the thermoplastic rubber molecule.

Solvent effect on drying time is shown in Fig. 11-14 for toluene and a blend of toluene with a typical aliphatic solvent of low boiling point and low solubility parameter (approximately 7.3). The illustration shows that the use of the aliphatic solvent greatly reduces evaporation time, due to its greater volatility and lower solubility parameter.

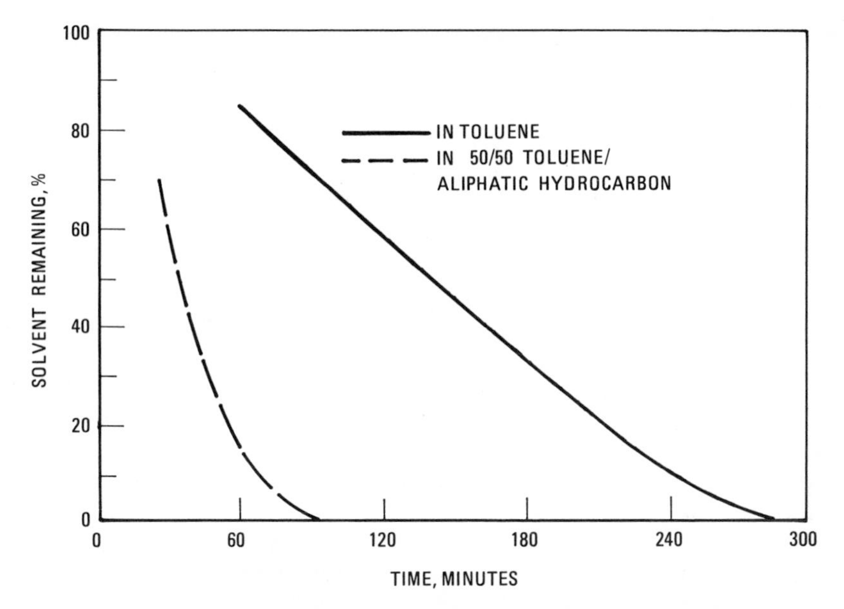


Fig. 11-14. Solvent evaporation rates for 10%W solutions of an S-I-S thermoplastic rubber (Kraton® D 1107 Rubber) in toluene or aliphatic hydrocarbon (Shellsol® BT66) solvent.

**Hot Melt Behavior of Thermoplastic Rubbers.** Heating the thermoplastic rubber above the glass transition temperature $T_g$ of the polystyrene endblocks (about 92°C) "unlocks" the physically crosslinked thermoplastic rubber network. Although the domains soften at temperatures above 92°C, the act of heating alone will not cause thermoplastic rubber to melt and flow, and the polymer remains very viscous. It will not flow unless mechanical energy is applied. It should be emphasized that phase interflow cannot be achieved without mechanical mixing even by greatly increasing the temperature. Instead, the rubber phase will begin to decompose at temperatures above about 220°C. Application of shear stress then makes mixing possible with other materials, as the melt viscosity decreases as the shear stress or shear rate increases (Figs. 11-15 and 11-16).

With commercial unsaturated midblock S-B-S and S-I-S thermoplastic rubbers, 120 to 180°C is the normal range for mixing and application. Temperatures above 180°C may lead to excessive oxidative degradation of the rubber network and above 220°C to thermal decomposition. At temperatures below 120°C, the viscosity becomes so high that mechanical breakdown of the thermoplastic rubber may occur. Since oxidative degradation is reduced by lowering temperature, the preferred mixing method is in high shear equipment at 135 to 160°C.

**High Shear Melt Mixing Precautions.** Thermoplastic rubbers, particularly those with the highest melt viscosities, may build up excessive temperatures in high shear internal mixers (e.g., Banbury mixers) if exposed to the high shear conditions in the absence of plasticizers and stabilizers. Certain precautions should be noted. While mass temperatures up to 180°C are normal in the mixing of thermoplastic rubber formulations, higher temperatures may indicate too high a melt viscosity or too vigorous a mixing action. If the temperature were allowed to rise uncontrolled in such a situation, varporization of plasticizer light ends and polymer decomposition could occur, creating a potential fire hazard. Excessive temperatures can be minimized during the mixing operation by adding a resin or plasticizer or by reducing mixer speed. It is often desirable to investigate the mixing characteristics of a new composition on a small scale before undertaking large-scale runs.

Saturated thermoplastic rubbers, because they have an olefin midblock, are substantially more resistant to degradation. They can be held at high temperatures for long periods of time with negligible oxidative degradation. In the absence of oxygen, they begin rapid thermal degradation at temperatures above about 275°C. A temperature of 280°C is considered the maximum for processing the saturated midblock thermoplastic rubber polymers.

Resin used in hot melt mixing can be premelted and in effect become solvents for the polymer. A resin which associates with the polystyrene endblock will therefore act as a solvent for this phase during hot melt mixing and enhance polymer dissolution. As mixing progresses, less mechanical energy is required to cause flow.

Successful mixing of hot melt pressure-sensitive adhesive based on unsaturated midblock thermoplastic rubbers depends on minimizing oxidative degradation by balancing the following variables in all types of mixing equipment and mixing processes:

1. Ensure adequate and uniform heating and shearing to soften and disperse polymers, resins, oils, and other ingredients to a homogeneous pressure-sensitive adhesive
2. Minimize the time and temperature in accomplishing dispersion
3. Minimize exposure of the hot melt to air (oxygen)

Manifestations of poor control and balance include gel and viscosity increase with S-B-S thermoplastic rubbers which tend to crosslink, and loss of network strength, i.e., lower viscosity, poorer holding power, and lower service temperature in finished adhesives through chain scission with S-I-S thermoplastic rubbers.

**Melt Rheology of Thermoplastic Rubbers: Effect of Shear Rate, Shear Stress and Temperature.** Viscosities of thermoplastic rubbers in the melt state change rapidly with variations in shear rate, shear stress or temperature. As noted previously, thermoplastic rubbers will not melt and flow at mixing or application temperatures if no shear stress is applied, due to the two-phase nature of the polymers.

The effect of shear rate on viscosity at 177°C is shown in Fig. 11-15 for an S-I-S polymer. This illustration shows the highly non-Newtonian behavior of thermoplastic rubbers by the rapid drop in melt viscosity with increasing shear rate. Note that at the shear rate of laboratory instruments (10 $sec^{-1}$) viscosity would be extrapolated to over 1,000 Pa · s, but at shear rates encountered in actual mixing and application ($10^2$–$10^3$ $sec^{-1}$), it drops substantially.

Thermoplastic rubbers other than the one used in Fig. 11-15 would produce similar melt viscosity curves. Generally, for a higher molecular weight thermoplastic rubber or for the saturated midblock thermoplastic rubbers, the entire curve would be moved upward; for a lower molecular weight thermoplastic rubber, it would be moved downward.

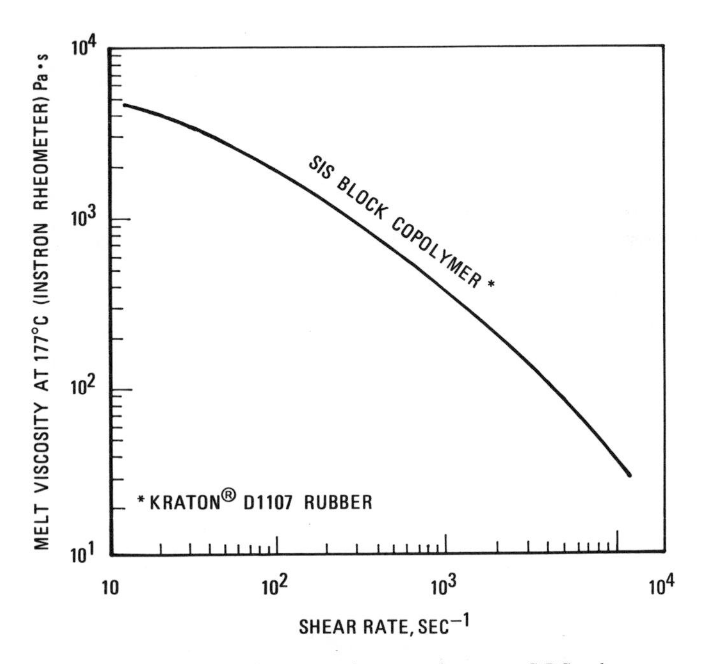


Fig. 11-15. Melt viscosity versus shear rate for a neat S-I-S polymer.

The effect of shear stress on melt viscosity is illustrated in Fig. 11-16 for an S-B-S thermoplastic rubber. Melt viscosity is shown to decrease nearly a hundredfold with increasing shear stress. This figure also shows the changes in melt viscosity effected by temperature. At a constant 0.7 MPa ($10^2$ psi) shear stress (the force exerted at melt flow condition *G* in a melt flow apparatus, ASTM Method D1238), a change of about 30°C will have a three-fold effect on viscosity.

At temperatures approaching the glass transition temperature of the polystyrene endblock, melt viscosity changes more rapidly with temperature, particularly at low shear rates. Figure 11-17 shows the effect of temperature on melt viscosity at constant shear rates for an S-B-S thermoplastic rubber.

As shown in Fig. 11-17, the rate of softening in the 80° to 120°C range

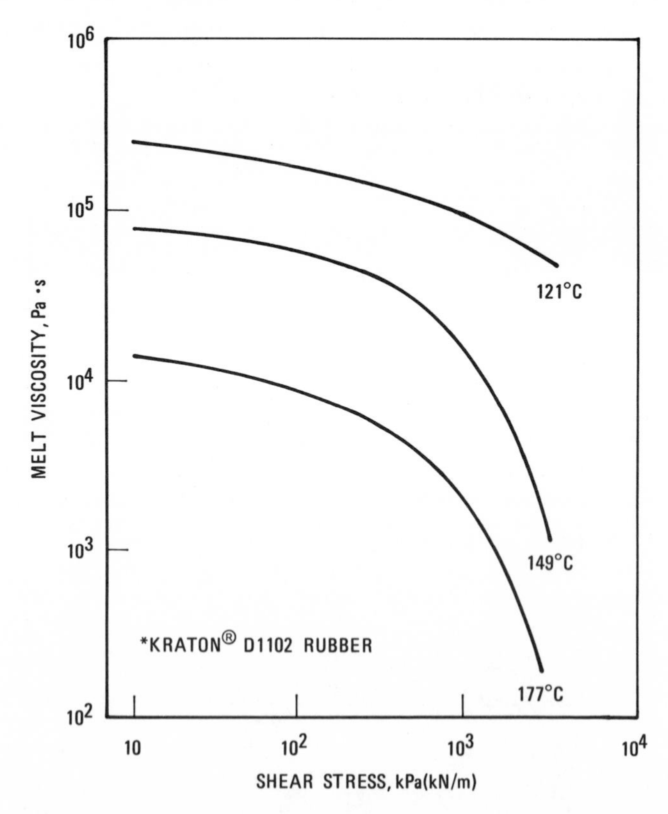


Fig. 11-16. Melt viscosity versus shear stress at constant temperature for a neat S-B-S* polymer.

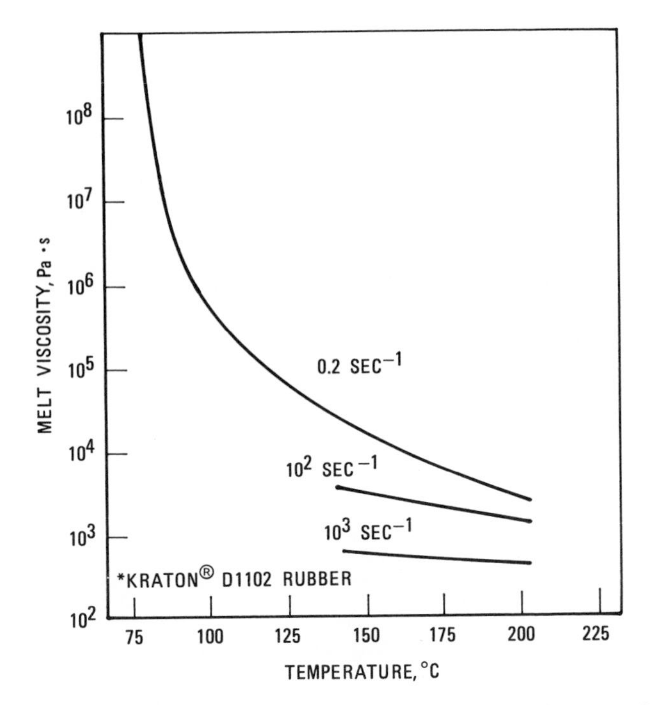


Fig. 11-17. Temperature versus melt viscosity at constant shear rates for a neat S-B-S* polymer.

is rapid as the polystyrene endblock domains become fluid. Conversely, on cooling, buildup of viscosity and strength is rapid.

**Effect of Resins on Melt Properties.** One function of resins is to reduce the viscosity of thermoplastic rubber melts. Even with resins added in, the behavior of thermoplastic rubber compounds is still non-Newtonian, but less so than for the polymer alone as described previously. Illustrative viscosities are shown in Fig. 11-18 for an S-B-S polymer/resin blend.

**General Approach to Melt Mixing.** Hot melt adhesive blends based on thermoplastic rubbers can be mixed by using a wide variety of batch or continuous mixing equipment. S-B-S and S-I-S polymers dissolve readily in hot resins, with only moderate agitation. Higher molecular weight thermoplastic rubbers require higher shear in mixing. Increasing the surface area of the polymer through granulation substantially reduces the dissolution time in batch mixing equipment. Using both polystyrene-associating resins and rubber-associating resins aids in dissolving both phases of the polymer, permitting faster cycles and reduced mixing temperatures.

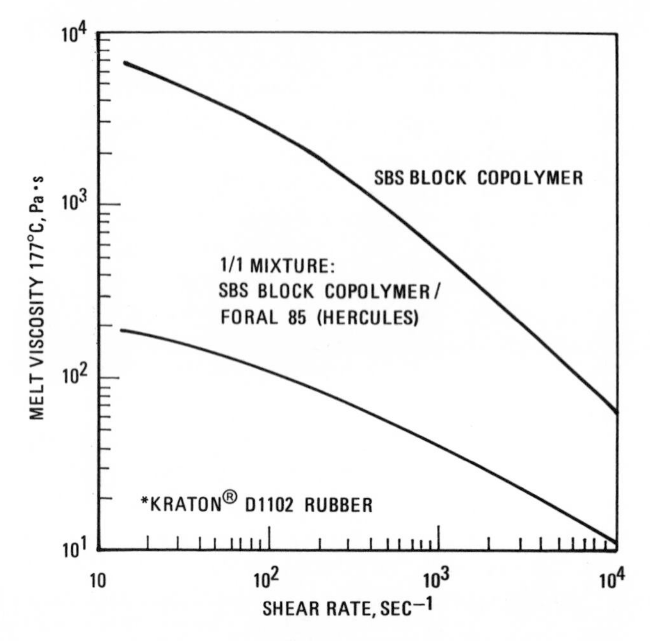


Fig. 11-18. Effect of a rosin-ester resin on the melt viscosity of an S-B-S* polymer at various shear rates.

As in hot melt processing of any type of hydrocarbon material, adhesive degradation can occur, with resultant changes in final properties through chain scission and chain crosslinking reactions of the polymer molecule. This is less of a concern when processing saturated midblock thermoplastic rubbers. The effectiveness of mixers should be judged by their ability to mix the adhesive uniformly and without degradation. Generally, high shear mixers such as twin screw extruders or sigma blade mixers are preferred for mixing because they can give homogeneous mixes in short periods of time, minimizing product degradation. For formulations in which degradation must be minimized, adhesives based on unsaturated thermoplastic rubbers must be protected against oxidative degradation by blanketing with an inert atmosphere.

**Degradation and Protection During Processing.** The manufacture of thermoplastic rubber-based melt formulations normally takes place in the range of 120 to 180°C. The primary factor in degradation is oxygen attack, which in melt mixing is accelerated by high temperature and high shear conditions. In addition, other hazards exist in melt processing and general handling of thermoplastic rubber crumb as previously discussed.

One approach to minimizing degradation is to reduce temperatures or shear rates where they are extreme. However, the most direct and most effective method of preventing degradation under melt mixing application conditions is to exclude oxygen. Contact with air is obviously greatest in open pot type mixers, less in closed equipment such as continuous feed extruders.

Blanketing or continuous purging of the mixing equipment or hot storage pots with $N_2$, $CO_2$, or other inert gas will significantly reduce oxidative degradation and should be considered mandatory. Even an initial purge of the mixer or storage vessel may be quite beneficial.

Figure 11-19 demonstrates the protection possible through inert gas blanketing. These data were derived from testing of an S-I-S based adhesive mixed in a Brabender Plasti-Corder (C. W. Brabender Instrument, Inc.) equipped with a high shear sigma blade mixing head. The illustration shows that with air present, viscosity decreases rapidly in proportion to shear applied and residence time in the mixer. Melt viscosity remains nearly constant at both shear rates when $N_2$ is purged through the mixing head.

Figures 11-20A and 11-20B demonstrate the beneficial effect $N_2$ blanketing has in preventing degradation of an S-I-S polymer-based pressure-sensitive adhesive when held in a melt holding tank for extended periods of time at elevated temperature. Severe decreases in melt viscosity and adhesive holding power are shown after less than 24 hours when the adhesive is in contact with air (Fig. 11-20A); the same adhesive when $N_2$ blanketed

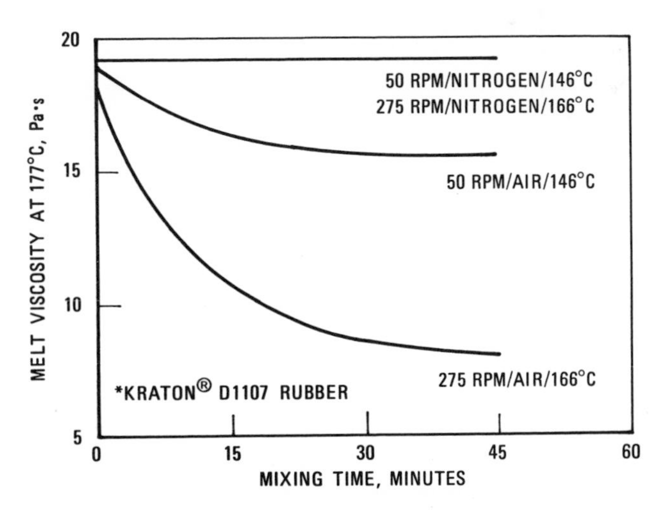


Fig. 11-19. Effect of mixing conditions on melt viscosity retention of a pressure-sensitive adhesive based on an S-I-S polymer.*

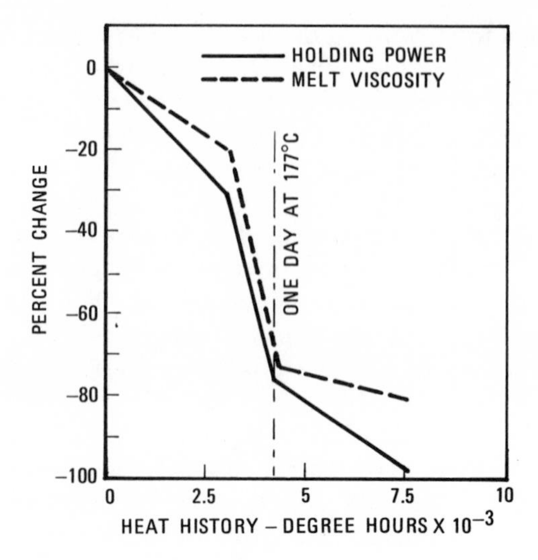


Fig. 11-20A. Hot melt pressure-sensitive adhesive pot life study in contact with air at 177°C.

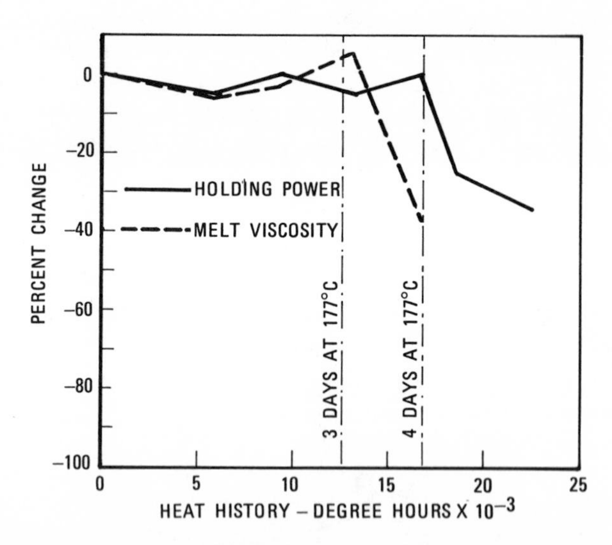


Fig. 11-20B. Hot melt pressure-sensitive adhesive pot life study under nitrogen blanket at 177°C.

shows little reduction in viscosity and holding power for over 72 hours under the same conditions (Fig. 11-20B).

Figure 11-21 shows general trends of adhesive properties for an S-I-S polymer system as degradation proceeds. Tack and adhesion values change

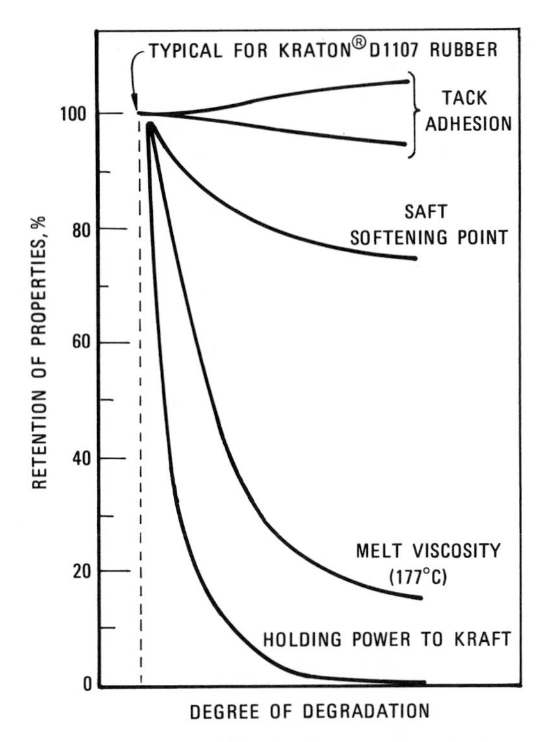


Fig. 11-21. Effect of degradation on retention of pressure-sensitive adhesive properties based on an S-I-S polymer.

little, with tack improving slightly. Thus, measurement of these properties of the finished product are not useful in measuring degradation. Reduction of melt viscosity is often a desirable goal in formulating. However, it is also a manifestation of chain scission of the S-I-S polymer when uncontrolled degradation has occurred in mixing, with resultant loss of service temperature (SAFT) and holding power to kraft paper adhesive properties.

Finally, qualitative degradation as a function of mixer type and heat history is shown in Fig. 11-22. Clearly, continuous mixers and high shear batch mixers are preferred (minimal heat histories to accomplish mixing). Where low shear batch mixers must be used, then residence times in mixing are longer, and the beneficial nature of $N_2$ blanketing of the system again becomes apparent.

## Adhesive Product Manufacture

Solvent applied thermoplastic rubber formulations employ equipment used to coat conventional solution rubbers.

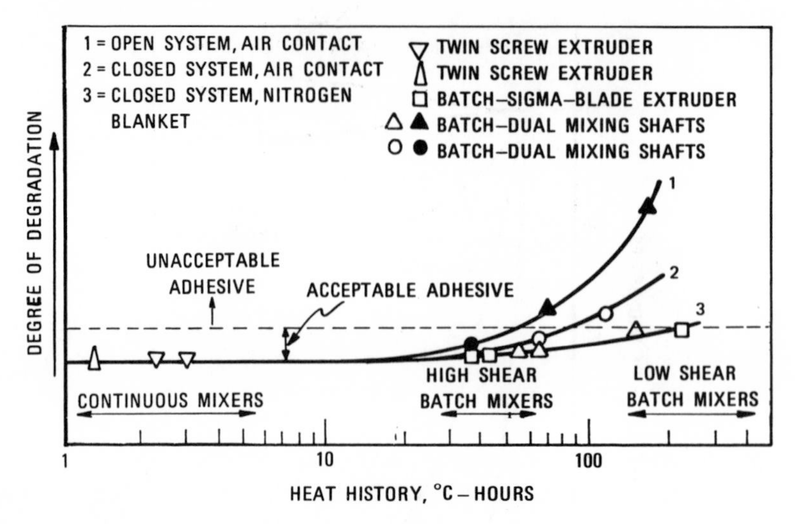


Fig. 11-22. Effect of heat history on degree of degradation for a pressure-sensitive adhesive based on an S-I-S polymer using various hot melt mixing equipment.

Dip roll, slot-die and extrusion types of equipment can be used to apply hot melt pressure-sensitive adhesives based on thermoplastic rubbers. Dip roll and slot-die coaters involve the use of a reservoir of molten adhesive as a supply for the coating head; the extrusion process is a continuous one, with adhesive mixed and fed to the coating head directly.

## FORMULATING TO MEET CONSTRAINTS

This section is devoted to discussion of examples of formulating technology and performance of thermoplastic rubber in a wide span of end-use target requirements. It serves the primary purpose of showing what can be done with thermoplastic rubbers as formulated pressure sensitive-adhesive compositions. It serves the secondary purpose of guiding initial thought in considering the kinds of tape and label applications and products in which thermoplastic rubbers can be useful as pressure-sensitive adhesives.

### General-Purpose Tapes

Having established the performance and end-use requirement targets for crepe paper and polyester tapes, the formulation shown in Table 11-8 is typical. This formulation has demonstrated a balance of pressure-sensitive properties when properly coated on a commercial type coater to commercially available backings. As the reference data in Table 11-8 show, this

**Table 11-8. General-purpose tape formulation and properties based on an S-I-S thermoplastic rubber.**

| FORMULATION NO. 7 | PARTS BY WEIGHT |
|---|---|
| S-I-S polymer[a] | 100.0 |
| Hydrocarbon tackifying resin[b] | 140.0 |
| Paraffinic/naphthenic oil[c] | 10.0 |
| Zinc dibutyl dithiocarbamate | 5.0 |

| | HOT MELT COATED TAPES | | | |
|---|---|---|---|---|
| | CREPE PAPER | | POLYESTER | |
| ADHESIVE PROPERTIES | TARGET | FORM NO. 7 | TARGET | FORM NO. 7 |
| Coat weight (g/m$^2$) | — | 43.0 | — | 36.0 |
| Rolling ball tack (cm) | 1.0 | 1.3 | 1.5 | 1.3 |
| Polyken probe tack (kg) | 0.3 | 0.6 | 1.0 | 1.3 |
| 180° peel strength (N/m) | 400 | 550 | 500 | 800 |
| SAFT to Mylar °C | 60 | 71 | 93 | 85 |
| SAFT to kraft °C | 54 | 61 | 65 | 67 |
| Holding power (shear) | | | | |
| to steel (min)[d] | 200 | 500 | 2000 | 2000 |
| to kraft (min)[d] | 175 | 300 | 2000 | 2000 |
| Box loop test (days) | 10 | >30 | >30 | >30 |

[a] Kraton D-1107 rubber; Shell Chemical.
[b] Such as: Escorez 1310, Exxon Chemical or Wingtack 95, Goodyear Chemical, or other 100°C softening point $C_5$ resins.
[c] Shellflex 371; Shell Chemical.
[d] 12.7 × 12.7 mm contact, 1 kg load.

formulation provides a good balance of properties for general-purpose crepe or polyester tapes.

Using the concepts of contour diagramming and investigating specific formulations, it is possible to adjust the resulting adhesive properties to meet a wide variety of performance properties. However, in all cases, care must be taken to prevent degradation of the adhesive to ensure uniform and reproducible adhesive properties. Hot melt viscosity, shear strength, and bleed and high-temperature performance are adversely affected by adhesive degradation. But with proper precautions, such as inert gas blanketing and minimal heat history of the adhesive, reproducible, high-quality tapes and labels can be produced as a 100% solids hot melt system.

## General-Purpose Permanent Labels

Adhesives for labels have different requirements than tapes and should be formulated accordingly. Labels have lower cohesive strength than tapes, but the adhesive should not bleed through the label stock when exposed to temperatures of 60°C for 2 to 3 weeks. Higher levels of resin and/or plasticizer can be used in label formulations to help lower the cohesive strength and reduce formulation costs.

New polymers, such as Kraton DX-1112 rubber, have been developed specifically for adhesives where low cohesive strength and lower viscosity is desired. The formulation shown in Table 11-9 demonstrates a balance of pressure-sensitive adhesive properties and die cuttability when using Kraton DX-1112 rubber in this formulation.

## General-Purpose Duct Tape

Duct tape is the familiar silver colored polyethylene/cloth laminated pressure-sensitive product. Its name is derived from the major intended use it serves—to wrap air conveying ducts, keeping insulation positioned in place. It is also a popular tape product with the general public for a variety of patching jobs around the home and at work.

**Table 11-9. Comparison of two linear S-I-S thermoplastic rubber polymers for permanent labels.**

| BASE POLYMER ADHESIVE PROPERTIES | KRATON D-1107 RUBBER FORMULATION[a] | KRATON DX-1112 RUBBER FORMULATION[a] |
|---|---|---|
| Rolling ball tack (cm) | 2.0 | 1.0 |
| Loop tack (N/m) | 500 | 550 |
| Polyken probe tack (g) | 400 | 500 |
| Holding power (shear) to steel (min)[b] | 200 | 170 |
| Bleed (3 weeks at 60°C) | None | None |
| Die cuttability | Fair-Good | Good |

[a] Formulation

| Formulation | Parts by weight |
|---|---|
| Kraton rubber | 100.0 |
| Hydrocarbon tackifying resin | 150.0 |
| Shellflex 371 oil | 50.0 |
| Zinc dibutyl dithiocarbamate | 5.0 |

[b] 12.7 × 12.7 mm contact, 500 g load.

Most duct tape is made using a calendering process, with natural rubber as the major base polymer in the pressure-sensitive adhesive. The adhesive mass is applied solventless, and is not further heat treated. This solventless method is used because the adhesive coat weight is high, and efficient solvent removal via forced drying at temperatures below the melting point of polyethylene is difficult and expensive. High coat weights are necessary in order to ensure good adhesive coverage of the open weave cloth scrim, and to form a functional, continuous layer of pressure-sensitive adhesive for bonding.

Properties of duct tape made via the hot melt process with block copolymer as the sole adhesive base polymer are shown in Table 11-10, and are compared in standard tests with four commercial calendered duct tape constructions.

## High-Temperature Tapes via Crosslinking

Pressure-sensitive adhesives based on thermoplastic rubbers perform well over the temperature range required in the majority of pressure-sensitive product applications. However, as the temperature at which the product will be used approaches the glass transition temperature of the endblock phase of the rubber (about 92°C), the polystyrene endblock domains begin to soften and unlock the physical crosslinks. When this occurs, the rubber midblocks are no longer crosslinked and the adhesive shows a drastic reduction in cohesive strength. Therefore, adhesives based on thermoplastic rubber cannot be used, for example, on high-temperature masking tapes which must withstand a significant load at temperatures in paint baking ovens of up to 170°C. This limitation can be overcome by chemically crosslinking the rubber in the adhesive so that it will no longer be thermoplastic.

The double bonds in the midblocks of the unsaturated S-B-S and S-I-S polymers provide sites for chemical crosslinks. These unsaturated thermoplastic rubbers can be crosslinked with the usual sulfur, sulfur donor, or peroxide curing systems traditionally used for crosslinking unsaturated rubbers. There are many combinations of ingredients which could be used in a sulfur-based crosslinking package. For example, the crosslinking package could contain 0.3–1.5 phr of elemental sulfur or a sulfur donor (e.g., Sulfads), 0.3–1 phr of an accelerator (e.g., Altax or Methyl Zimate) and 0.3–1 phr of ZnO and stearic acid. A peroxide crosslinking system could contain 1–4 phr of peroxide (e.g., Di Cup) and could be used in combination with 5–25 phr of a reactive acrylic monomer (e.g., Sartomer SR 351). Either the sulfur or the peroxide crosslinking package can be mixed directly into the adhesive solution. Crosslinking occurs thermally as the adhesive tape passes through the solvent evaporation ovens.

**Table 11-10. Thermoplastic rubber pressure-sensitive adhesive duct tape vs. commercial duct tape products.**

| | HOT MELT THERMO-PLASTIC RUBBER[a] | | COMMERCIAL DUCT TAPES | | | |
|---|---|---|---|---|---|---|
| IDENTIFICATION PROPERTIES | LOW COAT WEIGHT | HIGH COAT WEIGHT | A | B | C | D |
| Substrate | Joanna Western[b] | | PE film/cloth scrim | | | |
| Release coat | None | None | — | — | — | — |
| Adhesive coat weight ($g/m^2$) | 124 | 210 | — | — | 204 | 228 |
| Rolling ball tack (cm) | 1.2 | 0.5 | 0.9 | 0.8 | 1.7 | 1.2 |
| Polyken probe tack (kg) | 0.6 | 0.9 | 0.6 | 0.5 | 0.8 | 0.7 |
| SAFT to Mylar (°C) | 74 | 81 | 75 | 42 | 100 | 114 |
| SAFT to kraft (°C) | 77 | 77 | 61 | 40 | 100 | 102 |
| 180° peel to steel (N/m) | 1460 | 2100 | 370 | 390 | 520 | 580 |
| 180° peel to backing (N/m) | 790 | 1320 | 230 | 300 | 840 | 860 |
| Quick stick to steel (N/m) | 820 | 1020 | 230 | 300 | 320 | 370 |
| Holding power to kraft (min) | | | | | | |
| 12.5 × 12.5 mm, 2 kg load | 165 | 160 | 10 | 1 | — | — |
| 12.5 × 12.5 mm, 1 kg load | 4450 | 5470 | 90 | 10 | 31 | 68 |
| Holding power to steel (min) | | | | | | |
| 12.5 × 12.5 mm, 2 kg load | 120 | 55 | 40 | 5 | — | — |
| 12.5 × 12.5 mm, 1 kg load | 300 | 190 | — | — | 28 | 70 |
| Box loop test, days to failure | | | | | | |
| 25 mm test strip | >35 | >35 | 80 | <1 | — | — |
| 12.5 mm test strip | >35 | >35 | <1 | <1 | — | — |

[a] Formula: 100 parts Kraton D-1107 rubber, 140 parts Wingtack Plus, 10 parts Shellflex oil, 5 parts Butazate, 5 parts $TiO_2$.

[b] Prelaminated polyethylene/cloth scrim made by Joanna Western Mills Co., Industrial Products Division.

**Table 11-11. Pressure-sensitive adhesive cross-linked with phenolic resin.**

| | |
|---|---|
| S-I-S polymer[a] | 100 |
| $C_5$ Resin[b] | 50 |
| Phenol-formaldehyde resin[c] | 20 |
| Zinc resinate[d] | 10 |
| Zinc dibutyl dithiocarbamate | 2 |
| 2.5-di(*tert*-amyl) hydroquinone[e] | 1 |
| % solids in toluene | 50 |

[a] Kraton D-1107 rubber, Shell Chemical.
[b] Wingtack 95, Goodyear.
[c] Amberol ST-137, Rohm and Haas.
[d] Reichhold Chemical, Newport Division.
[e] Santovar A, Monsanto.

Another approach to improve the high-temperature cohesive strength of adhesives based on thermoplastic rubber is by establishment of a thermoset network extending throughout the rubber phase. This can be accomplished by the use of reactive phenolic resins in combination with a metal catalyst. An example of a formulation which is claimed to be effective is given in Table 11-11. This type of crosslinking system can be mixed directly into the adhesive solution. Crosslinking is initiated thermally in the solvent evaporation ovens.

A thermally-initiated crosslinking reaction would almost always be used to crosslink an adhesive applied from solvent because crosslinking can be conveniently initiated in the solvent evaporation ovens. However, thermally-initiated crosslinking systems present considerable handling problems for hot melt adhesives in that they may crosslink the adhesive in the hot melt processing equipment. A crosslinking system which can be initiated by exposure to radiation is a more practical route to crosslink a hot melt adhesive. The formulations shown in Table 11-12 are two which can be crosslinked by exposure to radiation. The reactive monomer, SR-350 (trimethylolpropane trimethacrylate), is needed in the formulation to reduce the radiation dose required to achieve crosslinking of the adhesive. Both adhesive formulations in Table 11-12 tend to crosslink when held at hot melt processing temperatures. However, both adhesives crosslink slowly enough that they can be extrusion coated onto the tape substrate before significant crosslinking has occurred. Formulation A in Table 11-12 can be crosslinked by exposure under a $N_2$ blanket to electron beam radiation at a dose of 2 to 6 megarads. Formulation B in Table 11-12 contains a photoinitiator, Irgacure 651, in addition to the reactive monomer. The function of the photoinitiator is to dissociate rapidly under exposure to UV light to generate free radicals which initiate the crosslinking reaction. Formulation

**Table 11-12. Pressure-sensitive adhesive formulations crosslinked by exposure to radiation.**

| | A | B |
|---|---|---|
| S-I-S polymer[a] | 100 | 100 |
| $C_5$ tackifying resin[b] | 100 | 80 |
| Paraffinic/naphthenic oil[c] | 25 | — |
| Multifunctional coupling agent[d] | 25 | 25 |
| UV initiator[e] | — | 6 |
| Zinc dibutyl dithiocarbamate | 2 | 1 |

[a] Kraton D-1107 rubber, Shell Chemical.
[b] Wingtack 95, Goodyear.
[c] Shellflex 371 oil, Shell Chemical.
[d] SR-350, Sartomer (trimethylolpropane trimethacrylate).
[e] Irgacure 651; Ciba-Geigy (acetophenone).

B in Table 11-12 can be crosslinked by exposure under a $N_2$ blanket to UV light at the dose achieved by passing the adhesive under two 200-watt/in. medium-pressure Hg lamps at a line speed of 20 m/min. However, it is required that the light be filtered to remove the portion of the spectrum whose wavelength is less than 320 nm in order to prevent degradation of the surface of the adhesive.

Because their midblock is saturated, the S-EB-S and S-EP-S polymers are much more difficult to crosslink. A formulation similar to formulation A in Table 11-12 containing an S-EB-S polymer rather than an S-I-S polymer cannot be crosslinked at commercially acceptable rates by exposure to UV light. It can be crosslinked by exposure to electron beam radiation but doses of at least 6 megarads are required.

It must be noted that while crosslinking an adhesive based on a thermoplastic rubber brings about the desired improvement in high-temperature cohesive strength, it also causes a reduction in aggressive tack of the adhesive. Unless the rubber has been designed specifically for use in crosslinked adhesives, chemical crosslinking of a thermoplastic rubber corresponds roughly to overcuring a conventional rubber vulcanizate. This reduction in aggressive tack can be partially offset by including a plasticizer in the adhesive formulation.

## Low-Temperature Tapes and Labels

A special type of formulation is required for pressure-sensitives which are used in cold environments, e.g., tapes and labels that are used for frozen food packaging and in some electrical insulating tapes.

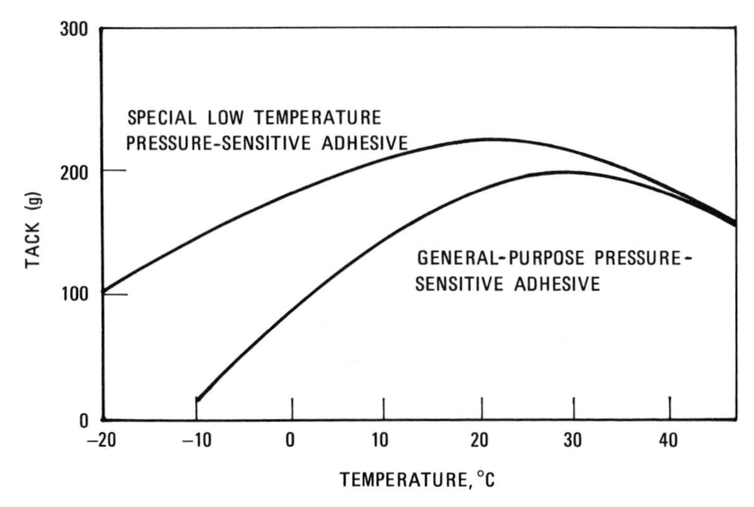


Fig. 11-23. Probe tack as a function of temperature.

Most adhesives lose their tack and adhesion properties as temperature decreases as depicted in Fig. 11-23, which presents the results of probe tack measurement as a function of temperature. If the temperature falls below 20°C, the tack decreases rather sharply. At 0°C, common adhesives are very low in aggressive tack, and at −10°C, they have no measurable tack at all.

This behavior of the adhesive upon lowering the temperature can be readily explained by considering its rheological properties. Tack of an adhesive is related to its modulus. The lower the modulus, the easier the adhesive will be deformed and flow, making good contact with the adherent. The rubber component in a pressure-sensitive adhesive alone has practically no tack. However, decreasing the modulus by addition of sufficient amounts of rubber-compatible resins and plasticizers raises the aggressive tack at room temperature.

The modulus value of the adhesive and thus the tack at any given temperature is highly dependent on the glass transition temperature of the rubber component $T_g$ and the glass transition temperature of the adhesive composition $T'_g$. These effects are shown schematically in Fig. 11-24. As the temperature decreases the modulus increases. The adhesive becomes less able to flow and make contact with the substrate. At or below $T'_g$, the adhesive behaves as a rigid solid and the tack will be nil. It is clear that the lower the $T'_g$ of an adhesive, the longer the tack properties will be maintained with a decrease in temperature.

The $T_g$ of elastomers is rather low. For polyisoprene and polybutadiene, midblock components in unsaturated thermoplastic rubbers, $T_g$'s

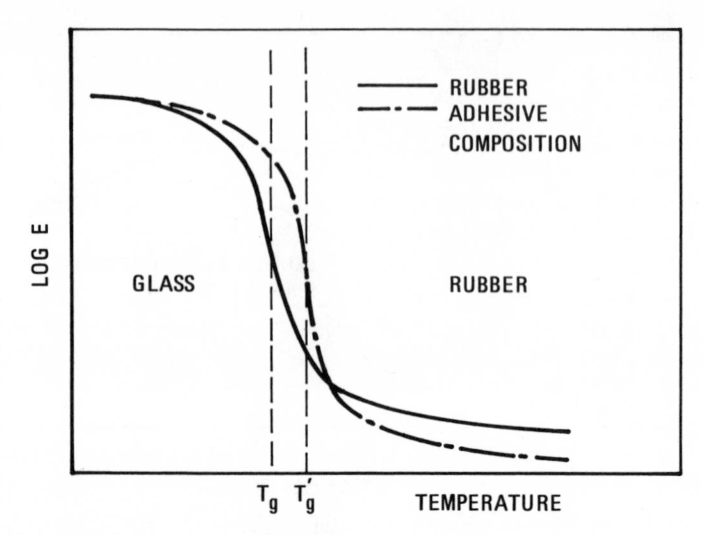


Fig. 11-24. Schematic presentation of modulus, $E$, as a function of temperature and pressure-sensitive adhesive glass transition temperature ($T_g$).

are −58 and −91°C, respectively. However, when midblock associating resins are added, to obtain tack at ambient temperature, the $T_g$ of the composition increases rapidly with increasing resin content ($T_g$ to $T'_g$ shift illustrated in Fig. 11-24). The extent of this shift depends on the type of resin and plasticizers incorporated, and the resultant $T'_g$ value of the ultimate adhesive blend.

By using materials with a relatively low $T_g$ and low rate of change in modulus with temperature, adhesives can be formulated with good low temperature tack. However, to maintain the ambient temperature properties, careful selection of all ingredients is necessary.

To predict the presence of tack at low temperature for a given adhesive formulation, it will be useful to know its $T'_g$. Equation (3) can be used to predict the $T'_g$ of a blend of fully compatible ingredients from the $T_g$ values of the separate components.[4] Hence,

$$\frac{1}{T'_g} = \sum_{i=1}^{n} \frac{W_i}{T_{gi}} \tag{3}$$

where $T'_g$ = glass transition temperature (°K) for the adhesive blend

$T_{gi}$ = glass transition temperature (°K) for component $i$

$W_i$ = weight fraction of component $i$

If the $T_{g_i}$ of the components are known, the $T'_g$ of the adhesive formulation can be calculated and the lowest temperature at which tack could be expected is then known. In Table 11-13, the calculated $T'_g$ for a number of adhesive formulations based on an S-I-S thermoplastic rubber are presented together with their low-temperature tack values.

S-I-S block copolymers show two $T_g$'s, i.e., one corresponding to their polystyrene phase (92°C) and one to their rubber phase (−58°C for polyisoprene). Because the low-temperature performance is determined by the latter, this value is used in the calculation.

A probe tack test was used to measure the tack at the low temperatures. In this test, a flat-ended stainless steel probe was contacted under a light weight with the adhesive surface for a very short time. The force required to withdraw the probe was measured. From the table, it can be seen that measurable tack is developed as long as the test temperature is above the $T'_g$ of the given adhesive formulation.

Contour diagrams presenting $T'_g$ for adhesive formulations as a function of type and amount of the ingredients can be very helpful to predict formulations with good low-temperature properties. In Fig. 11-25, an example is given for a combination of linear S-I-S polymer, Wingtack 76 as

**Table 11-13. Low temperature tack and $T'_g$ for a number of S-I-S based pressure-sensitive adhesives.**

| COMPONENTS | $T_g$(°C)[a] | COMPOSITION | | | | | |
|---|---|---|---|---|---|---|---|
| S-I-S Polymer[b] | −58 | 100 | 100 | 100 | 100 | 100 | 100 |
| Wingtack[c] 95 | 51 | 80 | — | — | 125 | — | 125 |
| Wingtack 76 | 32 | — | 80 | — | — | 125 | — |
| Wingtack 10 | −28 | — | — | 80 | — | — | — |
| Shellflex 371[d] | −64 | — | — | — | 25 | 25 | — |
| Indopol 150[e] | −80 | — | — | — | — | — | 50 |
| PROPERTIES | UNITS | | | | | | |
| $T'_g$ (calculated) | (°C) | −16 | −22 | −45 | −12 | −19 | −23 |
| Probe Tack at: | | | | | | | |
| −12°C | (g) | 150 | 180 | — | 35 | 160 | 225 |
| −18°C | (g) | 6 | 190 | 290 | 0 | 60 | 180 |
| −29°C | (g) | — | — | 310 | — | — | — |

[a] $T_g$ value measured by DSC.
[b] Kraton D-1107 rubber; Shell Chemical (isoprene midblock $T_g$).
[c] Hydrocarbon resin, Goodyear.
[d] Paraffinic/naphthenic processing oil, Shell Chemical.
[e] Low molecular weight polybutene, Amoco.

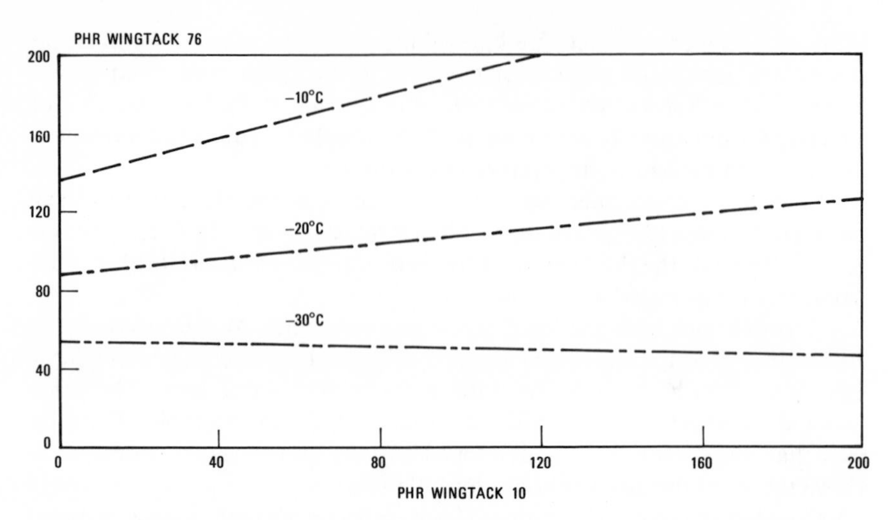


Fig. 11-25. Contour isopleths for calculated glass transition temperature ($T_g$) for two resins in an S-I-S polymer (Kraton® D 1107 Rubber) based system.

tackifying resin and Wingtack 10 as plasticizer. From this figure, it is easy to find formulations in which tack could be expected down to −30°C.

However, other property requirements, such as melt viscosity, ambient tack properties, peel adhesion, and holding power limit the area from which an adhesive for low-temperature tack can be chosen. Formulations with a relatively low amount of high softening resin and a relatively high amount of plasticizer may fulfill both the requirement for low-temperature tack and low melt viscosity. However, 180°C peel adhesion and holding power will be unacceptably low.

By superpositioning of the contour diagram for $T'_g$ with those for the other critical properties, it may be possible to choose a proper formulation with a satisfactory balance in properties, including low-temperature performance.

In Table 11-14, a number of adhesives, which were selected in that way, for low-temperature labels is presented together with the values of the most critical properties.

## Weatherable Tapes

Pressure-sensitive adhesives are acceptable for a wide variety of applications. However, only those polymers that are fully saturated can offer excellent resistance to degradation by air, ozone or UV light. Kraton G thermoplastic rubbers have a saturated rubber midblock and, when formulated with saturated resins, plasticizers and stabilizers, can obtain a balance of adhesive properties as well as resistance to degradation by oxidation or UV light.

The formulation shown in Table 11-15 demonstrates a good balance of adhesive properties in a formulation containing only saturated compounding ingredients. Although the tack of this formulation, as measured by standard tack tests, appears to be very good, its tack in a "thumb appeal" test (a qualitative estimate of tack when touched under mild pressure with one's finger) is only fair.

The UV resistance of two adhesives based on saturated formulating ingredients was compared in an accelerated weatherometer test with the resistance of an adhesive based on unsaturated formulating ingredients. The 180° peel strength of the adhesive cast on polyester backing and adhered to glass was measured as a function of exposure time in a Xe lamp weatherometer. As shown in Table 11-16, the performance of the adhesive based on unsaturated ingredients, formulation A, is poor. In less than six days exposure, the adhesive was so badly degraded that it failed cohesively when

**Table 11-14. Selected low temperature pressure-sensitive label formulations.**

| INGREDIENTS | UNITS | A | B | C | D | |
|---|---|---|---|---|---|---|
| S-I-S polymer[a] | (parts) | 100 | 100 | 100 | 100 | |
| Wingtack[b] 76 | (phr) | 75 | — | — | — | |
| Super Static 80[c] | (phr) | — | 62.5 | — | — | |
| Escorez 1310[d] | (phr) | — | — | 50 | — | |
| Wingtack 95 | (phr) | — | — | — | 50 | |
| Wingtack 10 | (phr) | 50 | 62.5 | 75 | 100 | |
| Zinc dibutyl dithiocarbamate | (phr) | 5 | 5 | 5 | 5 | |
| PROPERTIES | UNITS | | | | | REQUIREMENTS[e] |
| $T'_g$ (calculated) | (°C) | −21 | −24 | −26 | −25 | — |
| Probe tack at −18°C | (g) | 390 | 420 | 450 | 470 | > 70 |
| Rolling ball tack | (cm) | 1.2 | 1.0 | 1.1 | 1.0 | < 2.5 |
| Loop tack[f] | (N/m) | 540 | 620 | 580 | 700 | > 230 |
| 180° peel adhesion | (N/m) | 580 | 620 | 660 | 770 | > 390 |
| Holding power[g] | (min.) | > 1000 | > 1000 | > 1000 | 350 | > 60 |
| Melt viscosity at 175°C | (Pa·s) | 43 | 41 | 43 | 23 | — |

[a] Kraton D-1107 Rubber, Shell Chemical.
[b] Hydrocarbon resin, Goodyear Chemical.
[c] Hydrocarbon resin, Reichhold Chemical.
[d] Hydrocarbon resin, Exxon Chemical.
[e] As measured on available commercial labels.
[f] To steel, 25 × 25 mm contact area.
[g] Polyester film to steel, 1 kg load, 12.5 × 12.5 mm contact area.

## Table 11-15. Weather resistant pressure-sensitive adhesive tape based on a saturated S-EB-S thermoplastic rubber.

| FORMULATION | DRY WEIGHT (phr) |
|---|---|
| S-EB-S polymer[a] | 100 |
| Arkon P85[b] | 115 |
| Tuffalo 6056[c] | 15 |
| Irganox 1010[d] | 1 |
| Tinuvin 327[d] | 0.25 |
| ADHESIVE PROPERTIES | |
| Coat weight ($g/m^2$) | 35 |
| Rolling ball tack (cm) | 2 |
| Polyken probe tack (kg) | 0.9 |
| 180° peel strength (N/m) | 550 |
| SAFT to Mylar (°C) | 80 |
| Holding power to kraft (min) | 150 |
| Melt viscosity @177°C (Pa·s) | 60 |

[a] Kraton GX-1657 rubber, Shell Chemical.
[b] Saturated alicyclic hydrocarbon resin, Arakawa.
[c] Atlantic Richfield Co.
[d] Ciba-Geigy Co.

## Table 11-16. UV resistance: saturated polymer (S-EB-S) vs. unsaturated polymer (S-I-S) based pressure-sensitive adhesive tapes.

| EXPOSURE TIME[a] (days) | 180° PEEL STRENGTH (N/m) | | |
|---|---|---|---|
| | FORMULATION A[b] | FORMULATION B[b] | FORMULATION C[b] |
| 0 | 840 | 510 | 550 |
| 6 | 1000–2000 Coh[c] | 610 | 700 |
| 13 | 880–1600 Coh | 650 | 800 |
| 21 | 860 Coh | 770 | Tear[d] |
| 31 | 930 Coh | 1030 | Tear |
| 42 | 530 Coh | Tear[d] | Tear |
| 63 | 0 Coh | 1120 | Tear |

[a] 180° peel sample on glass, exposed through glass in Xe lamp weatherometer.

[b] Formulations

| | A | B | C |
|---|---|---|---|
| Kraton D-1107 rubber | 100 | — | — |
| Kraton GX-1657 rubber | — | 100 | 100 |
| Escorez 5380 | 80 | 80 | — |
| Arkon P85 | — | — | 115 |
| Tuffalo 6056 | — | — | 15 |
| Irganox 1010 | 1 | 1 | 1 |
| Tinuvin 327 | 0.25 | 0.25 | 0.25 |

[c] Coh indicates cohesive failure within adhesive layer.
[d] Tear indicates polyester (Mylar) backing becomes brittle and tears.

peeled from the glass. As exposure is continued, the peel sample continues to fail cohesively, but at lower and lower peel strengths as the cohesive strength of the adhesive is further reduced by degradation.

The UV resistance of the adhesives based on saturated ingredients, formulations B and C, is much better than that of formulation A in Table 11-16. In fact, these adhesives appear to weather better than the polyester backing since, after extended exposure, the backing tears during the peel strength measurement.

## REFERENCES


1. Craver, J. Kenneth, and Tess, Ray W. (eds.). *Applied Polymer Science*, Washington, D.C.: Organic Coatings and Plastics Chemistry Division of the American Chemical Society, p. 413, 1975.
2. Ceresa, R. J. (ed.). *Block and Graft Copolymerization, Volume 1*, London: John Wiley and Sons, Ltd., p. 151, 1973.
3. *Encyclopedia of Polymer Science and Technology, Supplement Volume 2*, New York: Wiley-Interscience pp. 417–418, 1977.
4. Nielsen, L. E. *Mechanical Properties of Polymers*, New York: Van Nostrand Reinhold Co., p. 27, 1963.
5. *Encyclopedia of Polymer Science and Technology*, Volume 3, p. 833, New York: John Wiley and Sons, 1965.
6. *Encyclopedia of Chemical Technology*, Supplement volume, p. 889, New York: Wiley-Interscience, 1971.
7. Solvent Properties Chart. Shell Chemical Bulletin.
8. Kraus, G., Rollmann, K. W., and Gray, R. A. *J. Adhesion*, **10**: 221–236 (1979).
9. Skeist, I. (ed.). *Handbook of Adhesives, Second Edition*, New York: Van Nostrand Reinhold Co., 1977, pp. 304–330.

# Chapter 12

# Butyl Rubber and Polyisobutylene

**J. J. Higgins**
*Exxon Chemical Americas*
*Florham Park, New Jersey*

**F. C. Jagisch**
*Exxon Chemical Company*
*Baton Rouge, Louisiana*

**N. E. Stucker**
*Exxon Chemical Americas*
*Baytown, Texas*

Butyl rubber and polyisobutylene are elastomeric polymers[1] used quite widely in pressure-sensitive adhesives both as primary elastomers and as tackifiers and modifiers. Butyl rubber is a copolymer of isobutylene with a minor amount of isoprene, while polyisobutylene is a homopolymer.

As shown in Fig. 12-1, the carbon-hydrogen backbone of butyl rubber is relatively long and straight, containing between 47,000 and 60,000 units. This regular structure, with few double bonds or reactive sites, renders butyl very stable and quite inert to the effects of weathering, age, and heat. It has good resistance to vegetable and animal oils and to attack by chemicals. Being an all-hydrocarbon material, the butyl polymer has very low water absorption and is soluble in typical hydrocarbon solvents. The many side groups attached to the polymer chain produce a high degree of damping. Since these side groups are not large in size and are regularly spaced, close, unstrained molecular packing results. To this is attributed the unique low air, moisture, and gas permeability of this polymer.

The polyisobutylenes are molecularly similar to butyl rubber but have only terminal unsaturation. Many of their characteristics are similar, e.g., age and chemical resistance, very low water absorptivity and low permeability. The polyisobutylenes are produced over a wider molecular weight range than butyl. The low molecular weight grades are soft, tacky, water-white semiliquids, while the higher molecular weight grades are strong, tough, elastic rubbers.

Polymers of the isobutylene family (i.e., butyl and polyisobutylene) have very little tendency to crystallize. They depend upon molecular entanglement or crosslinking for their strength, rather than upon crystallinity. The completely amorphous character of these polymers gives an internal mobility which imparts flexibility, permanent tack, and resistance to shock. Their low glass transition temperature, −60°C, indicates functional properties such as flexibility are maintained at temperatures considerably below ambient. While their tack is high, these polymers have no or very little polarity, and thus their chemical attraction to many surfaces is weak. Therefore, they are often mixed with resins and other materials which impart some polar character to the blend. Polyisobutylene polymers possess very good dielectric properties and electrical stability.

Butyl rubbers are also available in several halogenated grades, in a low molecular weight, semiliquid form, as a latex and in various modified and partially crosslinked grades. Grades of butyl rubber and polyisobutylene homopolymer currently available in commercial quantities are listed in Table 12-1. In addition to Exxon Chemical Company, other suppliers include Columbian Chemicals Division of Cities Service Company and Polysar Limited.

## BASIC PROPERTIES

### Butyl Rubber

All grades of regular butyl rubber are tacky, rubbery, light amber colored solids, manufactured by copolymerizing isobutylene with less than 3% iso-

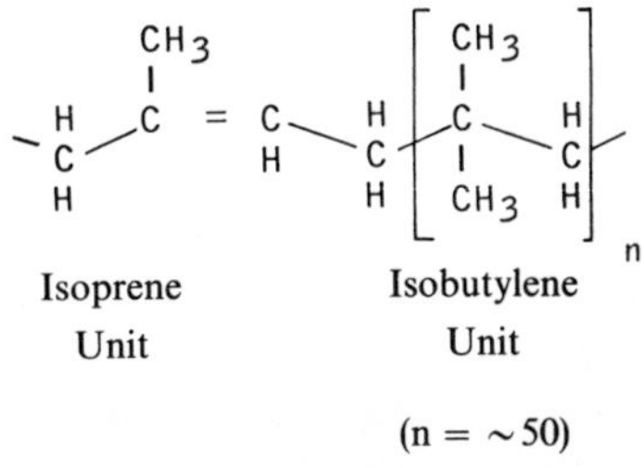

$$-CH_2-C(CH_3)=CH-CH_2-\left[C(CH_3)_2-CH_2\right]_n-$$

Isoprene Unit    Isobutylene Unit

(n = ~50)

Fig. 12-1. Butyl rubber molecule.

**Table 12-1. Commercial grades of Exxon butyl rubber and Vistanex polyisobutylene.**

| GRADE[a] | APPROX. VISCOSITY AVERAGE MOLECULAR WT. | APPROX. ISOPRENE UNITS/100 MONOMER UNITS (MOLE % UNSATURATION) | COMMENTS |
|---|---|---|---|
| Vistanex LM-MS | 44,000 | 0 | Semiliquid, tacky polymers used mainly as tackifiers |
| LM-MH | 53,000 | 0 | |
| Exxon Butyl 065 | 350,000 | 0.8 | Low MW grades with high age stability |
| 165 | 350,000 | 1.2 | |
| 268 | 450,000 | 1.5 | Used for higher cohesive strength |
| 365 | 350,000 | 2.0 | Used in curing applications |
| 077 | 425,000 | 0.8 | Special BHT-stabilized FDA grade |
| Chlorobutyl 1066 | 400,000 | 1.7 | Both grades contain approximately 1.2 wt. % chlorine |
| 1068 | 450,000 | 1.7 | |
| Bromobutyl 2244 | 425,000 | 1.5 | Contains approximately 2.0 wt. % bromine |
| Vistanex MM L-80 | 725,000–1,050,000 | 0 | Lowest viscosity of MM grades |
| MM L-100 | 1,050,000–1,450,000 | 0 | Widely used in pressure-sensitive adhesives |
| MM L-120 | 1,450,000 1,900,000 | 0 | Widely used in pressure-sensitive adhesives |
| MM L-140 | 1,900,000–2,350,000 | 0 | Highest viscosity of MM grades |

[a] Specific gravity: 0.93 : Bromobutyl 2244
0.92 : All other grades.

prene. Double bonds introduced into the macromolecule by isoprene permit the polymer to be crosslinked or vulcanized. The various butyl grades differ in mole percent unsaturation (number of isoprene units per hundred monomer units in the polymer chain), molecular weight, and the type of stabilizer incorporated during manufacture to prevent degradation. A minor amount of metallic stearate is also added to the rubber to prevent agglomeration of polymer particles during the manufacturing process.[2,3]

The stabilizer used in Exxon butyl rubber is "nonstaining", zinc dibutyl dithiocarbamate. Stabilizer content is in the 0.05–0.20 weight percent range. A special BHT-stabilized grade of butyl is also available for applications requiring broader FDA regulation.

A low molecular weight, semiliquid analog of butyl rubber is also commercially available. One such product is Kalene, marketed by Hardman Inc. (Belleville, N.J.). This tacky polymer can be partially cured through its unsaturation for cohesive strength benefits. It can be easily compounded into high solids adhesives, mastics, and coatings and will cure at room temperature. It can function as a curable tackifier/plasticizer in blends with higher molecular weight polymers.

## Polyisobutylene

Polyisobutylenes (PIBs) have the inertness of paraffinic hydrocarbons and cannot be cured or vulcanized using standard rubber technology. With adjustment for gross differences in molecular weight, polyisobutylenes can often be used interchangeably with butyl in cement and adhesives compositions that will not be cured.

The low molecular weight grades of polyisobutylene, Vistanex LM-MS and LM-MH, are permanently tacky, clear white to very light yellow semiliquids, containing no stabilizer.[4] They have broad acceptability in FDA-regulated applications. These grades will show some degree of flow at elevated temperature and can be pumped and handled as liquids at temperatures of 150–180°C. Primary uses are as permanent tackifiers in a variety of cements, pressure-sensitive adhesives and hot melt adhesives. These polymers provide tack, softness and flexibility and can assist in improving adhesion by "wetting out" various hard-to-adhere substrates. They are particularly useful for enhancing adhesion to polyolefin plastic surfaces.

The high molecular weight polyisobutylene grades, Vistanex MM L-80 through L-140, are white to light yellow, tough, rubberlike solids which contain less than 0.1% butylated hydroxytoluene (BHT) stabilizer.[5] These grades have fairly wide acceptance in FDA-regulated applications. They are used to impart strength and flow resistance to solvent cements and pressure-sensitive adhesive label stock. They are also used in certain hot melt compositions where they provide improved flexibility and impact resistance, particularly at low temperatures.

In addition to the Vistanex grades described, BASF (Badische Anilin und Soda Fabrik Co.) also supplies polyisobutylene homopolymers under the trade name Oppanol.

## Halogenated Butyl Rubber

Chlorobutyl rubber[6] is prepared by chlorinating the regular butyl polymer under controlled conditions so that reaction is primarily by substitution and little of the unsaturation originally present in the macromolecule is lost. The chlorine is believed to enter the molecule at the highly reactive allylic position, one carbon removed from a double bond. Approximately 1.2 weight percent chlorine is present in the commercial grades of Exxon Chlorobutyl. The chlorine tends to enhance the reactivity of the double bonds as well as supply additional reactive sites for crosslinking. As a result, a wide variety of different type cure systems can be used to vulcanize this elastomer.

Chlorobutyl can be readily blended and cured with more highly unsaturated elastomers. In the adhesives area, it has been blended with both regular butyl and natural rubber and then preferentially cured through the chlorine to improve strength. The reactive chlorine will also tend to increase adhesion to many polar substrates.

A bromobutyl polymer[7] is also available, marketed by both Exxon Chemical Company and Polysar Limited. Adhesives applications are under development.

## Butyl Rubber Latex

Butyl rubber can be emulsified to give a latex that might typically have a solids content of approximately 60 weight percent, a pH of 5.5 and a Brookfield viscosity of 7,500 cps. An anionic emulsifier is used to prepare butyl latex having an average particle size of 0.3 $\mu$m. BHT can be used as an antioxidant, with a small amount of formaldehyde sometimes added as a preservative. One supplier of butyl latex is Burke-Palmason Chemical Company (Pompano Beach, Florida). Vistanex polyisobutylenes have also been produced in latex form.

Butyl latex has excellent mechanical, chemical, and freeze thaw stability which allows for a wide latitude in compounding and blending with other ingredients. Because of its stability, it cannot be coagulated in a controlled manner using standard latex coagulants. When dried, it possesses the typical butyl characteristics of good aging, flexibility, low permeability, tack, etc.

Butyl latex is used in packaging adhesive applications and as a tackifying and flexibilizing additive in higher strength adhesives based on more brittle polymers. It is noted for its compounded adhesion to polyolefin film and fibers and is used in laminating and seaming adhesives and specialty binders and coatings for both polyethylene and polypropylene.[8]

## Modified Butyls

A number of partially crosslinked butyl rubbers and other modified forms are commercially available. Most of these are specifically produced for the sealants and adhesives markets.[9] Partially crosslinked butyls include Polysar's XL-20 and XL-50,[10] terpolymers in which DVB (divinylbenzene) is added during polymerization to impart a degree of cure, Columbian Chemicals' Bucar 5214,[11] a filled and plasticized, partially cured butyl rubber masterbatch crumb, and Bucar 5245, a more lightly crosslinked composition. These polymers have added toughness, strength, and flow resistance for use in sealing tapes for construction and automotive applications. These polymers are generally not used in standard solvent-based systems because their gel structure causes solution problems, although after compounding with fillers, etc., some measure of dispersion in hydrocarbon solvents is possible. An exception here is Bucar 5245, which dissolves more readily due to its lower crosslink density.

Various depolymerized butyls, butyl rubber solutions or cutbacks, butyl/plasticizer blends and highly plasticized/partially cured butyls are also available. These products are used by many adhesive companies where equipment limitations prohibit the use of the tougher, regular bale forms of butyl rubber. Principal suppliers of these materials include Rubber Research Elastomerics, Inc., Minneapolis, Minnesota; Adhesives Development and Chemical Operations, Inc. (ADCO), Michigan Center, Michigan; and A-Line Products Company, Detroit, Michigan.

## ADHESIVES COMPOUNDING

### Choice of Polymer

The family of isobutylene polymers (homopolymers and the butyl copolymers) is extremely broad. This wide selection of completely compatible polymers permits the knowledgeable compounder to achieve a wide range of properties. One example is the desire to increase cohesive strength while maintaining other valuable characteristics such as tack. There are a variety of ways in which cohesive strength can be modified by proper polymer selection in polyisobutylene systems. The following techniques are presently of commercial interest:

1. For strength, choose the highest molecular weight polyisobutylene (PIB) or butyl rubber grade consistent with application or processing requirements.
2. Blend butyl rubber with polyisobutylene and cure the butyl portion. Since the PIB does not cure, the extent of cure in the system and thus the cohesive strength is governed by the butyl content.
3. Partially cure butyl rubber. This can best be done by adding a carefully measured amount of curative and fully curing to the extent of the curative present, or by using one of the available precured grades.
4. Blend either PIB or butyl with Chlorobutyl and cure only the Chlorobutyl portion by utilizing the chlorine functionality. An example of a curative that will react with Chlorobutyl and not with butyl is zinc oxide. This technique is also of interest in blends of Chlorobutyl with either SBR or natural rubber for pressure-sensitive adhesives.

Another example of the compounding latitude afforded by polymer selection in the polyisobutylene family is tack. Tack is increased with Vistanex LM, butyl rubber can be depolymerized to any desired tacky state, or a wide range of compatible tackifiers can be incorporated.

In terms of the butyl polymers, the variable of greatest interest to the adhesives compounder is usually molecular weight. All grades of butyl rubber are of such low unsaturation that this characteristic does not affect most performance properties in uncured or partially cured formulations. However, in applications not requiring cure, the lowest unsaturation grades are usually chosen for optimum aging.

## Pigments and Fillers

The same pigments and fillers commonly used with other rubbers can be compounded with butyl rubber and polyisobutylene, and the general principles of selection are the same. Very fine pigments increase cohesive strength and stiffness, reduce cold flow, and also reduce tack. Platy pigments such as mica, graphite, and talc are preferred for acid and chemical resistance and low gas permeability. Some of the coarser pigments increase tack. Zinc oxide increases tack and cohesive strength (and also plays an important chemical role in the vulcanization of butyl). Aluminum hydrate, lithopone, whiting, and the coarser carbon blacks such as thermal blacks also increase tack with moderate increase in cohesivity. Clays, hydrated silicas, calcium silicates, silico-aluminates, and the fine furnace and thermal blacks increase cohesive strength and stiffness. Stiffness can also be increased by use of very fine silica pigment and magnesium oxide or carbonate.

Special situations call for special types of pigments and fillers. Several examples may make the point more clearly.

1. Where cost is a prime consideration, the calcium carbonates often produce adequate physical properties and are a good choice.
2. In the formulation of electrical tapes and mastics, the fillers must be chosen for their electrical properties, both original and after water immersion. Mistron Vapor Talc and Whitetex Clay are good selections.
3. In certain compositions, such as tacky sealing tapes for automotive glazing, the filler must contribute to the strength of the composition. In this case, a reinforcing black such as HAF performs best.

Filler selection can be critical to the performance of the product and should be given careful consideration.

## Tackifiers, Plasticizers, and Other Additives

As has been pointed out previously, a wide range of properties, especially rheological properties, can be obtained within the polyisobutylene family of polymers. In addition, a wide range of plasticizers and tackifiers are used to extend the viscosity range and to control the tack and cohesive strength level. A common plasticizer is polybutene.[12] This material is available in a number of molecular weights so the viscosity and volatility can be selected for the application. Other liquid materials used as plasticizers include paraffinic oils, petrolatum, and certain phthalates with long aliphatic side chains such as ditridecyl phthalate.

Resins are used in butyl and PIB-based pressure-sensitive adhesives to develop the desired level of tack. A wide range of resins can be used but the most common are probably the polyterpenes such as Piccolyte S 115, the terpene-phenolics such as Schenectady's SP-553, phenol-formaldehyde resins such as Schenectady's SP-1068, modified rosins and rosin esters like Staybelite and Pentalyn, and the hydrocarbons like Escorez 1102, Escorez 1304, and Escorez 1315. Usually, resins are used with plasticizers to obtain the desired balance of tack and cohesive strength.

Many other polymeric additives are used in butyl and PIB pressure-sensitive adhesives. Two worthy of mention are amorphous polypropylene, which can reduce cost and improve processing as well as impart thermoplastic character, and various waxes, which function in much the same way.

Special materials used as primers to obtain bonds of PIB-based adhesives to specific substrates such as glass are various organic silanes. An example is an epoxy silane like A-187. There is evidence that amino silanes like A-1100 and methacryoxy silanes like A-174 may also be of value. In certain textile applications, isocyanate primers or adhesion promoters can be used with butyl cements.

Antioxidants are often used in butyl and PIB adhesives to protect against severe environmental aging such as UV light or heat. Materials such as Ethanox 702, Butyl Zimate, AgeRite White, Irganox 1010, BHT, and even sulfur have been used.

## Curing Systems

There are four curing systems that are of general interest in cements, adhesives, sealants, and coatings based on butyl and Chlorobutyl.[13] These are the (1) "quinoid" cure, (2) cure with sulfur or sulfur donor compounds, (3) resin cure, and (4) for Chlorobutyl only, the zinc oxide cure. The introduction of crosslinking increases cohesive strength as stated previously but reduces tack and solubility in hydrocarbon solvents and thus finds only limited use in pressure-sensitive type adhesives. Therefore, only summary information on these cure systems will be provided here.

The quinoid cure is an old, easily controlled system that typically is formulated as a split batch two-part cement designed to cure at or even below room temperature. The resultant network is tightly cured, highly resistant to ozone, heat, chemicals and other environmental attack and provides good electrical properties. Dark color bodies are formed during the reaction, so quinoid cures cannot be used to make white products.

A commonly used system makes use of *p*-quinone dioxime (commercially known as QDO) or dibenzoyl *p*-quinone dioxime (DBQDO) combined with an oxidizing agent such as $MnO_2$, $PbO_2$, $Pb_3O_4$, or benzothiazyl disulfide. The DBQDO system is less active, and thus is easier to control at elevated temperatures.

Sulfur curing systems for butyl include elemental sulfur, thiuram or dithiocarbamate accelerators and thiazole or thiazyldisulfide activators. Zinc oxide or other metallic oxides are necessary to attain satisfactory cure. The relatively minor differences in unsaturation between grades of butyl are significant in regard to sulfur cures, the rate of cure and number of attainable crosslinks increasing with increasing polymer unsaturation. Sulfur cures find little commercial value in adhesives, however, due to the generally elevated temperatures required to generate crosslinks.

The resin cure of butyl normally uses one of a series of active brominated phenolic resins such as Schenectady's SP-1055. The resin cure of butyl can be made to proceed at temperatures from ambient to very high, depending on the specific resin chosen and on the concentration and type of cure activator. Features of the resin cure are high stability, heat resistance, and the ability to be compounded into light colored and white formulations.

The zinc oxide cure of Chlorobutyl is of interest in adhesives primarily because it allows Chlorobutyl to be cured in the presence of other elastomers without affecting the other elastomers. The zinc oxide cure is also of interest in certain FDA-sensitive applications where the elegant simplicity of a compound that contains only a polymer and zinc oxide has obvious appeal.

If it is desired to cure only a small fraction of the polymer base, this can be done in two ways. A small and carefully measured amount of curative could be used and reacted to the full extent of the curative present. Alternatively, a portion of Chlorobutyl could be blended with a polymer which does not cure via zinc oxide, for example, butyl; an excess of zinc oxide could be added and the dispersed Chlorobutyl phase totally cured. The result is butyl with higher cohesive strength for use as an adhesive (or sealant) base.[14] Similarly, Chlorobutyl can be blended with natural rubber and the Chlorobutyl phase cured with zinc oxide. The resultant material is a high cohesive strength base for pressure-sensitive adhesives.

## SOLVENTS AND SOLUTION PREPARATION PROCEDURES

The polyisobutylene family of polymers is soluble in hydrocarbon and chlorinated solvents, but not in the common alcohols, esters, ketones, and other low molecular weight oxygenated solvents. Volatile paraffinic solvents such as hexane, heptane, and naphtha are commonly used in cement and adhesive work. Cyclohexane and the chlorinated solvents like perchloroethylene give solutions of much higher viscosity on an equal weight percentage basis than the common paraffinic and aromatic solvents. A substantial part of this difference is due to the much higher specific gravity of chlorinated solvents. The resistance to solution in oxygenated solvents is often used since butyl and PIB films can withstand immersion in these common solvents.

A common consideration in solvent selection is air pollution requirements. For this reason solvents like Laktane (quite fast), VM&P Naphtha (medium evaporation rate) and Rule 66 Mineral Spirits (slow) are generally of interest. The presence of minor amounts of stabilizer and metal stearate in regular butyl and PIB may cause pure gum cements to be slightly cloudy even though the polymer is completely dissolved. Under certain circumstances, these will slowly settle out of low viscosity solutions of the polymers. The only deficiency is in appearance, and the settling can be avoided or retarded by working with high viscosity solutions.

The optimum solids level of cements to be applied by various methods is, of course, subject to broad variation depending on the viscosity given by various solvents, filler content, etc. In general, butyl rubber and polyisobutylene cements for application by spraying contain 5–10% solids, for dipping 10–30%, for spreading 25–55%, and for application by finger or spatula, 50–70%.

The relationship between solids and viscosity is logarithmic, as shown in Figs. 12-2 and 12-3.[15] A small change in solids can make a large difference in solution viscosity, especially at higher levels. Viscosity is also dependent on polymer molecular weight and a small change in the molecular weight of the polymer can result in a relatively large change in the viscosity of polymer solutions.[16]

The various butyl and high molecular weight PIB grades (e.g., Vistanex MM) are marketed in solid, bale form. To facilitate solvation for pressure-sensitive adhesives use, a common initial step is to increase surface area greatly by physical size reduction and thereby decrease solvation time. This

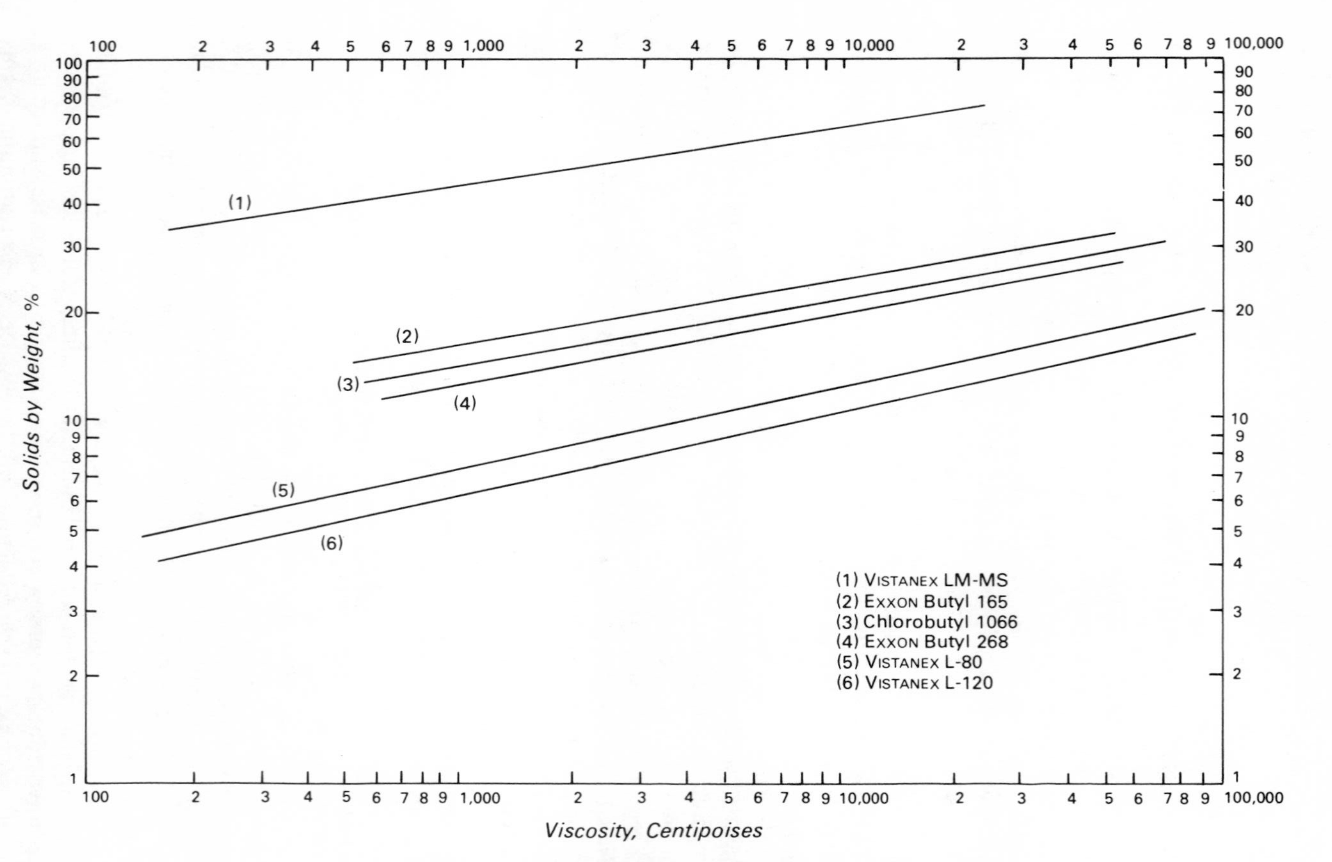


Fig. 12-2. Solution viscosities of isobutylene polymers in toluene.

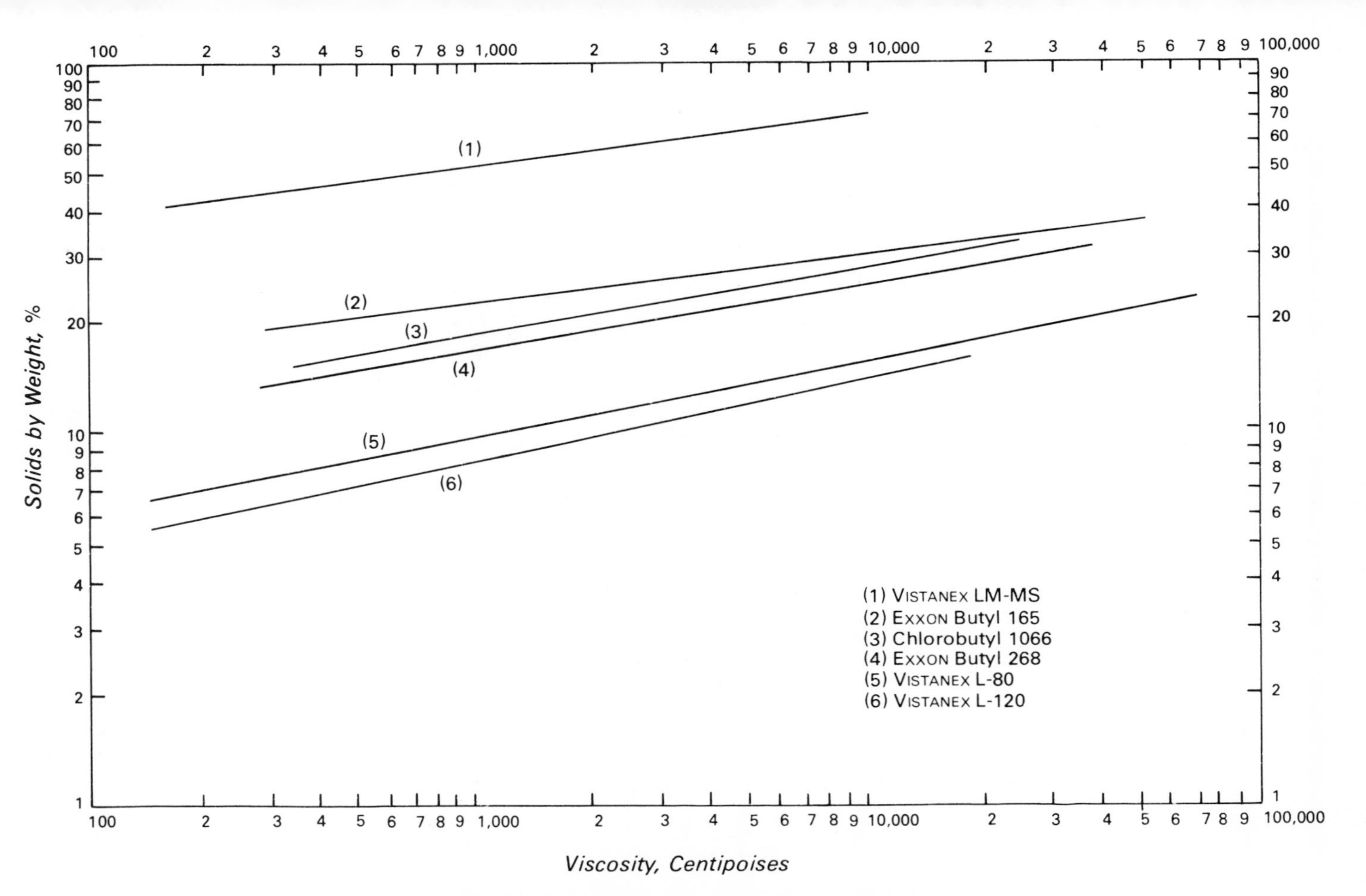


Fig. 12-3. Solution viscosities of isobutylene polymers in hexane.

can be accomplished using a (1) bale chopper or granulator, (2) sheet mill and slab chopper, or (3) Banbury, sheet mill, and slab chopper. It is important that the shredded polymer be stored for only a short time since it will reagglomerate.

The output of the shredder is most efficiently utilized by directing it into the subsequent solvation equipment. One clever and quite inexpensive unit uses a modified meat grinder mounted on the solvation tank to feed polymer particles from the grinder directly into agitated solvent.

The butyl polymers do not require the work input of a mill or Banbury to be solvated; simple chopping is quite adequate. Mills and Banburys, if available, can be used to incorporate other materials such as fillers into the polymer before solvation.

Low molecular weight polyisobutylene (e.g., Vistanex LM) and low molecular weight butyl, with their semiliquid forms, are much more rapidly solvated than the higher molecular weight analogs. It is advantageous to permit these low molecular weight polymers to flow gradually from their container directly into agitated solvent.[17] Fillers and other compounding ingredients are then added to the solution. If needed, the blend can then be processed through either a paint mill or ball mill to attain the desired degree of dispersion ("grind").

The choice of equipment for the production of solutions or mastics from butyl and PIB is based primarily on the viscosity of the product. Resins and other soluble ingredients are also often incorporated at this point. The following general guidelines may be useful:

1. For very low viscosity cements (i.e., probably below 50,000 cps), the high-speed, high shear rotating saw blade type mixers can solvate a small amount of polymer into a relatively large volume of solvent. Normally, the polymer would be introduced into the mixer in quite small particles.
2. For low viscosity cements (i.e., probably below 200,000 cps), high-speed, high shear, jacketed turbine mixers can be used.
3. For moderate viscosity cements and adhesives (i.e., viscosity about 200,000 to about 1,000,000 cps), adhesives churns or double arm planetary mixers are generally used.
4. Higher viscosity mixtures require the use of kneaders, mills or Banburys.

Butyl latex is normally compounded with quite simple equipment, commonly a propeller mixer and an open tank. When maximum filler dispersion is required, ball mills or similar equipment may be necessary for part of the mix.

Due to the low permeability of the butyl and PIB polymers, gum cements should be applied in thin coats to facilitate solvent release and avoid porosity. Adequate drying time should be permitted between coats when multiple applications are needed to increase thickness. Cements containing appreciable volumes of filler are less sensitive in this respect.

## APPLICATIONS AND FORMULATIONS

The inherent properties of polymers of the polyisobutylene family, particularly the chemical inertness, age and heat resistance, long-lasting tack, flexibility at low temperatures, and the favorable FDA position on selected grades, make these products commercially attractive in pressure-sensitive adhesives, as well as in other adhesives[18] and in the related areas of sealants[19] and coatings. An added dimension is achieved in the blendability of the polyisobutylene polymers with each other and with other adhesive polymers such as natural rubber, styrene-butadiene rubber, EVA, low molecular weight polyethylene, and amorphous polypropylene to achieve specific properties. They can, for example, be blended with the highly unsaturated elastomers to enhance age and chemical resistance. A description of polyisobutylene polymer use specifically in pressure-sensitive applications follows.

### Label and Tape Adhesives

Solution pressure sensitive adhesives for labels and tapes are primarily blends of rubber and tackifying materials such as resins. The low molecular weight polyisobutylenes find use as permanent tackifiers and modifiers. They enhance tack and softness and can contribute to adhesion by improving wetting of the substrate. The higher molecular weight products serve as the rubbery base materials.

Two important parameters in the formulation of pressure-sensitive adhesives are tack, and hold, which is the ability to resist creep under deadload. As noted, butyl and polyisobutylene are inherently tacky polymers. This tack can be increased with a wide variety of resins and other tackifiers. The cohesive strength is low compared to some other pressure-sensitive adhesive polymers, such as natural rubber, but can be increased if required by (1) incorporation of high molecular weight PIB or natural rubber, (2) the choice of the other ingredients, particularly resins and fillers, and (3) the partial or preferential curing techniques noted previously. The polyisobutylene polymers are primarily used in label pressure-sensitive adhesives and in certain tapes where high cohesive strength is not necessary.

The age resistance of polyisobutylene polymers manifests itself as lack of hardening and permanency of tack, and this has led to their use where long service life or aged adhesive integrity is required as, for example, in removable label pressure-sensitive adhesives. A removable label adhesive which ages poorly may lose cohesive strength and consequently will not peel as cleanly as desired. The higher molecular weight grades when used as the base polymer also die cut well, an important requirement for use in labels. Butyl and polyisobutylene are thus preferred materials in removable pressure-sensitive label and protective tape adhesives.

The low-temperature flexibility and tack of these polymers results in a service temperature range extending below ambient. Vistanex LM low molecular weight PIB is a particularly useful low temperature tackifier. The polyisobutylene polymers find application in solvent-borne, low-temperature pressure-sensitive adhesives such as freezer label adhesives.

The low order of toxicity, favorable FDA status and light color result in a variety of medical pressure-sensitive adhesive uses for selected polyisobutylene polymer grades. Applications include surgical tape, oral bandages, and ostomy appliances.

Adhesives formulated with butyl and polyisobutylene are used for adhesion to a variety of substrates including nonporous, hard to adhere to materials such as polyethylene. Other common substrates include paper, polyvinyl chloride, and polyester film.

Several typical or starting point solution pressure-sensitive adhesive formulations illustrating some of the uses described above are given in Table 12-2. They are provided for general guidance on materials selection and quantity but likely require modification to suit specific ingredients, manufacturing and finished property requirements. All concentrations in formulations are expressed in parts by weight.

## Hot Melt Pressure-Sensitive Adhesives

The polyisobutylene family of polymers can also be formulated in hot melt compositions and can impart many of the same property benefits noted for solvent-based systems. They improve flexibility particularly at low temperatures, provide good aging and chemical resistance, and serve to contribute toughness and strength in thermoplastic mixtures. Increases in hot tack, adhesion and overall seal strength have been widely documented.[20,21,22,23,24] Polyisobutylene grade choice and concentration are important considerations, however, in that these elastomeric products become more difficult to melt mix, significantly increasing melt viscosity as PIB molecular weight increases. Vistanex MM high molecular weight grades, in fact, are somewhat limited in hot melt applications.

For hot melt pressure-sensitive adhesives[25] butyl rubber is usually

## Table 12-2. Solvent-based pressure-sensitive adhesives.[a]

| | |
|---|---|
| **A. Simple Pressure-Sensitive Adhesive for Transparent Tape** | |
| Exxon Butyl 268 or Vistanex MM L-100 | 100 |
| Vistanex LM-MS | 75 |
| **B. Chlorobutyl Blend Pressure-Sensitive Adhesive for General-Purpose Tape** | |
| Chlorobutyl 1066 | 50 |
| SBR 1011 | 50 |
| Vistanex LM-MS | 30 |
| Hydrogenated rosin ester resin[b] | 30 |
| Escorez 2101 resin | 30 |
| Antioxidant[c] | 0.5 |
| **C. Removable Label Stock Pressure-Sensitive Adhesive** | |
| Vistanex MM L-100 | 100 |
| Vistanex LM-MS | 40 |
| Zinc oxide | 20 |
| Calcium carbonate | 20 |
| Escorez 5300 resin | 30 |
| Polyterpene resin (115°C S.P.)[d] | 30 |
| Polybutene (avg. MW 1200)[e] | 45 |
| Paraffinic process oil | 15 |
| Antioxidant | 0.5 |
| **D. Vinyl Floor Tile Pressure-Sensitive Adhesive** | |
| Exxon Butyl 268 | 100 |
| Vistanex LM-MS | 20 |
| Terpene-phenolic resin[f] | 70 |
| **E. Surgical Tape Pressure-Sensitive Adhesive** | |
| Vistanex MM L-100[g] | 100 |
| Vistanex LM-MS | 30 |
| Zinc oxide | 50 |
| Hydrated alumina | 50 |
| USP white oil | 40 |
| Phenol-formaldehyde resin[h] | 50 |
| **F. Paper Cement** | |
| Vistanex MM L-100 | 10% solids solution |

[a] Solvents (e.g., hexane, mineral spirits) added to achieve desired viscosity and solids range.
[b] Pentalyn H.
[c] Irganox 1010 or Ethanox 702 (all recipes where antioxidant indicated).
[d] Piccolyte S115.
[e] Indopol H-300.
[f] Schenectady SP-553.
[g] A minor percentage of the Vistanex MM L-100 can be replaced by natural rubber if increased cohesive strength is desired. In this case, an antioxidant may be required for age stability.
[h] Schenectady SP-1068.

blended with significant amounts of melt materials, for example, tackifying resin, petrolatum and amorphous polypropylene, to reduce viscosity to desirable levels. Low volatility hydrocarbon oils, polybutenes and paraffin and microcrystalline waxes also depress melt viscosity but excessive wax can adversely affect tack. Vistanex LM low molecular weight PIB is an effective butyl plasticizer and enhances tack. Applications for butyl hot melt pressure-sensitive adhesives often involve long service life, low temperature, and polyolefin plastic film adhesion, capitalizing on the inherent features of this polymer noted previously. An illustrative starting point formulation for adhesive-backed carpet tiles is shown in Table 12-3.

The Vistanex LM grades are perhaps the most functional polyisobutylene polymers for hot melt pressure-sensitive adhesives. They are easiest to melt mix and result in low viscosity blends due to their low molecular weight. Brookfield viscosity at 177°C (spindle SC4-27) is typically 30,000 to 60,000 cps depending on grade, and melt viscosity can readily be tailored to meet specific requirements by incorporation of thermoplastic ingredients.

Petrolatum and waxes are most effective as viscosity depressants. Resins, waxes and thermoplastics such as low density polyethylene (LDPE), ethylene vinyl acetate, and crystalline and amorphous polypropylene increase hardness and cohesive strength. LDPE and crystalline polypropylene particularly raise blend softening point, even at the 10 weight percent concentration level, thus extending the adhesive's service temperature range. Waxes and the thermoplastics must be used in moderation, however, since

**Table 12-3. Butyl rubber hot melt pressure-sensitive adhesive.**

| | |
|---|---|
| Exxon Butyl 065 | 100 |
| Escorez 1304 resin | 100 |
| Petrolatum[a] | 50 |
| Amber microwax[b] | 150 |
| Antioxidant[c] | 1 |
| Brookfield viscosity at 121°C | |
| Spindle SC4-29 (cps) | 80,000 |
| Surface | Tacky, firm |
| Adhesion[d] | |
| Vinyl foam to plywood | Good–excellent |
| Rubber foam to plywood | Good–excellent |

[a] 57 C.M.P.
[b] Be Square 175.
[c] Ethanox 702 or Irganox 1010.
[d] Adhesives coated on foam and subsequently manually pressed onto plywood.

they reduce the tack of the finished blend. Hydrocarbon oils and petrolatum plasticize and bring out aggressive tack in Vistanex LM, resin mixtures.

Blends containing low molecular weight PIB can be prepared with relative ease. Added incrementally, it dissolves quite effectively in the liquid and molten melt ingredients of an adhesive composition at 180°C with agitation.

Several formulations illustrating Vistanex LM use in hot melt pressure-sensitive adhesives are provided in Table 12-4.[26] Vistanex LM functions as an age-resistant tackifier, adhesion promoter and flexibilizer, and its favorable FDA regulatory position extends use to medical applications. Again, these formulations should be considered as starting points only. All concentrations are expressed in parts by weight.

## Pipe Wrap and Electrical Tape

The polyisobutylene polymers have a history of excellent performance in pipe wrap tape and in a variety of electrical tape areas. These applications exploit the age resistance, low water absorptivity and permeability, inherent tack and electrical insulative properties of these polymers.

Pipe wrap tape is most commonly an adhesive based on butyl rubber or possibly a blend of butyl and ethylene propylene rubber[27] on a polyethylene or polyvinyl chloride backing. A formula for the adhesive is shown in Table 12-5. High molecular weight grade Exxon Butyl 268 is used for maximum cohesive strength at high loading levels. Low unsaturation grade Exxon Butyl 065 can be used if maximum age resistance is desired. Resins are chosen for tack and cohesive strength balance, and high resin levels combined with polybutene provide a tacky, pressure-sensitive adhesive type pipe wrap tape. These adhesives are either calendered onto the backing, extruded or solvated and coated from solution.

Electrical tapes based on butyl are actually bulk mastics compounded for good electrical, heat aging and ozone resistant characteristics.[28] Tapes for splicing and terminating wires and cables are self-fusing (i.e., when wrapped upon itself the tape must fuse, forming a permanent bond). Additionally, PIB tapes, particularly those tackified with Vistanex LM, provide good adhesion to polyethylene and thus are often the products of choice for splicing crosslinked polyethylene insulated cables. Splicing tapes are unsupported (as used they do not have a fabric or film backing) and are typically available in roll form on a release liner.

There are numerous pressure-sensitive tapes with various adhesive and backing combinations for specific electrical applications. The nonpolarity of the polyisobutylene polymers makes them quite resistant to attack and

## Table 12-4. Low molecular weight polyisobutylene hot melt pressure-sensitive adhesives.

A. General Recipes

| | 1 | 2 | 3 |
|---|---|---|---|
| Vistanex LM-MS | 25 | 35 | 45 |
| Escorez 1310 resin | 50 | 50 | 15 |
| Petrolatum[a] | 15 | — | 40 |
| High Molecular Wt. LDPE[b] | 10 | — | — |
| Paraffinic Process Oil | — | 15 | — |
| Brookfield Viscosity at 177°C | | | |
| Spindle SC4-27 (cps) | 1,900 | 1,200 | 850 |
| Surface | Tacky, firm | Very tacky, soft | Very tacky, soft |
| Softening Point | | | |
| Ring and Ball (°C) | 101 | 49 | 55 |

B. Low-Temperature Label Pressure-Sensitive Adhesive

| | |
|---|---|
| Kraton 1107[c] | 50 |
| Solprene 1205[d] | 50 |
| Vistanex LM-MS | 90 |
| Escorez 2101 resin | 130 |
| Naphthenic process oil | 58 |
| Antioxidant[e] | 2 |
| Rolling Ball Tack (cm) | 2.5 |
| 180° Peel (g/cm) | |
| Stainless steel | 880[g] |
| Treated LDPE | 570 |
| 180° Peel at 0°F (g/cm)[f] | |
| Stainless steel | 1,250[g] |
| Treated LDPE | 90 |
| Loop Tack at 0°F (g/cm)[f] | |
| Stainless steel | 950 |
| Corrugated board | 290[h] |
| Waxed board | 410[g] |

C. EVA-based Pressure-Sensitive Adhesive

| | |
|---|---|
| Elvax 40 | 100 |
| Vistanex LM-MS | 20 |
| Escorez 2101 resin | 80 |
| Antioxidant[e] | 1 |

[a] 57°C M.P.
[b] Exxon LD 600; a low MW LDPE grade such as Epolene C-10 would give lower viscosity and softening point.
[c] Styrene-isoprene-styrene block copolymer.
[d] Styrene-butadiene block copolymer; improves polyisobutylene compatibility with Kraton 1107.
[e] Irganox 1010.
[f] Adhesive and adherend conditioned at −18°C before contact; adhesion determined at −18°C.
[g] Some adhesive transfer from polyester backing noted.
[h] Fiber tearing noted.

**Table 12-5. Butyl mastic for pipe wrap tape.**

| | |
|---|---|
| Exxon Butyl 268 | 100 |
| FEF carbon black | 100 |
| Mistron Vapor Talc | 200 |
| Polybutene (avg. MW 900)[a] | 100 |
| Paraffinic process oil | 50 |
| Escorez 1304 resin | 75 |
| Amorphous polypropylene[b] | 50 |

[a] Indopol H-100
[b] A-Fax 600-HL-5

swelling by the commonly used vinyl plasticizers. Thus, they are less susceptible to softening when used on vinyl film backings.

Typical nonconductive splicing tape and pressure-sensitive electrical tape formulations are given in Table 12-6. The splicing tape mix is subjected to brief heat treatment in a Banbury mixer to maximize cohesive strength and electrical properties.

**Table 12-6. Electrical tape formulations.**

| | |
|---|---|
| A. Butyl Nonconductive Splicing Tape[a] | |
| Exxon Butyl 268 | 100 |
| Vistanex MM L-100 | 10 |
| AgeRite Resin D | 1 |
| Zinc oxide | 5 |
| Mistron Vapor Talc | 60 |
| Whitetex Clay | 60 |
| MT carbon black | 10 |
| Low density polytheylene | 5 |
| Escorez 1315 resin | 5 |
| Alkyl–Phenol resin[b] | 5 |
| Poly DNB | 0.5 |
| QDO | 0.2 |
| Tensile strength (psi) | 650 |
| Ultimate elongation (%) | 750 |
| Dielectric strength (volts/mil) | 745 |
| Volume resistivity (ohm-cm $\times 10^{-14}$) | 140 |
| B. PSA for Plastic Film-Backed Electrical Tape | |
| Vistanex MM L-100 | 100 |
| Escorez 2101 resin | 35 |
| Liquid hydrogenated rosin ester[c] | 35 |
| Polybutene (avg. MW 900)[d] | 35 |

[a] This compound is prepared in a Banbury mixer and held for about three minutes at 163°C to allow the promoters (Poly DNB and QDO) to couple the polymer and fillers. All properties measured on 0.030-inch thick uncured pads.
[b] Schenectady SP-1077
[c] Hercolyn D
[d] Indopol H-100

## Suppliers of Trade Named Formulating Ingredients

| MATERIALS | SUPPLIER |
|---|---|
| A-Fax 600-HL-5 | Hercules Inc. |
| AgeRite Resin D | R. T. Vanderbilt Co. |
| AgeRite White | R. T. Vanderbilt Co. |
| Be Square 175 | Petrolite Corp., Bareco Div. |
| Bromobutyl 2244 | Exxon Chemical Co. |
| Bucar Butyl | Cities Service Co., Columbian Chemicals Div. |
| Butyl Zimate | R. T. Vanderbilt Co. |
| Chlorobutyl 1066 and 1068 | Exxon Chemical Co. |
| DBQDO | Hughson Chemicals Lord Corp. |
| Elvax | E. I. DuPont, Inc. |
| Exxon Butyl | Exxon Chemical Co. |
| Exxon LD 600 | Exxon Chemical Co. |
| Epolene | Eastman Chemical Products Inc. |
| Escorez Resins | Exxon Chemical Co. |
| Ethanox 702 | Ethyl Corp. |
| Hercolyn D | Hercules Inc. |
| Indopol | Amoco Chemicals Corp. |
| Irganox 1010 | Ciba-Geigy Corp. |
| Kalene 800 and 1300 | Hardman Inc. |
| Kraton 1107 | Shell Chemical Co. |
| Laktane | Exxon Corp. |
| Mistron Vapor Talc | Sierra Talc Co. |
| Oppanol | BASF |
| Pentalyn Resins | Hercules Inc. |
| Piccolyte S115 | Hercules Inc. |
| Poly DNB | Hughson Chemicals Lord Corp. |
| Polysar Butyl | Polysar Ltd. |
| QDO | Hughson Chemicals Lord Corp. |
| Schenectady Resins | Schenectady Chemicals, Inc. |
| Silanes | Union Carbide Corp. |
| Solprene 1205 | Phillips Petroleum Co. |
| Vistanex | Exxon Chemical Co. |
| Whitetex Clay | Freeport Kaolin Co. |

## REFERENCES

1. Buckley, D. J. *Rub. & Chem. Technol.*, **32**(5): 1475 (December 1959).
2. Dunkel, W. L., Neu, R. F., and Zapp, R. L. Morton, M. (ed). *Introduction to Rubber Technology*, p. 309, New York: Van Nostrand Reinhold Co., 1959.
3. Thomas, R. M., and Sparks, W. J. (Whitby, G. S., ed.) *Synthetic Rubber*, p. 838, New York: John Wiley and Sons, 1954.

4. Exxon Chemical Company USA. *An Introduction to Vistanex® LM Low Molecular Weight Polyisobutylene*, Brochure SC-79-131, Houston, 1979.
5. Exxon Chemical Company USA. *Vistanex Polyisobutylene, Properties & Applications*, Brochure SYN-76-1434, Houston, 1974.
6. Exxon Chemical Company USA. *Chlorobutyl Rubber, Compounding and Applications*, Brochure SYN-76-1290, Houston, 1976.
7. Harianawala, A., and Wilson, G. J. "The Cure Rate and Vulvanizate Properties of Bromobutyl Compounds," Presented at ACS Division of Rubber Chemistry, Cleveland, Ohio, October 15, 1971.
8. Gunner, L. P. *J. Adhesive and Sealant Council*, **1**(1): 23 (1972).
9. Berejka, A. J., and Lagani, A., Jr. U.S. Patent 3,597,377 (to Esso Research and Engineering Co.) (August 3, 1971).
10. Paterson, D. A. *Adhesives Age*, **12**(8) 25 (August 1969).
11. Scheinbart, E. L., and Callan, J. E. *J. Adhesive and Sealant Council*, **1**(1): 82 (1972).
12. Amoco Chemical Corp *Indopol Polybutenes for High Quality Caulks and Sealants*, Bulletin 12-46, pp. 8–14, Chicago, Illinois, 1970.
13. Smith, W. C. "The Vulcanization of Butyl, Chlorobutyl and Bromobutyl Rubber," in Alliger, G., and Sjothun, I, J. (eds.). *Vulcanization of Elastomers*, p. 230, New York: Van Nostrand Reinhold Co., 1964.
14. Eby, L. T., and Thomas, R. T. U.S. Patent 2,948,700 (to Esso Research and Engineering Co.) (August 9, 1960).
15. Exxon Chemical Company USA. *Viscosities of Solutions of Exxon Elastomers*, Bulletin SC-75-108, 1975.
16. Stucker, N. E., and Higgins, J. J. *Adhesives Age*, **11**(5): 25 (May 1968).
17. Exxon Chemical Company USA. *Vistanex LM Polyisobutylene, Handling Suggestions*, Bulletin SC-79-130A, 1979.
18. Stucker, N. E., and Higgins, J. J. "Butyl Rubber and Polyisobutylene in Adhesives and Sealants," in Skeist, I. (ed.). *Handbook of Adhesives*, p. 255, New York: Van Nostrand Reinhold Co., 1977.
19. Jagisch, F. C. *Adhesives Age*, **21**(11): 47 (November 1978).
20. Tyran, L. W. U.S. Patent 3,321,427 (to E. I. DuPont) (May 23, 1967).
21. Moyer, H. C., Karr, T. J., and Guttman, A. L. U.S. Patent 3,338, 905 (to Sinclair Research Inc.) (August 29, 1967).
22. Cox, E. R. U.S. Patent 3,396,134 (to Continental Oil Co.) (August 6, 1968).
23. Kremer, C. J., and Apikos, D. U.S. Patent 3,629,171 (to Atlantic Richfield Co.) (December 21, 1971).
24. Hammer, I. P. U.S. Patent 3,322,709 (to Mobil Oil Corp.) (May 30, 1967).
25. Trotter, J. R., and Petke, F. D. U.S. Patent 4,022,728 (to Eastman Kodak Company) (May 10, 1977).
26. Exxon Chemical Company USA. *Escorez® Resins and Vistanex LM Polyisobutylene in Low Temperature Pressure Sensitive Adhesives*, Bulletin R-79-54, 1979.
27. Harris, G. M., "Plastic Tapes—Twenty Years of Underground Corrosion Control," Paper at 26th NACE Conference, March 2–6, 1970.
28. Federal Specification HH-1-553, *Insulation Tape, Electrical*, Grade A–Ozone Resistant.

# Chapter 13

# Acrylic Adhesives

**Donatas Satas**
*Satas & Associates*
*Warwick, Rhode Island*

Acrylic polymers are important, although relatively recently introduced, materials for pressure-sensitive adhesive applications.

The acrylic acid was first synthesized in 1843. By 1901, research was carried out on acrylic acid esters.[1] The first commercial production of poly(methyl methacrylate) took place in 1927 by Röhm and Haas AG in Germany and acrylic dispersions have been produced by Badische Anilin und Soda Fabrik AG since 1929. The suitability of polyacrylates for pressure-sensitive adhesives was recognized as early as 1928.[2] Despite all this early work, the polyacrylates found an extensive use for pressure-sensitive adhesives only in the 1950's and secured its present prominent position in the 1960's.

Rubbery polyacrylates can be compounded to pressure-sensitive products by addition of tackifying resins, plasticizers, and other ingredients generally used for such purposes. This approach has not found many applications in the pressure-sensitive product industry.

Polyacrylates of a proper monomer composition are inherently pressure-sensitive without any compounding. Only polyvinyl ethers have this property besides polyacrylates. This single component feature has some advantages over the compounded adhesives.[3] Low molecular weight ingredients that can migrate to the surface of an adhesive coating are absent. Adhesive bond is a surface phenomenon, and minimizing the compositional variations at the surface is highly desirable. Some variation is difficult to avoid in multiphase systems, while uniformity is easier achieved in single component adhesives.

Polyacrylates possess some inherent properties superior to many other polymers used for pressure-sensitive adhesives. The polymer is saturated and resistant to oxidation. It is water white and does not yellow on exposure to sunlight. The resistance to oxidation surpasses that of most polymers used for pressure-sensitive adhesives, except silicones. Monomers with various functional groups can be introduced during polymerization, and an adhesive with various degrees of thermosetting properties can be prepared.

Acrylic adhesives are available as solutions, aqueous emulsions, hot melts, and 100% reactive solids. The latter two forms require further development to gain a wider acceptance, but basically polyacrylates have the versatility to be applied in many different ways, unlike most of the other polymers used in pressure-sensitive adhesive applications.

Acrylic adhesives have caused some changes in the pressure-sensitive product business. The demand for compounded pressure-sensitive adhesives was very small. Every tape or label producer was compounding his own adhesives and was able to claim some degree of uniqueness in his product. The largest tape and label manufacturers are producing their own proprietory acrylic adhesives, but the rest of the manufacturers of pressure-sensitive products depend upon the adhesives sold by polymer producers on the open market. The skills required to produce acrylic adhesives in-house are lacking in the case of most smaller tape manufacturers. This ready availability of good quality adhesives on the open market made it much easier to enter the manufacturing of pressure-sensitive products. Thus, the polyacrylates are responsible for the proliferation of pressure-sensitive product manufacturers.

## MONOMERS

The basis of pressure-sensitive acrylic adhesives is acrylic esters which yield polymers of low glass transition temperatures $T_g$ and can be copolymerized with acrylic acid and many other functional monomers in either emulsion or solution polymerization. The patent literature generally claims alkyl acrylates and methacrylates of 4–17 carbon atoms as suitable monomers for pressure-sensitive adhesives. The most commonly used monomers are 2-ethylhexyl acrylate, butyl acrylate, ethyl acrylate, and acrylic acid. Some nonacrylic monomers such as vinyl acetate are also used quite frequently to modify the polymer properties or to decrease the cost of raw materials.

In addition to the above mentioned monomers, many other compounds have been used in the synthesis of pressure-sensitive acrylic adhesives. Some are used in substantial quantities, others in small amounts for some special effect related to the functional groups carried by such monomers. Almost any conceivable monomer that is capable to undergo vinyl polymerization has been mentioned in the voluminous patent literature. Table 13-1 lists

**Table 13-1. Monomers used for pressure-sensitive acrylic adhesives.**

| Comonomer | Reference |
|---|---|
| Acrylates | |
| methyl | 16,30,57,69,70,71,80 |
| ethyl | many references |
| *n*-propyl | 72,80 |
| butyl | many references |
| pentyl | 63,68 |
| isopentyl | 7,80 |
| 2-methylbutyl | 42 |
| *n*-hexyl | 42,54,72 |
| 2-ethylbutyl | 7,70 |
| methyl pentyl | 25,78 |
| heptyl | 63 |
| 2,4-(methyl)pentyl | 50 |
| octyl | 35,40,41,42,70 |
| 2-ethylhexyl | many references |
| *n*-decyl | 17,42,54,76 |
| *n*-undecyl | 42,63 |
| *n*-dodecyl | 54,63,72 |
| tridecyl | 17 |
| lauryl | 16,27,53 |
| stearyl | 53,71 |
| 10-cyclohexyl undecyl | 62 |
| fusel oil | 78 |
| 6-methoxy | 50 |
| methoxyethyl | 68 |
| ethoxyethyl | 17,33,68,70 |
| methoxybutyl | 59 |
| hydroxyethyl | 48,56 |
| hydroxypropyl | 48,56,67 |
| 4-hydroxybutyl | 38 |
| butanediol | 42 |
| tetrahydrofurfuryl | 77 |
| abitol | 17 |
| cyanoethyl | 17,23,39 |
| dimethylaminoethyl | 63,80 |
| diethylaminoethyl | 52,63 |
| glycidyl | 44,52,55,67 |
| benzophenone glycidyl | 67 |
| 3-chloro-2-hydroxypropyl | 42 |
| 3-(3,4-dichlorophenoxy)-2-hydroxypropyl | 69 |
| 3-(2,4,6-trichlorophenoxy)-2-hydroxypropyl | 69 |
| 3-(2,3,4,5-tetrachlorophenoxy)-2-hydroxypropyl | 69 |
| 3-(pentachlorophenoxy)-2-hydroxypropyl | 69 |
| Diacrylates | |
| dimethylaminoethyl | 67 |
| 1,6-hexanediol | 67 |
| diethylene glycol | 67 |

**Table 13-1.** *Continued*

| Comonomer | Reference |
|---|---|
| triethylene glycol | 67 |
| tetraethylene glycol | 67 |
| glycol | 70 |
| Triacrylates | |
| diethylene glycol | 37 |
| trimethylolpropane | 37,67 |
| Methacrylates | |
| methyl | 28,47,54,55,58,61,68,69,80 |
| butyl | 53,54,55,63,68,72 |
| pentyl | 63 |
| hexyl | 17,27,42,63,70,72 |
| octyl | 40,70,72 |
| 2-ethylhexyl | 40,53,55,56,63 |
| *n*-nonyl | 42 |
| *n*-decyl | 42,54,63 |
| *n*-dodecyl | 17,50,54,63,72 |
| lauryl | 7 |
| isobornyl | 68,80 |
| hydroxyethyl | 35,49,56 |
| hydroxypropyl | 66,67 |
| 2-hydroxypropyl | 26 |
| 3-hydroxypropyl | 56,67 |
| cyanoethyl | 23,39 |
| dimethylaminoethyl | 18,20,40,52,63,80 |
| *t*-butylaminoethyl | 52,80 |
| glycidyl | 16,32,34,38,44,55,57,67 |
| benzophenone glycidyl | 67 |
| 3-(3,4-dichlorophenoxy)-2-hydroxypropyl | 69 |
| 3-(2,4,6-trichlorophenoxy)-2-hydroxypropyl | 69 |
| 3-(2,3,4,5-tetrachlorophenoxy)-2-hydroxypropyl | 69 |
| 3-(pentachlorophenoxy)-2-hydroxypropyl | 69 |
| 10-chlorodecyl | 50 |
| sodium-2-sulfoethyl | 61 |
| trimethoxysilylpropyl | 41 |
| 2-sulfoethyl | 80 |
| Dimethacrylates | |
| ethylene glycol | 42,67 |
| 1,3-butylene glycol | 67 |
| triethylene glycol | 20,66,67 |
| polyethylene glycol | 23,67 |
| dimethylaminoethyl | 67 |
| Trimethacrylates | |
| trimethylolpropane | 37 |
| Tetramethacrylates | |
| pentaerythritol | 67 |
| Acrylamides | |
| acrylamide | 8,23,25,41,54 |

**Table 13-1.** *Continued*

| Comonomer | Reference |
|---|---|
| *N*-methylol | 36,42,46 |
| *N*,*N*-dimethyl | 44 |
| *N*-*t*-butyl | 18,22,30,74,77 |
| *N*,*N*-diacetonyl | 56,60 |
| *n*-octyl | 22 |
| *t*-octyl | 69 |
| *N*-*t*-$C_9$ | 17 |
| *N*-*t*-$C_{12}$ | 17 |
| *N*-[(2-ethylhexoxy)methyl] | 59 |
| diacetone | 29,30,40,41,67 |
| diacetophenone | 67 |
| 2-isocyanate | 44 |
| 2-acrylamido-2-methylpropane sulfonic acid | 64 |
| Methacrylamides | |
| methacrylamide | 8,22,23,42,54 |
| *N*-methylol | 42 |
| *N*,*N*-diacetonyl | 56,60 |
| *N*-(*n*-butoxymethyl) | 36,42 |
| *N*-*t*-$C_{12}$ | 17 |
| 2-isocyanate | 44 |
| Acrylimides | |
| trimethylamine | 50,61 |
| Methacrylimides | |
| trimethylamine | 50,61 |
| triethylamine | 50,61 |
| tributylamine | 50,61 |
| 1,1-dimethyl-1-(2-hydroxypentyl)amine | 50 |
| 1,1-dimethyl-1-(2-hydroxydecyl)amine | 50 |
| 1,1-dimethyl-1-(2,3-dihydroxypropyl)amine | 50 |
| 1,1-dimethyl-1-(2-hydroxy-3-phenoxypropyl)amine | 50 |
| 1,1-dimethyl-2-(2-hydroxy-3-isopropoxypropyl)amine | 50 |
| Methacrylic acid | 8,42,52,80 |
| Acrolein | 44 |
| Acrylonitrile | 9,10,11,12,17,23,27,54,61,64,71,72 |
| Methacrylonitrile | 12,23,54,64 |
| Maleic acid derivatives | |
| maleic acid | 52 |
| maleic anhydride | 16,19,30,41,44,45,52,53,58,67 |
| methyl maleate | 67 |
| dibutyl maleate | 70 |
| di(2-ethylhexyl)maleate | 15 |
| bis(2-hydroxyethyl)maleate | 56 |
| maleic acid amide | 54 |
| maleic acid diamide | 54 |
| maleic nitrile | 54 |
| maleic dinitrile | 54 |
| Fumaric acid derivatives | |
| fumaric acid | 10,52 |
| diisopropyl fumarate | 38 |

**Table 13-1.** *Continued*

| COMONOMER | REFERENCE |
|---|---|
| di-*n*-butyl fumarate | 7,38 |
| di-*sec*-butyl fumarate | 38 |
| diamyl fumarate | 70 |
| *n*-hexyl fumarate | 70 |
| di-2-ethylbutyl fumarate | 69 |
| diisoamyl ethylene difumarate | 70 |
| di-*n*-octyl fumarate | 38 |
| di-2-ethylhexyl fumarate | 38 |
| didodecyl fumarate | 38 |
| di-"Cellosolve" fumarate | 38 |
| bis(2-hydroxyethyl) fumarate | 56 |
| polypropylene glycol fumarate | 78 |
| fumaric acid amide | 54 |
| fumaric acid diamide | 54 |
| fumaric acid nitrile | 54 |
| fumaric acid dinitrile | 54 |
| Crotonic acid derivatives | |
| crotonic acid | 10,11,51,55 |
| glycidyl crotonate | 55 |
| Itaconic acid derivatives | |
| itaconic acid | 8,17,42,52,55,80 |
| itaconic anhydride | 52 |
| half esters of itaconic acid | 24 |
| Citraconic acid derivatives | |
| citraconic acid | 52 |
| citraconic anhydride | 52 |
| half esters of citraconic acid | 24 |
| Maleamic acid derivatives | |
| butyl maleamic acid | 17 |
| *t*-octyl maleamic acid | 17 |
| dibutyl maleamic acid | 17 |
| di-(2-ethylhexyl) maleamic acid | 17 |
| Primene maleamic acid | 17,21,23,31 |
| Vinyl compounds | |
| vinyl chloride | 27,64,72 |
| vinylidene chloride | 70,72 |
| vinyl acetate | 13,15,17,19,26,27,28,29 30,34,37,38,39,40,41,45 48,49,52,53,55,57,61,64,69,70,78 |
| vinyl propionate | 70,72 |
| vinyl butyrate | 38 |
| vinyl 10-phenylundecanoate | 62 |
| vinyl toluene | 64,72 |
| vinyl benzoate | 70 |
| sodium vinyl sulfonate | 42 |
| methyl vinyl ketone | 70 |
| vinyl naphthalene | 70 |
| *N*-vinyl succinimide | 70 |
| *N*-vinylimidazole | 42 |

**Table 13-1.** *Continued*

| Comonomer | Reference |
|---|---|
| 2-vinyl pyridine | 63 |
| 4-vinyl pyridine | 63 |
| *N*-vinyl pyrrolidone | 38,42,43,63,77 |
| *N*-vinyl piperidone | 43 |
| *N*-vinyl caprolactam | 42,43 |
| Vinyl ethers | |
| vinyl methyl ether | 72 |
| vinyl ethyl ether | 72 |
| vinyl butyl ether | 38,70 |
| vinyl octyl ether | 61 |
| divinyl ether | 70 |
| 2-chloroethyl vinyl ether | 61 |
| Allyl compounds | |
| tetraallyloxyethane | 67 |
| diallyl succinate | 70 |
| tetraallyl silicate | 70 |
| allyl glycidyl ether | 44,57 |
| triallyl cyanurate | 67 |
| triallyl isocyanurate | 67 |
| Styrene derivatives | |
| styrene | 17,38,55,61,64,70,72 |
| $\alpha$-methylstyrene | 61,72 |
| *t*-butylstyrene | 61 |
| Lactones | |
| $\beta$-propiolactone | 14 |
| $\gamma$-butyrolactone | 14 |
| $\delta$-valerolactone | 14 |
| $\varepsilon$-caprolactone | 14 |
| diketene | 44,73 |
| Betaines | |
| 3-[(2-acryloxyethyl)dimethylammonium]propionate betaine | 50,61 |
| 3-[(2-acryloxyethyl)dimethylammonium]propane sulfonate betaine | 50 |
| 3-[(2-methacryloxyethyl)dimethylammonium]propionate betaine | 50 |
| Isocyanates | |
| methylenebisphenyl-4-4′-diisocyanate | 66 |
| acrylic acid-2-isocyanate ester | 44 |
| polyisocyanate prepolymers | 26 |
| Miscellaneous | |
| 1,2-bis(dimethylamino)-2-hydroxypropane | 65 |
| 3-methacryloxypropyltrimethoxysilane | 29,30,66 |

some of the monomers along with the reference to the patent literature. The best source of detailed technical information in this field is the patent literature.

In some cases, the presence of a large amount of a nonacrylic monomer might dilute the properties generally associated with polyacrylates, so that even the classification of such copolymers as acrylates might become questionable. A large quantity of a comonomer of poorer aging properties and chemical stability can change the characteristics of the adhesive substantially. Vinyl acetate is one monomer used in large quantities in pressure-sensitive adhesives.

The polymerization of acrylates proceeds by a chain transfer mechanism described briefly in Chapter 30. The acrylic polymerization is discussed in detail in several books.[4,5,6] The literature on vinyl polymerization is quite voluminous because of its commercial importance.

Because of the early discovery of the suitability of polyacrylates for pressure-sensitive adhesive applications, it was not possible for any single patent to have a dominant position in this field. Such condition was responsible for the growth of a large number of compositional patents, different in the use of some comonomer, sometimes of questionable importance, in obtaining the specific adhesive properties. A patent by Ulrich[8] was as close to a dominant position as possible under the circumstances. The patent claims the use of small quantities of various functional monomers such as acrylic, methacrylic, or itaconic acids and their amides. Introduction of carboxylic groups is believed to enhance the adhesion and functional groups in general are useful for crosslinking purposes.

## COPOLYMERIZATION

Polymer properties can be varied by copolymerization with other monomers. The physical properties of a copolymer are roughly intermediate of those of the homopolymers and might not be drastically different from the properties of corresponding homopolymer blends. In case of a long side-chain polymer, which exhibits a tendency to crystallize, copolymerization interferes with the crystallization and has a plasticizing effect on the copolymer.

Pressure-sensitive polymers are soft, capable of wetting the adherend surface, and capable of sufficient cold-flow to fill the surface irregularities. Such properties are found in polymers of a low glass transition temperature. Table 13-2 lists the glass transition temperatures of some acrylic and methacrylic homopolymers. Homopolymers are rarely used for pressure-sensitive adhesive applications. Copolymerization with other monomers is the universally used technique to vary the physical properties of the adhesives.

The glass transition temperature is neither an accurate measure of polymer stiffness at room temperature, nor is it an accurate measure of pressure-sensitive properties. It is an easy and convenient method to predict the suitability of a polymer for pressure-sensitive adhesive application and to predict the effect of a comonomer on the copolymer properties.

**Table 13-2. Glass transition temperatures of some homopolymers.**

| HOMOPOLYMERS | $T_g$ (°C) |
|---|---|
| Poly(acrylic acid) | 106 |
| Poly(vinyl acetate) | 30 |
| Poly(vinyl pyrrolidone) | 54 |
| Poly (ethylene glycol dimethacrylate) | 132 |

| ESTER GROUP | ACRYLATE | METHACRYLATE |
|---|---|---|
| methyl | 8 | 105 |
| ethyl | −22 | 65 |
| isopropyl | −5 | 81 |
| *n*-propyl | −52 | 33 |
| isobutyl | −40 | 48 |
| *n*-butyl | −54 | 20 |
| *t*-butyl | 43 | 107 |
| *n*-pentyl | — | 10 |
| *n*-hexyl | — | −5 |
| 2-ethylhexyl | −85 | — |
| *n*-octyl | −80 | −20 |
| *n*-decyl | — | −70 |
| *n*-dodecyl | — | −65 |
| *n*-tetradecyl | 20 | −9 |
| *n*-hexadecyl | 35 | — |
| isobornyl | — | −114 |

It has been suggested[79] that some of the pressure-sensitive properties such as peel adhesion can be correlated with the glass transition temperature. It is true that a polymer which is obviously too soft and fails cohesively when peeled can be improved by copolymerization with a monomer that raises its $T_g$ and, conversely, a polymer that is too hard and fails in a peel test in a stick-slip fashion can be improved by copolymerization with a monomer which will lower the $T_g$. Any finer adjustment and correlation of adhesive performance with $T_g$ or any other single physical property is not possible.

In order to possess sufficient tack required for pressure-sensitive adhesives, the homopolymer must have a fairly low $T_g$. Both 2-ethylhexyl acrylate ($T_g = -54$°C) and *n*-butyl acrylate ($T_g = -54$°C) are sufficiently tacky to serve as pressure-sensitive adhesives. Homopolymers of higher $T_g$ generally are not tacky enough. The tackiness can be improved by copolymerization.

The polymer flexibility and tackiness increases with increasing side-chain length until a certain chain length is exceeded, and the chains start to form crystalline regions which cause the stiffening of the polymer. In case of straight side-chain polymers, this maximum softness occurs in *n*-octyl acrylate ($T_g = -80°C$) for the acrylic polymer series and in *n*-decyl or *n*-dodecyl methacrylates in the methacrylic polymer series.[81,82] The maximum softness occurs higher in the methacrylic series because of the steric hindrance by the methacrylate group.[83] The softness and tackiness maximum is reached at a higher chain length if these properties are tested at temperatures above 20°C.

Branched side-chains stiffen the polymer in the range from propyl to hexyl. In case of longer chains, branching decreases the tendency to form crystalline regions and therefore branching makes the polymer softer. Thus, 2-ethylhexyl acrylate exhibits a lower $T_g$ than *n*-octyl acrylate.

The effect of introducing a comonomer of a different side-chain length is not dissimilar from the effect of branching. It increases the randomness of the polymer and decreases its tendency to form crystalline regions. It could be expected that the comonomer distribution could be important in some cases. The distribution can be affected by the order of monomer addition. If the comonomer is added at the beginning of the reaction, its distribution will depend on the ratio of monomer reactivities. Withholding a part of the comonomer and introducing it after the polymerization reaction has proceeded will delay its incorporation into the polymer chain. An investigation of the effects of comonomer distribution on the pressure-sensitive adhesive properties did not yield any interesting results. Apparently, the pressure-sensitive polymers were sufficiently random to be significantly affected by some differences in comonomer distribution.

Mao and Reegen[84] have shown that the peel adhesion of a nonpressure-sensitive copolymer of methylmethacrylate with various alkyl methacrylates and alkyl acrylates in a mole ratio 9 : 1 increased with increasing length of the comonomer side-chain. These data are shown in Fig. 13-1.

## MOLECULAR WEIGHT

Molecular weight has a profound effect on the mechanical properties of polymers. Therefore, it is expected that the adhesive performance will be influenced by molecular weight variations. Low molecular weight polymers have poor mechanical properties and, although they might exhibit good tack, their adhesive properties, especially resistance to carry a load, are poor. Such polymers are not suitable for adhesive applications, unless they can be crosslinked at some point. It is generally considered that a polymer

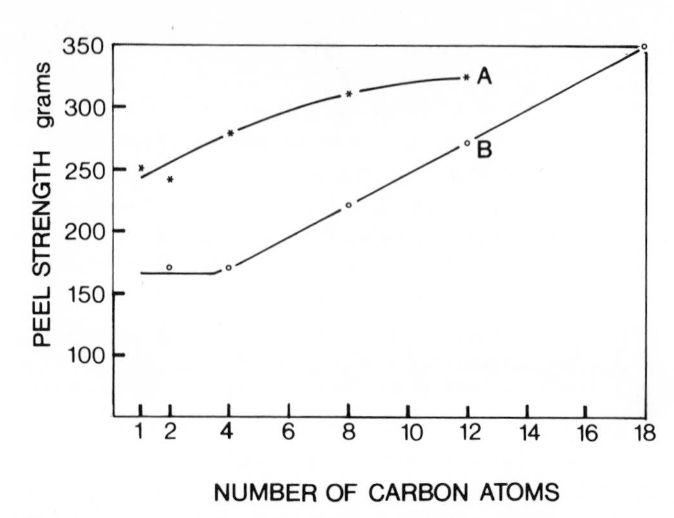


Fig. 13-1. Peel strength of methyl methacrylate copolymerized with various alkyl methacrylates and acrylates. Mole ratio 9 : 1. Curve A = acrylic comonomer; curve B = methacrylic comonomer.

must have a degree of polymerization of at least 300 before its physical properties are developed to a useful level. The molecular weight should be considerably higher to develop a reasonably high resistance to creep, unless the level of secondary bonding is high.

Table 13-3 shows the effect of molecular weight on the properties of pressure-sensitive adhesive polymers. The polymer series A through G are of identical monomer composition. The molecular weight was varied by varying the rate of polymerization reaction; lower molecular weight polymers are obtained at higher reaction rates. The intrinsic viscosity was determined by regular viscometric techniques and the number average molecular weight by an osmometer. Adhesive properties correlate better with intrinsic viscosity than with number average molecular weight. Number average molecular weight is sensitive to the presence of low molecular weight materials, while most mechanical properties are mainly determined by the larger molecular weight fractions. Tack was determined by the Polyken Probe Tack Tester. Resistance to creep was measured by applying a shear force to an adhesive bond and observing the time required for a complete failure. The splitting temperature is defined as the temperature at which the transition of mode of failure in the 180° peel test takes place from adhesive to cohesive.[85]

Figure 13-2 shows the dependence of peel force on peel rate and Fig. 13-3 on the temperature for polymers of various molecular weights as shown in the Table 13-3. The solid lines indicate an adhesive failure; the

**Table 13-3. Effect of molecular weight on some of the adhesive properties.**

| POLYMER | INTRINSIC VISCOSITY | NUMBER AVG. MOL. WT. | PROBE TACK (g)[a] AT PRESSURE 10 g | 100 g | 500 g | CREEP (hr) 24°C | 71°C | SPLITTING TEMP. (°C) |
|---|---|---|---|---|---|---|---|---|
| A | 4.50 | 1,500,000 | 163 | 493 | 545 | >200 | 16 | >93 |
| B | 3.04 | 770,000 | 133 | 513 | 557 | 120 | 2 | 88 |
| C | 2.65 | 251,000 | 180 | 537 | 637 | 10 | 0.2 | 37 |
| D | 2.17 | 250,000 | 157 | 550 | 670 | 9 | 0.2 | 29 |
| E | 1.71 | 353,000 | 143 | 453 | 473 | 1.8 | 0.04 | 20 |
| F | 1.67 | 276,000 | 170 | 310 | 373 | 1.3 | 0.03 | 21 |
| G | 1.62 | 360,000 | 260 | 563 | 593 | 1.2 | 0.03 | <20 |

[a] Contact time 1 sec, rate of separation 1 cm/sec, temperature 23°C.

broken lines, a cohesive failure where a visually detectable layer of adhesive remains on the steel panel after the test.

From Fig. 13-2, it is evident the molecular weight has a large effect on the behavior of these adhesives. High molecular weight adhesives A and B show an adhesive failure at the rates tested. Intermediate molecular weight polymers C and D show a transition from adhesive to cohesive failure. Low molecular weight adhesives E, F, and G, show cohesive failure.[86] Extension of the testing to higher peel rates should also show a transition to adhesive

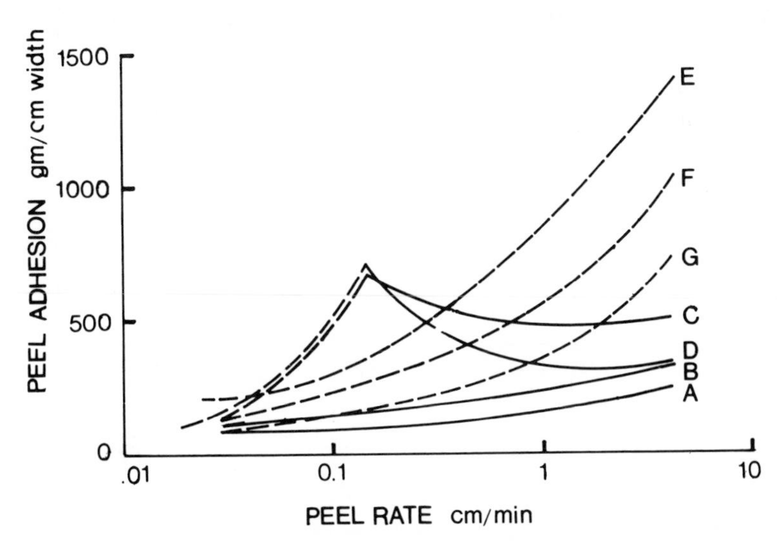


Fig. 13-2. Dependence of peel adhesion on peel rate for various molecular weight polymers shown in Table 13-3.

failure in these polymers. The increase of cohesive strength with increasing molecular weight as indicated by increased resistance to creep is, of course, expected. The transition from cohesive to adhesive failure is the result of increased modulus and cohesive strength with increasing strain rates. At higher peel rates, the cohesive strength exceeds the bond strength and transition of failure locus takes place.

Figure 13-3 shows the peel force of the same adhesives as a function of temperature. Force-temperature and force-time curves are qualitatively similar and superimposable. Comparison of Fig. 13-2 and 13-3 indicates this similarity. Adhesives E and F exhibit the transition of mode of failure at the lower end of the temperature scale. Similar behavior would be expected by extension of the test conditions to higher peel rates.

The changes of pressure-sensitive adhesive properties as a function of molecular weight are shown in the generalized way in Fig. 13-4.[3] Both tack and resistance to peel increase with increasing molecular weight until a maximum is reached. The maximum is at a fairly low molecular weight and the transition of the mode of failure from cohesive to adhesive failure takes place in this region. A further increase in molecular weight causes a decrease and leveling of these properties at some value suitable for a functional pressure-sensitive adhesive. Commercial adhesives would be offered in this range of molecular weight. A good pressure-sensitive adhesive will show only minor variations of tack with increasing molecular weight in the region past the transition area. Resistance to shear increases with increasing molecular weight and levels off at a fairly high molecular weight.

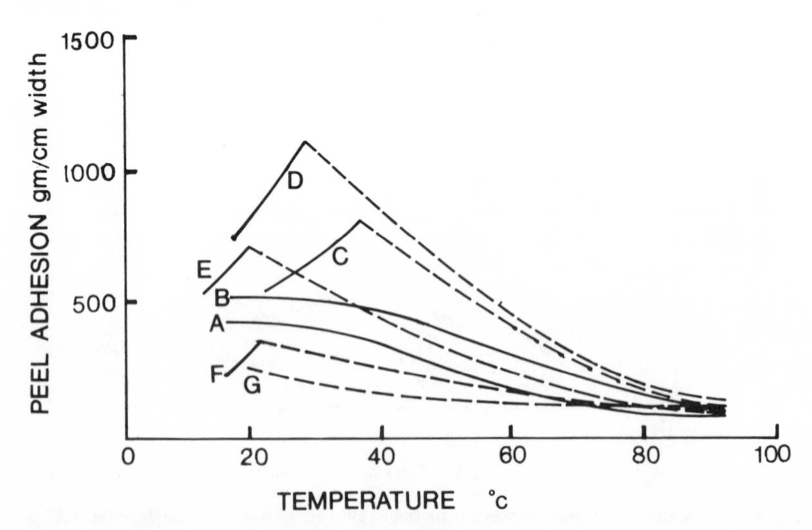


Fig. 13-3. Dependence of peel adhesion on temperature for various molecular weight polymers shown in Table 13-3.

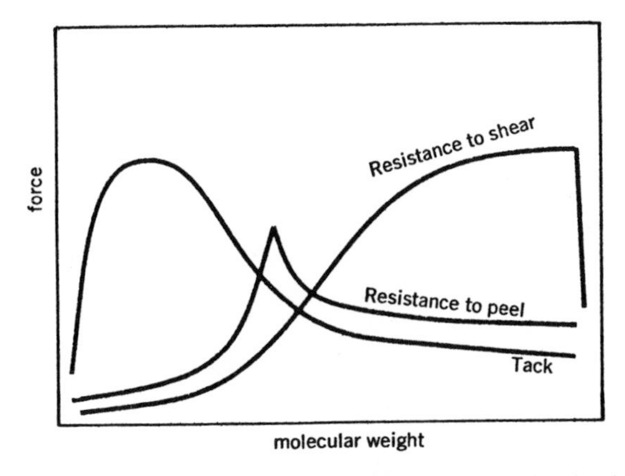


Fig. 13-4. Pressure-sensitive adhesive properties as a function of molecular weight.

Aubrey, Welding, and Wong report the relationship between the peel strength and peel rate for various molecular weight poly(butyl acrylate) polymers.[87] In the region of slow peel rates, the peel strength increased with increasing molecular weight, as is expected, since the peel strength is controlled by viscous flow of the adhesive. The peel rate at which a transition to stick-slip failure occurs, increases with decreasing molecular weight. In the region of high peel rates, no indication of appreciable dependence of peel strength on the peel rate was observed. This indicates that the adhesive is elastically and reversibly deformed at the high peel rates and that there is no flow in the adhesive.

The effect of molecular weight distribution on the pressure-sensitive adhesive properties is more difficult to assess. Tack and resistance to peel at low peel rates are expected to be sensitive to the presence of low molecular weight fractions. The resistance to shear is mainly controlled by the high molecular weight fraction of the adhesive.

Molecular weight distribution of acrylic pressure-sensitive adhesives is fairly broad with a considerable low molecular weight fraction. A typical molecular weight distribution is shown in Fig. 13-5. For some applications, it has been found that blending of two polymers gives improved properties. Such blends exhibit two peaks on a molecular weight distribution curve as illustrated in Fig. 13-6.

## CROSSLINKING

The main purpose of crosslinking is to improve the shear resistance of pressure-sensitive adhesives, especially at elevated temperatures. The improvement of resistance to creep can be quite significant even at low crosslinking densities. Crosslinking decreases the free movement of polymer

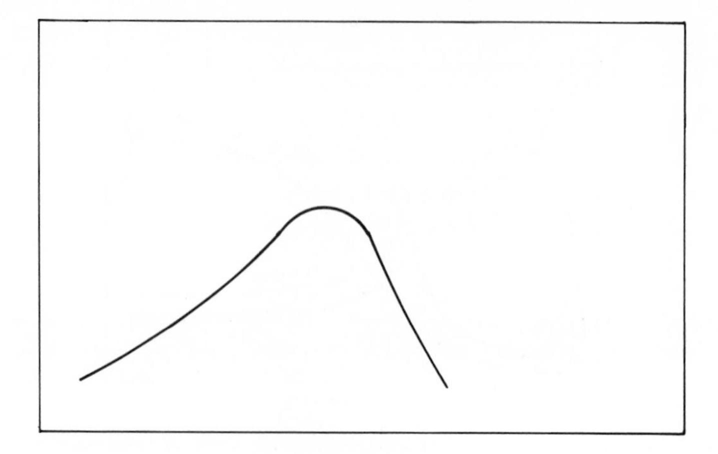


Fig. 13-5. Molecular weight distribution of a typical acrylic pressure-sensitive adhesive.

molecules and a decrease in tack is normally observed. Adhesive polymers are viscoelastic materials and the crosslinking tends to increase the effect of the elastic component at the expense of the viscous one. The peel adhesion might decrease with the increasing degree of crosslinking. The transition from a cohesive to an adhesive failure during peeling advances to lower peel

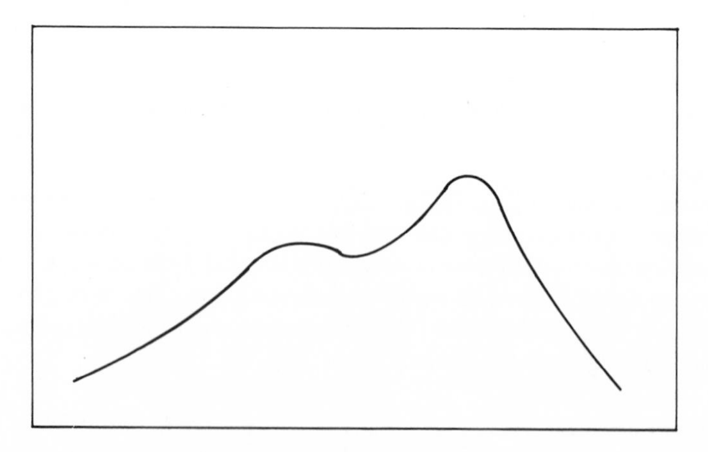


Fig. 13-6. Molecular weight distribution of a typical blend of two acrylic pressure-sensitive adhesives.

rate and higher temperature with increasing crosslinking density. Crosslinked adhesive might not exhibit this transition at all, but fail adhesively under all conditions.

In the polymerization of acrylates, multifunctional monomers or reactive groups carrying monomers can be introduced into the polymer chain for crosslinking purposes. There are many different ways to crosslink such polymers and the main problem is not the crosslinking itself, but maintaining of a low crosslinking density and producing a polymer of the same properties batch to batch.

Introduction of divinyl monomers during polymerization is a well-known method to produce crosslinked polymers. Low crosslinking densities are possible at low concentration of such monomer. Divinyl benzene, ethylene glycol dimethacrylate, and similar monomers are used for this purpose. Ethylene dimethacrylate is reported to improve the tensile strength of acrylic rubbers[88] and to crosslink pressure-sensitive acrylic adhesives.[23] Latent crosslinking can be introduced in the polymers by incorporating ethylene glycol dimethacrylate or its low molecular weight polymers. Loshaek and Fox[99] have proposed an interesting explanation of the crosslinkability of ethylene glycol dimethacrylate containing copolymers on heating. Ethylene glycol dimethacrylate and especially its dimer, trimer, or low molecular weight polymer consists of a long flexible chain with the methacrylate groups at the ends. The reactivities of both methacrylate groups are equal until one group becomes attached to a polymer chain. At this time, the reactivity of the pendant group is decreased, because of its restricted movement. Thus, at the end of the reaction, a number of pendant methacrylate groups are left unreacted. These are available for crosslinking among themselves at temperatures higher than the polymerization temperature because their mobility, as well as the mobility of the main polymer chain, increases with increasing temperature and these groups can be brought into contact to accomplish the crosslinking.

Various vinyl monomers carrying functional groups can be easily copolymerized with acrylic monomers providing sites for crosslinking. The crosslinking should take place through long and flexible chains in order to retain the flexibility and the high stress relaxation rate. A minimum decrease in flexibility minimizes the effect on adhesive properties. Two basic approaches can be used in crosslinking:

1. Functional groups are introduced along the polymer chain and a multifunctional additive capable of reacting with the pendant functional groups is introduced later.
2. Multiple functionalities capable of interaction are introduced into the polymer chain providing the sites for intermolecular crosslinking.

The polymer can crosslink during the polymerization or after, on heating or aging. The reactivity of the crosslinking reaction can be adjusted by the types and concentrations of the reactive groups and by their freedom of movement.

The use of various monomers for crosslinking purposes has been most widely explored for surface coating applications. The degree of crosslinking required for coatings is much higher than for pressure-sensitive adhesives, but the same reactants can be used. Various functional group containing monomers for coating applications have been reviewed by Miranda[100] and by a group of authors in the symposium on thermosetting acrylic resins.[101]

Functional groups often used for crosslinking are listed below.

- Carboxylic
- Hydroxyl
- Epoxy
- Allylic double bond
- Amide
- Amine

Carboxylic groups are the most versatile and react with many other groups as well as with multifunctional additives. They react with epoxides,[32,47] amides,[56] and amines such as hexamethylene diamine,[33] which can form salts with metallic ions. Zirconium and zinc alalinate,[47] zinc acetate,[47] zinc ammonium glycinate,[60] titanium compounds,[50,52] and many other salts of multivalent metals have been used for crosslinking purposes. Ionic crosslinking makes the polymer stronger, stiffer, and less tacky. The solvent resistance develops gradually over a period of several days indicating that it takes a long time for the metal ions to diffuse into the polymer when added to aqueous copolymer latex.[47] Ionic crosslinking through Zn is shown in the structural formula below.

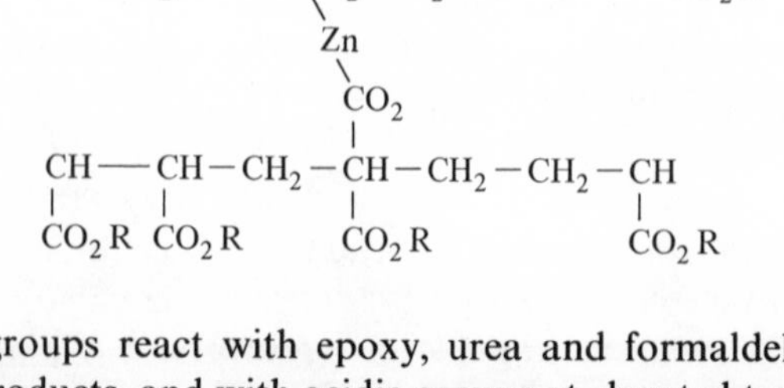

```
CH2-CH-CH2-CH-CH-CH2-CH2-CH
    |       |  |           |
    CO2R    CO2 CO2R       CO2R
            \
             Zn
              \
               CO2
               |
CH---CH-CH2-CH-CH2-CH2-CH
|    |       |          |
CO2R CO2R    CO2R       CO2R
```

Hydroxyl groups react with epoxy, urea and formaldehyde-melamine condensation products, and with acidic groups at elevated temperatures.[46]

Epoxy groups react with acids, anhydrides, and amines. Room temperature self-curing pressure-sensitive adhesive has been prepared by incorporating 5 parts of acrylic acid and 1 part of glycidyl methacrylate.[16]

Amide groups react with condensation products of formaldehyde and melamine and with urea. Methylolamides are useful crosslinking agents with epoxy and carboxylic groups. Crosslinking of amide containing polymer by formaldehyde is shown in the structural formula below.

$$
\begin{array}{l}
CH_2-\underset{\displaystyle CO_2R}{\underset{|}{CH}}-CH_2-\underset{\displaystyle \underset{\displaystyle CH_2-NHCO}{\underset{|}{CONH}}}{\underset{|}{CH}}-CH_2-\underset{\displaystyle CO_2R}{\underset{|}{CH}}-CH_2-\underset{\displaystyle CO_2R}{\underset{|}{CH}}-CH_2 \\
\\
\underset{\displaystyle CO_2R}{\underset{|}{CH}}-CH_2-\underset{\displaystyle CO_2R}{\underset{|}{CH}}-CH_2-\underset{\displaystyle CO_2R}{\underset{|}{CH}}-CH_2-\underset{\displaystyle CO_2R}{\underset{|}{CH}}
\end{array}
$$

Allylic double bonds can be crosslinked by peroxide catalysts.[72] Free radical-initiated curing can be used for crosslinking of pressure-sensitive acrylic polymers. Use of peroxides, especially those peroxides with various coagents, has been discussed by Mendelsohn.[102] The polymers can also be cured by high-energy irradiation such as accelerated electrons and also UV light.[66]

Polyfunctional aziridines are good room temperature crosslinking agents for carboxylic group-containing latexes. Diisocyanates are highly reactive with compounds containing active hydrogen and serve as crosslinking agents for solvent-based adhesives.

## SECONDARY BONDING

The cohesive strength of a polymer is also influenced by other bonds besides the covalent bonding. The most common bonds are the ones caused by dispersion forces which act between all atoms and are largely responsible for intermolecular cohesion. These forces act over a short distance: the interaction energy decreases with the sixth power of the distance.

Cohesive properties of a polymer can be affected by introduction of polar groups that interact forming secondary bonding. Such interactions are classified as hydrogen bonding, dipole–dipole and dipole–induced dipole bonding. It is not always possible to distinguish among various types of bonds, since the same groups might be involved in different types of bonding. While the effect of such bonding is similar to covalent crosslinking, such polymers are soluble, and the strength of such bonds decreases rapidly with increasing temperature because the bond strength decreases with increasing distance between atoms.

**Table 13-4. Polar groups useful in pressure-sensitive acrylic adhesives.**

| Group (positive end on left side) | Dipole Moment (Debye units) |
|---|---|
| H−F | 1.9 |
| H−Cl | 1.1 |
| H−O | 1.6 |
| H−S | 0.9 |
| H−N | 1.6 |
| ≡C−Cl | 1.7 |
| H−C≡ | 0.4 |
| >C=O | 2.5 |
| ≡C−F | 1.5 |
| >C=NH | 1.9 |
| −NO₂ (−N(=O)=O) | 3.9 |
| −C≡N | 3.8 |
| ≡C−O−C≡ | 0.9 |
| ≡C−N=O | 1.9 |
| >C=S | 3.0 |

A large number of vinyl monomers containing polar groups is available to tailor the properties of acrylic adhesives by introducing secondary bonding. Various useful groups have been reviewed by Mark[88,89] and by Rutzler.[90,91] Some of the polar groups of interest for pressure-sensitive acrylic adhesives are shown in Table 13-4.[89]

A dipole-dipole bond is the interaction of two dipoles. The bond energy is equal to the product of the strength of the dipoles. It decreases with the sixth power of the distance between the atom centers. The range of these forces is 4.0–5.0 Å.

Dipole-induced dipole bond takes place between one end of a strong dipole and an easily polarizable area of another molecule. A secondary dipole is induced and the bond is similar to the one between two permanent dipoles. Some easily polarizable groups are the following:

−C≡C− ; >C=C< ; >C=C−C=C< ; >C=C−C=O

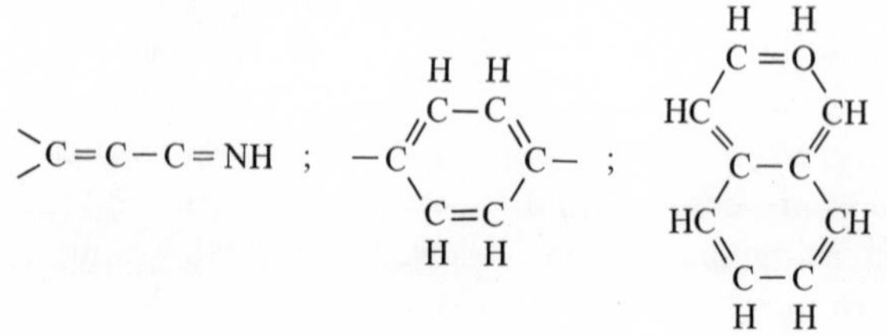

>C=C−C=NH ; −C₆H₄− (p-phenylene, −C(HC=CH)(HC=CH)C−) ; naphthalene ring (HC, CH, C−C, H H)

Hydrogen bonding takes place between the negative pole of a dipole and the positively charged H atom of another dipole as shown below.

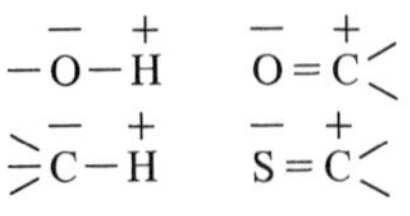

$$-\overset{-}{O}-\overset{+}{H} \qquad \overset{-}{O}=\overset{+}{C}\lt$$
$$\gt\overset{-}{C}-\overset{+}{H} \qquad \overset{-}{S}=\overset{+}{C}\lt$$

The range of hydrogen bonding is 2.6–3.0 Å, and the bond is very strong.

Acrylonitrile has a very strong dipole $-C\equiv N$ capable of forming hydrogen bonds, or interacting with other dipoles. Introduction of acrylonitrile as a comonomer, even in small amounts, has a great effect on the pressure-sensitive properties of acrylic polymers. It serves well as an example of the importance of secondary bonding. Other polar groups generally have a lesser effect. Table 13-5 shows the effect of acrylonitrile.

The glass transition temperature increases with an increasing amount of polar monomer such as acrylonitrile. The tack remains constant, except at higher concentrations of the polar monomer. Sample L, containing 10% acrylonitrile, shows a decrease in tack. The creep resistance increases with increasing acrylonitrile content. The increase is less pronounced at elevated temperatures, but considerable at room temperature.

Figures 13-7 and 13-8 show the peel force as a function of peel rate and temperature. The transition from cohesive to adhesive failure is observed in some polymers (H, I, and J in Fig. 13-7). In the range of adhesive failures, the peel force decreased with increasing acrylonitrile content. The decreasing compliance of the adhesive distributes the stress over a smaller area that accounts for the observed decrease of the peel force. Figure 13-9 shows

**Table 13-5. The effect of acrylonitrile comonomer on the properties of pressure-sensitive adhesives**

| | | PROBE TACK[b] | | | CREEP RESISTANCE (hr) | | |
|---|---|---|---|---|---|---|---|
| POLYMER[a] | $T_g$ (C°) | 10 g | 100 g | 500 g | 21°C | 43°C | 66°C |
| H | −60 | 113 | 400 | 447 | 0.9 | 0.04 | 0.01 |
| I | −60 | 102 | 433 | 473 | 1.2 | 0.07 | 0.02 |
| J | −55 | 133 | 405 | 440 | 2.0 | 0.06 | 0.01 |
| K | — | 103 | 432 | 492 | 48 | 0.3 | 0.03 |
| L | −40 | 57 | 337 | 338 | 48 | 48 | 0.5 |

[a] Polymers are listed in order of increasing acrylonitrile content. Polymer L has 10% by weight of acrylonitrile.

[b] Probe tack measured on Polyken Probe Tack Tester, contact time 1 sec, rate of separation 1 cm/sec, temperature 83°C.

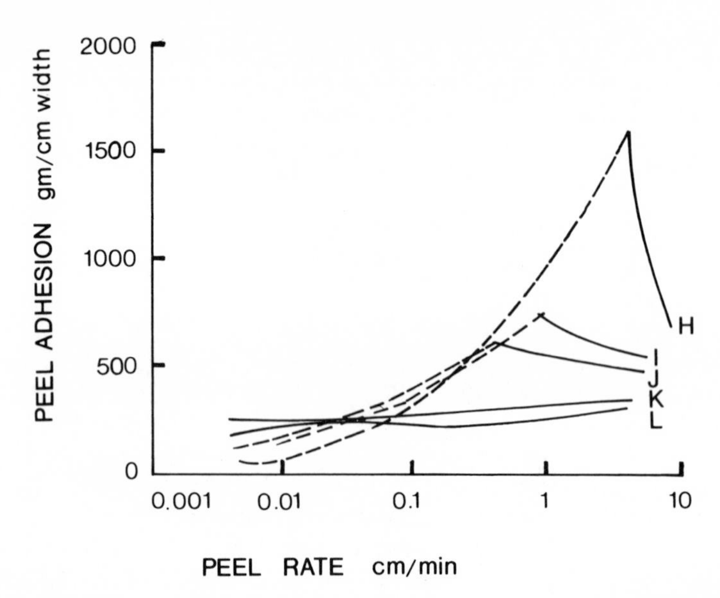


Fig. 13-7. Dependence of peel adhesion on peel rate for various acrylonitrile containing polymers shown in Table 13-5.

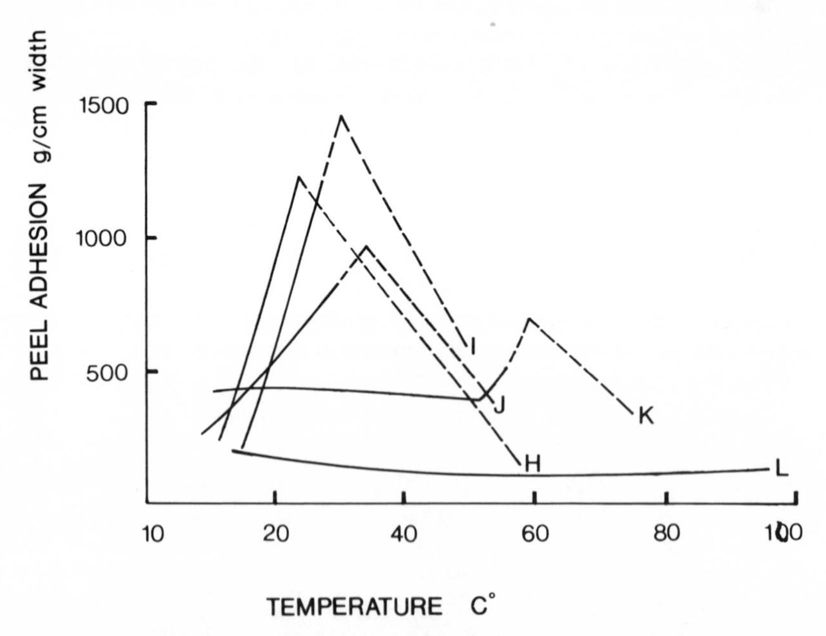


Fig. 13-8. Dependence of peel adhesion on temperature for various acrylonitrile containing polymers shown in Table 13-5.

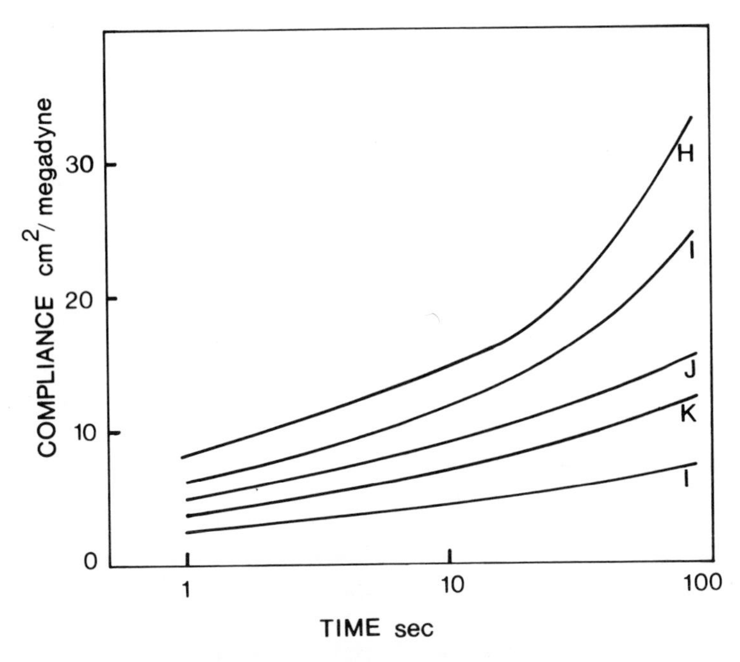


Fig. 13-9. Penetrometer compliance as a function of temperature for various polymers shown in Table 13-5.

penetrometer compliance data of these polymers, illustrating the stiffening effect of acrylonitrile because of hydrogen bonding or dipole interaction.

## IONIC BINDING

Carboxylic group containing polymers can be crosslinked by multivalent metallic ions. Introduction of monovalent metallic ions can also affect a change in properties similar to crosslinking.

This ionic binding has been discussed by Satas and Mihalik.[85] A significant difference between ionically bound and crosslinked polymers is that ionic binding, while effective at room temperature, is weakened at elevated temperatures. This property obviously makes such polymers more easily processible.

Ionic binding has been imparted by soaking the polymer in a solution of an appropriate base. Reasonably uniform properties can be obtained if the polymer is in a form of porous crumbs. In the case of a polymer prepared by emulsion polymerization, the emulsion can be introduced into the alkaline solution, thus providing a good penetration of metal ions into the polymer. Milling of solid hydroxide into the polymer mass on a hot two-roll mill is also an available method.

The pressure-sensitive polymers investigated are easily soluble in ketones and aromatic solvents. Introduction of ionic binding changed the solubility behavior considerably. The polymers with the higher densities of ionic binding did not go into solution but only swelled in these solvents. Addition of isopropyl alcohol was usually sufficient to remove this effect of the ionic binding, with the result that the polymers were soluble in solvent blends containing alcohol.

The effect of sodium binding on the adhesive properties is shown in Table 13-6. The modulus of these adhesives increases with increasing ionic character. It is also expected that the tack should decrease with increasing modulus and rigidity of the adhesive. This is supported by the data in Table 13-6, showing that the tack decreases rapidly with increasing binding. The expected decrease of peel adhesion with increasing binding density is also supported by the data in Table 13-6. The failure in the creep tests can occur either in the adhesive mass or at the bond interface, the former being the case in these tests. The time required for the tape to fail increased with increasing cohesive strength of the polymer. It is obvious from the data in Table 13-6 that ionic binding increased the cohesive strength.

The resistance to creep is temperature-dependent. The ionic binding, although quite effective at room temperature, loses its effectiveness at elevated temperatures. This indicates that ionically bound polymers could be processed at elevated temperatures on equipment designed for thermoplastic materials. The data shown in Fig. 13-10 indicate that resistance

**Table 13-6. Effect of bound sodium on pressure-sensitive adhesive properties of the polymer.**

| SAMPLE | BOUND SODIUM (wt %) | TACK ($g/cm^2$ at 10 g CONTACT PRESSURE) | PEEL ADHESION (g/cm Width) | RESISTANCE TO CREEP (hr) |
|---|---|---|---|---|
| 1 | 0.0 | 163 | 524 | 0.6 |
| | 0.06 | 123 | 312 | 4.5 |
| | 0.08 | 77 | 480 | 1.7 |
| | 0.32 | 40 | 368 | > 100 |
| 2 | 0.0 | 607 | 971 | 0.2 |
| | 0.34 | 247 | 513 | 4.2 |
| | 0.50 | 142 | 502 | 7.5 |
| | 0.52 | 133 | 558 | 2.9 |
| | 0.59 | 92 | 279 | 68 |
| | 0.69 | 0 | 279 | 64 |

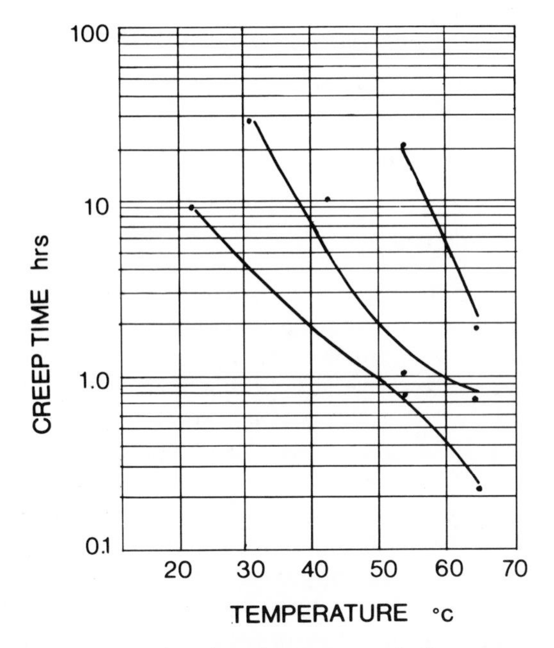


Fig. 13-10. Creep resistance as a function of temperature. Sodium content: top curve 0.8%; middle curve 0.08%; bottom curve 0%.

to creep falls off rapidly with increasing temperature and that at elevated temperatures, the effect of ionic binding is minor. In its temperature sensitivity, the heat-labile ionic binding is basically different from covalent cross-linking.

The effect of ionic binding on splitting temperature is shown in Table 13-7. Splitting temperature is defined as the temperature at which the mode

**Table 13-7. Effect of bound sodium on splitting temperature.**

| Bound Sodium (wt%) | Splitting Temperature (°C) |
|---|---|
| 0 | 21 |
| 0.18 | 32 |
| 0.34 | 68 |
| 0.50 | 77 |
| 0.52 | 77 |
| 0.59 | 82 |
| 0.69 | 88 |

of failure changes from adhesive to cohesive. Instead of a clean peel as obtained in the case of adhesive failure, the polymer splits at or above this temperature, leaving a residue on the steel plate. The data shown are obtained from the tests of samples from the same polymer batch differing only in the amount of bound sodium.

## FUNCTIONAL GROUPS

Functional groups, in addition of providing sites for chemical reactions such as crosslinking, are also known to improve adhesion. Carboxylated polymers are widely used for adhesive applications.

It is difficult to assess whether the observed effect of the functional groups is due to a change of polymer bulk properties or to the true adhesion effects at the interface between the polymer and the adherend. A change in bulk properties has an effect on peel force, tack and other properties obtained by standard testing methods. The same properties can also be affected by interfacial changes: improved wettability and increased contact or formation of bonds between the adhesive and the substrate. Aubrey[79] on the basis of work by Ginosatis[92] shows that the carboxylic acid groups on the surface of a poly(butyl acrylate) pressure-sensitive adhesive increase the peel force without affecting the bulk properties of the polymer. Lewis and Forrestal[93] have shown that a small amount of functional groups, deposited by grafting of 2-chlorovinyl phosphonate onto polypropylene, have improved adhesion without a change in bulk properties. In general, it is claimed that acidic groups and other electron acceptors improve adhesion to metals; electron donors, such as hydroxyl groups, are more effective on steel than on copper, and tertiary amines are less effective than the primary ones.

Mao and Reegen[84] reported their findings on the effect of copolymeriz-

**Table 13-8. Effect of various comonomers on the peel adhesion.**

| | Peel Adhesion (g) | | | |
|---|---|---|---|---|
| Comonomer (mole %) | Acrylic Acid | Acrylamide | Methacrylic Acid | Methacrylamide |
| 0 | 160 | 160 | 160 | 160 |
| 2 | 220 | 200 | 170 | 140 |
| 5 | 250 | 270 | 170 | 170 |
| 10 | 310 | 280 | 100 | 100 |
| 15 | — | 140 | — | — |
| 20 | 420 | — | — | — |

| Comonomer (10 mole %) | Peel Adhesion (g) |
|---|---|
| $CH_2=C(CH_3)CO_2CH_2CF_3$ | 90 |
| $CH_2=C(CH_3)CO_2CH_2CH_3$ | 170 |
| $CH_2=C(CH_3)CO_2CH_2CH_2CN$ | 170 |
| $CH_2=C(CH_3)CO_2CH_2Si(CH_3)_2OSi(CH_3)_3$ | 210 |
| $CH_2=C(CH_3)CO_2CH_2CH_2Cl$ | 220 |
| $CH_2=C(CH_3)CO_2CH_2CH_2Br$ | 260 |
| $CH_2=C(CH_3)CO_2CH_2CH(OH)CH_3$ | 260 |
| $CH_2=C(CH_3)CO_2CH_2CH_2OH$ | 270 |
| $CH_2=C(CH_3)CO_2CH_2CH_2-C_6H_5$ | 290 |
| $CH_2=C(CH_3)CO_2CH_2-C_6H_5$ | 350 |
| $CH_2=C(CH_3)CO_2CH_2CH_2OCH_2CH_3$ | 360 |

ing methyl methacrylate with acrylic acid, acrylamide, methacrylic acid, and methacrylamide. The peel adhesion of various copolymers is shown in Table 13-8. Acrylic acid was the most effective comonomer. The decrease of peel adhesion in case of higher concentration of acrylamide is attributed to the increase of crystallinity.

Mao and Reegen[84] also prepared methyl methacrylate copolymers with various functional monomers. The results on peel adhesion are shown in Table 13-9. The authors explain the difference in the effect on adhesive properties on the basis of available electrons to form the bonds between the adhesive and the substrate.

The patent literature has many examples of the beneficial effects of relatively small quantities of modifying monomers. It is not clear whether such results are to be attributed to bulk or to surface effects. Pietsch and Curts[43] describe the use of modifying monomers *N*-vinyl pyrrolidone, *N*-vinyl piperidone, and *N*-vinyl caprolactam to improve the adhesive properties of an acrylic pressure-sensitive polymer. The effect of *N*-vinyl pyrrolidone is shown in Fig. 13-11. These polymers are especially useful for electrical applications. They do not contain reactive groups which contribute to electrolytic corrosion.

Water-dispersable and water-soluble acrylic pressure-sensitive adhesives can be prepared by incorporating a higher concentration of hydrophilic groups containing monomers. Such adhesives are of interest for repulpable tapes used in splicing of paper. Acrylic acid and vinyl pyrrolidone containing repulpable adhesives are described by Peterson.[33]

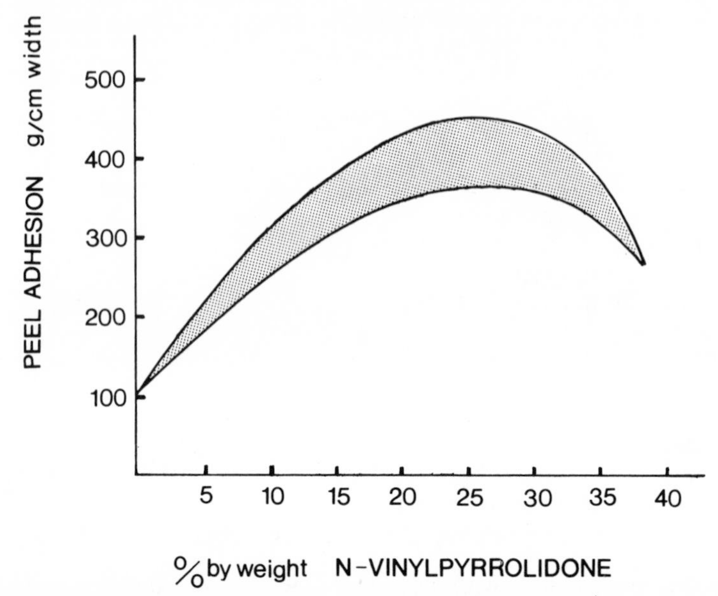


Fig. 13-11. The effect of *N*-vinyl pyrrolidone comonomer on the peel adhesion of an acrylic pressure-sensitive adhesive.

## COMPOUNDING

Acrylic pressure-sensitive adhesives do not require compounding in order to develop the optimum combination of properties. The adhesive properties, however, can be changed and varied by compounding with various ingredients just like in case of other elastomer-based adhesives.

Incorporation of tackifying resins improves tack and peel adhesion. Table 13-10 shows the effect of addition of Cellolyn 21 (a phthalate ester of

**Table 13-10. Effect of tackifying resin on adhesive properties.**

| Plasticizer Content (phr) | Peel Adhesion (g/cm width) | Creep Resistance (hr) | Tack[a] (g/cm$^2$) at a Pressure of | | |
|---|---|---|---|---|---|
| | | | 10 g | 100 g | 500 g |
| 0 | 275 | 32 | 117 | 530 | 515 |
| 5 | 319 | 25 | 183 | 540 | 548 |
| 10 | 341 | 11 | 200 | 540 | 555 |
| 20 | 352 | 5 | 322 | 548 | 555 |
| 30 | 462 | 2.7 | 315 | 570 | 622 |

[a] Polyken Probe Tack Tester, contact time 1 sec, rate of separation 1 cm/sec, temperature 23°C.

**Table 13-11. Effect of plasticizer on pressure-sensitive adhesive properties.**

| Amount of Plasticizer (phr) | Peel Adhesion (g/cm width) | Tack | |
|---|---|---|---|
| | | Qualitative | Rolling Ball (cm) |
| 0 | 429 | Good | 3.0 |
| 4 | 407 | Good | 3.6 |
| 10 | 363 | Excellent | 2.8 |

technical hydroabietyl alcohol by Hercules, Inc.) to a typical acrylic pressure-sensitive adhesive. The adhesive thickness over a polyester film is kept constant at 0.05 mm.

Aqueous emulsions of acrylic adhesives can be tackified by addition of pre-emulsified tackifying resins.

Addition of plasticizer has an effect similar to that of tackifying resin. Table 13-11 shows the results of adding a phosphate plasticizer to an acrylic pressure-sensitive adhesive.[94]

Fillers can be used to extend the polymer or to change its properties. The effect of some of the fillers is shown in Table 13-12.[95]

Larger amounts of nonreactive pigments such as clay and calcium carbonate can be tolerated before a noticeable effect on peel adhesion is observed. Zinc oxide and colloidal silica are active pigments and much smaller amounts can be tolerated before the adhesive properties are lost.

**Table 13-12. Effect of some fillers on the peel adhesion of aqueous acrylic adhesives.**

| Filler | Peel Adhesion (g/cm Width) — Parts of Filler per 100 Parts of Latex[a] | | | | | | |
|---|---|---|---|---|---|---|---|
| | 10 | 20 | 30 | 40 | 50 | 100 | 150 |
| Hydrasperse clay (70% dispersion) | 380 | 380 | 413 | 436 | 436 | 447 | 0 |
| Calcium carbonate | 402 | 413 | 402 | 402 | 413 | 424 | 405–525 |
| Zinc oxide (50% dispersion) | 405 | 447 | 424 | 447 | 559 | 469 | 0 |
| Colloidal silica AM (30% dispersion) | 536 | 447–559 | 380 | — | — | — | — |
| Colloidal silica HS (40% dispersion) | 458 | 536 | — | — | — | — | — |

[a] Hycar 2600 × 146 by B. F. Goodrich Co.

Flame-retardant properties are incorporated into an acrylic pressure-sensitive adhesive by compounding with phosphate plasticizers and antimony oxide.[25]

Acrylic pressure-sensitive adhesives have been compounded with other polymers for special effects. Compounding with a noncompatible synthetic elastomer gave an adhesive especially suitable for vibration dampening over a wide temperature range.[96]

## COATING METHODS

Pressure-sensitive acrylic adhesives are available in various forms for coating. Solutions in hydrocarbon solvents are commonly used, although the method is under pressure to be replaced by other adhesive coating techniques, because of the environmental and economical reasons. Many large coaters are choosing to remove the solvent vapor from the exhaust by activated carbon adsorption and then to recover the adsorbed solvent. Despite a large investment, this method is economically feasible for controlling vapor emissions to the atmosphere and conserving on the solvent usage.

Adhesive polymer solutions are usually made by solution polymerization, but sometimes the polymer is emulsion polymerized, coagulated, washed and then dissolved. The normal range of polymer solids in solution coatings is 25–40%. Some high solid content (50–70%) nonaqueous dispersions have also been reported.[97]

Liquid acrylic oligomers or monomers have been polymerized *in situ* to pressure-sensitive adhesives. The reaction can be initiated by radiation or by thermal decomposition of a catalyst.

Liquid acrylic polymers containing hydroxyl groups and isocyanate-terminated prepolymer are combined and reacted to form a pressure-sensitive adhesive. Such system is available on the market[98] and is processed as a high solid solution (80% solids) or as a 100% solid coating. At 150°C, complete cure, using dibutyl tin dilaurate catalyst, is obtained in 30–45 sec; at 93°C, in 2.5–4 min. The adhesive properties depend on OH/NCO ratio which is adjusted prior to coating.

Acrylic latexes are widely used for pressure-sensitive applications. They are compounded with thickeners, antifoaming agents, wetting agents, and other ingredients to render the latexes easily coatable. The additives must be carefully chosen and evaluated because of their possible effect on the adhesive properties.

Acrylic adhesives are also available as solids suitable for application by the hot melt technique. Hot melt adhesives and higher molecular weight polymers, which are too viscous for hot melt application, can be calendered.

## REFERENCES


1. Peckmann, H. V., and Roehm, O. *Berichte der Deutschen Chemischen Gesellschaft* **34:** 429 (1901).
2. Bauer, W. Ger. Patent 575,327(1933) (assigned to Röhm and Haas AG).
3. Satas, D. *Adhesives Age* **15**(10): 19–23 (October 1972).
4. Ham, G. E. *Vinyl Polymerization.* New York: Marcel Dekker, 1969.
5. Ham, G. E. *Copolymerization.* New York: Wiley-Interscience, 1964.
6. Luskin, L. S., and Myers, R. J. Acrylic Ester Polymers. *Encyclopedia of Polymer Science and Technology*, Vol. **I,** 246–328. New York: John Wiley and Sons, 1964.
7. Kugler, J. H. U.S. Patent 2,544,692 (1951) (assigned to Minnesota Mining and Manufacturing Co.)
8. Ulrich, E. W. U.S. Patent 2,884,126 (1959) and Re 24,906 (assigned to Minnesota Mining and Manufacturing Co.).
9. Bergstedt, M. A., and Young, K. C. U.S. Patent 2,953,475 (1960) (assigned to Johnson and Johnson).
10. Young, H. C., and Toy, W. W. U.S. Patent 2,976,203 (1961) (assigned to Rohm and Haas Co.).
11. Young, H. C., and Toy, W. W. U.S. Patent 2,976,204 (1961) (assigned to Rohm and Haas Co.).
12. Ulrich, E. W. U.S. Patent 3,008,850 (1961) (assigned to Minnesota Mining and Manufacturing Co.).
13. Cantor, H. A., Horback, W. B., Vona, J. A., and Kuczynski, W. J. U.S. Patent 3,258,443 (1966) (assigned to Celanese Corporation of America).
14. Vona, J. A., and Wyart, J. W. U.S. Patent 3,258,454 (1966) (assigned to Celanese Corporation of America).
15. Alexander, R. R., and Urjil, A. J. U.S. Patent 3,275,589 (1966) (assigned to W. R. Grace and Co.).
16. Knapp, E. C. U.S. Patent 3,284,423 (1966) (assigned to Monsanto Co.).
17. Samour, C. M. U.S. Patent 3,299,010 (1967) (assigned to Kendall Co.).
18. Gander, R. J. U.S. Patent 3,321,451 (1967) (assigned to Johnson and Johnson).
19. Brooks, B. A., and Jubilee, B. D., Jr. U.S. Patent 3,371,071 (1968) (assigned to National Starch and Chemical Corp.).
20. Gander, R. J. U.S. Patent 3,475,363 (1969) (assigned to Johnson and Johnson).
21. Samour, C. M. U.S. Patent 3,483,155 (1969) (assigned to Kendall Co.).
22. Popa, L., and Lydick, H. K. U.S. Patent 3,485,896 (1969) (assigned to B. F. Goodrich Co.).
23. Samour, C. M., and Satas, D. U.S. Patent 3,492,260 (1970) (assigned to Kendall Co.).
24. Samour, C. M. U.S. Patent 3,509,111 (1970) (assigned to Kendall Co.).
25. Tomita, J., and Strahan, T. W. U.S. Patent 3,515,578 (1970) (assigned to 3M Co.).
26. Zang, D. H., and Lader, W. U.S. Patent 3,532,652 (1970) (assigned to PPG Industries, Inc.).
27. Davis, I. J., and Sirota, J. U.S. Patent 3,535,295 (1970) (assigned to National Starch and Chemical Corp.).
28. Burke, R. L., Jr. U.S. Patent 3,547,852 (1970) (assigned to United Merchants and Manufacturers, Inc.).
29. Crocker, G. J., and Doehnert, D. F. U.S. Patent 3,551,391 (1970) (assigned to Johnson and Johnson).
30. Doehnert, D. F. U.S. Patent 3,558,574 (1971) (assigned to Johnson and Johnson).
31. Newman, N. S., and Satas, D. U.S. Patent 3,562,088 (1971) (assigned to Kendall Co.).

32. Lehmann, G. W. H., and Curts, H. A. J. U.S. Patent 3,563,953 (1971) (assigned to Beiersdorf AG).
33. Peterson, R. L. U.S. Patent 3,575,911 (1971) (assigned to 3M Co.).
34. Kordzinski, S., and Horn, M. B. U.S. Patent 3,579,490 (1971) (assigned to Ashland Oil and Refining Co.).
35. Aranyi, C., Gutfreud, K., Hawrylewicz, E. J., and Wall, J. S. U.S. Patent 3,607,370 (1971) (assigned to U.S. Secretary of Agriculture).
36. Coffman, A. M. U.S. Patent 3,637,615 (1972) (assigned to B. F. Goodrich Co.).
37. Brookman, R. S., Grib, S., and Pearson, D. S. U.S. Patent 3,661,618 (1972) (assigned to The Firestone Tire and Rubber Co.).
38. Guse, G., and Pietsch, H. G. U.S. Patent 3,690,937 (1972) (assigned to Beiersdorf AG).
39. Maska, R. U.S. Patent 3,701,758 (1972) (assigned to National Starch and Chemical Corp.).
40. Christenson, R. M., and Anderson, C. C. U.S. Patent 3,725,115 (1973) (assigned to PPG Industries, Inc.).
41. Fournier, A. A. U.S. Patent 3,725,121 (1973) (assigned to Johnson and Johnson).
42. Reinhard, H., Mueller, A., and Dotbauer, B. U.S. Patent 3,725,122 (1973) (assigned to Badische Anilin and Soda Fabrik AG).
43. Pietsch, H., and Curts, J. U.S. Patent 3,728,148 (1973) (assigned to Beiersdorf AG).
44. Lehmann, H., and Curts, J. U.S. Patent 3,729,338 (1973) (assigned to Beiersdorf AG).
45. Sirota, J., and Davis, I. J. U.S. Patent 3,733,292 (1973) (assigned to National Starch and Chemical Corp.).
46. Coffman, A. M. U.S. Patent 3,738,971 (1973) (assigned to B. F. Goodrich Co.).
47. Sanderson, F. T., and Zdanowski, R. E. U.S. Patent 3,740,366 (1973) (assigned to Rohm and Haas Co.).
48. McKenna, L. W., Jr., and Gardner, D. M. U.S. Patent 3,763,117 (1973) (assigned to Monsanto Co.).
49. Anderson, C. C., Maska, R., and Das, S. K. U.S. Patent 3,769,254 (1973) (assigned to National Starch and Chemical Co.).
50. Knoepfel, H., and Silver, S. F. U.S. Patent 3,770,708 (1973) (assigned to 3M Co.).
51. Fukukawa, S., Shimomura, T., Ijichi, I., Yokishawa, N., and Murakami, T. U.S. Patent 3,772,063 (1973) (assigned to Nitto Electric Industrial Co., Ltd.).
52. Samour, C. M. U.S. Patent 3,790,533 (1974) (assigned to Kendall Co.).
53. Kosaka, Y., Uemura, M., Fujiki, T., and Saito, M. U.S. Patent 3,838,079 (1974) (assigned to Toyo Soda Mfg. Co., Ltd.).
54. Dalibor, H., Gutte, R., and Stenzel, G. U.S. Patent 3,842,133 (1974) (assigned to Reichhold-Albert Chemie AG).
55. Yamada, F., Fukuda, Y., and Uraya, T. U.S. Patent 3,849,840 (1974) (assigned to Kanebo, Ltd.).
56. McKenna, L. W., Jr. U.S. Patent 3,886,126 (1975) (assigned to Monsanto Co.).
57. Gardner, D. M., and McKenna, L. W., Jr. U.S. Patent 3,893,982 (1975) (assigned to Monsanto Co.).
58. Kosaka, Y., Uemura, M., Fujiki, T., Kimura, M., and Saito, M. U.S. Patent 3,896,067 (1975) (assigned to Toyo Soda Mfg. Co., Ltd.).
59. Dowbenko, R., and Christenson, R. M. U.S. Patent 3,897,295 (1975) (assigned to PPG Industries, Inc.).
60. McKenna, L. W., Jr. U.S. Patent 3,900,610 (1975) (assigned to Monsanto Co.).
61. Silver, S. F., Winslow, L. E., and Zigman, A. R. U.S. Patent 3,922,464 (assigned to 3M Co.).

62. Gobran, R., Knoepfel, H., and Silver, S. F. U.S. Patent 3,924,044 (1975) (assigned to 3M Co.).
63. Davis, I., Skoultchi, M. M., and Fries, J. A. U.S. Patent 3,925,282 (1975) (assigned to National Starch and Chemical Co.).
64. Baatz, J. C., and Corey, A. E. U.S. Patent 3,931,087 (1976) (assigned to Monsanto Co.).
65. McKenna, L. W., Jr., and Gardner, D. M. U.S. Patent 3,931,444 (1976) (assigned to Monsanto Co.).
66. Douek, M., Schmidt, G. A., Malofsky, B. M., and Hauser, M. U.S. Patent 3,996,308 (1976) (assigned to Avery Products Corp. and Loctite Corp.).
67. Mowdood, S. K., and Given, D. A. U.S. Patent 3,998,997 (1976) (assigned to Goodyear Tire and Rubber Co.).
68. Guerin, J. D., Hutton, T. W., Miller, J. J., and Zdanowski, R. E. U.S. Patent 4,045,517 (1977) (assigned to Rohm and Haas Co.).
69. Pastor, S. D., and Skoultchi, M. M. U.S. Patent 4,052,527 (1977) (assigned to National Starch and Chemical Corp.).
70. British Patent 624,764 (1949) (assigned to Minnesota Mining and Manufacturing Co.).
71. Dahlquist, C. A., and Zenk, R. E. British Patent 870,022 (1961) (assigned to Minnesota Mining and Manufacturing Co.).
72. British Patent 930,761 (1963) (assigned to Farbenfabriken Bayer AG).
73. British Patent 938,217 (1963) (assigned to Beiersdorf and Co., AG).
74. British Patent 1,086,262 (1967) (assigned to Johnson and Johnson).
75. Aubrey, D., and Saville, R. W. Canadian Patent 670,164 (1963) (assigned to Adhesive Tapes, Ltd.).
76. Bailey, F. E., and Stickle, R., Jr. Canadian Patent 676,096 (1963) (assigned to Union Carbide Corp.).
77. Brown, F. W., and Keck, F. L. Canadian Patent 769,780 (1967) (assigned to 3M Co.)
78. Bond, H. M., and Tomita, J. Canadian Patent 778,731 (1968) (assigned to 3M Co.).
79. Aubrey, D. W. *Developments in Adhesives*—1, (William C. Wake, ed.). Barking, England: Applied Science Publishers, Ltd., pp. 127–156, 1977.
80. Guerin, J. D., Hutton, T. W., Miller, J. J., and Zdanowski, R. E. U.S. Patent 4,152,189 (1979) (assigned to Rohm and Haas Co.).
81. Rehberg, C. E., and Fisher, C. H. *J. Am. Chem. Soc.* **66**: 1203 (1944).
82. Rehberg, C. E., and Fisher, C. H. *Ind. Eng. Chem.* **40**: 1429 (1948).
83. Nuessle, A. C. *Chemistry and Industry*, 1580–1594, 1966.
84. Mao, T. J., and Reegen, S. L. *Proceedings of the Symposium on Adhesion and Cohesion.* (Philip Weiss, ed.) Amsterdam: Elsevier Publishing Co., 209–217, 1962.
85. Satas, D., and Mihalik, R. *J. Appl. Polym. Sci.* **12**: 2371–2379 (1968).
86. Satas, D. *Adhesive Tapes.* Polymer Conference Series, Wayne State University, June 12–16, 1967, Detroit.
87. Aubrey, D. W., Welding, G. N., and Wong, T. J. *J. Appl. Polym. Sci.* **13**: 2193–2207 (1969).
88. Mark, H. *Proceedings of the Symposium on Adhesion and Cohesion.* (Philip Weiss, ed.). Amsterdam: Elsevier Publishing Co., 240–269, 1962.
89. Mark, H. F. *Adhesives Age*, **22**(9): 45–50 (1979).
90. Rutzler, J. E. *Adhesives Age*, **2**(6): 39 (1959).
91. Rutzler, J. E. *Adhesives Age*, **2**(7): 28 (1959).
92. Ginosatis, S. M. Sci. Thesis, 1977, The City University, London.
93. Lewis, A. F., and Forrestal, L. J. ASTM Preprint, Atlantic City Meeting, June 1963.
94. Sanderson, F. T. Pressure-Sensitive Tape Council, Technical Meeting on Water-Based

Systems, June 21–22, 1978, Chicago, pp. 13–27.
95. Latexes in Adhesive Systems, Bulletin L-14, B. F. Goodrich Co.
96. Knutson, A. T., and Kallenberg, M. O. U.S. Patent 3,092,250 (1963) (assigned to 3M Co.).
97. Marchessault, R. G., and Plummer, A. P. *Adhesives Age* **20**(4): 34 (1977).
98. Hycar Pressure-Sensitive Polymers. PSP I-3AA-2, revised 11/10/79. B. F. Goodrich Co.
99. Loshaek, S., and Fox, T. G. *J. Am. Chem. Soc.*, **75**: 3544 (1953).
100. Miranda, T. J. *J. of Paint Tech.*, **38**: 469–477 (1966).
101. *Official Digest*, 679–746 (June, 1961).
102. Mendelsohn, M. A. *Ind. and Eng. Chem., Product Research and Development*, **3**: 67–72 (1964).

# Chapter 14

# Vinyl Ether Polymers

**Helmut W. J. Mueller**
*BASF AG, Ludwigshafen/Rhein*

Polymers and copolymers of vinyl ethers are used for various applications. Homopolymers with the general formula

$$-\left[CH_2-\underset{\displaystyle OR}{\underset{|}{CH}}\right]_n-$$

| | | |
|---|---|---|
| where R is | $-CH_3$ | (methyl) |
| | or $-CH_2-CH_3$ | (ethyl) |
| | or $-CH_2-\underset{\displaystyle CH_3}{\underset{|}{CH}}-CH_3$ | (isobutyl) |

and copolymers with acrylates are used as raw materials for producing pressure-sensitive adhesive compounds. Homopolymers of the aforementioned vinyl ethers are most widely used. These polymers may be of oily to rubberlike consistency.

Vinyl ether was first converted to a balsamlike polymer more than 100 years ago.[1] Between 1920 and 1930, Reppe chemistry made vinyl ethers readily accessible so that they became technically attractive. Systematic research work was carried out to investigate their polymerizability. Production of polymers on an industrial scale commenced in 1938 in the Ludwigshafen works of the then IG–Farbenindustrie (now the main works of BASF AG). After 1940, production was taken up by GAF Corporation, and later by Union Carbide Corporation, which discontinued their production in 1976.

## MONOMERS

The addition of alcohols to acetylene, discovered by Reppe, yields vinyl ethers.

$$HC{\equiv}CH + H{-}OR \longrightarrow CH_2{=}CH{-}OR$$

This is the only method of preparation of vinyl ethers that has achieved importance in industrial scale production. The reaction is highly exothermic. The process is carried out at 120–180°C and 5–20 bar in the presence of alkalis or methoxides as catalyst, depending on the type of alcohol used.[2]

Vinyl ether monomers are produced by the aforementioned manufacturers of polymers and the People's Republic of China. The monomers used for polymerizing intermediates for pressure-sensitive adhesives are shown in Table 14-1.

## PRODUCTION OF POLYMERS

### Homopolymers

Vinyl ethers belong to the group of readily polymerizable compounds. Cationic initiators such as $BF_3$ and $AlCl_3$ allow rapid polymerization.

Anionic and free-radical initiators have no significance for polymerizing vinyl ethers on an industrial scale. The high sensitivity of the cationic polymerization reaction imposes very stringent requirements on the purity of the monomers.

Polymerization can be carried out batchwise or continuously in bulk or in solution. On account of the considerable reaction heat, polymerization

**Table 14-1. Monomers used for producing raw materials for adhesives.**

| PRODUCT | STRUCTURE | MOLECULAR WEIGHT | DENSITY AT 20°C (g/cm³) | FLASH POINT (°C) | MELT POINT (°C) | BOILING POINT (°C) |
|---|---|---|---|---|---|---|
| Vinyl methyl ether | $CH_2{=}CH{-}O{-}CH_3$ | 58.1 | 0.747 | −60 | −122 | ~6 |
| Vinyl ethyl ether | $CH_2{=}CH{-}O{-}CH_2{-}CH_3$ | 72.1 | 0.754 | −45 | −115 | ~36 |
| Vinyl isobutyl ether | $CH_2{=}CH{-}O{-}CH_2{-}\underset{CH_3}{CH}{-}CH_3$ | 100.2 | 0.769 | −15 | −112 | ~83 |

must be carried out with great care and elaborate equipment. Monomers and initiators are metered continuously into the reactor. Polymerization sets in within a few minutes. The reactor is closed after the charging cycle and polymerization continues under pressure. The course of the temperature of the reaction is determined by the boiling temperature of the monomer or solvents. The reaction heat is removed by means of a reflux condenser and/or cooling of the vessel. The continuous low-temperature polymerization process is another method of producing high molecular weight polyvinyl isobutyl ethers. In this process, a mixture of vinyl isobutyl ethers, liquefied propane and initiator is applied on a conveyor belt. The reaction, which sets in immediately, lasts only a few seconds. The reaction heat is removed by evaporation of the liquid propane.[3]

### Copolymers

Copolymers of vinyl ethers and acrylates are obtained by free-radical emulsion polymerization. Polymerization takes place after the introduction of the monomer mixture into the slightly alkaline water at 70–90°C containing a dispersing agent. Potassium peroxodisulfate is used as initiator.

## COMMERCIAL PRODUCTS

Table 14-2 shows a list of conventional vinyl ether polymers.

**Table 14-2. Conventional polyvinyl ether pressure-sensitive adhesive raw materials.**

| POLYMER BASED ON | COMMERCIAL NAME | MANUFACTURER | K VALUE | NATURE OF POLYMER |
|---|---|---|---|---|
| Vinyl methyl ether | Lutonal M 40 | BASF | ~ 50 | Soft resin |
| | Gantrez M | GAF | | Soft resin |
| Vinyl ethyl ether | Lutonal A 25 | BASF | ~ 12 | Viscous oil |
| | Lutonal A 50 | BASF | ~ 60 | Soft resin |
| | Lutonal A 100 | BASF | ~ 105 | Tacky rubber |
| Vinyl isobutyl ether | Lutonal I 30 | BASF | ~ 25 | Viscous oil |
| | Lutonal I 60 | BASF | ~ 60 | Soft resin |
| | Lutonal IC | BASF | ~ 125 | Rubber |
| | Lutonal I 60 D | BASF | ~ 60 | Soft resin |
| | Lutonal I 65 D | BASF | ~ 60 | Soft resin |
| Methylacrylate Vinyl isobutyl ether Acrylic acid | Acronal 550 D | BASF | ~ 65 | Soft resin |

The K value is widely used as a criterion for indicating the degree of polymerization. This value indicates the average molecular weight of the polymer.[4]

The polymers are supplied in various forms, depending on the requirements imposed. They are free from solvents, or dissolved in conventional solvents to facilitate handling. The Lutonal types I 60 D and I 65 D are 55% high-viscosity secondary dispersions in water, produced from an 80% solution of Lutonal I 60 in mineral spirit (b.p. 60–140°C) and various protective colloids. Acronal 550 D is an aqueous, anionic, low-viscosity polymer dispersion with a solids content of approximately 45%.

## PROPERTIES

Depending on the degree of polymerization, polyvinyl ethers are viscous oils, tacky soft resins or rubberlike substances. Commercial products are normally colorless. However, side reactions of the initiators may cause a pale yellow to brown discoloration. This has usually no influence on the pressure-sensitive adhesive properties. Vinyl ether polymers are very resistant to hydrolysis and are tasteless and odorless. Products with low to medium molecular weight (K value $< 65$) may have a weak odor which is caused by residual monomers and oligomers. Tests for determining the primary skin irritation caused by vinyl ether polymers on the dorsal skin of white rabbits showed no or very slight irritation. The instructions of the manufacturers must be observed in using the products for medical applications.[5] The solubility of the vinyl ether polymers is indicated in Table 14-3.

**Table 14-3. Solubility of vinyl ether polymers.**

| SOLVENT / POLYMER | Aliphatic Hydrocarbons | Aromatic Hydrocarbons | Chlorinated Hydrocarbons | Lower Alcohols | Higher Alcohols | Esters | Lower Ketones | Higher Ketones | Water |
|---|---|---|---|---|---|---|---|---|---|
| Polyvinyl methyl ether | – | + | + | + | + | + | + | + | + |
| Polyvinyl ethyl ether | + | + | + | + | + | + | + | + | – |
| Polyvinyl isobutyl ether | + | + | + | – | + | + | – | + | – |

The solubility is influenced by the alkyl group. The solubility of polyvinyl methyl ether in water is due to hydrogen bonding between the water molecules and ether oxygen. The intrinsically insoluble polyvinyl methyl ether is surrounded by water molecules and is thus solvated. The bridge bonds are broken on application of energy and the nonhydrated polymer precipitates. In the case of Lutonal M 40 solutions, for instance, this process takes place at approximately 28°C. The precipitation point is influenced by the concentration of the aqueous solution and by any solvents present.

## STABILIZATION

The influence of oxygen, heat and light causes chain cleavage, crosslinking and reactions with oxygen. These change the molecular weight, molecular weight distribution, adhesiveness, color, etc., of the polymer. The changes are usually the result of free-radical reactions. The first step is cleavage of a carbon-hydrogen bond by uptake of energy, leading to the formation of free radicals. The energy can be introduced as heat, light (UV radiation), or mechanical action. By oxidation, new radicals are formed via peroxide intermediates. The reaction thus initiated is continued either by chain cleavage (causing softening of material) or crosslinking (causing hardening of material.[6]

These aging processes affect vinyl ether polymers in an undesirable way as well. Antioxidants must therefore be used in order to obtain adhesive compounds with constant properties. Numerous products are available on the market. Products that have proved very effective include Ionox 330 (Shell), Irganox 1010 (Ciba-Geigy) and Antioxidant ZKF (Bayer Leverkusen).

Figure 14-1 shows the aging behavior of a Lutonal I pressure-sensitive adhesive applied from a hot melt. The adhesive compound was tested in unstabilized form and with an addition of approximately 1.5% of Antioxidant ZKF. The bonding strength was measured in the original condition, after two, eight and 24 hr aging at 150°C in a circulated air oven. Within a few hours, the nonstabilized adhesive degraded to such an extent that it became completely unusable. The stabilized adhesive, however, changed insignificantly for practical application. First, a slight hardening took place, and a negligible tendency to chain fracture was observed thereafter. The results of this extremely stringent short-term test confirm the good experience that has been gained in practice with polyvinyl ethers in the course of decades. The adhesion values were determined by internal BASF methods with due regard to the PSTC standards.

The diagrams in Figure 14-2 show the viscosity behavior of stabilized and unstabilized 35% solutions of Lutonal A 50 and I 60 in mineral spirit.

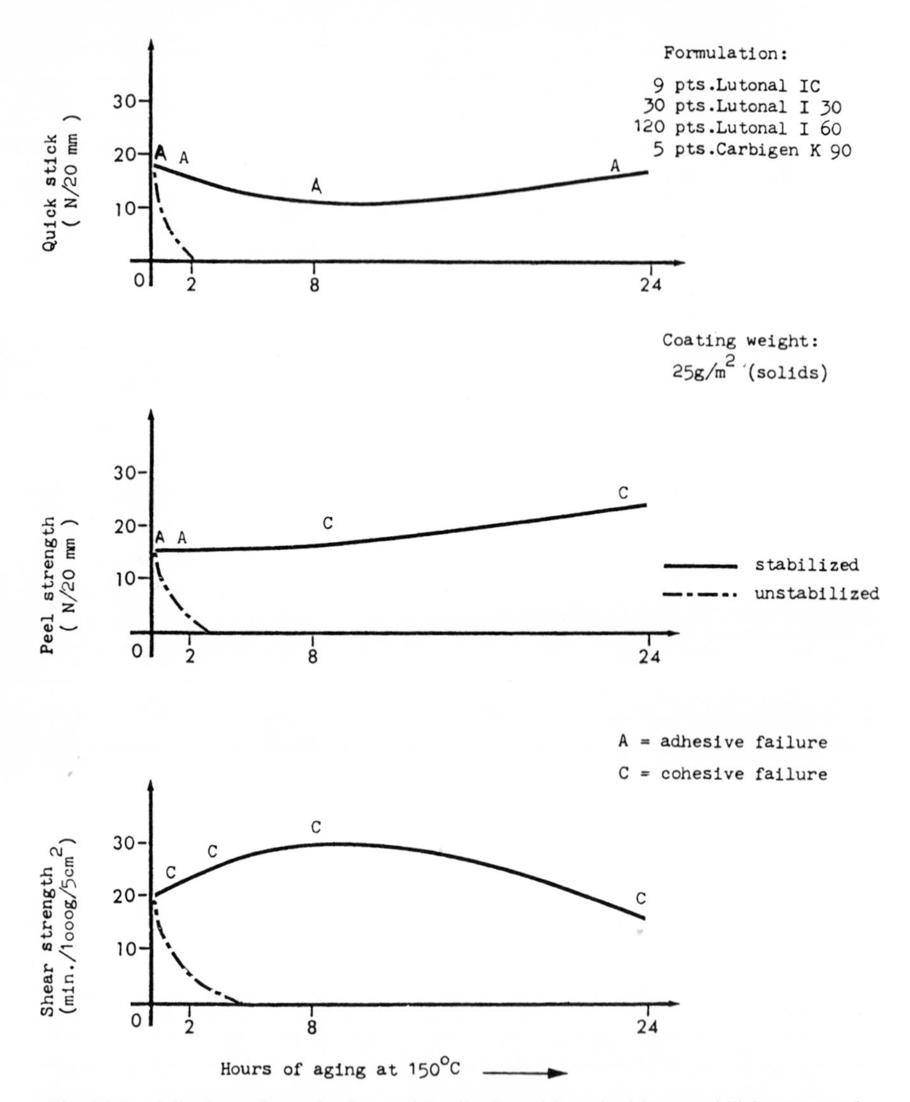


Fig. 14-1. Adhesion values of a Lutonal I adhesive with and without stabilizing agent after exposure to heat.

The solutions were stored in colorless, transparent glass bottles for nine months at room temperature. In this case, too, the unstabilized solution showed distinct chain degradation. Even an addition of 0.1% of Antioxidant ZKF is so effective that no change in viscosity of any practical significance takes place. The curves show the flow time determined with DIN

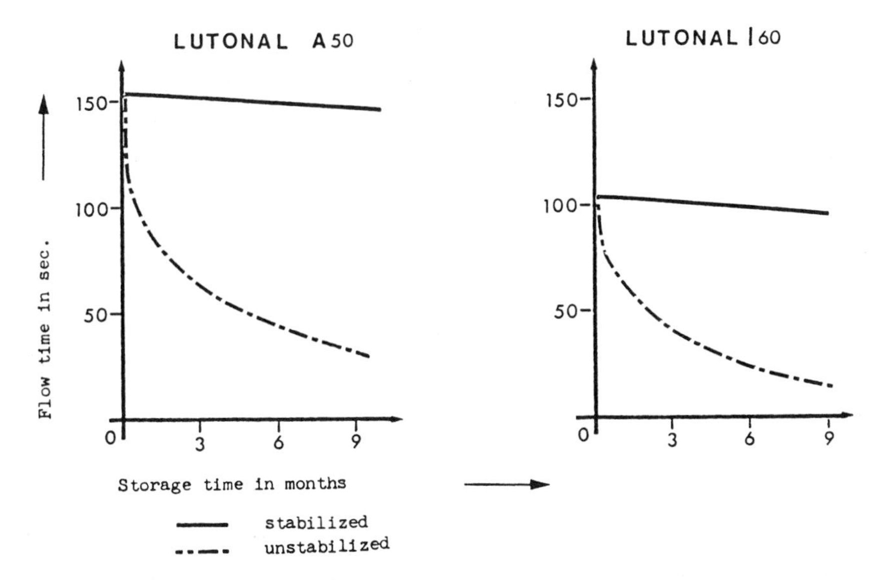


Fig. 14-2. Behavior of stabilized and unstabilized 35% Lutonal A 50 and Lutonal I 60 solutions in mineral spirit during storage.

Cup, 4 mm orifice, as a function of storage time. No appreciable change in viscosity takes place in opaque containers even when no antioxidant has been incorporated.

## APPLICATION

### General

In addition to rubber and polyisobutylene, vinyl ether polymers belong to the classical intermediates for pressure-sensitive adhesive compounds. They are suitable for producing adhesive compounds for pressure-sensitive medical articles, pressure-sensitive labels, films, etc. Vinyl ether polymers are usually mixed with polymers of the same type or with other substances. When vinyl ether polymers of the same type but of different degree of polymerization are mixed with one another, the desired adhesive properties are obtained by varying the mixing ratio of the polymers. The constant quality of the polyvinyl ether polymers allows the production of pressure-sensitive articles with excellently reproducible properties. The commercial products are available either in solvent-free form or as solutions in conventional solvents. When the base materials for pressure-sensitive adhesives must be transported over long distances, it is advisable to use products free from solvent. In such cases, it is advantageous that polyvinyl ether solutions can be made up not only in solution kneaders and planetary mixers but also in dissolvers with a high shearing effect. Nowadays, vinyl ether polymers are used to advantage in the following pressure-sensitive products:

1. For producing pressure-sensitive articles.
2. For formulating adhesive compounds with accurately reproducible properties.
3. In cases when compounding and dissolving equipment for rubber is not available.
4. As a blending component for adhesive compounds based on other raw materials in order to achieve special properties.

Adhesive compounds based on vinyl ether polymers can be processed on all conventional coaters. These compounds were considered already in the early stage of developing solvent-free coating processes. Apart from other substances, vinyl ether polymers were suggested for use as raw materials for hot melt adhesives in 1953, for processing on conventional coaters[7,8] in 1957 and for the extrusion process[9] in 1960. The extrusion process allows short exposure of the adhesive to heat and in-line melting and coating. The formulation of adhesive compounds that can be applied as hot melts is subject to certain limitations. To achieve the required coating behavior, it is often necessary to subordinate quality requirements to the processing requirements.

The formulations suggested in paragraphs below should be applied at a rate of approximately 25 $g/m^2$. The optimum coating weight depends on the purpose of the finished article and the nature of the adhesive compound. A coating weight of 20–30 $g/m^2$ is usually applied for articles used on smooth surfaces, such as glass, lacquered wood or metal surfaces. Articles for use on uneven surfaces, such as pressure-sensitive tapes for laying carpets, require coating weights of up to 100 $g/m^2$. Moreover, there is a certain relationship between adhesion and coating weight, as shown in Fig. 14-3. With low coating weights, the adhesion normally increases with the coating weight, provided that adhesive failure takes place on separation. The adhesion is largely independent of the coating weight when the bond separates in the adhesive layers, i.e., when cohesive failure occurs. The optimum coating weight must be determined by trials in each individual case.

## Vinyl Methyl Ether Polymers

Vinyl methyl ether polymers are applied as blending components for pressure-sensitive adhesive compounds based on other raw materials—

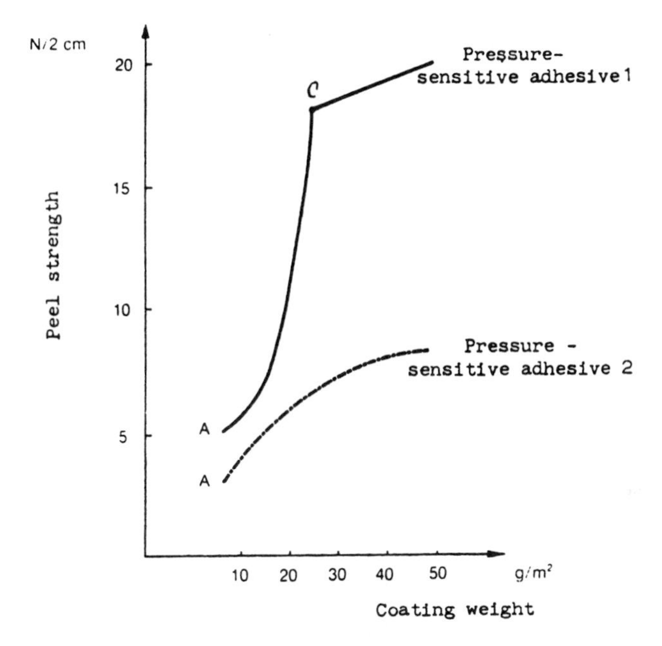


Fig. 14-3. Adhesion as a function of coating weight.

usually aqueous dispersion-type adhesives. They are used primarily for achieving defined hydrophilic properties and increasing the adhesiveness.

Tapes with adhesive compounds containing polyvinyl methyl ether are used, for instance, for bonding paper roll to paper roll in continuous printing of paper webs. The bonded areas are cut out after printing. They must be dispersible in water so that the paper can be repulped.[10,11,12]

One of the methods of increasing the water-vapor permeability of adhesive compounds is to add vinyl methyl ether polymers. This possibility of modification can be exploited, for instance, in the production of pressure-sensitive medical articles.[13]

Labels that can be removed from the substrate with the aid of water can be produced by coating paper with an adhesive according to the typical formulation given blow. When stuck to glass plates, these labels are detached cleanly by soaking in cold water for 15 minutes.

50 parts Acronal 880 D
30 parts Acronal 500 D
20 parts Lutonal M 40 ~50% in water
1 part Lutensol AP 6
5 parts toluene

A very thin film of condensation water may form on deep-frozen articles before the labels are applied. Adhesives for deep-frozen articles must therefore have good adhesion to moist surfaces as well. A typical formulation for obtaining an adhesive that meets this requirement is given below.

70 parts Acronal V 205
30 parts Lutonal M 40, 50% in water
40 parts water
5 parts toluene

In the following typical formulation for plasticized PVC decorative films, Lutonal M 40 is used for increasing the adhesion.

70 parts Acronal 80 D
30 parts Lutonal M 40 ~50% in water
5 parts ethyl acetate

At temperatures above 28°C, mixtures of Lutonal M 40 and polymer dispersions are unstable on account of the solubility inversion of the vinyl methyl ether polymers. A slightly gritty to completely coagulated mixture is formed, depending on the temperature and duration of exposure. These mixtures can be rendered stable even at temperatures above the precipitation point of Lutonal M 40 by adding small quantities of solvent, if necessary, in combination with an emulsifier. Acronal V 205 has a solids content of approximately 70%. Mixtures of this polymer dispersion, of exceptionally high solids content, and Lutonal M 40 cannot be stabilized solely by the means described above. The solids content of the typical formulation containing Acronal V 205 was, therefore, reduced to approximately 50% by adding 40 parts water.

When used in adhesive compounds for coating plasticized PVC, vinyl ether polymers fulfill another function. They reduce the decrease in adhesive strength caused by plasticizer migration. This phenomenon is probably due to the good compatibility of the vinyl ether polymers with a great variety of solvents and plasticizers. Vinyl ether polymers take up the plasticizer migrating from the carrier film and form a stable mixture. The usual accumulation of plasticizer on the surface of the adhesive compound and the resultant decrease in adhesive strength can thus be diminished. However, it was found that this effect is largely dependent on the nature of the film, so that investigations should be carried out in each individual case.

## Vinyl Ethyl Ether Polymers

Vinyl ethyl ether polymers are very similar to the substantially more widely used vinyl isobutyl ether polymers with regard to properties and processibility. In principle, the formulations and application data given below are also applicable to polyvinyl ethyl ethers. One difference is that the water-vapor permeability of these polymers is about equivalent to the average loss of water through the human skin.[13] On account of this property, vinyl ethyl ether polymers are particularly suitable for use in the medical field.

### Vinyl Isobutyl Ether Polymers

Vinyl isobutyl ether polymers belong to the oldest and most widely used group of polyvinyl ethers. They are employed as adhesive compounds applied from a solution or melt, and as blending components for increasing the adhesion. The latter are usually applied as secondary dispersions together with aqueous acrylic polymer dispersions. Adhesive compounds with different properties can be formulated by varying the mixing ratio of polymers of different molecular weights (K value). Adhesive compounds with optimum shear strength for vinyl ether polymers, and adequate adhesion can be obtained with the following formulation.

30 parts Lutonal IC
40 parts Lutonal I 60
10 parts Lutonal I 30
0.5 part Antioxidant ZKF
~150 parts mineral spirit (b.p. 65–95°C)

On the other hand, an adhesive compound with optimum adhesion and adequate shear strength can be obtained according to the following formulation.

30 parts Lutonal IC
60 parts Lutonal I 60
30 30 parts Lutonal I 30
0.5 part Antioxidant ZKF
~150 parts mineral spirit (b.p. 65–95°C)

Both adhesives can be applied only from a solution. Application from a melt requires a larger proportion of additives with thermoplastic behavior. A typical formulation for an adhesive compound for roll coater application developed according to this principle is as follows.

9 parts Lutonal IC
30 parts Lutonal I 30
120 parts Lutonal I 60
5 parts Carbigen K 90

This hot melt pressure-sensitive adhesive was already mentioned. It can be used, for instance, for double-faced tapes with high coating weight on a textile carrier, e.g. for securing edges (carpet laying tape).

Widely differing requirements are imposed on the adhesive behavior of pressure-sensitive medical articles. Formulations must, therefore, be worked out in each individual case with due regard to the required properties. The following formulation for an adhesive compound applied from a solution may serve as a guide. The developed adhesive should in any case be examined for its skin irritating effect.

10 parts Lutonal IC
60 parts Lutonal I 60
5 parts Lutonal I 30
0.5 part Antioxidant ZKF
25 parts zinc oxide
~80 parts mineral spirit (b.p. 65–95°C)

Pressure-sensitive tapes and films produced with hot melt adhesive compounds consisting of 40–60 parts Lutonal I 60 and 60–40 parts bitumen can be used as "bituminous adhesive off the roll" for a great variety of applications. They can, for instance, be used on building sites where cookers for hot bituminous adhesives are undesirable because of the danger of fire and accidents, or for other reasons.

Vinyl isobutyl ether polymers of lower degree of polymerization are also used for increasing the adhesiveness of adhesive compounds based on other raw materials. They are mainly applied as secondary dispersions in admixture with aqueous acrylic polymer dispersions. In the following typical formulation for an adhesive compound for plasticized decorative PVC films, Lutonal I 65 D also brings about a reduction of the aforementioned decrease in adhesive strength

80 parts Acronal 80 D
20 parts Lutonal I 65 D

## Outlook

The significance of vinyl ether polymers has changed in the course of time. However, these polymers still have a firm foothold as products for special applications in the adhesives field. The fact that polyvinyl ethers can also be used for adhesive compounds suggested for new technologies indicates that they will maintain this position in the future as well. They are recommended, for instance, for use as an adhesion-increasing component in solvent-free, heat-crosslinking pressure-sensitive polyurethane systems[14] and solvent-free, radiation-curing pressure-sensitive polyacrylate adhesives.[15]

The properties of acrylic copolymers can be varied to a great extent by judicious selection of the monomers. In view of this possibility and the growing demand for new products, it is to be expected that acrylic/vinyl ether copolymers will be applied on an increasing scale.

The formulations suggested here are based on practical experience and extensive laboratory trials. Trials must be carried out if these formulations are to be used as a guide for developing adhesive systems. This is necessary because many factors which play a role in the production of pressure-sensitive adhesives (e.g., storage stability, adhesion to the various carrier materials, etc.) cannot be considered in laboratory trials. Patent rights should be observed.

## REFERENCES

1. Wislicenus, I., *Justus Liebigs Ann. Chem.* **192**: 106 (1878).
2. Hofmann, E. "Vinylether" in Ullmanns *Enzyklopaedie der technischen Chemie*, Third Ed., Vol. **18** (1963).
3. Schroeder, G. "Vinylether-Polymerisate" in Ullmanns *Enzyklopaedie der technischen Chemie*, 4th ed., Vol. **18** (1980).
4. Pruefmethoden Polymerdispersionen/Polymerloesungen; Bestimmung des K-Wertes. BASF AG.
5. Technische Information Haftkleber, Part 3, Sheet 9. BASF. AG.
6. D'Ianni, Jame D., and Widmer, Hans, Stabilisierung von synthetischen Kautschuken und Latices, GAK 9/1973.
7. Salditt, F. Swiss Patent 327,731 (1953) (assigned to Lohmann KG).
8. Salditt, F. U. S. Patent 2,861,006 (1957) (assigned to Scholl Mfg. Co., Inc.).
9. Sahler, W. DAS 1,113,777* (1960) (assigned to Beiersdorf AG).
10. Kline, M. M., and Kwok, M. C. T. U. S. Application 95,024 (1961) British Patent 941,276 (1962) (assigned to Norton Company).
11. Sorell, H. P. U. S. Patent 3,556,835 (1968) (assigned to Borden, Inc.).
12. Hauber, R. DOS 2,142,770** (1973) (assigned to Beiersdorf AG).
13. Seymour, D. E., Da Costa, N. M., Hodgson, M. E., and Dow, J. British Patent 1,280,631 (1968) (assigned to Smith and Nephew Ltd.).
14. Hagenweiler, K. and Scholz, K. DOS 2,328,430** (1973) (assigned to BASF AG).
15. Steuben, K. C. *Adhesives Age*, **20**(6): 16 (June 1977).

---

* DAS is an abbreviation for *Deutsche Auslegeschrift*. This is the last stage in the procedure for granting a patent.

** DOS is the abbreviation for *Deutsche Offenlegungsschrift*. This is a patent application which has been published by the German Patent Office without previous investigation of the subject.

# Chapter 15

# Silicone Pressure-Sensitive Adhesives

**Duane F. Merrill**
*General Electric Company*
*Waterford, New York*

Silicone pressure-sensitive adhesives exhibit the flexibility of silicone rubber and the high temperature resistance of silicone resin. They are chemically inert, providing extended service life. They can be utilized at temperatures between −73 and 250°C without becoming brittle or drying out. They can bond both low- and high-energy surfaces, including etched tetrafluoroethylene (TFE) fluorocarbon polymers, unetched polyolefins, untreated fluorohalocarbon films, and silicone release coatings. Organic adhesives will not stick to such surfaces, while silicone adhesives are effective bonding agents for these materials.

## MANUFACTURING AND PRODUCTS

Like most pressure-sensitive adhesives, silicone adhesives are based on a gum and a resin. There are two chemical types of silicone gum: the all methyl based and the phenyl modified. The resin is used as a tackifier and for adjusting the physical properties of the pressure-sensitive adhesive. By adjusting the resin-to-gum ratio, pressure-sensitive adhesives are made in a wide range of tack and peel adhesion properties. The polymers are dissolved in a solvent and chemically reacted in such a way that condensation takes place. The condensation reaction of SiOH groups proceeds readily as water is removed. The physical properties of the adhesive, especially shear strength, are controlled through intra- and intermolecular condensation. High resin content adhesives are not tacky (see Table 15-1) at room temperature and are tackified with heat and pressure. High gum content adhesives are extremely tacky at room temperature.

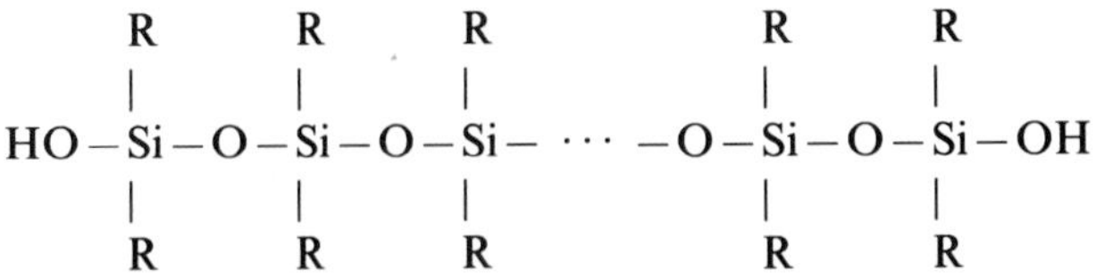

$$HO-\underset{R}{\overset{R}{Si}}-O-\underset{R}{\overset{R}{Si}}-O-\underset{R}{\overset{R}{Si}}-\cdots-O-\underset{R}{\overset{R}{Si}}-O-\underset{R}{\overset{R}{Si}}-OH$$

R = Methyl ($CH_3$) or Phenyl ($Ph_2SiO$) groups

Gum Structure

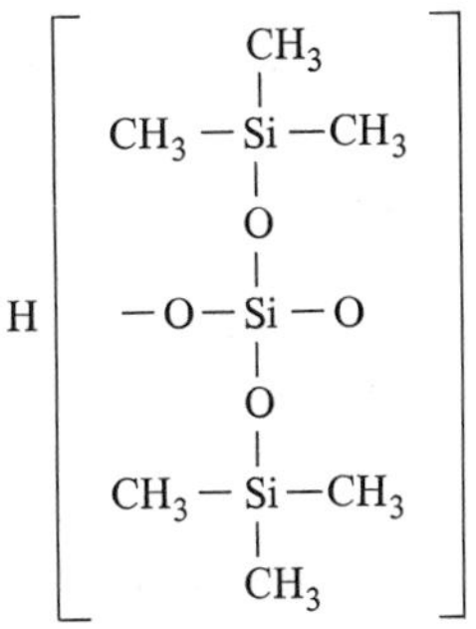

$$H\left[\begin{array}{c} CH_3 \\ | \\ CH_3-Si-CH_3 \\ | \\ O \\ | \\ -O-Si-O \\ | \\ O \\ | \\ CH_3-Si-CH_3 \\ | \\ CH_3 \end{array}\right]H$$

Resin Structure

$$\left[-O-\underset{R}{\overset{R}{Si}}\right]_n - O(H\ HO-) \text{ (} H_2O \uparrow \text{) } Si-O-Si-O-Si-O-Si-$$

(Resin network: Si–O–Si–O–Si–O–Si– chain with O bridges to RSiR, Si—, R groups; lower Si–O–Si–OR–R–SiR; O–RSiR–R)

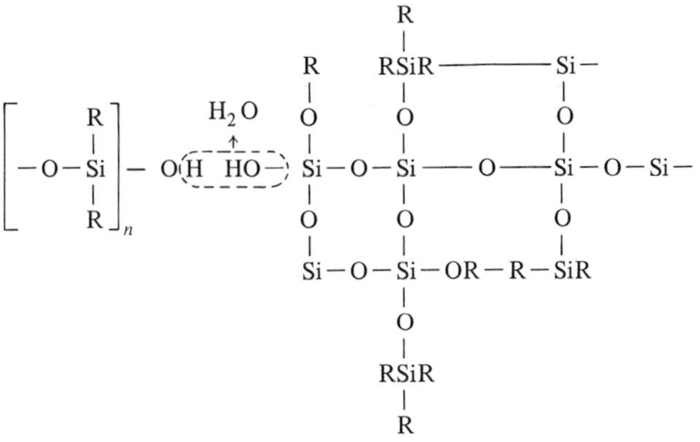

Condensation Reaction of Resin with Silanol Stopped Gum

**Table 15-1. Adhesive properties of typical methyl-based adhesives.**

| VISCOSITY (cps) | PEEL ADHESION ASTM D1000 (g/cm) | LAP SHEAR STRENGTH[a] @ 25°C (kg) | TACK[b] |
|---|---|---|---|
| 65,000 | 445 | 33 | Medium |
| 3000 | 535 | 60 | Low |
| 9000 | 1000 | 68 | Tack-free |
| 4800 | 714 | 50 | High |

[a] Force required to separate a bonded area 1 × 1 in.
[b] As measured on Polyken Probe Tack Tester

The methyl-based adhesives are produced in a wide range of viscosity and physical properties. They typically have a solids content of 55 ± 1% and range in viscosity from as low as 1000–5000 centipoise to as high as 40,000–90,000 centipoise giving the formulator and coater wide parameters for filler loading and selection of tape backing.

The methyl-based silicone adhesives are compatible and can be blended for customizing to specific requirements.

A dry type (nontacky) adhesive having a peel adhesion of 1100 g/cm can be used to increase the peel adhesion of a tacky adhesive having a peel adhesion of 440 to 770–880 g/cm with no loss of quick stick properties as shown in Fig. 15-1.

The shear strength of the tacky adhesive increases with the addition of the dry type adhesive. Some blends have a holding power greater than 1 hr under the shear load of 500 g at 150°C on a 1.27 × 1.27 cm (0.5 × 0.5 in.) lap of 0.05 mm (2 mil) coating on 0.05 mm polyester film bonded to a steel test panel.

There are two types of phenyl-modified silicone adhesives (see Table 15-2). One type is referred to as the low (6 mole %) phenyl made in a viscosity range of 50,000 to 100,000 centipoises. The other is a high (12 mole %) phenyl adhesive having a viscosity range of 6,000 to 25,000 centipoises. The phenyl adhesives have excellent holding power at high (250°C) and low (−73°C) temperatures and are used in specialty and general tape applications. The major feature of phenyl adhesives is that they have a unique combination of high viscosity, high peel strength and high tack. Another unique property of the phenyl adhesives is that they are not compatible with methyl adhesives and other methyl polymers. When applied and cured over a cured film of methyl-based silicone paper release coating, the high phenyl adhesive can be transferred to other surfaces. The high phenyl transfer adhesive has been found excellent for various bonding applications:

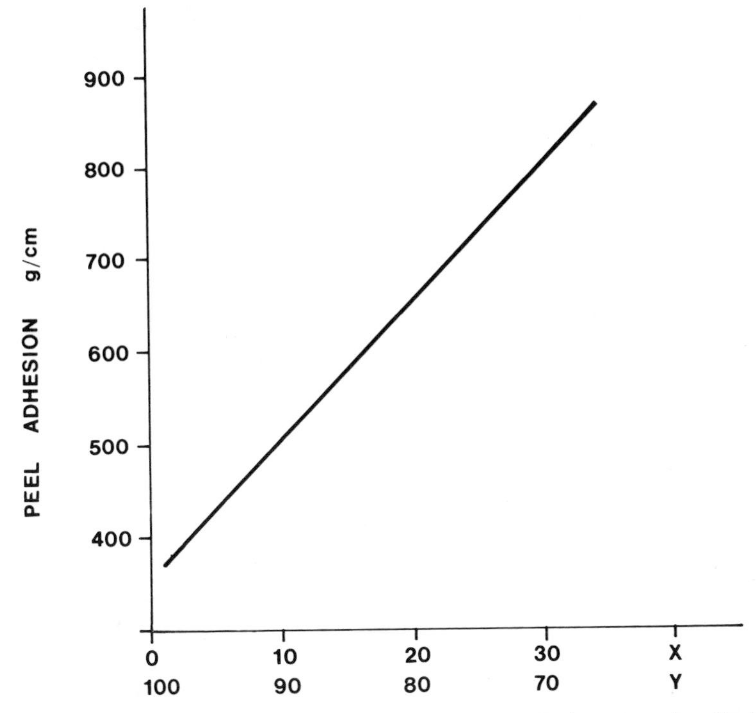


Fig. 15-1. Blends of methyl-based adhesives. (x = dry type adhesive; y = tacky adhesive).

from the automotive and aerospace industries to the electrical insulation and appliance markets. For example, it is used to bond warning and information labels to hot engine parts, thermal blankets in electrical appliances, solar energy panels in collectors, and has been used to bond friction pads in subzero temperatures of the Alyeska Pipeline.

**Table 15-2. Adhesive properties of typical phenyl-based adhesives.**

| VISCOSITY (cps) | PEEL ADHESION ASTM D1000 (g/cm) | LAP SHEAR STRENGTH[a] @ 25°C (kg) | TACK[b] |
|---|---|---|---|
| 75,000 | 500 | 39 | High |
| 15,000 | 890 | 45 | High |

[a] Force required to separate a bonded area 1 × 1 in.
[b] As measured on Polyken Probe Tack Tester

Not only are the silicone adhesives made in a wide range of viscosities, they have a wide range of physical properties. High tack for quick stick in repair and masking tapes, to tack free heat-activated adhesive for making laminates from dissimilar materials. Tack ranges from 0–1500 g/cm$^2$ measured on a Polyken Probe Tack Tester. Peel adhesion for pressure-sensitive adhesives is 450 g/cm minimum to 725 g/cm maximum in the methyl adhesives and 480–890 g/cm in the phenyl adhesives. Tack-free methyl adhesives become aggressively adhesive under heat (93°C) and pressure (7 kg/cm$^2$) and have been found to be excellent for making laminates of films (polyester to aluminum) and foils (aluminum to aluminum). Peel adhesion of benzoyl peroxide crosslinked adhesives is as high as 1100 g/cm.

## CROSSLINKING

Two different catalyst systems are used for curing silicone pressure-sensitive adhesive. The most common catalyst used is benzoyl peroxide. 2,4-dichlorobenzoyl peroxide can be substituted for the benzoyl peroxide where lower temperature (140°C) curing conditions and/or higher cohesive strength of the adhesive is desirable. Curing of the adhesive improves the high temperature shear properties with only a slight loss of peel adhesion compared with noncured adhesive. When using benzoyl peroxide, caution should be taken to insure that the curing temperature is above 150°C for complete decomposition of the catalyst. At temperatures below 150°C, the benzoyl peroxide can volatilize and condense in exhaust pipes building up to explosive levels.

The second type of catalyst system is an amino silane, which acts as a room temperature curing catalyst. This catalyst is not recommended for making tapes or in other applications where materials are stored before use.

```
        CH3    CH3                       CH3     CH3
         |      |                         |       |
 ···—O—Si—O—Si—O—···             —O—Si——O—Si—O—···
         |      |                         |       |
        CH3  H—C—H                       CH3  H—C—H
                |                                 |
              (H)                                 |
               ¦         ————————————→            |
              (H)                                 |
                |                                 |
        CH3  H—C—H                       CH3  H—C—H
         |      |                         |       |
 ···—O—Si—O—Si—O—···             —O—Si—O—Si—O—···
         |      |                         |       |
        CH3    CH3                       CH3     CH3
```

Crosslinking with a Peroxide

The amino silane continues to react with the reactive sites in the adhesive at room temperature, resulting in complete cure and loss of pressure sensitivity. It is used in applications where materials are bonded or laminated soon after solvent removal.

When silicone adhesives are crosslinked with an amino silane, they develop peel strengths as high as 5000 g/cm. Highly crosslinked silicone adhesives are used in bonding of silicone rubber to metal, fabrication of lightweight building structures and bonding of other dissimilar materials. Because of the low flammability of silicone pressure-sensitive adhesives, they are also used in aerospace and aircraft structures.

## TEST METHODS

Silicone pressure-sensitive adhesives are tested by standard ASTM test methods. Some adhesives are tested uncrosslinked, others crosslinked. The peel adhesion test of uncrosslinked adhesives (ASTM D1000) run on low tack adhesives differs only in the way the adhesive is dried and applied to the test panel. Following application of the adhesive to the polyester film, the tapes are hung in a $150 \pm 3$ C oven for 10 minutes. The panel is preheated to $90 \pm 5$ C before applying the tape. High tack adhesives are tested catalyzed with 1.8% benzoyl peroxide based on adhesive solids. A $0.05 \pm 0.005$ mm dry film of the catalyzed adhesive is applied to a strip of 0.05 mm polyester. The coating is cured for 90 seconds at $90 \pm 5$°C and 120 seconds at $165 \pm 5$ C. The cured adhesive tape is applied to a standard steel test panel with a 2 kg rubber roller and allowed to stand for 20 minutes before running the peel adhesion test.

Not only are silicone pressure-sensitive adhesives tested for physical properties as made, but after long-term aging under adverse storage conditions, the tests simulate high-temperature storage conditions of finished rolls of tape. The cured adhesive tapes are protected by release paper and placed under a steel weight equal to the internal pressure of a given roll of tape. It is then exposed to 60°C for eight weeks. Tapes are removed each week and tested for peel adhesion from standard steel panel at room temperature and from aluminum at 200°C. There is no loss of peel adhesion under these conditions.

Silicone pressure-sensitive adhesives are now being used in the restoration of valuable works of art. Deteriorated canvas is strengthened with new backing material coated with silicone pressure-sensitive adhesive. Loose and cupped paint is held in place and flattened by infusion of adhesive through and around the paint with the aid of electrostatic charge. A conservator, Robert E. Fieux, Fieux Restoration Laboratory, Barnstable,

Massachusetts has found that after 100 years of accelerated light exposure, the silicone adhesive does not discolor or lose its original adhesive properties. Total restoration is accomplished without heat or damaging pressures.

## PRIMERS

Silicone paper release coatings, because of their chemical similarity to methyl-based adhesives, act as primers for the methyl adhesives. When properly coated and cured on polyester film, the primer reduces transfer of the adhesive to the tape backing, allowing for reduced legging (pulling away from the substrate) and edge rolling of the adhesive during tape slitting and cutoff. Solvent solutions of the release coating have been found most satisfactory for priming polyester film. Less than 0.8 $g/m^2$ of coating is required to provide excellent anchorage of the adhesive to the film. A short heat cure of the coating is required to develop primer properties. A typical cure cycle is 15–30 seconds at 150°C.

## TAPES

Coated polyester film is used for making platers masking tape and silicone-coated paper splicing tape. Coated polyimide film and etched tetrafluoroethylene films are used in electrical insulation. Many other low-energy film type substrates can be coated for similar uses. High viscosity adhesives facilitate the making of tape composites in which strike through (wetting through to the back side of substrate) is undesirable, e.g., on glass cloth and mica insulation tape.

Glass cloth tapes are used in high-temperature masking, coil and support ring insulation and holding and tying down wiring. Tapes made of fluorocarbon film provide release surfaces on wrapping equipment and heat sealer bars. Silicone rubber tapes are used in splicing of silicone rubber cable and taping of electrical connections. Polyester tapes are used in masking printed circuit boards during plating operation. Because of their unique high temperature and chemical resistance, silicone pressure-sensitive adhesives have outperformed other adhesives in this application. The silicone-coated tape gives straight tin line stopoff and remains in place (no adhesive ooze from edges). The tape is easy to use in hand application and there is less stretch in automatic equipment with no waste.

High molecular weight polyethylene has the greatest resistance to abrasion of all the plastic films. Coated with silicone pressure-sensitive adhesive, it can be used in tape form or die cut for a variety of applications to prevent wear, reduce soil and grime buildup, and increase the life of the substrate.

Fluorocarbon films are used for plasma masking tapes to resist the intense heat of plasma spraying, wire coating, and sandblasting. Through control of the catalyst and curing, silicone adhesive will resist a shear load of 300–500 g at 250°C (bonded area 1″ × 1″).

## LAMINATION

Silicone pressure-sensitive adhesives have been suggested by O'Malley[2] for lamination of flexible printed circuits and flexible flat cables. O'Malley has developed various data with 2,4-dichloro-benzoyl peroxide-cured adhesives (SR573 and SR574 adhesives by General Electric Company).

The data illustrate the properties of silicone pressure-sensitive adhesives in general and is presented in Table 15-3.

**Table 15-3. Dependence of lap shear strength on temperature.[a]**

| TEMPERATURE (°C) | LAP SHEAR STRENGTH (kg) |
|---|---|
| 38 | Break |
| 93 | Break |
| 149 | 15 |
| 204 | 10 |
| 260 | 2.3 |

[a] 1 oz. copper foil laminated to 0.05 mm polyimide film over a bonded area 1 × 1 in. by a 0.05 mm-thick SR573 adhesive coating.

The adhesive exposed to the high temperature remains flexible and aggressively pressure-sensitive. At room temperature, this adhesive has a lap shear strength of at least 70 kg.

Resistance to adverse conditions was tested by aging a glass fabric tape at 250°C and measuring the unwind force according to a military specification (MIL-I-19166A). The results are shown in Table 15-4.

The chemical inertness of silicone pressure-sensitive adhesives is further demonstrated by exposing the adhesive to the atmosphere at 95°C and 100% relative humidity. After a 10 day exposure to these conditions, the adhesive showed no change in lap shear strength. Exposure of laminates to 150°C for seven days in a forced air oven had no effect on the adhesive performance.

**Table 15-4. Resistance of silicone adhesive to aging.**

| UNWIND FORCE (g/cm WIDTH) | | | | |
|---|---|---|---|---|
| CONTROL, CONDITIONED FOR 7 DAYS @ 25°C | AGED @ 250°C FOR 7 DAYS | 14 DAYS | 21 DAYS | 28 DAYS |
| 500 | 740 | 650 | 560 | 580 |

## REFERENCES

1. Merrill, D. F. *Adhesives Age*, **22**(3): 39–41 (March 1979).
2. O'Malley, W. J. *Adhesives Age*, **18**(6): 17–20 (June 1975).
3. McGregor, R. R. *Silicones and Their Uses.* New York: McGraw-Hill, 1954.
4. Rochow, E. G. *Chemistry of the Silicones.* Waterford, N.Y.: General Electric Co., 1946.
5. Dexter, J. F. U.S. Patent 2,736,724 (February 28, 1956) (assigned to Dow Corning Corp.).
6. Goodwin, J. T. U.S. Patent 2,857,356 (October 21, 1958) (assigned to General Electric Co.).

# Chapter 16

# Tackifier Resins

**James A. Schlademan**
*Neville Chemical Company*
*Pittsburgh, Pennsylvania*

The properties of a pressure-sensitive adhesive depend primarily on the viscoelastic nature of the adhesive mass. In formulating a pressure-sensitive adhesive, an elastomer or rubbery polymer provides the elastic component while a low molecular weight tackifying resin constitutes the viscous component. Thus, for any given elastomer system, it is the tackifying resin which ultimately determines the viscoelastic behavior and the final properties of the finished adhesive.

The adhesive properties known as tack, resistance to peel, and shear depend on three different aspects of the viscoelastic nature of the tackifier-elastomer blend. Tack is essentially a measure of viscous flow under conditions of fast strain rates and low stress magnitudes while shear adhesion measures viscous flow at low strain rates and intermediate stress magnitudes. Peel adhesion measures resistance to flow at intermediate strain rates and moderate to high stress magnitudes as well as the cohesive strength of the adhesive.

It is obvious that in any given system these properties can tend to be mutually exclusive and a balance must be found based on cost and the desired performance requirements. The proper tackifier is not necessarily the one which gives the best tack values or the best shear values. Furthermore, no matter what tackifier one is using, it is usually possible to find another one having a lower price or better availability which can serve as an adequate replacement. Thus, to understand and formulate pressure-sensitive adhesives requires a basic understanding of tackifier resins including their raw material sources, methods of manufacture, properties, uses, and performance characteristics.

Tackifier resins may be broken up into two major groups. These are rosin and its derivatives and hydrocarbon resins. The rosin group, as the name implies, consists of rosin, modified rosins, and their various derivatives such as esters. The hydrocarbon resin group consists of polyterpenes, synthetic hydrocarbon resins, and a collection of modified or special resins primarily phenolics.

## ROSIN AND ROSIN DERIVATIVES

Rosin is a thermoplastic acidic resin obtained from pine trees. There are three major methods for obtaining rosin and the resulting product is named according to the process used.

Gum rosin, along with turpentine, is obtained as oleoresin from the living tree. This harvesting procedure, also called turpentining, involves wounding the tree and collecting the oleoresin which exudes from the wound. Sulfuric acid is usually used to enhance exudation and increase the yield of oleoresin. More recently, other materials such as the herbicide paraquat have been found to further increase the yield of oleoresin as well as the resinous content of the wood in the lower portion of the tree.[1] The oleoresin obtained from this process is then distilled to give gum rosin and turpentine.

Wood rosin is obtained from aged pine stumps which are pulled up, washed, and mechanically reduced to chip form. These chips are then extracted with solvent followed by distillation and further refining to give wood rosin.

Tall oil rosin is the third type of rosin and it is obtained from tall oil which is a by-product of the paper industry. During the pulping operation, wood chips are cooked under heat and pressure in an alkaline medium. The pulp is removed by filtration and washing, and the liquid portion is concentrated, resulting in a precipitate which is removed and acidified to give crude tall oil. This crude tall oil consists mainly of fatty acids and rosin acids which are separated by distillation.

Rosin from any of these three sources is a mixture of organic acids called rosin acids. Minor components consist of rosin esters and anhydrides, unsaponifiable matter, and fatty acids.

The rosin acids can be divided into two different types. These are the abietic acid type and the pimaric acid type, and unmodified rosin consists primarily of abietic type acids. Figure 16-1 shows some of the more common members of these two types. It can be seen that the difference in the two types is the substitution pattern at the $C_{13}$ position. Abietic acids

ABIETIC TYPE ACIDS

$CO_2H$

ABIETIC ACID

$CO_2H$

NEOABIETIC ACID

$CO_2H$

PALUSTRIC ACID

$CO_2H$

DIHYDROABIETIC ACID

$CO_2H$

DEHYDROABIETIC ACID

PIMARIC TYPE ACIDS

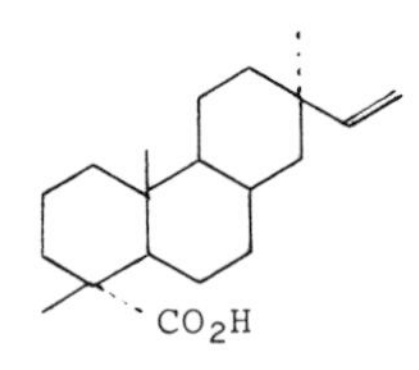


PIMARIC ACID

$CO_2H$

ISOPIMARIC ACID

Fig. 16-1. Rosin acids.

have an isopropyl group while the pimaric acids have a methyl and a vinyl group.

Rosin by itself is generally unsuited for use in modern adhesive systems because it is subject to oxidation and other reactions as well as crystallization. Because of this, most rosins for adhesive use have been subjected to one or more of a series of reactions to form modified rosins or rosin derivatives. Among the more important reactions for the production of materials for use in adhesives are disproportionation, polymerization, hydrogenation, and esterification.

Disproportionation takes place when rosin is heated to about 270°C. The reaction may be accelerated by the use of various catalysts such as platinum or iodine. The result is a hydrogen exchange between the rosin acids where, for instance, two abietic acids are converted into a dihydroabietic acid and a dehydroabietic acid.

Polymerization is affected by treating a solution of rosin in an inert solvent with an acid catalyst such as boron trifluoride or sulfuric acid. Removal of the catalyst and solvent results in a product composed largely of rosin dimer and having a higher softening point than unmodified rosin.

Hydrogenation using Raney nickel or a similar catalyst can be tailored to give products which are only partially saturated such as dihydroabietic acid, or completely saturated such as tetrahydroabietic acid.

Rosin acids react with a variety of alcohols to form esters. Because of steric hindrance, however, the esterification requires rather high temperatures, but the resulting esters are quite stable and resistant to hydrolysis. Usually, polyhydric alcohols are used for esterification in order to give products with higher softening points and higher molecular weight. Ethylene glycol, glycerol, and pentaerythritol are the most common alcohols employed for this purpose. When special stability is desired, rosin esters are produced from hydrogenated rosins.

A few examples of rosin-based tackifier resins used in pressure-sensitive adhesives are shown in Table 16-1.

## HYDROCARBON TACKIFIER RESINS

Hydrocarbon tackifier resins are low molecular weight polymers derived from crude monomer streams. These streams are obtained from wood, coal,

**Table 16-1. Rosin tackifiers.**

| TRADE NAME | DESCRIPTION | MANUFACTURER |
|---|---|---|
| Floral 85 | Glycerine rosin ester | H |
| Pentalyn H | Hydrogenated pentaerythritol ester | H |
| Staybelite Ester 10 | Hydrogenated glycerine ester | H |
| Sylvatac RX | Modified tall oil rosin | S |
| Sylvatac 95 | Polymerized rosin | S |
| Zonester 85 | Rosin ester | A |

A: Arizona Chemical Co.
H: Hercules, Inc.
S: Sylvachem Corp.

or petroleum sources and their composition dictates the polymerization method used as well as the properties of the resins that are obtained.

Chemically, hydrocarbon resin streams may be classified as containing primarily aromatic, aliphatic, and diene (cyclic olefin) monomers. These are also referred to as C-9, C-5, and $(C\text{-}5)_2$, respectively, due to the average number of carbon atoms per monomer molecule. Polymerization of these streams is carried out using a Lewis acid catalyst or by a free-radical process using heat and pressure. These streams, polymerization methods, and the general designations of the resins produced are shown in Fig. 16-2.

Because of the variety of feed streams available and the various polymerization techniques employed, a wide variety of hydrocarbon resins is available with a broad range of properties and prices. Furthermore, it is this diversity that has made hydrocarbon resins useful in so many tackifier applications, especially those involving pressure-sensitive adhesives. The following discussion will briefly treat each of the main types of hydrocarbon

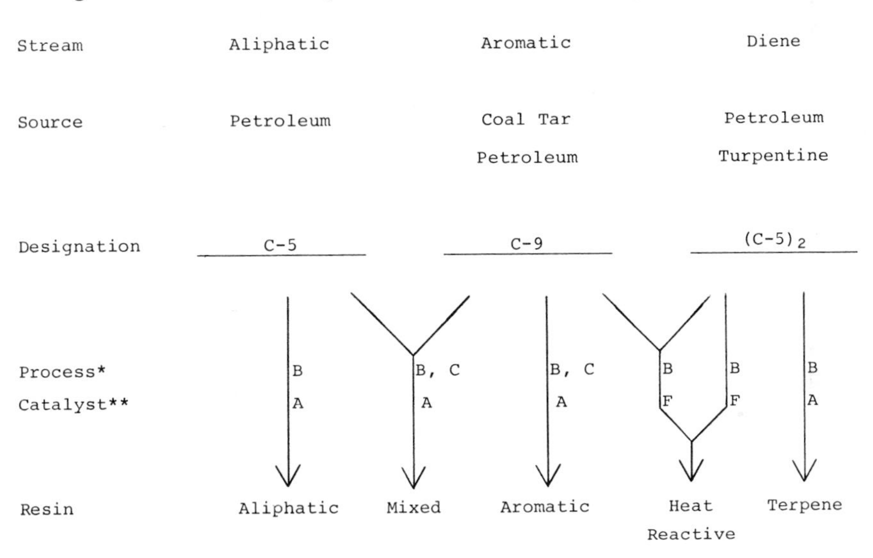


Fig. 16-2. Hydrocarbon resins.

resins and the typical range of properties available will be outlined. Tables 16-2 through 16-7 contain representative lists of the materials available from various manufacturers. It should be remembered that each trade name used here represents only one resin out of a series or family that may be available under that general name.

## Aromatic Hydrocarbon Resins

The aromatic resin group consists of aromatic petroleum resins and resins from coal tar, commonly called coumarone-indene resins. These resins are produced by the cationic solution polymerization of aromatic crude streams containing indene as the principal polymerizable monomer along with varying minor percentages of styrene, methylindenes, and methylstyrenes. Coumarone or dicyclopentadiene may also be present, depending upon the raw material source (Table 16-2).

**Table 16-2. Aromatic resins.**

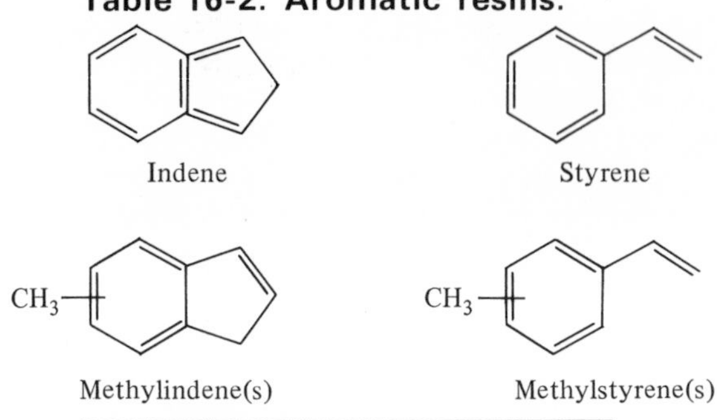


| TRADE NAME | MANUFACTURER |
|---|---|
| Betaprene AC-100 | R |
| Chemfax 5A88-100 | CF |
| Cumar R-7 | N |
| Nevchem 100 | N |
| Nevex 100 | N |
| Petro-Rez 100 | L |
| Picco 6100 | H |
| Piccoumaron 100 | H |
| Zecorez 700 | Z |

CF: Chemfax, Inc.
H: Hercules, Inc.
L: Lawter Chemical, Inc.
N: Neville Chemical Co.
R: Reichhold Chemicals, Inc.
Z: Ziegler Chemical and Mineral Corp.

Aromatic resins can be obtained in varying softening points from about 10° to over 150°C and in colors ranging from pale straw to dark amber. The iodine number, generally taken as a measure of reactivity or unsaturation, is relatively low in these resins, ranging from 30 to 60. The specific gravities of aromatic resins are greater than 1.0, usually falling in the 1.05 to 1.15 range. Finally, number average molecular weights generally vary with the softening point and fall in the range of 290 to 1150.

Aromatic resins have a wide utility in a number of end-use areas. Because of their excellent compatibility with most synthetic elastomers, they can be used as tackifiers in a variety of contact cements, mastics, and construction adhesives. The high softening point grades are finding increased use in adhesives based on the thermoplastic elastomers. They associate with the styrene end blocks and can thus be used to modify the strength, viscosity, and upper service temperature of the adhesive blend.

Aromatic resins are also used in the production of ink and various coatings, as well as in rubber compounding where they serve as process aids, plasticizers, and tackifiers.

## Aliphatic Hydrocarbon Resins

Aliphatic resins are produced from light, so-called C-5 petroleum fractions. The principal monomers are *cis*- and *trans*-piperylene. Also found in these streams are varying amounts of isoprene, 2-methylbutene-2, and in some cases, dicyclopentadiene (DCPD). These streams are polymerized using aluminum chloride and the softening point is controlled by adjusting the composition of the diluent or monomers. This differs from the production of aromatic resins using $BF_3$ where the softening point is controlled by changing the reaction conditions (Table 16-3).

Aliphatic resins are generally available only in softening point grades from 80 to 115°C because of production and demand limitations. The specific gravities of these resins are below 1.0, usually in the range of 0.95 to 0.98, and molecular weights run from 1000 to 1500, which is quite high when compared to other types of resins of equivalent softening points.

The aliphatic or C-5 resins have been available for many years. They were used primarily in rubber compounding, floor tile production and some coating applications. The high quality grades that are currently available, however, are a relatively recent innovation. These new generation C-5 resins are sometimes referred to as synthetic polyterpene resins to give an indication of their outstanding properties. They possess excellent heat stability and very light colors. They also have excellent compatibility with paraffin waxes, low density polyethylenes, and amorphous polypropylene. They are also outstanding tackifiers for natural rubber and *cis*-polyisoprene. These properties combine to make these resins useful in the formulation of a wide

**Table 16-3. Aliphatic resins.**

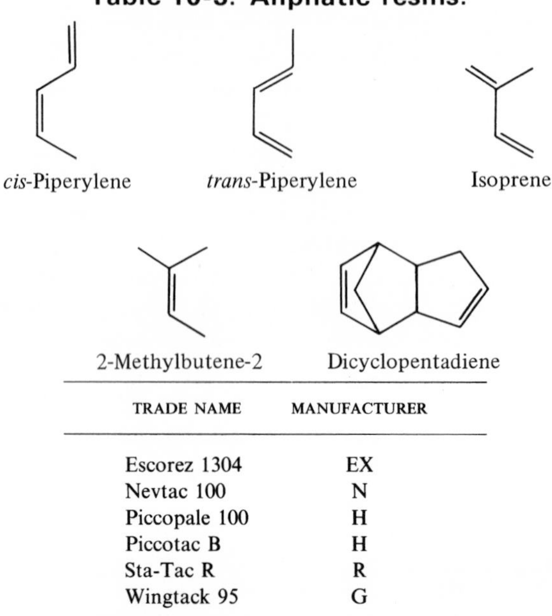


| TRADE NAME | MANUFACTURER |
|---|---|
| Escorez 1304 | EX |
| Nevtac 100 | N |
| Piccopale 100 | H |
| Piccotac B | H |
| Sta-Tac R | R |
| Wingtack 95 | G |

EX: Exxon Chemical Co.
G: The Goodyear Tire and Rubber Co.
H: Hercules, Inc.
N: Neville Chemical Co.
R: Reichhold Chemicals, Inc.

variety of hot melt adhesives and coatings as well as solvent-applied and hot melt pressure-sensitive adhesives. They can also be used as light colored tackifiers and plasticizers in compounding a variety of types of elastomers.

## Mixed Aliphatic/Aromatic Resins

The mixed C-5/C-9 resins are hybrid resins produced by blending C-5 and C-9 streams in varying proportions and then polymerizing with either $AlCl_3$ or $BF_3$ as required. These resins are usually produced in only one or two softening point grades and the properties obtained are generally intermediate between those of either straight aromatic or aliphatic resins. These resins are the result of research efforts by various manufacturers to fine tune the compatibility and tackifying characteristics of their C-5 resins to produce products with outstanding performance in the hot melt and pressure-sensitive adhesives (Table 16-4).

**Table 16-4. Mixed aliphatic aromatic resins.**

| TRADE NAME | MANUFACTURER |
|---|---|
| Escorez 2101 | EX |
| Hercotac AD-1100 | H |
| Nevpene 9500 | N |
| Super Nevtac 99 | N |
| Super Sta-Tac 100 | R |
| Wingtack Plus | G |

EX: Exxon Chemical Co.
G: The Goodyear Tire and Rubber Co.
H: Hercules, Inc.
N: Neville Chemical Co.
R: Reichhold Chemicals, Inc.

## Heat-Reactive Hydrocarbon Resins

The next general class of hydrocarbon resins are the so-called heat-reactive or diene resins. These are petroleum resins with a mixed aromatic and aliphatic composition. The streams used to produce these resins contain indene and styrene, mentioned before, along with substantial amounts of dicyclopentadiene, dimethyl-dicyclopentadiene, and various other C-5 dimers, sometimes better known as diene monomers. These resins are generally polymerized by the action of heat and pressure in a free radical process. This method of polymerization is generally less costly than the use of a Friedel-Crafts catalyst, and the resins are therefore usually less expensive than the catalyst resins (Table 16-5).

Colors of the heat-reactive resins can vary from light pale yellow to dark amber. The iodine numbers of these reactive resins are quite high, ranging from about 120 to 220 depending on the feed blend and the actual polymerization conditions employed. Specific gravities are similar to those of aromatic resins ranging from 1.08 to 1.10 and molecular weights generally fall between 500 and 800. Softening points of various grades usually fall between 95° and 140°C, although some resins with higher values are available for special applications.

Because of the lower cost and excellent compatibility characteristics of these resins, they are widely used in compounding low cost adhesives such as cements, mastics, sealants, and even some solvent-applied pressure-sensitive adhesives. Because of their reactivity, they are subject to oxidation, however, and an adequate stabilizer system should be used where long-term oxidation resistance is required.

**Table 16-5. Heat-reactive resins.**

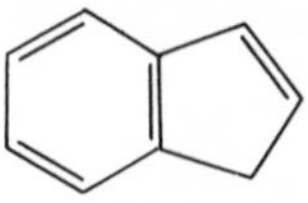

Indene

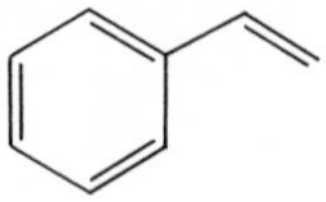

Styrene

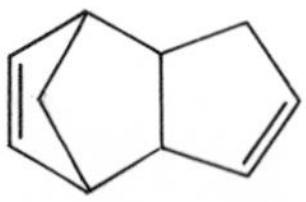

$CH_3$

$CH_3$

Dicyclopentadiene

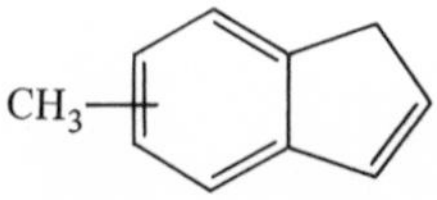

Methyldicyclopentadiene

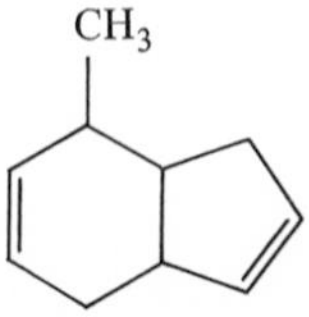

C–5 Dimers

| TRADE NAME | MANUFACTURER |
|---|---|
| Betaprene BR-100 | R |
| Chemprene 210 | CF |
| Neville LX-685-125 | N |
| Neville LX-1000 | N |
| Piccodiene 2215 | H |
| Zecorez 650 | Z |

CF: Chemfax, Inc.
H: Hercules, Inc.
N: Neville Chemical Co.
R: Reichhold Chemicals, Inc.
Z: Ziegler Chemical and Mineral Corp.

## Terpene Resins

Along with the heat-reactive resins, the terpene resins may be classified as diene or $(C\text{-}5)_2$ resins since the monomers used in their production may be considered to be dimers of isoprene. Unlike the heat-reactive resins, terpene resins are produced by cationic solution polymerization of raw materials obtained from turpentine and other natural sources, including citrus peels. The actual monomers are usually one or a combination of the terpenes $\alpha$-pinene, $\beta$-pinene, and dipentene or limonene. The catalyst is usually $AlCl_3$, although other Lewis acids can be used (Table 16-6).

Terpene resins are light in color and can be obtained in softening points from 10 to 140°C. The specific gravities range from 0.97 to 1.00 and molecular weights from about 300 to almost 2000. Generally speaking, at equiva-

**Table 16-6. Terpene resins.**

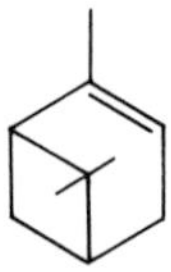

α-Pinene 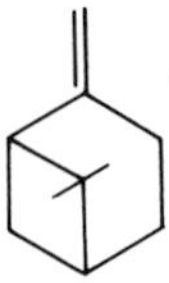 β-Pinene 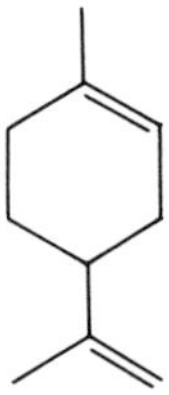 Dipentene

| TRADE NAME | MANUFACTURER |
|---|---|
| Croturez 100 | CR |
| Gammaprene A-100 | R |
| Piccolyte S-100 | H |
| Zonarez B-100 | A |

A: Arizona Chemical Co.
CR: Crosby Chemical, Inc.
H: Hercules, Inc.
R: Reichhold Chemicals, Inc.

lent softening points, β-pinene resins have the highest molecular weights and lowest gravities while dipentene resins have the highest gravities and lowest corresponding molecular weights.

Terpene resins, especially β-pinene resins, have been the resins of choice in the production of pressure-sensitive adhesives for many years. These resins have outstanding tackifying abilities when used with natural rubber, a wide range of softening points and extremely broad FDA clearances. Over the last ten years, however, due to rising costs and uncertain availability, the terpene resins and rosin derivatives have lost considerable ground to the growing number of less expensive synthetic hydrocarbon resins designed specifically to replace them.

## Modified/Special Resins

The final group of resins to be discussed are the so-called special or modified resins. This is a catchall group used for resins having special properties which do not quite fit into the general classes of resins already mentioned. These resins are prepared by various techniques and by blending selected feed streams or pure monomers to obtain resins with special solubility, compatibility, or performance characteristics. Table 16-7 lists several of these resins along with a general description of the type of resin. Table 16-7 is far from complete and is only offered here to give the reader a brief glimpse of the types of special resins available.

## EVALUATING TACKIFIER RESINS

Virtually all adhesive applications and many of the more important fabricated rubber applications depend on that property of rubber known as tack. Rubber tack is a perverse and misunderstood property possessed to some degree by natural rubber but usually lacking in synthetic elastomers and other polymers. It must be formulated into the adhesive compound by the use of tackifier resins. This discussion will attempt to describe tack in terms of what it is and how resins can be evaluated to promote and maintain it.

As anyone in the area of rubber and adhesives knows, tack is not a single phenomenon. Tack is actually a generic term used to describe one or more of a number of phenomena relating to the speed of bond formation. While most everyone is aware of this, all too often people can end up comparing apples and oranges.

The types of tack most often dealt with in rubber and rubber-based adhesive applications are wet tack, pressure-sensitive tack, and building or green tack. Wet tack is the term used in conjunction with adhesive systems such as mastics where the bond is formed while the adhesive still contains appreciable quantities of a carrier solvent or vehicle. Wet tack is a function of the rheology of the total system and will not be dealt with here, although most of the discussion to follow will apply at least indirectly to wet tack.

**Table 16-7. Modified/special resins.**

| TRADE NAME | DESCRIPTION | MANUFACTURER |
|---|---|---|
| Eastman Resin H-100 | Hydrogenated resin | ES |
| Escorez 5300 | Hydrogenated resin | EX |
| Piccofyn A-100 | Terpene phenolic | H |
| Piccotex 100 | Vinyltoluene/$\alpha$-methylstyrene | H |
| Piccovar AP-25 | Alkylated aromatic | H |
| Nebony 100 | Petroleum tar resin | N |
| Nevillac Hard | Phenol-modified aromatic resin | N |
| Nevroz 1520 | Acid-modified resin | N |
| SP-553 | Terpene phenolic | SC |

ES: Eastman Chemical Products Corp.
EX: Exxon Chemical Co.
H: Hercules, Inc.
N: Neville Chemical Co.
SC: Schenectady Chemicals, Inc.

Considering the remaining two types of tack as total dry rubber tack, we can now break it down into three components.

- Adhesion
- Autohesion
- Cohesion

Adhesion is pressure-sensitive or sticky and is nonspecific. Autohesion is the specific adhesion of the rubber compound or adhesive to itself and is referred to as the true tack or building tack. Finally the new term, cohesion, refers to the internal strength of the rubber or adhesive and is often referred to as green strength.

## Pressure-Sensitive Tack—Adhesion

Pressure-sensitive tack or adhesion is that property which allows an adhesive film to adhere tenaciously to any surface with which it comes in contact. The strength of the bond is influenced by the contact pressure used during application which, of course, gives rise to the term pressure-sensitive adhesive.

Most rubbery polymers having a glass transition temperature $T_g$ well below room temperature possess at least some small degree of pressure-sensitive tack, but for the production of adhesives a tackifier is usually needed. While quite a bit of work has been done in this area, the actual function of tackifier resins is not fully understood from a theoretical standpoint. Thus, it still remains for an adhesive formulator to evaluate the various tackifiers in the systems in which he is interested to determine what will work best for him. Therefore, a few comments on tackifiers and the evaluation of pressure-sensitive adhesives may prove instructive.

Pressure-sensitive adhesives are usually evaluated by measuring tack, peel, and shear adhesion. These properties measure, in a very empirical and end-use-oriented way, three different aspects of the viscoelastic nature of the adhesive.

Tack measures viscous flow under conditions of fast strain rates and low stress magnitudes. A good tackifier should modify the rubber in such a way that it can flow easily to wet the surface of the material to which it is bonding. This flow should take place quickly under the influence of very slight pressures.

Peel measures the resistance to flow at intermediate strain rates and high stress magnitudes. It also measures the cohesive strength of the adhesive.

Finally, shear adhesion or creep measures viscous flow at very low strain rates and intermediate stress magnitudes. It is highly sensitive to the molecular weight distribution and any crosslinking or network structure in the system.

It is obvious that in any given system these properties tend to be mutually exclusive and a balance must be found. The proper tackifier is not necessarily the one which gives the best tack values or the best shear values.

Not all tackifiers will tackify all elastomers. In fact, the best tackifiers for promoting pressure-sensitive tack are often fairly specific. Table 16-8 shows the results of evaluating three tackifiers in two different elastomers. As can be seen, the best tackifier for natural rubber is essentially dead in SBR, and likewise, the best tackifier for SBR is worthless in natural rubber. The resin with intermediate characteristics has some activity in both elastomer systems but is of limited utility.

Why certain resins will tackify one elastomer and not another while some can tackify a variety of elastomers is not precisely known at this time. Some studies indicate that rubber-resin blends exhibiting good adhesion have a two-phase morphology which disappears along with adhesion on heat aging.[2] Just what this might mean is unclear because the best tackifiers for a given elastomer tend to have compatibility and solubility characteristics similar to the elastomer itself. Furthermore, it is well known that within a given series of tackifiers, as the softening point goes up, the optimum resin-to-rubber ratio decreases as shown in Fig. 16-3. This suggests that the tack resin may be acting to modify the $T_g$ of the base elastomer and that there is an upper limiting value, probably room temperature, above which tack falls off sharply.

## Evaluating Tackifier Resins

At this point, we have seen that there are a wide variety of tackifier resins available to the adhesive compounder and a scarcity of published general information on the performance of resins with the various polymers used to

**Table 16-8. Preliminary tackifier evaluation.**

| | | NATURAL RUBBER | | | SBR 1011 | | |
|---|---|---|---|---|---|---|---|
| RESIN @ 80 PHR | TYPE | TACK | PEEL | SHEAR | TACK | PEEL | SHEAR |
| Nevex 100 | Aromatic | > 15 | — | — | 1.3 | 1.06 | > 120 |
| Nevpene 9500 | Mixed | 3.8 | 0.73 | 25 | 1.8 | 0.99 | 75 |
| Nevtac 100 | Aliphatic | 1.3 | 0.61 | 18 | > 15 | — | — |

Tack: ASTM D3121-73 rolling ball tack (cm)
Peel: PSTC-1 180° peel adhesion (kg/cm)
Shear: PSTC-7 shear adhesion 1 lb/in$^2$ @ 60°C (min)

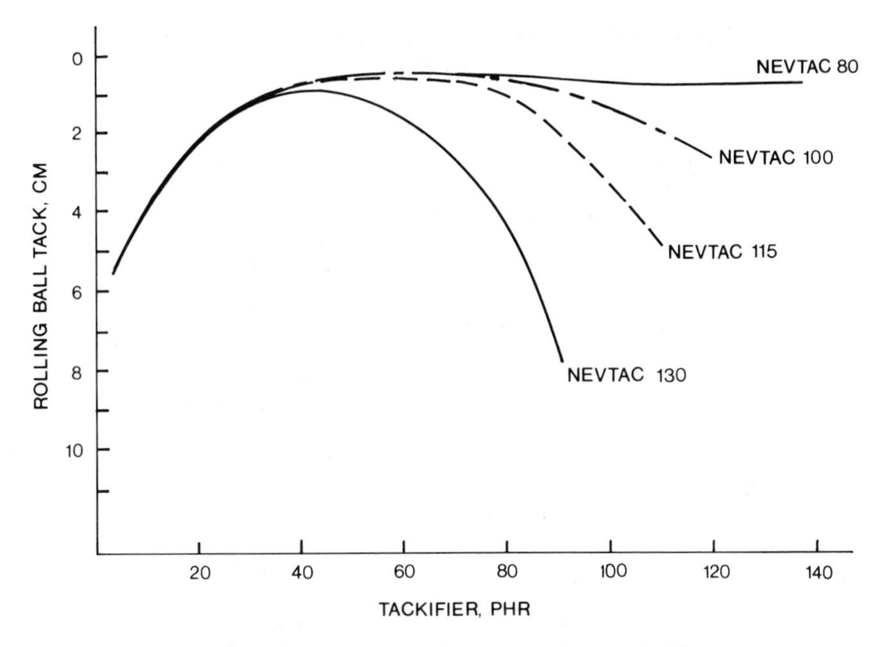


Fig. 16-3. Tack versus resin softening point in natural rubber.

produce pressure-sensitive adhesives. In fact, there are over 13 tackifier resin producers in the United States alone offering approximately 110 different resins in a variety of grades. This means the adhesive compounder has well over 300 resins to choose from. Actually, less than 50 probably have any real utility in pressure sensitive adhesive formulations. And, depending on the base polymer and the performance characteristics desired, there may be as few as two or as many as 20 resins which will fit any particular need.

Finding the proper resin for a given system is not too difficult, however, if a systematic approach is taken. Table 16-9 groups the more common base polymers used in pressure-sensitive adhesives along with the types of tackifiers that can normally be used with them. Using this as a starting point, the first step is to contact the suppliers of these types of tackifiers for information on the specific resins they supply, their properties and prices, and any specific recommendations they might make. This information is then evaluated to eliminate resins with properties and prices which do not lend themselves to the application being studied.

The second step in tackifier evaluation is to screen the remaining candidates to determine their suitability as tackifiers for the system being studied. This screening test can be quite simple, and may only involve dissolving roughly equal quantities of the base polymer and tackifier in a suitable

**Table 16-9. General tackifier recommendations.**

| BASE POLYMER | TACKIFIERS |
|---|---|
| 1. Natural rubber (NR)<br>*cis*-Polyisoprene (IR)<br>Styrene/isoprene block copolymers | Aliphatic resins<br>Mixed C-5/C-9 resins<br>Rosin derivatives<br>Special resins<br>Terpene resins |
| 2. Styrene-butadiene rubber (SBR)<br>Styrene/butadiene block copolymers | Aromatic resins<br>Heat reactive resins<br>Mixed C-5/C-9 resins<br>Rosin derivatives<br>Special resins<br>Terpene resins |
| 3. Ethylene-vinylacetate copolymers | Aromatic resins<br>Rosin derivatives<br>Special resins |
| 4. Atactic polypropylene | Aliphatic resins<br>Mixed C-5/C-9 resins<br>Terpene resins |
| 5. Miscellaneous elastomers | Various, depending on the polarity of the system |

solvent and casting a film which may be evaluated, upon drying, by touching it and observing its clarity. A refinement of this procedure was illustrated in Table 16-8 where actual tack, peel, and shear values were determined at a resin loading of 80 phr using ASTM and Pressure Sensitive Tape Council methods. Based on these preliminary results, the list of tackifier resin candidates is reduced to some manageable number based on time and manpower for further evaluation.

The third step involves determining the overall performance characteristics of the individual resins in the pressure-sensitive system being studied. This involves the measurement of tack, peel, shear, and other desired properties over a wide range of polymer/tackifier compositions. This results in a performance profile for the tackifier as shown in Fig. 16-4.

It can be seen that the optimum value for each property tested usually occurs at a different resin loading level. Different tackifiers can give widely different performance profiles. These profiles, therefore, are carefully compared and the resin which gives the best balance of compounded properties

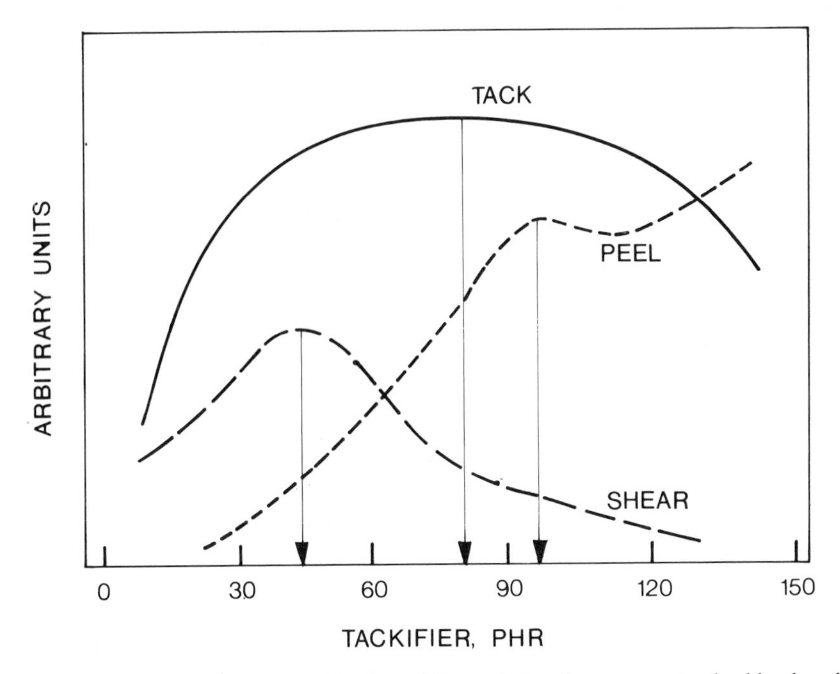


Fig. 16-4. Typical performance profile of a tackifier resin in solvent cast natural rubber-based system.

at a given loading level also considering the economics of the system is the resin of choice.

This, then, is how tackifier resins are chosen and evaluated for pressure-sensitive adhesives. This has been a general treatment with no mention of the special problems encountered in preparing hot melt applied adhesives[3,4] or water-based adhesives from rubber latices[5,6] Tackifiers for these systems are chosen in the same manner outlined above. The only differences encountered are associated with the actual formulation of the test adhesive. This is especially true in the case of the latex-based adhesives. (The reader is referred to the references cited above.)

## REFERENCES

1. Rothrock, C. W., Jr., and Rhyne, J. B. *TAPPI* **60** (6): 54–57 (1977).
2. Wetzel, F. H., and Alexander, B. B. *Adhesives Age*, **7**(1): 28–35 (1964).
3. Brinker, K. C. *Adhesives Age*, **20**(8): 38–40 (1977).
4. Coker, G. T., Jr., Lauck, J. E., and St. Clair, D. J. *Adhesives Age*, **20**(8): 30–35 (1977).
5. Jahn, R. G. *Adhesives Age*, **22**(12): 35–44 (1979).
6. Wherry, R. W. "Resin Dispersions for Water Base Pressure-Sensitive Adhesives," paper presented at Pressure-Sensitive Tape Council seminar on Water-Based PSA Systems, June, 1979.

# Chapter 17

# Release Coatings

**Donatas Satas**
*Satas & Associates*
*Warwick, Rhode Island*

Release coatings are important components of pressure-sensitive products. Tape, label, or sheet material with a pressure-sensitive adhesive must be protected from unwanted contact with other surfaces. Protection with a release sheet or, in the case of tapes, winding upon its own backing are the methods used to prevent the tacky mass from accidental sticking to other surfaces.

The low adhesion to the release-coated surface may be caused by several factors or a combination of them. A release coating may be poorly wetted by the adhesive, so that a good contact is not established and the separation is easy. Release properties may be imparted by an agent which splits cohesively at a low level of force, leaving a light deposit on the adhesive. A sufficient release may be obtained from a smooth and hard plastic surface without any release agent. In such cases, it is important to prevent adhesive transfer from one to the other side of the backing, because the release level might not be very low. Prime coating may be used to improve the anchorage of the adhesive to the other side and to assure that the separation takes place in the desired location.

The surface over which the release coating is applied should be reasonably nonporous to prevent the flow of soft pressure-sensitive adhesive into the pores and irregularities, causing an increase in the unwind force. Backsize coating, which might or might not have the release characteristics, is often used on fabric or paper surfaces to provide a smooth and impermeable surface for the release.

In some cases, a rough surface may serve as a release plane. The adhesive contacts the surface only in a limited area because of its roughness. This allows for an easy separation.

A release coating is expected to meet the following requirements:

1. It should provide an adequate release. The release level should be sufficiently low to provide an easy unwind, but it should be high enough to prevent the flagging when the tape is wrapped over its own backing. Higher release level also minimizes the possibility of telescoping in the roll. The release agent should exhibit the "shingling" failure. The release should be slow at a low peel angle and rapid at 90° peel, normally used for the tape unwind.
2. The level of the release should be easily reproducible. This requires that the release level is not very sensitive to the amount of release agent used.
3. The release agent should be firmly anchored and should not transfer to the adhesive surface causing a decrease in tack.
4. The aging properties should be good and the unwind should not change much on prolonged aging.

There might be other requirements for specific applications. Masking tape requires that the release coating has a good solvent resistance and that the paint adheres reasonably well to the release coating. Otherwise, the paint chips might fall off and remain adhered to the freshly painted surface.

The release coating is applied over the backing from a dilute solution or dispersion. Light coatings of 2–3 g/m$^2$ or even less are sufficient. The anchorage of the release coating to the backing is not always sufficiently good and the release agents can be incorporated into the backsize coating. This improves the anchorage, and it dilutes the release agent. This method can be used to control the release level. In such cases, the coating weight is higher.

There are several groups of compounds which are useful as release agents for pressure-sensitive tapes. Silicones, long chain alkyl branched polymers, chromium complexes, fluoro chemicals, and various hard polymers have been used for release purposes.

## POLYMER COATINGS

The first release coatings for pressure-sensitive tapes were natural and synthetic polymers such as shellac, starch, zein, casein, and nitrocellulose. Used on porous backings, such as saturated paper used for masking tapes, these coatings function as barriers preventing the adhesive from penetrating the backing. Such release coatings do not exhibit an easy release: they fall into 50–400 g/cm width release level.

Formaldehyde modified starch has been used as a release coating.[1] Cellulose acetate butyrate and alkyd resins have been found useful as release coatings or antiblocking additives. Polyvinyl chloride resins, such as Geon 352 or a PVC-nitrile rubber blend (Geon 552) have been used as release coatings for paper tapes.

Many acrylic polymers have found use in this application. Solvent-based acrylates are most commonly used for release applications, but aqueous emulsion systems are offered as well. Rhoplex R-225 (Rohm and Haas) is offered specifically for masking tape application at a coating weight of 5–10 $g/m^2$. Ethyl acrylate-acrylonitrile copolymer (65 : 35) 100 parts blended with 10 parts of heat-reactive formaldehyde resin was utilized as a release coating.[2] Acrylic acid-ethyl acrylate copolymer was also used in the same way.

Blum[3] describes a release agent that is printable i.e., exhibits a good receptivity of flexographic inks and a good ink adhesion, while still acting as a release agent for pressure-sensitive adhesives. Usually, it is required to print first and then apply the release coating over the print. The proposed release coating consists of a vinyl film former, a hard thermoplastic acrylic resin, and a polyamide plasticizing resin which acts as an adhesion promoter for inks. An example of this release coating is given below.

| | |
|---|---|
| Vinyl polymer film former (Vc-171, Borden) | 10 |
| Hard acrylic resin a copolymer of methyl methacrylate and ethyl acrylate (Acryloid A-11, Rohm and Haas) | 40 |
| Polyamide plasticizer (Versamid 950, Henkel) | 50 |

The coating is applied as a dilute solution in organic solvents.

Dabroski[4] gives an interesting release agent for printable tapes. The ink is printed on release-coated film. The release coating migrates through the print, providing a continuous release surface as if the coating were applied over the print. Such coatings consist of water dispersible aldehyde or imide resin, emulsifying agent, and an acrylic film forming polymer, such as ethyl acrylate, acrylonitrile, or methacrylic acid terpolymer (63 : 35 : 2).

Polyvinyl *N*-octadecyl carbamate made by reacting polyvinyl alcohol and octadecyl isocyanate is described by Dahlquist *et al.*[5] for use as a release agent for cellophane and cellulose acetate film tapes. The bond to cellulose film surface was very good and a small amount (1 $g/m^2$) was sufficient to provide an adequate release. Poly-*N*-acyl imine is also proposed for release coating applications by Bartell *et al.*[6] Poly(octadecyl vinyl ether/maleic anhydride) (Gantrez AN-8194 by GAF) is useful as a release coating.

Settineri and O'Brien[7] describe the use of acrylic acid-acrylonitrile oligomers polymerized in the presence of a mercaptan to a molecular weight of 200–5000.

It is well known that polar pressure-sensitive adhesives do not adhere well to nonpolar surfaces. This is utilized in the construction of polyethylene and polypropylene film tapes. Oriented polypropylene film tape can be made by applying the adhesive on the corona discharge treated side of the film. This results in sufficiently good adhesive anchorage and the untreated film side provides the release surface. Similarly, release coating is not required for polyethylene-coated tapes.

## WAXES

Various waxes are used as additives to polymer coatings to improve the slip, blocking resistance, and release. The effectiveness of these waxes is based on their ability to migrate to the surface of the polymeric coating. They should be carefully evaluated for application as pressure-sensitive adhesive release coatings. Waxes have an inclination to transfer to other surfaces and can easily affect the tack and adhesion properties. For this reason, the waxes are not widely used in pressure-sensitive applications.

Polyethylene waxes and modified polyethylenes can be used as release coatings for polar adhesives, especially polyacrylates.

Polyamide waxes are used as slip, antitack, antiblocking, and release agents. When added to the polymer coating, these compounds exude to the surface providing a surface more characteristic of the amide than of the base polymer. Various amide waxes are used for such applications: oleamide, erucamide, *N,N'*-ethylenebis stearamide, bis-oleorylamide.

Holtz[8] has discussed waxes with film formers as release coatings for pressure-sensitive adhesives. Film formers such as nitrocellulose, vinyl chloride-vinyl acetate copolymers, and polystyrene are used with waxes of melting point above 90°C. About 10–15% of wax is used in the polymer coating.

## SILICONES

Silicones are the most widely-used materials for release coating applications. They give a low level of release and are useful for coating the release paper used in pressure-sensitive adhesive applications. While widely used for release sheets, silicone release coatings are not used that widely in tapes because of the extremely easy release. Generally, the force required to separate a pressure-sensitive adhesive from a silicone-treated surface by peeling is typically in the range of 4 to 15 g/cm width. That is too low for use as low adhesion backsize in the tapes.

Silicone release coatings are available in solvent solution, aqueous

**Table 17-1. The release force of organopolysiloxane mixtures with vinyl methyl ether-maleic anhydride copolymer.**

| Silicone (wt %) | Copolymer (wt %) | Release Force (g/cm Width) |
|---|---|---|
| 100 | 0 | 20 |
| 90 | 10 | 8 |
| 75 | 25 | 7 |
| 50 | 50 | 7 |
| 25 | 75 | 18 |
| 10 | 90 | 58 |

emulsion or as 100% solids, usually added as monomers to be polymerized during the coating process.

Mestetsky[9,10] describes the use of polysiloxane release agents with alkyl ether-maleic anhydride copolymer as the carrier for the release agent. The higher the amount of polysiloxane in the blend, the easier is the release. Same author also discloses the use of a cured mixture of aqueous organopolysiloxane emulsion with water-soluble polyvinyl methyl ether homopolymer or with vinyl methyl ether-maleic anhydride copolymer.[10] Table 17-1 shows the variation of release level with the varying amount of poly-

**Table 17-2. Unwind force of some tapes with polyurethane-silicone based release coating.**

| Composition of Release[a] Coating (pbw) | | Unwind Force (g/cm Width) | | | |
|---|---|---|---|---|---|
| | | Acrylic Adhesive | | Rubber-Resin Adhesive | |
| NCO-terminated Prepolymer | Silicone | Initial | After Aging | Initial | After Aging |
| 75.4 | 24.6 | 24 | 120 | 24 | 120 |
| 75.2 | 24.8 | 24 | 24 | 24 | 67 |
| 64.4 | 35.6 | 24 | 120 | 67 | 110 |
| 64.4 | 35.6 | 24 | 24 | 67 | 200 |
| 48.5 | 51.5 | 24 | 24 | 67 | 110 |
| 48.5 | 51.5 | 24 | 24 | 67 | 190 |
| 77.2 | 22.8 | 225 | split | 110 | 380 |
| 57.5 | 42.5 | 24 | 24 | 67 | 67 |
| 64.9 | 35.1 | 24 | 24 | 110 | 110 |
| 52.0 | 48.0 | 24 | 24 | 67 | 110 |
| 74.8 | 25.2 | 24 | 24 | 110 | 110 |

[a] The composition of NCO-terminated prepolymer varied in a manner not shown in the table. Some of the variation of the unwind force is due not only to the polyurethane-silicone ratio, but also to the composition of the former compound. (See Reference (13) for details.)

siloxane in the coating. The interesting part is that the release force goes through a minimum at about 75% of silicone.

Hockemeyer *et al.*[11] discuss the use of diorganopolysiloxanes for release coatings. Sorell *et al.*[12] describe the reaction product of polysiloxane having hydroxy groups with an organic titanate. This release agent is used with a film former: poly(vinyl N-octadecylcarbamate).

Schurb and Evans[13] combine organosilicone release agents with isocyanates to obtain a series of compounds with varying release levels. Table 17-2 shows some of the data. Phenolic crosslinking resin was used with some of the adhesives.

Release agents for vinyl floor tiles are described by Boranian and Timm.[14] Such tiles can be stacked without using release paper. The release agents consist of a mixture of a crosslinkable acrylic emulsion (Rhoplex TR-407 by Rohm and Haas), silicone release agent (Syl-off 22 by Dow Corning Corp.), catalyst (organotin salt), and colloidal silica that serves as an antislip agent. The release coating is applied at a level of 5 g/m$^2$. A typical formulation is shown below.

| Ingredient | Parts added | Percent solids |
|---|---|---|
| Organopolysiloxane emulsion | 30 | 40 |
| Organotin salt catalyst | 6 | 20 |
| Acrylic emulsion polymer | 63 | 45 |
| Colloidal silica | 0.25 | 40 |
| Glacial acetic acid | 0.75 | 100 |

Caimi and Schlauch[15] disclose an aqueous release agent consisting of a crosslinkable vinyl acetate polymer (vinyl acetate-N-methylol acrylamide polymer 94 : 6) to which release agent is added. Various release agents can be used: organosilicone compounds (dimethyl polysiloxanes L-522 and L-45 by Union Carbide), chrome complexes (Quilon by DuPont), and fluorocarbons (FC-805 and FC-808 by 3M). Water-based release coatings for masking tapes are described by Young.[16] The acrylic coating contains a release agent that migrates to the surface during the drying and becomes immobilized by the reaction between the release coating and the acrylic polymer.

The release level of silicone coatings can be controlled by the corona discharge treatment. Hurst[17] describes a corona discharge or flame treatment to decrease the release. Oxidative treatment disturbs the molecular structure of the silicone surface and changes its release purposes. The polymer molecules have a tendency to realign back to the original structure. If the surface is in contact with the pressure-sensitive adhesive, realignment does not take place and the higher release is maintained.

## LONG CHAIN BRANCHED POLYMERS

Polyvinyl esters of long chain aliphatic acids are waxy compounds exhibiting sharper melting points, quite unlike the corresponding esters of short chain acids. Apparently, the long chains show a tendency to crystallize. Such compounds are suitable as release coatings of medium release value, which are especially desirable for pressure-sensitive tapes. These polymers are used alone or as a part of the backsize. Many release coating patents describe the use of such long chain compounds.

Hendricks[18] proposes the copolymers of alkyl acrylate and acrylic acid (3:2) for tape release coatings. The alkyl side chain of 16–20 carbon atoms is the most suitable; side-chains shorter than 12 carbon atoms do not have sufficient low adhesion. Octadecyl acrylate is the preferred comonomer in this case. The acrylic acid helps adhesion to cellulosic films and the coating gives a good release for cellophane and cellulose acetate film tapes. Crocker[19] describes stearyl methacrylate-acrylonitrile (50 to 80% stearyl methacrylate) for release coating applications. This release agent can also be used with other polymer coatings: alkyd with melamine formaldehyde and a curing agent. Polyvinyl esters such as vinyl stearate, palmitate, arachidate, and behenate give release coatings with polyvinyl alcohol and polyvinyl acetate.[20] Stearic acid polyester with formaldehyde resin and proper catalysts is useful for a release coating, which is cured after application as backsize.

Collins[21] describes a solvent-based release coating consisting of a release agent such as polyvinyl stearate or laurate, copolymers of vinyl stearate with vinyl acetate and maleic anhydride, or copolymers of octadecyl acrylate and of a filmformer such as nitrocellulose or polystyrene. A typical formulation is given below.

| | |
|---|---|
| Nitrocellulose (R.S. 5–6 sec) | 18.1 |
| Di-octyldecyl phthalate | 5.9 |
| Release agent, such as solution copolymer of vinyl stearate-maleic anhydride-vinyl acetate (0.8 : 0.2–1.0) 4.8 parts in 60.4 parts toluene | 65.2 |
| Isopropyl alcohol | 33.5 |
| Methyl ethyl ketone | 42.5 |

Solution polymerization of other release agents is described in this patent. The release agent concentration is higher at the surface of such coating and lower at the coating-substrate interface. This distribution helps to obtain a good adhesion to the backing and a good release. Dabroski[2] also describes the use of other copolymers with long side chains in combination with various polymer coatings such as shown in the examples below.

| | |
|---|---|
| Nitrocellulose | 100 |
| Stearyl maleate–vinyl acetate copolymer | 7 |

or

| | |
|---|---|
| Vinyl chloride–vinyl acetate copolymer (88 : 12) | 100 |
| Sorbitan monostearyl–toluene diisocyanate reaction product | 7 |

Demmig and Rehnelt[22] describe release surface coatings consisting of polyethylene imines acylated with higher fatty acids, such as palmitic, stearic, behenic, and myristic acids. These coatings were especially useful for cellulosic backings. Solution polymers of vinyl stearate, allyl stearate, or vinyl octadecyl ether with maleic anhydride are described by Dahlquist *et al.*[23] Lavanchy[24] discusses *N*-stearyl polyacrylamide release coating. Applied from a dilute solution in toluene over nitrocellulose coated masking tape the release coating exhibited good anchorage and adequate release. The unwind force at a speed of 50 m/min was 500 g/cm width after aging for 24 hr at room temperature and practically did not change after aging at 50 and 65°C for 24 hr. This release agent can be compounded with phenol-, urea-, or melamine-formaldehyde resins to give a cured coating with improved solvent and heat resistance.

Solution polymerized stearyl itaconate[25] was useful as a release agent over nitrocellulose coated kraft paper backing. Poly(monocetyl itaconate) and poly(monobehenyl itaconate) were also useful for the same purpose. Copolymers of *N*-substituted long straight chain alkyl maleamic acids and vinyl comonomers formed release agents for products such as creped masking tapes with rubber-resin adhesive.[26] The release agent was applied from a 2% solution in toluene-isopropanol blend over the backsize coating or added to the backsize coating in the amount of 5 parts per hundred on solids basis. The backsize consisted of short oil alkyd (80 parts), butylated melamine formaldehyde (10 parts), and carboxylated polyvinyl chloride (10 parts). The performance of this release agent is shown in Table 17-3.

Grossman and Webber[27] give a release agent composition specific for silicone pressure-sensitive adhesives. Solution of stearic acid is treated with vanadium oxytrichloride, molybdenum pentachloride, or antimony pentachloride in an alcohol solution of morpholine or other secondary amine.

The most unusual way of constructing a pressure-sensitive tape of an easy unwind is suggested by McGarry and Weidner.[28] They suggest to add

**Table 17-3. The release properties of octadecyl maleamic acid and ethyl acrylate copolymer.**

| Aging Conditions | Release Force (g/cm) | | | Peel Adhesion to Glass (g/cm Width) |
|---|---|---|---|---|
| | Release Over the Backsize | Release Added to the Backsize | No Release | |
| No aging | 80 | 125 | 240 | 510 |
| 66°C, 16 hr | 87 | 138 | 315 | 510 |
| 66°C, 7 days | 100 | 150 | 475 | 490 |
| 66°C, high humidity, 7 days | 110 | 160 | 490 | 475 |

a small amount of glyceryl monostearate or a similar compound to the adhesive mass. While this influences the unwind, it does not affect the adhesion to other surfaces.

## AMINES

Various alkyl substituted amines show some release properties, especially to paper surfaces. It is suggested that they are adsorbed on the paper surface and orient by the mechanism similar to that of the chromium complexes. Tristearoyl tetraethylene pentamine[30] applied from 5% solution in an organic solvent at a weight of 2.5 g/m$^2$ serves as a release agent for a saturated paper tape. The release agent is prepared by reacting the polyalkylene amine with the fatty acid. Cristmas[31] describes a release agent prepared by

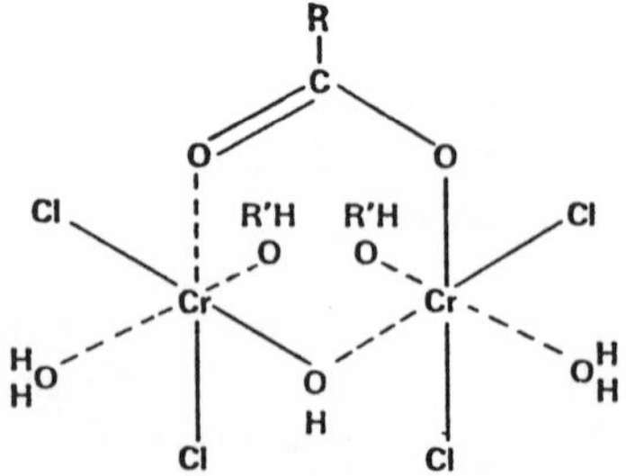


Fig. 17-1. Configuration of chrome complexes used for release coatings. *R* represents a fatty acid group and *R′* the alkyl group of the alcohol.

reacting stearylamine with ethylene-maleic anhydride (1:1) copolymer. Crone and Pike[32] used *p*-phenylene diamine distearamide as the active release agent to be added to the backsize. Webber[33] has found that octadecylamine salt of monooctadecyl acid orthophosphate and similar salts are effective release agents.

## CHROMIUM COMPLEXES

A class of chromium complex compounds have been disclosed to impart water repellency to paper.[34] These are dark blue-green Werner-type complexes where $C_{14}$–$C_{18}$ fatty acid is coordinated with chromium. The configuration of the complex is shown in Fig. 17-1. The surfaces which contain polar groups, such as paper and textiles, which are negatively charged, form a bond with the chromium portion of the complex. The fatty acid groups then become oriented outward and provide the water repellant and adhesive release properties to the surface. When the complex is heated, it polymerizes, forming an insoluble layer of crosslinked material. The cure is fast at 60–93°C. Light coatings of 0.5–1.5 g/m$^2$ are sufficient to impart the release properties. Properly crosslinked coatings rarely transfer to the adhesive side causing a decrease in tack.

The compound is manufactured by DuPont and is available in several grades: Quilon M is the myristic acid ($C_{14}$) complex and is more stable than Quilon S which is the complex of stearic acid ($C_{18}$). Quilon C, the most commonly used grade for pressure-sensitive adhesive release coatings, is also a stearic acid complex, but of improved stability. Quilon H and L are higher strength grades. Quilon C can also be used with film formers such as polyvinyl alcohol, starch and many other polymers. Quilon C applied over a glassine surface improved its release properties as shown in Table 17-4.

**Table 17-4. Release properties of stearato chromic chloride.**

| | | Initial Test | | Pressure Test[a] | |
|---|---|---|---|---|---|
| | Coating Weight (g/cm$^2$) | Peel Adhesion (g/cm) | Percent Retained Adhesion | Peel Adhesion (g/cm) | Percent Retained Adhesion |
| Control | none | 125 | 83 | 161 | 80 |
| Chrome complex | 0.6 | 36 | 91 | 72 | 105 |
| | 1.2 | 18 | 97 | 18 | 63 |

[a] In the pressure test, the release coated paper and the adhesive tape were compressed for 60 sec at a pressure of 175 kg/cm$^2$.

Reaction product of hydroxyethyl cellulose and chromium stearate was used as a release coating for kraft paper tape.[35] Stearato chromic chloride with phenolic aldehyde binder gave an improved release coating for polyester film tapes. This curable coating was also suitable for masking tapes.[36]

## MISCELLANEOUS

Fluorocarbon Werner-type complexes have been used for release coatings for pressure-sensitive adhesives.[37] Nichols[38] describes an unusual release coating. A gelled cellulose triacetate film containing glycerine or some other plasticizer releases the oil under pressure. A pressure-sensitive adhesive adheres to such film if the pressure applied is light. An increase of pressure forces the oil into the interface and the bond is lost. Removal of pressure causes the reabsorption of plasticizer and the adhesive bond can be made again.

## TESTING OF RELEASE QUALITY

The level of release is measured as a peel force at 180° or at 90°. It is also of interest to measure the peel force at various peel rates, since the tape might

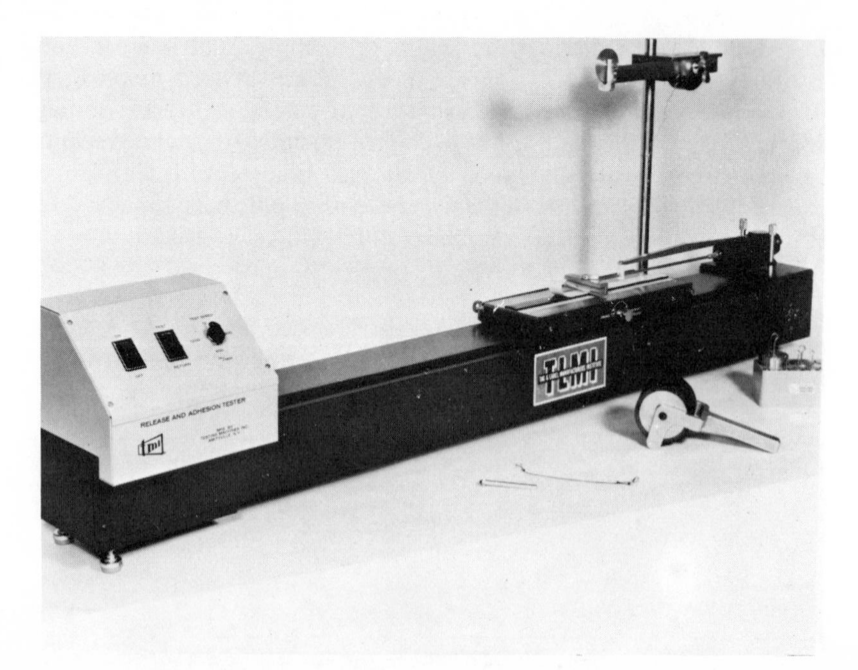


Fig. 17-2. TLMI Release and Adhesion Tester. (*Courtesy Testing Machines Inc., Amityville, New York*).

be unwound at higher speeds. This can be especially important for high-speed application of tapes or labels.

It is important to check the change of release level on aging at various conditions including pressure that might be experienced in a roll of tape. A specific pressure test is used for measuring the release from silicone-coated papers. The tape release paper sandwiches are compressed in a hydraulic press for 2 min at a pressure of 28 kg/cm$^2$ (400 lbs/in.$^2$). The samples are tested immediately and after aging for 7 days at 66°C. It is believed that aging 5–7 days is sufficient for a tape to develop maximum adhesion to a release surface.

Figure 17-2 shows a tester suitable for measuring the peel force at 90°, 135°, and 180° at various speeds (10–3000 cm/min). The apparatus is constructed to meet the specifications of Tag and Label Manufacturers Institute (TLMI Test No. VIILD 4-68) for measuring the force required to strip

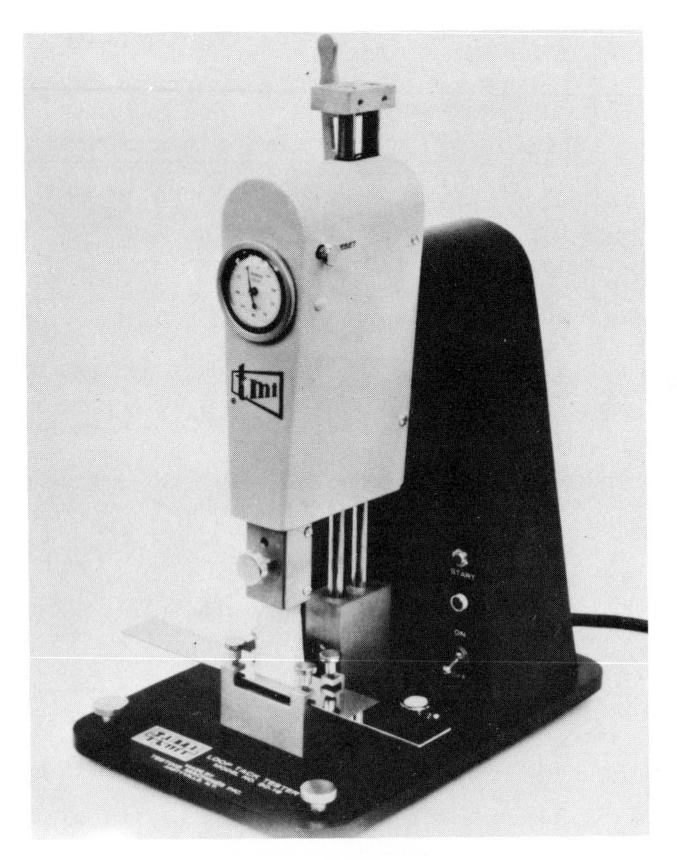


Fig. 17-3. TLMI Loop Tack tester. (*Courtesy Testing Machines Inc., Amityville, New York*).

away label stock from release liners. The Keil test is also frequently used for measuring the release level and subsequent adhesion.[29] It is a peel test at 180° at a peel rate of 25 cm/min. The adhesion after contact with release surface is called subsequent adhesion. A standard stainless steel steel surface is used for these tests.

Figure 17-3 shows a loop tack tester also designed to meet the specifications of the Tag and Label Manufacturers Institute for measuring the tack of pressure-sensitive adhesives. The instrument is suitable to check the effect of release coating on the adhesive.

## REFERENCES


1. Vassel, B., Furtner, V., and Feige, H. I. U.S. Patent 3,300,330 (1967) (assigned to Johnson and Johnson).
2. Dabroski, W. C. U.S. Patent 3,285,771 (1966) (assigned to Johnson and Johnson).
3. Blum, A. U.S. Patent 3,978,274 (1976) (assigned to Borden, Inc.).
4. Dabroski, W. C. U.S. Patent 3,067,057 (1962) (assigned to Johnson and Johnson).
5. Dahlquist, C. A., Hendricks, J. O., and Sohl, W. E. U.S. Patent 2,532,011 (assigned to Minnesota Mining and Manufacturing Co.).
6. Bartell, C., Milutin, I. C., Porsche, J. D., and Rolih, R. J. U.S. Patent 3,475,196 (1969) (assigned to Borden, Inc.).
7. Settineri, R. A., and O'Brien, J. T. U.S. Patent 3,928,690 (1975) (assigned to W. R. Grace and Co.).
8. Holtz, A. U.S. Patent 2,914,167 (1959) (assigned to Johnson and Johnson).
9. Mestetsky, T. S. U.S. Patent 3,770,687 (1973) (assigned to GAF Corp.).
10. Mestetsky, T. S. U.S. Patent 3,823,025 (1974) (assigned to GAF Corp.).
11. Hockemeyer, F., Hechtl, W., Marwitz, H., and Hittmair, P. U.S. Patent 4,154,714 (1979) (assigned to Wacker-Chemic GmbH).
12. Sorell, H. P., Pomazak, E., and Rygg, S. U.S. Patent 3,679,458 (1972) (assigned to Borden, Inc.).
13. Schurb, F. A., and Evans, J. L. U.S. Patent 3,957,724 (1976) (assigned to Minnesota Mining and Manufacturing Co.).
14. Boranian, A. G., and Timm, W. C. U.S. Patent 4,041,200 (1977) (assigned to GAF Corp.).
15. Caimi, R. J., and Schlauch, W. F. U.S. Patent 3,933,702 (1976) (assigned to National Starch and Chemical Corp.).
16. Young, J. E. *Pressure-Sensitive Tape Council Technical Meeting on Water Based Systems*, June 21–22, 1978, Chicago, Illinois, p. 100–105.
17. Hurst, A. R. U.S. Patent 3,632,386 (1972) (assigned to Arhco, Inc.).
18. Hendricks, J. O. U.S. Patent 2,607,711 (1952) (assigned to Minnesota Mining and Manufacturing Co.).
19. Crocker, G. J. U.S. Patent 3,502,497 (1970) (assigned to Johnson and Johnson).
20. Williams, P. L. U.S. Patent 2,829,073 (1958) (assigned to Adhesive Tapes, Ltd., England).
21. Collins, W. C. U.S. Patent 2,913,355 (1959) (assigned to Johnson and Johnson).
22. Demmig, H. W., and Rehnelt, K. U.S. Patent 3,510,342 (1970) (assigned to Henkel and Cie GmbH).
23. Dahlquist, C. A., Albrecht, A. H., and Dixon, G. M. U.S. Patent 2,876,894 (1959) (assigned to Minnesota Mining and Manufacturing Co.).

24. Lavanchy, P. U.S. Patent 3,051,588 (1962) (assigned to Johnson and Johnson).
25. Smith, R. M. U.S. Patent 3,052,566 (1962) (assigned to Johnson and Johnson).
26. Grossman, R. F., and Webber, C. S. U.S. Patent 3,342,625 (1967) (assigned to Norton Co.).
27. Grossman, R. F., and Webber, C. S. U.S. Patent 3,508,949 (1970) (assigned to Norton Co.).
28. McGarry, A., and Weidner, C. L. U.S. Patent 2,790,732 (1957) (Industrial Tape Corporation).
29. Mosher, R. H., and Davis, D. S. (eds.) *Industrial and Specialty Papers*, Vol. **III**, New York: Chemical Publishing Co., 1969.
30. Ritson, D. D., and Reynolds, W. F. U.S. Patent 3,394,799 (1968) (assigned to American Cyanamid Co.).
31. Cristmas, H. F. U.S. Patent 3,240,330 (1966) (assigned to Adhesive Tapes, Ltd., England).
32. Crone, J. W., Jr., and Pike, C. O. U.S. Patent 3,282,727 (1966) (assigned to Shuford Mills, Inc.).
33. Webber, C. S. U.S. Patent 2,822,290 (1958) (assigned to Norton Co.).
34. Iler, R. K. U.S. Patent 2,273,040 (1942) (assigned to E. I. DuPont de Nemours and Co., Inc.).
35. Martin, J. B., and Funk, C. S. U.S. Patent 2,803,557 (1957) (assigned to Crown Zellerbach Corp.).
36. Bradstreet, J. A. U.S. Patent 3,236,677 (1966) (assigned to Johnson and Johnson).
37. Estes, P. W. U.S. Patent 3,690,924 (1972) (assigned to W. R. Grace and Co.).
38. Nichols, L. D. U.S. Patent 3,846,404 (1974) (assigned to Moleculon Research Corporation).

# Chapter 18

# Silicone Release Coatings

**Mary D. Fey and John E. Wilson**
*Dow Corning Corporation*
*Midland, Michigan*

Without silicone release coatings, the tag and label segments of the pressure-sensitive adhesives industry as we know it today would not exist. In the United States, silicone release coatings are used to produce products having an annual market value of over one billion dollars. Worldwide, products employing silicone release coatings have an annual market value of over two billion dollars.[1] These products include product identification labels, price tags, care and warning labels, name tags, display banners, bumper stickers, all kinds of packaging tapes, self-adhesive wall coverings, floor and carpet tiles, and transfer tapes used in manufacturing assembly operations, to name just a few.

Silicones were first produced commercially during World War II to protect electrical systems on aircrafts. Silicones are based on organosilicon chemistry where both the carbon-silicon bond and the silicon-oxygen bond contribute to their unique properties. Some of these properties are high temperature stability, good electrical insulating properties, weather resistance, and low intermolecular forces which result in low surface tension, low density, low heats of fusion, high compressibility (high molar volumes), low freeze point, and smaller changes in viscosity with temperature than other materials.

One of the basic properties of silicones are their chemical inertness which results in low toxicity and inherent incompatibility with other materials. This incompatibility based on silicone's low polarity and low surface energy led to several of the first nonmilitary "release" applications for silicones after World War II. Silicones were used to help keep automobile tires from sticking in their molds and fresh loaves of bread from sticking to their baking pans.

Silicone release coatings for the pressure-sensitive industry are an extension of this early technology. The first silicone coating designed especially to provide stable, pressure-sensitive release was made commercially available in 1954.[2] Since then, new, improved products have kept pace with the industry's increasingly sophisticated requirements. During these 25 years, silicones have remained the industry standard for achieving controlled pressure-sensitive adhesive release, primarily in the tag and label industry.

Figure 18-1 shows the basic construction of most pressure-sensitive laminates used for labels. The release liner, coated with a very thin film of thermoset silicone polymer, permits the laminate to be rolled up or stacked flat. For many applications, without the liner, the adhesive would stick to anything it touched and be impossible to store, ship or handle. Sufficient adhesion exists between the adhesive and the silicone coating to hold the laminate together during handling. But when it is time to use it, the label and its adhesive peel easily from the protective, silicone-coated release sheet.

The major component of a silicone release coating is polydimethylsiloxane. Thermosetting this polymer with a crosslinker and a catalyst produces an irreversibly crosslinked film that resists penetration by the organic adhesive with which it is laminated. Without crosslinking, the adhesive molecules would readily penetrate through the thin silicone film because of the flexibility of the siloxane polymer chain.

## PHYSICAL PROPERTIES

One of several important physical properties contributing to the release characteristics of the polydimethylsiloxane polymer is its low surface energy, only 22 to 24 dyne/cm compared to a range of 30 to 50 dyne/cm for most organic adhesives.[3] This differential helps prevent the two from bonding tightly by preventing the adhesive from intimately wetting the silicone surface. The result is enough adhesion to keep the adhesive in place, yet not enough to prevent easy parting of the silicone-adhesive interface.

Being nonpolar, silicone coatings exhibit low attractive forces for other molecules. Also, the polydimethylsiloxane is incompatible with organic polymers, including those on which most pressure-sensitive adhesives are based. However, the polydimethylsiloxane polymer is compatible with aliphatic and aromatic solvents. This makes solvent-dilutable silicone coatings easy to apply from a coating bath. The polymer requires little energy to activate viscous flow, making it relatively easy to apply in very thin films.

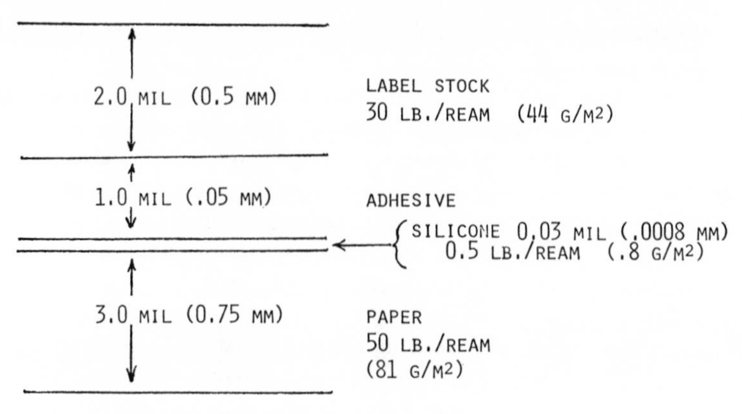


Fig. 18-1. Schematic cross section of a pressure-sensitive laminate.

This means that the penetration of silicone polymers into porous substrates is not as much a function of temperature as it is a function of viscosity, i.e., molecular weight.

Pressure-sensitive adhesive release is a very complex phenomenon. Table 18-1 lists the many factors that affect release characteristics.

**Table 18-1. Factors affecting release level.**

| Nature of the Adhesive | Laminate Characteristics |
|---|---|
| Chemical type | Paper age |
| Thickness | Laminate age |
| Modulus | Thickness and modulus |
| Diluents | Mode of adhesive application |
| **Nature of the Silicone Coating** | **Stripping Operation** |
| Chemical composition | Speed |
| Coating weight | Angle |
| Film continuity | Physical dimensions |
| Degree of cure | |
| Crosslink density | |
| **Substrate Variations** | |
| Roughness | |
| Porosity | |

One of the most important factors is crosslink density, the number of polymer crosslinks per unit volume in the cured film. Crosslink density is a function of the molecular weight of the polymer and the amount and structure of the crosslinker. This relationship provides the ability to vary the release characteristics of a silicone coating from very low, or easy release to very high, or tight release.

Figure 18-2 illustrates how the molecular weight of the polymer determines crosslink density and release performance. All other parameters being equal, the higher the molecular weight, the higher the force required to release a given adhesive, and vice versa.[4] Higher molecular weight polymers means longer polymer chains between crosslinks which means greater flexibility of the polymer chains. This greater flexibility allows more penetration of the adhesive polymer chains into the silicone than a more highly crosslinked silicone coating.

The ability to vary the release characteristics of a silicone release coating makes it possible to provide different release values on opposite sides of the same substrate. Controlled differential release represents the most soph-

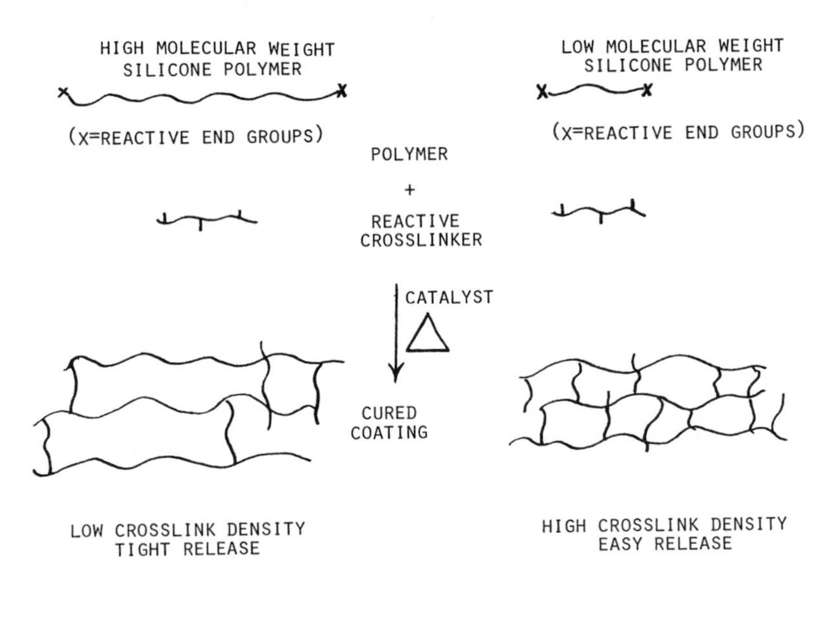


SUPPORTING RELEASE DATA

| SILICONE COATING | RELEASE |
|---|---|
| LOW CROSSLINK DENSITY (E.G., SYL-OFF® 23 PAPER COATING) | 125 G/IN (49 G/CM) |
| HIGH CROSSLINK DENSITY (E.G., SYL-OFF® 291 PAPER COATING) | 50 G/IN (20 G/CM) |

Fig. 18-2. Effect of crosslink density on release values.

isticated application of silicone release technology. It is critical to the proper functioning of many complex labeling systems and transfer tapes used throughout industry.

## CHEMISTRY

Thermoset silicone release coating systems consist of a reactive polymer, a crosslinker, and a catalyst. Two cure systems are available, one based on the condensation reaction and another on an addition reaction. The condensation reaction is the older of the two and has the most fully developed technology behind it.

The addition system was developed in Europe and Japan primarily for use with solvent-based coatings. Today, however, solventless[5,6,7] and emulsion[8] coatings can also be cured by the addition system.

Although more expensive than the condensation system, addition cure offers several advantages. It provides greater control over release characteristics. There is no post cure; once the coating is properly cured, there is little tendency to block. Differential release can be achieved using different polymers and additives, and, as there is no post cure reaction, release equalization is not a serious problem.

At the present time, the addition system provides less processing flexibility than the condensation reaction. However, the technology is relatively

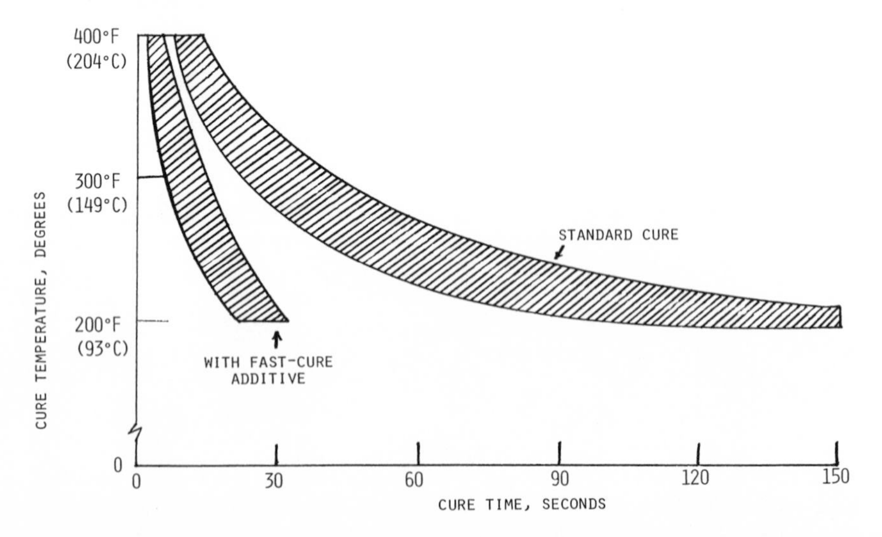


Fig. 18-3. Typical cure time of solvent-based (Syl-OFF 23–30%) silicone release coatings as a function of temperature.

new and increased processing latitude is one of the goals of continuing development work.

The greater processing latitude of the condensation system is the result of refinements developed throughout its 25-year history. The condensation reaction takes place between a silanol-functional polymer and a silane-functional crosslinker in the presence of a catalyst. Moisture must be present for the reaction to take place at the normal rate. Normal atmospheric humidity is usually sufficient, except in extremely dry climates or at very low temperatures.

The reaction will take place at room temperature, so the catalyst must remain separate from the polymer system until it is ready for use. Heat accelerates the reaction, and the silicone cures to a nonmigrating state within 20 to 30 sec at 120 to 150°C. (Fig. 18-3)

The most commonly used catalyst is dibutyltin diacetate, although other dialkyltin esters may be used. They are not true catalysts. The esters hydrolyze during the reaction, eventually ending up as a diorganotin oxide.

Silanol-functional polymer

Silane-functional crosslinker

Thermosetting coating

$$H-\left[O-\underset{CH_3}{\overset{CH_3}{Si}}-\right]_x OH + CH_3-\underset{CH_3}{\overset{CH_3}{Si}}-\left[O-\underset{H}{\overset{CH_3}{Si}}-\right]_y O-\underset{CH_3}{\overset{CH_3}{Si}}-CH_3 \xrightarrow[H_2O]{\text{Catalyst}}$$

$$\left[H-\left[O-\underset{CH_3}{\overset{CH_3}{Si}}\right]_x - \left[\begin{matrix} (CH_3)_3Si-O-\underset{|}{\overset{|}{Si}}(H)-CH_3 \\ O \\ O-\underset{|}{\overset{|}{Si}}-CH_3 \\ O \\ CH_3-Si-H \end{matrix}\right]_z - O-Si(CH_3)_3\right] + H_2\uparrow$$

$x \gg y$

$z < y$

Various additives are available to increase bath life and to control the reaction. In both solvent- and water-based emulsion systems, the condensation reaction gives greater flexibility of cure times and temperatures than the addition reaction.

In a roll of coated substrate in the presence of heat, moisture, and pressure, the reaction can reverse direction. This phenomenon is more pronounced in the presence of excess tin catalyst. Fortunately, the extent of the positive, forward reaction is always greater than the reverse reaction.

$$\equiv SiOH + HOSi\equiv \overset{Sn}{\rightleftharpoons} \equiv Si-O-Si\equiv + H_2O$$

In the case of one-side coated webs, the continuing reaction acts as a post cure for the coating. However, on two-side coated papers, the silicone can migrate across the interface between the two coatings after the paper is rolled up, especially if it is still hot. This can create adhesion between the two coatings and cause blocking. If the two coatings are designed to provide differential release, migration tends to equalize the differential.

In the United States, the addition reaction was utilized specifically to cure the solvent-free silicone release coatings introduced in 1975. Since that time, the system has been extended to encompass solvent and water-based emulsion systems as well.

In the addition reaction, a vinyl-functional polydimethylsiloxane reacts with a silane-functional crosslinker in the presence of a noble metal catalyst. In this case, the catalyst is a metal complex of platinum or rhodium. Used in small quantities of from 1 to 100 ppm, it promotes the addition reaction and as a result cure.

Addition reaction systems cure in 30 sec at 120 to 175°C, forming a three-dimensional polymer crosslinked by silicon-carbon-carbon bonds, in contrast to the silicon-oxygen-silicon bonds of those formed by the condensation reaction. Vinyl-functional polydimethylsiloxane polymer as shown below

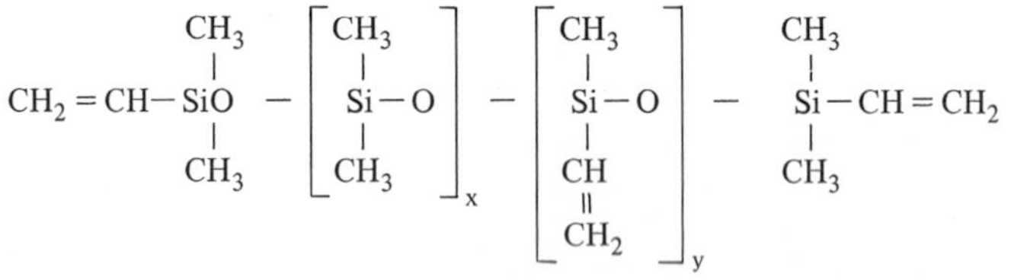

$$CH_2=CH-\underset{CH_3}{\overset{CH_3}{\underset{|}{\overset{|}{Si}}}}O - \left[\underset{CH_3}{\overset{CH_3}{\underset{|}{\overset{|}{Si}}}}-O\right]_x - \left[\underset{\underset{CH_2}{\overset{\|}{CH}}}{\overset{CH_3}{\underset{|}{\overset{|}{Si}}}}-O\right]_y - \underset{CH_3}{\overset{CH_3}{\underset{|}{\overset{|}{Si}}}}-CH=CH_2$$

reacts with silane-functional crosslinker in the presence of Group VIII noble metal catalyst

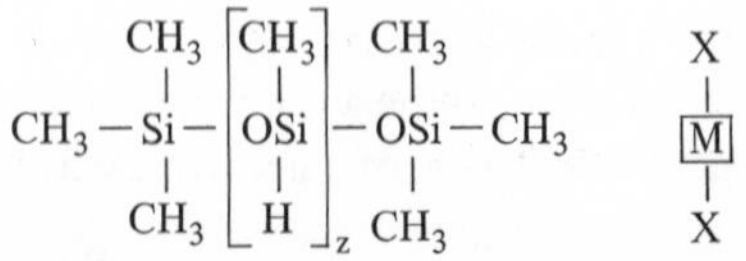

$$CH_3-\underset{CH_3}{\overset{CH_3}{\underset{|}{\overset{|}{Si}}}}-\left[\underset{H}{\overset{CH_3}{\underset{|}{\overset{|}{OSi}}}}\right]_z-\underset{CH_3}{\overset{CH_3}{\underset{|}{\overset{|}{OSi}}}}-CH_3 \qquad \underset{X}{\overset{X}{\underset{|}{\overset{|}{\boxed{M}}}}}$$

$$\equiv Si-CH=CH_2 + HSi\equiv \xrightarrow{\text{Catalyst}} \equiv Si-CH_2-CH_2-Si\equiv$$

## COATING TYPES

The three basic types of silicone release coatings used in the pressure-sensitive industry can be categorized as in Table 18-2 by the form in which they are applied: solvent-borne, water-borne emulsions, and solvent-free coatings applied at 100% solids.

Solvent systems are the most flexible, easiest to apply and account for approximately 65% of the pressure-sensitive release market in the United States and 70% globally. However, usage of solvent systems is declining because of air pollution regulations and the increasing cost of hydrocarbon solvents and energy.

Emulsion systems are as old as solvent systems, but the problems associated with applying water-based systems to paper substrates has limited their application to date to about 15% of the pressure-sensitive release market in the United states and 20% globally. For the same reasons that solvent systems are losing favor, the use of emulsion release coatings is growing.

Although introduced only five years ago, solventless systems today represent about 20% of the United States and 10% of the global pressure-sensitive markets. As might be expected, this market share is increasing globally for reasons of ecology and economy.

Table 18-2 summarizes the basic properties and characteristics of the three types of silicone release coatings.

## SOLVENT-BORNE RELEASE COATINGS

Solvent-based silicone release coatings can be formulated to produce a wide range of release characteristics. In addition, a variety of additives are available to ease processing by extended bath life and increasing the rate of cure.

**Table 18-2. Comparison of the properties of the three release coating systems.**

| | CURE RATE | RELEASE LEVEL | COMMENTS |
|---|---|---|---|
| Solvent-based Coatings | Ranges from fast to slow | Ranges from easy to tight | Most versatile curing Temperatures. Dull to glossy coatings. |
| Emulsion Coatings | Ranges from medium to slow | Ranges from easy to medium | Some coatings require post cure. |
| Solventless Coatings (100% solids) | Ranges from fast to medium | Ranges from easy to tight | Glossy coatings. Two products blendable to achieve a range of release values. |

**Table 18-3. Summary of silicone pressure-sensitive release coating application and performance properties.**

| TYPE OF COATING | SOLVENT-BASED | WATER-BASED EMULSION | SOLVENTLESS |
|---|---|---|---|
| Primary substrates | S2S kraft,<br>Clay-coated kraft,<br>PE coated kraft | Machine finished (MF) kraft | Two-side size Super calendered (S2S) kraft |
| Primary methods of application | Direct gravure,<br>Wire-wound rod | Wire-wound rod | Offset gravure |
| Application solids (%) | 2–7 | 5–10 | up to 100 |
| Typical cure cycle | | | |
| Temp (°C) | 80–150 | 120–150 | 120–175 |
| Time (sec) | 20–30 | 30–40 | 20–30 |
| Additives | Bath-life extender<br>Fast cure<br>Adhesion aid | Hold-out agent<br>Wetting agent<br>Defoamer | None |
| Processing latitude | Wide | Narrow | Medium |
| Typical release values (g/cm) | | | |
| SBR adhesive | 10–63 | 10–18 | 8–63 |
| Acrylic adhesive | 10–47 | 12–20 | 12–69 |

A typical solvent-based silicone release system will consist of:

1. Silicone polymer and crosslinker
2. Fast cure additive
3. Bath life extender
4. Catalyst
5. Solvent

Both high and low molecular weight silicone polymers are used in solvent systems. All other conditions being equal, a high molecular weight, long chain polymer will produce a coating with low crosslink density and high release values. The lower the molecular weight, the higher the crosslink density and the easier the release. As an example of the range of release possible with solvent systems, release of a typical commercial SBR adhesive may vary from 10 to 63 g/cm. Values from 10 to 47 g/cm may be achieved with a typical commercial acrylic adhesive. When choosing a solvent-based release coating, the initial polymer selection is based on the desired release level.

## Additives

The most widely used additives are bath life extenders, as the moisture-sensitive coating can gel in the bath. Typical bath life extenders include methyl ethyl ketone, which also reduces static buildup at the coating head during coating, isopropyl alcohol, acetic acid, and an organofunctional silane. The latter also promotes adhesion of the coating to the substrate.[9]

Although not absolutely necessary, most users employ fast cure additives. These additives permit faster coating speeds at a given temperature, or, conversely, lower cure temperatures at a given speed. They may also improve the ability to coat polyethylene-coated paper and film, and some fast cure additives improve adhesion of the coating to the substrate.

## Catalysts

Catalyst selection also affects bath life and cure time. In general, catalysts which increase cure rates decrease bath life, because of their greater activity.

## Solvents

Silicone polymers are compatible with both aliphatic and aromatic solvents. The latter, such as toluene, provide the longest bath life. Aliphatic solvents, such as heptane, improve wetting. Usually, a blend of aliphatic and aromatic solvents is used. To obtain good wet-out with a low molecular weight silicone on a hard-to-wet substrate such as polyethylene, the coating bath must contain a high concentration of aliphatic solvent. When using a high level of aliphatic solvent, the addition of methyl ethyl ketone in the solvent blend to reduce static buildup at the coating head is important. Solvent-based systems are usually applied at only 5% silicone solids, so substrates must have good solvent hold out.

## Substrates

Solvent-borne silicone release coatings are applied to a variety of substrates. Major usage is on supercalendered, two-side sized (S2S) semibleached kraft, clay-coated kraft, and polyethylene-coated kraft. Other substrates being coated with solvent systems include machine finished (MF) kraft, glassine, kraft-glassine, foil, and polytheylene, polyester, and polypropylene film.

Air pollution regulations and the cost of hydrocarbon solvents and energy are causing a reduction in solvent system use. Because the solvent concentration in the exhaust must be kept at a low level, energy requirements are high. It is possible to recover solvent by adsorption on activated carbon, but the effect of the silicone and additives on the adsorbent should be investigated. Generally, solvent recovery or incineration requires more energy than the coating operation itself.

## Water-Borne Systems

Historically, aqueous emulsion-type silicone release coatings have been used primarily for non-pressure-sensitive applications. Typical applications include releasing sticky products such as rubber or asphalt from containers and as release sheets for processing sticky foods such as baked goods and candy. However, the development of faster, more stable cure systems and more shear stable water-based systems for release coatings has led to some limited use in pressure-sensitive products. Some tapes, pressure-sensitive floor tiles and vinyl wall coverings have emulsion coated release liners.

Emulsion-based release coatings are limited to low or medium molecular weight silicone polymers and are manufactured by mechanical emulsification or emulsion polymerization techniques. As a result, the polymers, the catalysts and the range of release values attainable are more limited than with solvent-based systems. Also, there is less latitude in terms of cure time and temperature. Typically, emulsion systems are cured for 30 to 40 sec at 120 to 150°C. Bath life extenders and fast cure additives are not available for emulsion coatings.

The major problem in applying water-based systems is to prevent saturation of the substrate with water. Water swells paper fibers, destroying the release liner's dimensional stability. This cannot be tolerated with die cut label stocks. One method of controlling this problem is with hold-out agents, such as sodium carboxymethylcellulose, sodium alginate and other water-soluble organic compounds. These compounds adhere well to both the silicone polymer and the paper and have a higher specific gravity than the coating. They fill the pores of the paper to prevent water from penetrating and form a thin layer between the coating and the substrate.

Although normally applied at 5% solids, emulsion coatings can be applied from baths containing as much as 20 to 40% solids by gravure or trailing blade coaters. This reduces cure times because less water must be evaporated and reduces dimensional stability problems with the paper.

Water-borne silicone emulsion release coatings are used on a limited number of substrates. By far the most common substrate for pressure-sensitive applications is machine finished (MF) kraft. Bag kraft and parchment are coated for non-pressure-sensitive applications. Emulsion coatings are also used to a limited extent on glassine, kraft-glassine, and polyethylene-coated kraft. Surface wetting problems have to be addressed for application on films.

The technology of emulsion-type silicone release coatings continues to evolve. Environmental and energy considerations are giving impetus to further development, and the use of water-based silicone release coatings for pressure-sensitive applications can be expected to increase in the future. Up to 95% of oven air used to cure solvent-free silicone emulsion coatings can be recirculated. This provides a definite economic advantage over solvent-based systems.

## Solventless Silicone Coatings

The most recent development in silicone release coating technology is a solventless coating which is applied at 100% solids. Introduced in 1975,[1] these coatings have, in only four years, gained approximately 20% of the U. S. pressure-sensitive release market. Solventless silicone release coatings were designed specifically, and are used almost entirely, to provide pressure-sensitive release.

Solventless silicone release coatings consist of a vinyl-functional polydimethylsiloxane polymer, a silane-functional polymethyl hydrogen siloxane crosslinker, and a noble metal catalyst. The catalyst is incorporated with the polymer. Addition of the crosslinker activates the cure mechanism.

Polymer, crosslinker, and catalyst are matched to achieve optimum cure times. The greater the heat input, the faster the cure. The system is designed to give a minimum bath life of 8 hr at temperatures below 40°C, yet provide short cure times. At the present time, there are no bath life extenders or fast cure additives for solventless coatings.

The major initial problem with solventless release coating was to apply thin films on the order of 0.03 to 0.04 mils (0.0008 to 0.001 mm) thick. This translates to approximately 0.5 lb of silicone per 3000 $ft^2$, or 0.8 $g/m^2$ of paper. Low molecular weight polymers in solventless coatings are of low viscosity that allows low coating weights to be achieved.

As previously stated, low molecular weight polymers give low release with most adhesives. Adding a silicone resin to the low molecular weight polymer produces higher release values. Intermediate release is obtained by blending the two polymer systems, the low-release polymer and a high-release silicone resin. In this manner, solventless coatings can provide virtually the entire range of release values given by solvent release coatings.

For example, an unmodified solventless silicone release coating gives release values of approximately 20 g/in. with a typical SBR adhesive and 37 g/in. with a typical acrylic adhesive. Blending three parts of high-release silicone resin with the basic solventless coating polymer gives a release value of about 60 g/in. with the SBR adhesive and 140 g/in. with the acrylic adhesive. Figure 18-4 shows release values for various blends with typical commercial adhesives.

Blending the two coatings affects cure parameters. As shown in Fig. 18-5, the basic unmodified coating cures in 20 sec at 143°C.

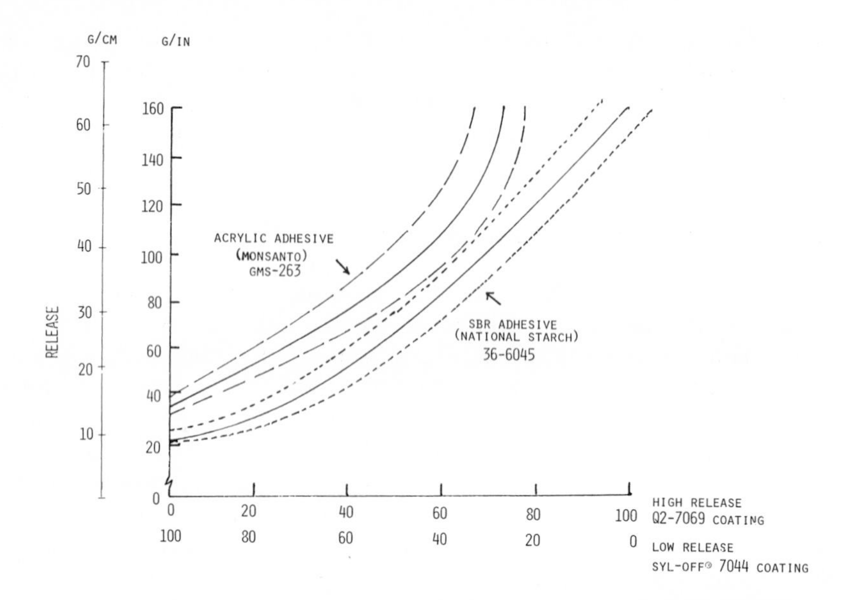


Fig. 18-4. Release as a function of solventless silicone release coating composition.

A tight-release coating containing 100% of the high-release silicone resin requires a temperature of 175°C for 20 sec cure. Cure parameters for blends of the two coatings fall between these extremes.

Solventless silicone release coatings require modification of standard coating equipment. Usually only the coating head need be altered. However, the coatings can be blended with aliphatic solvent for application as a high solids solvent system with standard equipment. This alternative can significantly reduce solvent emissions with little or no equipment modification.

The lack of solvents makes solventless release coatings economically, as well as environmentally, beneficial. Oven air can be recirculated, drastically reducing process energy costs. Compared to solvent systems requiring solvent recovery and/or incineration, solvent-free silicone release coatings can usually be applied at lower cost.

Other advantages of solventless coatings include more reliable process control. There is little tendency for the coating to post cure on the roll or block in two-side coating operations.

Solventless coatings have their limitations. The noble metal catalysts are easily inhibited by strongly electron-donating atoms such as sulfur, nitrogen, or phosphorus in the form of mercaptans, amines or phosphines. Such materials should not be introduced into the coating or the oven air.

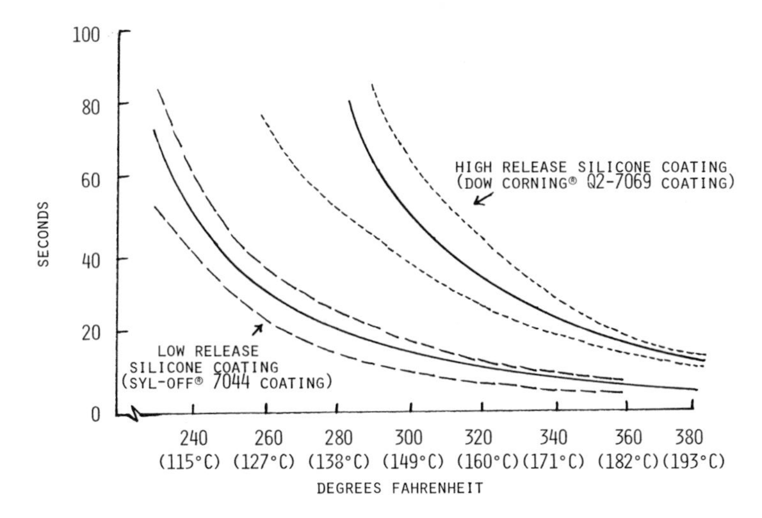

Fig. 18-5. Cure time of solventless release coatings as a function of temperature and composition.

Solventless release coatings are applied almost entirely to two-side sized, supercalendered (S2S) kraft substrates. The coating has a tendency to soak into clay-coated krafts. However, it is being used to some extent to coat better quality clay-coated stocks.

Although it is difficult to obtain adhesion and cure of solventless silicone coating to films, some films are being coated with the material. Solventless silicone coatings are used on few other substrates, either because of substrate porosity, or because the coating does not cure on the substrate or because the substrate cannot withstand the cure temperature.

## COATING APPLICATION

Silicone release coatings are applied by a converter. The converter may be a paper company which also manufactures the base substrate or it may be a company that specializes in converting paper or film manufactured by others. The converter may or may not then go on to produce finished labels or tape.

There are two basic types of converting operations, off-line and in-line. In an off-line operation, only the silicone coating is applied (Fig. 18-6). Adhesive and label stock are mated to the coated release roll in a subsequent operation.

In-line converting equipment has two or three coating heads and a laminating station (Fig. 18-7). The entire pressure-sensitive laminate is made

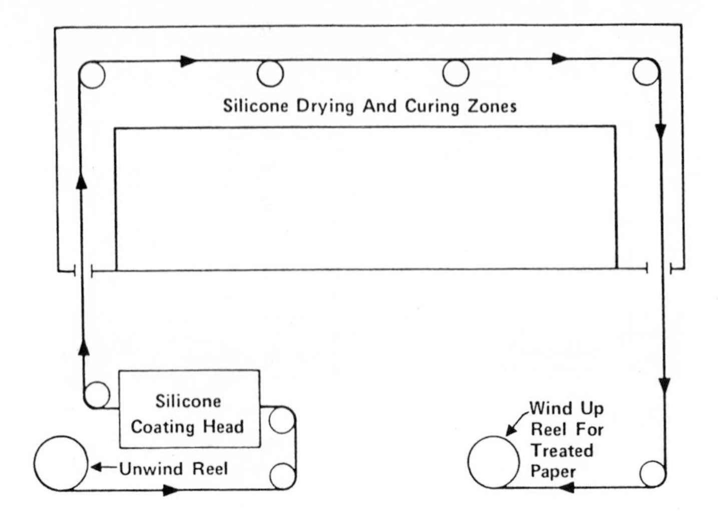


Fig. 18-6. Off-line silicone coating.

in one continuous operation. First, the silicone release coating is applied and cured. Then the adhesive is applied to the silicone coated stock and dried. Finally, the label stock is laminated to the adhesive.

Line speeds for off-line converting are usually faster than in-line. Speeds of in-line operations are usually limited by the adhesive drying rate. The increased use of hot melt adhesives may change this equation, with the silicone cure rate becoming the limiting factor. Line speeds may vary from 31.5 to 315 m/min. Most are in the range from 95 to 190 m/min.

The adhesive is usually transfer coated. That is, the adhesive is applied to the silicone coated sheet and transferred to the label stock in the final laminating step. In this way, a more efficient use of the adhesive can be realized. Transfer coating also simplifies the design of in-line machines.

Die cutting labels and rewinding tape is a separate operation. It should be pointed out that die cut labels require a dimensionally stable release sheet so that the label can be cut cleanly from its matrix without penetrating the release sheet. Release values must also be closely controlled for die cutting. If release is too tight, the matrix may break when it is stripped away. If release is too low, the label may come off with the matrix.

Silicone release coatings are applied by several types of coating heads. The three most commonly used are direct gravure, wire-wound rod, and offset gravure. Other coaters used include reverse roll, air-knife, trailing blade, and flexographic gravure.

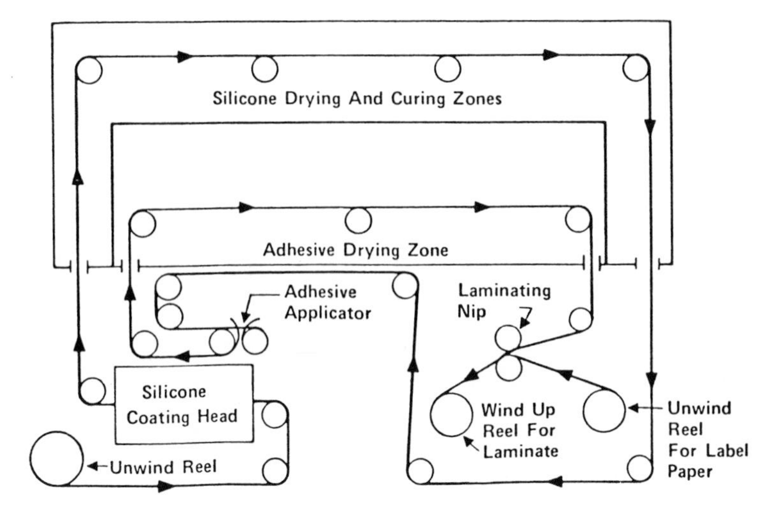


Fig. 18-7. In-line silicone/adhesive coating and lamination.

The majority of solvent-based silicone coatings are applied by direct gravure. The gravure rolls used usually have a quadrangular or pyramidal pattern. The amount of coating applied depends upon pattern cell size and shape. A typical 3% solids coating bath might use a 25Q (25 cells per inch) cylinder, and a 5% bath a 45Q cylinder, to deposit 0.5 lb of coating solids per 3000 $ft^2$ ream (0.8 $g/m^2$). These figures are given only as a guide. Actual results depend upon a number of variables, such as line speed, bath viscosity, substrate finish, and tension.

Although most solvent-based silicone release coatings are applied at less than 7% solids, some success has been achieved with 10% solids coatings using very finely etched gravure rolls and in-line metering.

Some solvent-based silicone coatings and nearly all emulsion coatings for pressure-sensitive applications are coated on a Mayer, or wire-wound, rod coater. Versatile and easy to maintain, these coaters work well with low viscosity coatings.

Rod coating is a low shear process and thus is ideal for shear-sensitive emulsion coatings. Excessive shear forces can cause emulsion coatings to gel in the bath and clog the rod.

Virtually all solventless silicone release coatings are applied by three- or four-roll offset gravure or smooth roll coaters. Attention must be paid to line speed, paper tension, roll pressure, and the condition of the coater frame and bearings to achieve optimum results. With proper care, solventless coatings can be applied at the same weight as solvent systems.

Reverse roll coaters may be used to apply both solvent and aqueous systems. This type of coater handles a wide range of viscosities, and coating thickness is independent of both viscosity and speed. Air-knife coaters are recommended only for applying emulsion coatings. They offer many adjustments for controlling coating weight, and can apply a uniform thickness on rough substrates. Trailing blade coaters, are used to a limited extent to apply emulsion coatings.

In some cases, silicone release coatings are not applied as a continuous film, but are laid down in a pattern. This is done by flexographic gravure coating. The etched roll transfers the coating to a patterned roll which applies the coating to the substrate.

The coating is usually cured in a forced-air oven. Oven performance influences the rate of cure and selection of the catalyst for condensation cure systems.

Jets direct hot air at the coated side of the web, and sometimes at the back as well. Lineally directed air flow at the jets is frequently a limiting factor. Ovens with air velocity above 230 m/min are preferred although they may prohibit at oven temperatures above 150°C the use of volatile, low molecular weight catalysts, e.g. dibutyltin diacetate.

Infrared heating may be used to supplement oven curing. Heaters are usually located after the oven tunnel.

## COATING EVALUATION

The methods of evaluating silicone release coatings can be classified into two broad categories:

1. Quick tests are used on the coating line to check coating quality during the run, to determine the end of the useful life of the coating bath, and to check the degree of cure.
2. Control tests quantify release values, subsequent adhesion and coating weight.

### Quick Tests

As silicone coatings are almost invisible on many substrates, it is difficult to judge coating uniformity and state of cure by casual observation. The following quick tests serve these purposes.

Coating adequacy and uniformity can be judged by applying lengths of cellophane tape to the coated surface, rubbing it with the finger and then pulling it off. The amount of force required to lift the tape is a quick indication of coating quality.

When the tape has been removed, it is doubled back on itself so the sticky sides just touch. The tape is then pulled apart. Any loss of tack, as compared to that of a fresh piece of tape, indicates that silicone has transferred to the tape, and may be evidence of incomplete cure. With experience, an operator can grade tack loss as "none," "very slight," "slight," "some, and "gross."

Dye tests are used to check the state of cure and film uniformity of release coatings. They also form a permanent record of a particular run. Coated test strips are smeared with a dye solution and wiped dry with absorbent paper. One frequently used dye solution is 1% Rhodamine 6 GDN extra (DuPont) in diacetone alcohol.

The quality of the coating is judged by the amount of color retained on the test strip and the amount striking through to the substrate. Very little color is retained by a good silicone coating. The dye test is less effective for testing coatings containing water-soluble extenders and coatings applied to nonstaining substrates, such as polyethylene-coated paper.

Simply making a single firm stroke across the coating with a finger is an effective cure test. Fully cured coatings will show little or no mark. Poorly cured coatings will show a greasy streak. This and the rub test are very useful for a fast in-line evaluation of silicone coatings.

Sometimes a coating passes the smear test indicating full cure, but has not yet developed full adhesion to the substrate. This can be detected by rubbing the coating with the fingers. If the adhesion is poor, the coating will ball up as it is rubbed. The degree of ruboff is usually reported as "none," "some," or "gross."

## Control Tests

Standard tests exist for measuring release force. These tests are described in TAPPI useful method 502. "Quality of Release Coatings," TAPPI useful method 504, "Quick Adhesion Test for Release Papers" and Pressure-sensitive Tape Council Method PSTC-4, "Adhesion to Liner of Pressure-Sensitive Tapes at 180° Angle." Several test methods are described in the "Manual of Standard Specifications" by the Label Division of the Tag and Label Manufacturers Institute.

Although these tests are useful, many converters today have adapted them to more nearly approximate actual end-use applications. The essential ingredients of a good test of release force are a standard pressure-sensitive adhesive tape, a means of applying the tape uniformly to the release coating and a method of determining the force required to peel the tape from the release coating at a controlled angle and rate.

Test instruments commonly used for peel testing include the Keil Tester, which operates at a 180° peel angle and 12 in./min. (30.48 cm/min) the Scott Horizontal Tester and TLMI Tester, each of which provides peel angles adjustable from 90° to 180° at speeds variable to 400 in./min (10.2 m/min). Release force is usually expressed in grams per inch of tape width.

Subsequent adhesion is a measure of the adhesion of a pressure-sensitive adhesive after it has been in contact with a release coating. The test is used to check for any migration of silicone onto the adhesive. Properly applied and cured silicone release coatings should be stable under normal storage conditions for from six months to two years or longer.

Requirements for testing subsequent adhesion are similar to those used for testing release force: a standard tape, a means of applying the tape uniformly to the release coating, a standard surface to which the tape is applied after it is removed from the release coating, and an instrument for measuring peel force. The same testers used for measuring release force are used to test subsequent adhesion and values are reported in ounces per inch or grams per centimeter.

Release tests for high-speed label stock are usually designed to duplicate the actual stripping conditions as closely as possible.

There are several methods for determining the weight of silicone coating solids deposited on the substrate.

**Material Balance.** One of the most commonly used methods is by material balance. The amount of silicone used is divided by the amount (area) of paper coated. Errors are introduced by uncoated paper, waste paper, and unused bath unless accounted for in the material balance.

**Actual Pick-Up Weight Method.** An untreated paper is subjected to the same cure conditions as a coated paper, and its weight per unit area is determined. The difference in weight from that of the same unit area of coated paper is the weight of the applied coating. The error in this method can be quite large (±100%) due to variations in the basis weight.

**Aluminum Foil Pick-Up Weight Method.** Pass a piece of aluminum foil of known weight through the coating system. Determine its unit weight after processing. Any increase is the weight of the coating applied. There can be ±50% error in this method due to differences in transfer to aluminum foil versus paper.

## SUMMARY

As the cost of petroleum-based solvents increases and the cost of energy increase, the trend in the silicone coating industry will be change to solvent-free systems and lower energy or more efficient cure systems. Solventless (100% solids) silicone and emulsion systems will be the systems of choice. New cure technology will allow the systems of the future to be more energy-efficient.

## REFERENCES

1. Lewis, P. W. "Commercial Importance of Solventless Silicone Coatings," Solventless Silicone Coatings Short Course, Technical Association of the Pulp and Paper Industry (November 1978).
2. Dennett, F. L., U. S. Patent 2,588,367 (1952) (assigned to Dow Corning Corporation).
3. Gordon, D. J., and Colquhoun, J. A. *Adhesives Age*, **19** (6): 21–30, (June 1976).
4. Bey, A. E., *Adhesives Age*, **15** (10): 19–32 (October 1972).
5. Chandra and Rowland, U. S. Patent 3,928,629 (1975) (assigned to Dow Corning, Limited).
6. Chandra and Rowland, U. S. Patent 3,960,810 (1976) (assigned to Dow Corning, Limited).
7. Grenoble, M. E. U. S. Patent 4,162,356 (1979) (assigned to General Electric Company).
8. Grenoble, M. E. U. S. Patent 3,900,617 (1975) (assigned to General Electric Company).
9. Ault and Salisbury, U. S. Patent 3,076,726 (1963) (assigned to Dow Corning Corporation).

## ADDITIONAL READING

Campbell, J. K., and Weber, C. D., "Silicone Coatings: How They Add Release Properties to Substrates," *Package Engineering.* **12** (9): (September 1967).

Noll, Walter. *Chemistry and Technology of Silicones*, 2nd Ed. New York: Academic Press (1968).

# Chapter 19

# Saturated Paper and Saturated Paper Tapes

**Charles Bartell**
*Mystik Corporation*
*Northfield, Illinois*

Raw paper is saturated with many different types of rubber-like materials to improve its physical properties such as tensile strength, elongation, tear strength, bursting strength, grease and water resistance, internal strength, porosity, and many other properties. These saturated papers are used for many different types of coating bases such as synthetic leathers, gaskets, and tapes. Use of saturated paper has been reviewed by J. F. Hechtman and J. E. Jayne.[1] Saturated papers for pressure-sensitive adhesive tapes and especially for general-purpose creped masking tape will be covered in this chapter.

## PHYSICAL PROPERTIES

The properties of saturated paper, also called impregnated paper (these two terms will be used interchangeably), are the result of the combining saturating grade raw paper with the saturant. The physical properties which are affected are tensile strength, wet tensile strength, elongation, tear resistance, delamination resistance, stiffness, porosity, sensitivity to solvents, and color. These properties will be discussed below.

### Tensile Strength

Tensile strength is usually defined as the strength of a 2.5 cm (1 in.) wide sample about 10 cm long pulled apart on a pendulum tensile tester at

30 cm/min (12 in./min). The tensile strength is also tested in strain gauge type tensile testers at speeds of 30 cm/min. Actually, the test speed of a pendulum type tester is lower, perhaps about 5 cm/min. The pendulum type tester is actually slower than the set speed, because of variable jaw separation during testing, whereas the strain gauge tensile tester operates at a constant speed.

For creped masking tapes, a tensile strength range of 3–4.5 kg/cm width (18–25 lb/in.) is used, with a more common range of 3.5–4 kg/cm (20–22 lb/in.). The tensile strength values vary with the supplier, depending on the raw paper source and saturation formula. Generally, the tensile strength is not critical, if it is within the range stated.

### Wet Tensile Strength

It is measured the same way as the dry tensile strength. It is not important for most applications, but it might be important for some packaging uses where the item might get wet. In general, 50% retention of the dry tensile strength is considered to be good and 75% is excellent. The wet tensile strength is also indirectly important in that it can reflect the degree of cure of impregnant and can be useful as a control test.

### Elongation

It is measured at the same time one measures the tensile strength. It depends on the amount of crepe in the raw paper, type of raw paper and type of saturant. Elongation is especially important in those paint masking applications where the tape is applied around curved surfaces. A tape of low elongation may tear at the edge and yield a jagged paint edge, whereas one of higher elongation will yield a smooth edge. Elongation below 6% is considered too low, 8% is acceptable and 10–12% very good.

### Tear

The tear resistance can be measured by a number of methods: Elmendorf, edge tear, trapezoid, Finch, etc. All of these indicate the resistance the tape offers to tearing. The most critical operation requiring a good tear strength is when the tape is pulled around curved surfaces. The tear resistance tends to be a function of tensile strength and especially of elongation.

### Delamination Resistance

This very important test was developed by Bartell and Goun and is described in detail in Dunlap's paper.[2]

The test sample is prepared by placing a strip of impregnated paper to be tested between two strips of rug binding tape (Bondex BT7). The sandwich is put in a press and bonded by applying a pressure of 0.4 kg/cm$^2$ for 30 sec at 135°C. The bonded laminate is cut into strips 2.5 cm wide. The ends are then separated by hand until the paper splits. Then the ends are placed in a tensile machine and the force measured to continue the paper splitting. The rate of jaw separation is 25 cm/min.

If one considers the three dimensions of an impregnated paper sheet, the surface area of the paper would be the XY plane with Y in the machine direction and X in the cross direction. The direction perpendicular to this plane would be the Z direction. The delamination test measures the strength in the Z direction.

This property is a good indication of the resistance of paper to delamination under stresses of high unwind force and under stresses of peeling from surfaces to which the tape is applied. The delamination value can vary with the tapes from various suppliers and the proper backing is chosen depending on the stresses the tape is expected to encounter, both internally, from adhesive-release coating combinations, and externally from the forces on tape removal.

The delamination resistance of most creped masking tapes varies depending on the level and type of impregnation. Some such studies are presented in Goodrich's Latex Manual[3] and in a paper by Finnegan and Miller.[4] In general, the higher the rubber pickup, the higher the delamination resistance and the higher the cost.

There is another test, the brass block test, that measures resistance to delamination in the Z direction, but it is generally used for thick coating bases intended primarily for synthetic leather. Two brass blocks 2.5 × 2.5 × 1 cm thick are bonded to the opposite sides of the backing by a hot melt adhesive. The sample block is put into a tensile machine and the brass blocks are pulled apart. The force is measured in kg/cm$^2$.

## Stiffness

This test is not very often used. It is of value along with the elongation in determining the conformability of tape. Also, stiff tapes tend to flag and tear more easily. There is a number of ways of measuring stiffness. One often used in the industry is the Gurley stiffness test.

## Porosity

The most important value of this test is in measuring the resistance the paper offers to subsequent release, or to coating penetration into the paper. It indicates the ability of the adhesive coating to penetrate into or through impregnated paper. If the porosity is too high, the adhesive and release coatings can penetrate through the backing. If the adhesive penetrates the backing during the processing, the opposite side of the tape becomes sticky, can stick to the rollers and cause web breaks. This property can be measured by Gurley's Densometer.

### Sensitivity to Solvents

This property is not usually measured but could be of concern in masking applications. During painting of masked surfaces, the solvent in paint may solubilize the impregnant by soaking in at the edges. Usually, a good release coating will help to prevent such solvent attack if impregnant is of the type affected by the solvent.

### Color

The color of creped masking tapes is usually natural. It can vary from a very light, almost white, for bleached papers, to a dark tan. The color depends largely on the color of raw paper: bleached paper being white, semibleached being light tan, and unbleached being darker tan. The impregnant can also influence the shade. If a darker raw paper is used and a lighter shade of the tape is desired, some $TiO_2$ may be incorporated into the impregnant formula.

### Interaction between Adhesive and Saturant

After coating the impregnated paper with pressure-sensitive adhesive, sometimes a lowering of tack and adhesion on aging is observed. In most cases, this is due to the resin migrating from adhesive into the impregnant. There are several ways to avoid or minimize this phenomenon. One simple way is to use an impregnant into which the resins will not migrate, such as a butadiene acrylonitrile copolymer, although such impregnants tend to be more expensive. Another way is to use a primer barrier coat. A third way is to design the adhesive compensating for some loss of resin. A fourth way is to include resin in the impregnant as described by Hechtman and Bonnslaeger.[5] In any case, this migration must be minimized or eliminated. Otherwise, the tape will fail in the field due to drop-off of adhesion, tack, and holding power.

## SATURATION

Most impregnated papers are made by dry saturation or off-machine saturation. In this process, saturation is carried out in a separate operation.

Many tape companies saturate their own paper. There are also some companies that specialize in paper saturation and sell the product to tape manufacturers. Figure 19-1 shows the equipment generally used for dry saturation.

Another process of making impregnated paper is saturation on machine. In this setup, impregnation takes place on the paper machine in line with the paper making operation. The paper is formed, dewatered, partially dried, impregnated with latex saturant, and then completely dried. This operation combines two operations into one and is less costly.

A third process that is not used very often for tape backings is saturation by beater addition of saturating ingredients. In this system, the latex saturant is added to the paper slurry (1–5% paper fibers in water) and the latex is precipitated onto the paper fibers. Then the paper-latex slurry is formed into a sheet on the paper machine. Thicker sheets used for book covers and synthetic leather may use this system and have certain advantages. But the process is rarely used for thin pressure-sensitive tape backings.

## RAW PAPER

The physical properties of impregnated sheet are largely determined by the raw paper. The impregnant will improve upon the properties of the paper, but the degree of improvement is determined by the properties of raw paper. Creped raw paper is made by a small number of paper companies. Under a microscope, the paper looks like a bundle of flattened straws with relatively high amount of void space. It is believed that adhesion between fibers is primarily due to hydrogen bonds. Raw paper is a weak structure somewhere in the area of 1.3–1.8 kg/cm in tensile strength. On saturation, the rubberlike material fills the voids and much of it tends to clump around fiber-to-fiber contact points, gluing them together. It also forms a network of rubber between fibers. This increases the tensile strength to 3.2–4.4 kg/cm as well as improving other physical properties, especially the delamination

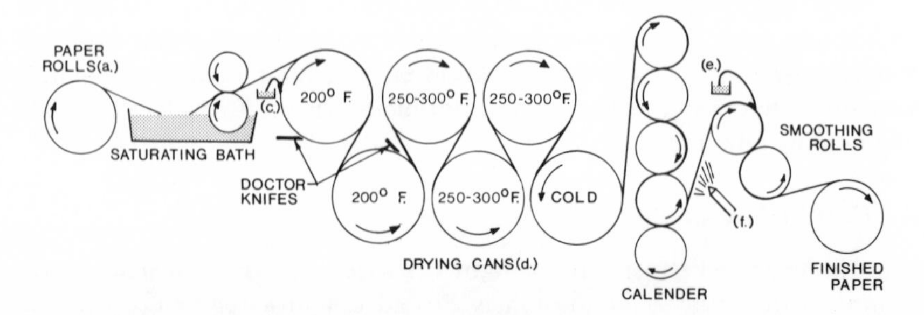


Fig. 19-1 A schematic diagram of a typical paper saturating equipment.

resistance. To improve the properties at a minimum cost is a difficult task, and the properties of the raw paper become important.

On continued refining of the pulp, the straw like fibers making up the paper are split and hairlike fibrils extend from them. These increase the number of contact points between fibers and increase the tensile strength. Simultaneously, the sheet becomes denser with less void space; it becomes stiffer and more difficult to saturate. The paper also becomes less stretchy and tears easier. Thus, there is an optimum amount of refining necessary to give a sheet of optimum properties which are well-balanced. The effect of different pulps and treatment is discussed by Page *et al.*[6] and by Dunlap.[2] The correct pulp or pulp blend must be chosen. Relatively long, softwood pulp fiber is preferred in order to obtain the properties discussed: sufficient void space for saturation and resulting good elongation and tear properties. The fiber length of hardwood pulp is too short and does not give the best properties.

The paper discussed above is primarily used for masking tapes. Usually it is a semibleached color—light tan. Besides creped papers, flat, and special papers are also used. These can be bleached white for printable tapes and other products, semibleached or unbleached, depending on the end-use. The paper may be considerably refined to give a relatively dense stiff sheet, or it may be lightly refined to give a less dense, soft, and stretchy sheet.

Usually, the paper is oriented so that the strength ratio in machine/cross direction is approximately 2/1–2/1.5. There is a type of paper highly oriented in the machine direction to give a higher tensile strength. A 30 lb/ream paper normally has 3.2–4.4 kg/cm. If made with special fibers (hemp or high tensile kraft) and highly oriented, such paper could have a tensile strength of 8–9.2 kg/cm. In certain papers, the fibers before paper formation are mechanically treated or treated with caustic, to impart a greater curl to individual fibers giving a sheet with a greater stretch. There are also other special papers treated in various ways to obtain specific properties. Heavier papers, 33 lb/ream and up, give stronger sheets and are used for special applications.

Cost is a very important factor in such a competitive product like creped masking tapes, and it is important to be able to make the least expensive sheet that still meets the requirements. The usual creped paper used to be 30 lb/ream weight. It has come down in the past 25 years to 28 lb and 27 lb paper as a standard, with even a lower weight used occasionally.

## SATURATED PAPERS

Two main types of saturants are used: solvent-based and water-based latexes. The solvent-based saturants were the earliest developed as illustrated in a series of patents.[7,8,9] Water-based latexes were developed later and are by far the most extensively used today.

### Solvent-based Impregnants

Solvent-based saturants coat all fibers uniformly. This, together with the fact that there are no water-sensitive materials in the saturant, as there might be in water-based latexes, tends to give a sheet of better water resistance. However, the sheet tends to be weaker, as compared to a saturated one by water-based latex. Also the number of impregnating polymers that can be used is limited. Natural rubber and SBR are the preferred polymers. Pollution problems and the high price of solvents are the other reasons favoring water-based impregnants. However, the solvent impregnated papers are necessary for electrical paper tapes.

### Water-based Impregnants

Latexes are by far most commonly used by tape manufacturers as well as by independent saturators for tape backings. There is a wide variety of latexes available which are used alone or blended. Latexes are much simpler to use as compared to solvent-based impregnants which require breaking down, compounding and dissolving the rubber.

Some of the synthetic latexes used are enumerated below.

**SBR.** It improves the overall properties of paper and is most widely used. A significant advantage is its low price. A disadvantage is that the adhesive used on such backing may lose tack and adhesion on aging due to migration of resins from adhesive into SBR impregnant.

**NBR.** It was widely used from 1940–1960 because it produced excellent backing properties and stable adhesive properties: no loss of adhesion and tack due to migration of resins into NBR on aging. Its disadvantage is its high price.

**Acrylics.** Like NBR, these latexes produce excellent properties and give stable adhesives. Furthermore, they have a light color and are heat and light stable. They were expensive and therefore not widely used during the development of masking tapes (1940–1060). Later, prices came down and the use of acrylics for paper saturation has increased. The price is still higher than SBR.

**Neoprene.** It was widely used in coating base sheets and gaskets, but not for tape backings. It needs to be compounded with ZnO and antioxidants to prevent tenderizing of backing on aging.

**Carboxylated SBR.** These polymers are improvements over SBR and tends to resist drop-off of adhesion after aging. These are a little more costly than SBR but are less expensive than other saturants.

**Other Types.** A number of other latexes are used,[4] such as various acrylic copolymers, carboxylated NBR, and different butadiene-styrene copolymers.

The effect of various saturants on the physical properties is given in a number of papers.[2,3,10,11] Two important broad patents on latex impregnated backings were issued to Engel *et al.*[12] and especially to Eger and Engel.[13] The latter one teaches the impregnation of a thin web formed of paper fibers with an aqueous dispersion of polymeric particles containing 40–60% solids. The latex may consist of copolymers of butadiene with styrene, acrylic esters, nitriles, isobutylene, or methacrylates. This covers most latexes used at that time. It claimed a backing of sufficient internal strength, so that it would not delaminate when subjected to forces exerted by adhesive on unwinding.

## Delamination Resistance

As noted earlier, this is a very important property that largely determines the cost of the backing, which in turn is the most costly part of the tape. Research in latex saturation, concentrating heavily on this property, was initiated in the early 1950's and resulted in a series of patents issued to Bartell.[14,15,16,17,18] The work leading to these patents was made possible by the development of the delamination test, described earlier, initiated by Goun and perfected by Bartell.

## Effect of Product Variables on Physical Properties

Most physical properties improve with higher level of impregnation. The most critical properties affecting the tape performance are delamination resistance and edge tear. The effect on delamination resistance by the impregnation level is illustrated in Fig. 19-2 by Weschler and Bartell.[17] In this case, which uses a conjugated diene-acrylic acid copolymer emulsion as saturant, an additional increase in delamination resistance is obtained on heating at 163–193°C. For a creped masking tape of 500 g/cm width peel adhesion, 80–100% saturation is recommended.

In general, the firmer the saturating polymer, the higher the delami-

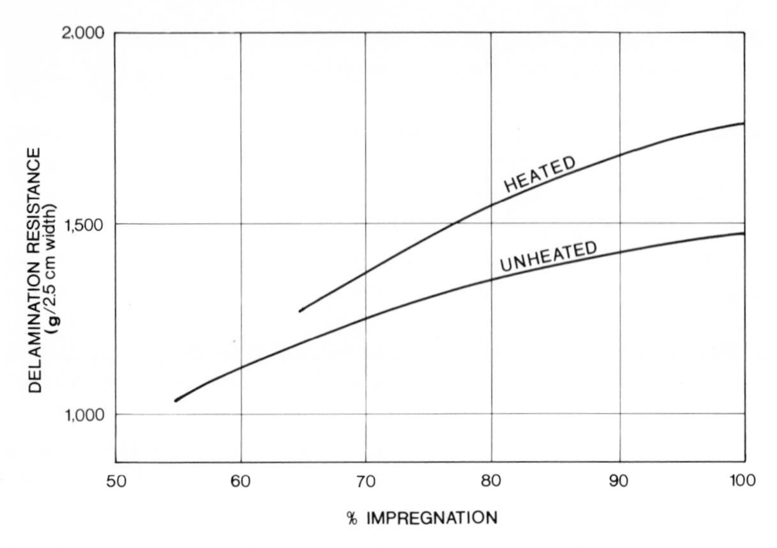


Fig. 19-2 Delamination resistance as a function of impregnation.

nation resistance and tensile strength, but this gives a stiffer backing and lower edge tear. Butadiene/styrene 75/25 copolymer is soft and gives products of lower delamination resistance, lower tensile strength and higher edge tear than butadiene/styrene copolymer in the range 35/65. Same considerations are valid for butadiene/methacrylic ester copolymers as described by Weschler and Bartell[17] and by Finnegan and Miller.[19] In the latter reference, incorporation of polycarboxylic acid containing copolymers were shown to develop higher delamination values.

Early work to increase the delamination resistance at the lowest possible levels of impregnation led to curable impregnants as illustrated in Bartell's patents.[14,15] Curing effect increased with increasing concentration of curing agent and increased heat treatment. The higher the degree of cure, the higher the delamination resistance, but the lower the edge tear and elongation, so that a compromise between these properties is needed. This is illustrated in Table 19-1 based on Bartell's patent.[15] Note that although an excellent delamination resistance was obtained at a curing resin concentration of 13%, the elongation and edge tear were poor. The edge tear was acceptable up to a curing agent concentration of 7%, but the best results were obtained in the range of 2–3%. This patent illustrates the effect of a phenolic curing resin incorporated into *N*-containing copolymer emulsion such as butadiene-acrylonitrile latex.

Another patent[14] teaches that carboxylated SBR emulsions can be effectively cured to increase the delamination resistance by water-dispersible heat reactive resins such as phenol-formaldehyde, urea-formaldehyde and melamine-formaldehyde, whereas standard SBR could not. Other aspects of saturated papers are discussed in patents dealing with various emulsions, beater addition techniques, etc.[20,21,22,23,24,25,26,27] Cluett[28] describes a unique treatment of impregnated papers that produces unusual properties. It uses special Clupak equipment to compact fibers together but still producing a sheet with a smooth surface. This treatment was tried on saturated papers and led to a patent by Dunlap for paper tape backings[29] and to a patent by Sonnichsen and Bartell[30] on heavier impregnated paper for coating bases. For tape backings the treatment produces a thinner sheet with a higher elongation, increased softness (lower stiffness), increased delamination resistance and increased edge tear. In thicker backings, it produced a sheet with excellent piping resistance, higher flex endurance as measured by the MIT fold test, and a soft, leatherlike feel.

## PAPER TAPES

### General-Purpose Creped Paper Masking Tape

The tape meets Military Specification PPP-T-42C (UUT-106D) Type 1. It has many uses: masking for painting when the paint is dried at ambient or relatively low temperatures of up to 80°C, light holding, bundling, packaging and other applications. A relatively low cost paper is sufficient with a minimum delamination resistance of 330 g/cm, though higher delamination resistance might be desired depending on the adhesive-release coat system and aging conditions. Tensile strength of 3.4–4 kg/cm and elongation of 8–12% are adequate. Impregnant and the adhesive must be compatible to avoid adhesion drop-off. This is true for all impregnated papers.

In just about all commercial products a release coat is used, which, in general, is considered confidential. One such release coating is described.[18] The backing is backsized with a mixture of 100 parts of an urea formaldehyde resin, 13.6 parts of copolymer of vinyl acetate and stearyl maleate, and 0.45 part of phosphoric acid. The backsized sheet was dried at 93°C and cured at 177°C. A water-based release coating is described by Hechtman.[31] Various other release coatings are covered in a separate chapter in this book. A number of companies are offering polymer emulsions for use as release coatings (Rohm and Haas Co., National Starch and Chemical Corp., Reichhold Chemicals, Inc.).

This tape may or may not have primer depending partly on the paper impregnant. A primer can be made up of a solution of 320 g of a mixture composed of a 25% solution of pale creep rubber, 40 g of a 50% solution of *p,p'*-diphenylmethane diisocyanate, and 210 g of toluene. Various blends of SBR and NBR, with and without resins, are also used.

Upon the primed side of backing an adhesive coating of 40–50 g/m$^2$ is applied. Such an adhesive based on natural rubber is described by Collins.[21]

| Ingredient | Parts |
|---|---|
| Crude rubber | 100 |
| Dehydrogenated rosin | 90 |
| Aluminum hydrate | 90 |
| Mineral oil | 30 |
| Antioxidant | 2 |

Adhesive formulas based on block copolymers have been proposed. These are suitable for hot melt and high solids adhesives. A possible hot melt formula for a general-purpose masking tape is given below.

| | |
|---|---|
| Kraton 1107 | 100 |
| Escorez 5280 | 125 |
| Shellflex 371 | 25 |
| Butyl Zimate | 2 |

## High-Temperature Paper Masking Tape

This tape is used mainly by automotive companies that mask the chrome, glass and rubber parts before painting. The paper and release coat must prevent paint solvent from soaking into and softening rubber adhesive. Softening of the adhesive could leave a residue on paint. The impregnant must also resist heat during the paint baking and should not soften or harden too much. Otherwise, the tape will tear on removal. This requires a higher quality paper of higher delamination values, so that the paper does not split on removal in case the adhesive adheres tightly to the masked surface.

The release coat and primer can be the same or similar as those used for the general-purpose masking tape. The adhesive must resist high temperatures during baking cycle, should not stain paint, and should not leave a residue on removal, especially when removed warm. An example of such adhesive which is cured by phenolic resins is given below.[18]

| Ingredient | Percent by weight |
|---|---|
| Pale crepe rubber | 36.5 |
| Copolymer of butadiene and styrene (SBR 1022) | 12.2 |
| Polyterpene resin (Piccolyte S115) | 36.1 |
| Oil soluble octyl phenol-formaldehyde resin (Amberol ST 137) | 4.2 |
| Modified phenol-formaldehyde resin (Amberol M88) | 4.2 |
| Zinc resinate | 4.2 |
| M.tolylene diisocyanate | 1.0 |
| 2,5-di-*tert*-amyl hydroquinone (Santovar B) | 0.8 |
| 2,6-di-*tert*-butyl *p*-cresol (Ional) | 0.8 |
| | 100.0 |

There are many variations of the above formula employing different phenolic resins and different tackifiers and their levels.

## Medium Tensile Strength Flatback Paper Tape

The tape meets Military Specification PPP-T-42C (UUT-106D) Type II. These tapes are used for general packaging applications where moderate tensile strength requirements are sufficient. Tensile strength varies from approximately 5.7–6.8 kg/cm. Delamination stresses seem to be less important, because more efficient release coatings are used. Release as low as 0.3 kg/cm have been used. Elongations are low in the range 4–8%. Both, natural (light tan) and dark tan (kraft brown) colored backings are used.

Release coating, primer and adhesive systems similar to general-purpose masking tape may be used. In general, good adhesion, tack and hold to kraft boxboard are desired.

## High Tensile Strength Flatback Paper Tapes

Regular kraft paper tape meeting the Military Specification PPP-T-76C (Carton Sealing) is a special packaging tape with stringent government specification requirements. The tensile strength of at least 8 kg/cm and usually in the range of 9-10 kg/cm is offered. A minimum wet cross-sectional tensile strength of 2.1 kg/cm and a minimum Elmendorf tear resistance of 75 g after UV and accelerated aging are required. Higher weight properly designed papers are necessary to meet these specifications. The adhesive also has to meet certain strict military specification requirements. Minimum peel adhesion as received and after aging must be at least 390 g/cm. It must have at least 110 g/cm adhesion to its own backing before and after aging, as well as holding power of at least 96 hr.

Another variety of these tapes is high tensile strength rope paper tapes. These tapes are also used for packaging and require tensile strength of 8-10 kg/cm in the machine direction. Tensile strength in the cross-machine direction is low and unimportant. These requirements are met by oriented rope (hemp) paper. This paper also has the advantage of being thinner, about 0.1 mm vs. 0.15 mm for standard high tensile flatback paper, and also less stiff. The adhesive must have a high holding power. In addition, the white goods market requires it to be nonstaining. A noncured adhesive with very high Mooney viscosity rubber, or a phenolic resin cured one with a lower Mooney viscosity are suitable. Release coats and primers are similar to the ones used in other constructions.

Since hemp used in the rope paper is in short supply and is expensive, a considerable amount of work has been done by paper manufacturers in developing all-kraft or kraft-hemp blend substitutes to provide a similar backing material. These have in some measure replaced all hemp papers. The kraft paper tends to be somewhat thicker and stiffer than all-rope paper.

### Protective Flatback Paper

These papers are used to cover and protect stainless steel, chrome plated steel and lacquered brass. These metal stocks are sold in roll form to companies who cut, drill and bend them into various articles, such as handles for furniture, ashtrays, various kitchen utensils, light switch plates, etc. These protective tapes are also used to cover plastics such as sheets of polymethyl methacrylate or high pressure laminates. The purpose is to protect the surface from scratches and dents until it is used. Then the customer has to be able to remove the tape easily and cleanly, leaving a highly polished and scratch-free surface. These are relatively inexpensive backings with low level of latex impregnation. Delamination resistance values are under 0.17 kg/cm. The color is usually natural or white.

The adhesive has to be of a low adhesion level 70-110 g/cm and must not build up higher than 180 g/cm. The tape may be in contact with stainless steel, or other surface, for many months. The adhesive should release easily and should not leave any residue of "ghosts." Most of these adhesives are based on proprietary formulas involving natural rubber latex with some water soluble or dispersible tackifying resin. Some solvent and latex acrylic adhesives have been also used for this application.

## Printable and Colored Tapes

Low tensile strength (3.5–4.3 kg/cm) crepe or flatback tapes of various colors are used for identification and packaging.

High tensile strength (7.9-9.8 kg/cm) printable tapes in various colors are used for identification and packaging applications where high strength is needed, especially where narrow widths are used. These are made from rope paper. These tapes require a printable release coat: the release coating must be printable by various inks and the bond should be firm so that the ink is not pulled off by the adhesive upon tape unwinding. One class of such tapes printable release coats is given by Hechtman.[31]

## Electrical Paper Tapes

Since they are used for insulation, holding and other applications in contact with conductors, they must have acceptable electrical properties. The latex adhesives usually have ionic compounds that conduct electricity and can be corrosive. Thus solvent impregnated papers are used. Both lower tensile strength of 3.5–4.3 kg/cm and higher tensile strength of 7.9–9.8 kg/cm tapes are used.

The adhesive, primer, and release coating must be of electrical grade. As with impregnant, the emulsion type coatings are avoided. The adhesive must not contain contaminants such as chlorides or sulfur which could stain and corrode copper wires with which it might be in contact. Most systems use rubber-resin combinations in solvents and are cured with phenolic resins. Curable solvent type acrylic adhesives are also used.

**Table 19-1. Effect of curing resin on the physical properties of saturated paper tape backing.**

| CURING RESIN CONTENT (%) | DELAMINATION RESISTANCE | ELONGATION (%) | EDGE TEAR (g) |
|---|---|---|---|
| 0 | Fair | 12.6 | 2225 |
| 3 | Excellent | 12.6 | 2000 |
| 13 | Excellent | 6.2 | 950 |

## REFERENCES

1. Hechtman, J. F., and Jayne, J. E. *Industrial and Specialty Papers*, Vol. II (Mosher and Davis, eds.). New York: Chemical Publishing Co., Inc., 1968.
2. Dunlap, I. R. *TAPPI*, **40**(8) (August 1957).
3. Latex Manual HL-2. B. F. Goodrich Chemical Co.
4. Finnegan, L. P., and Miller, V. A. *TAPPI*, **41**(12) (December 1958).
5. Hechtman, J. F., and Bonnslaeger, S. R. U. S. Patent 3,464,848.
6. Page, D. H. *TAPPI*, **62**(9) (September 1979).
7. Drew, R. G. U. S. Patent 2,236,527 (1941) (assigned to 3M Co.).
8. Kellgren, W. U. S. Patent 2,410,078 (1946) (assigned to 3M Co.).
9. Tierney, H. J. U. S. Patent 2,438,195 (1948) (assigned to 3M Co.).
10. Thommen, E. K., and Stannett, V. *TAPPI*, **41**(11) (November 1958).
11. Heyse, W. T., Sarkanen, K., and Stannett, V. *J. Appl. Polym. Sci.* **3**(9) (1900).
12. Engel, E. W., Grolito, C. J., McGarry, J. A., Morris, V., and Werdner, C. L. U. S. Patent 2,592,550 (1952).
13. Eger, L. W., and Engel, E. W. U. S. Patent 2,726,967 (1955).
14. Bartell, C., and Weschler, J. R. U. S. Patent 2,848,105 (1958).
15. Bartell, C. U. S. Patent 2,848,355 (1958).
16. Bartell, C., Weidner, C., and Smith, R. M. U. S. Patent 2,937,109 (1960).
17. Weschler, J. R., and Bartell, C. U. S. Patent 2,963,386 (1960).
18. Bartell, C. U. S. Patent 3,197,330 (1965).
19. Finnegan, L. P., and Miller, V. A. U. S. Patent 3,156,581 (1964).
20. Hechtman, J. F., and Greenman, E. G. U. S. Patent 3,026,217 (1962).
21. Collins, W. C. U. S. Patent 3,012,913 (1961).
22. Bailin, M. M. U. S. Patent 3,127,284 (1964).
23. Weschler, J. R. U. S. Patent 3,068,121 (1962).
24. Hechtman, J. F., and Greenman, E. G., U. S. Patent 3,066,043 (1962).
25. Engel, E. W., and Dunlap, I. R. U. S. Patent 2,899,353 (1959).
26. Kellgren, W., Birchwood and Marschall, U. S. Patent 2,633,430 (1953).
27. Gustafson, K. H., Excelsior and Picard, L. E., U. S. Patent 3,503,495 (1968).
28. Cluett, S. L. U. S. Patent 2,624,245 (1953).
29. Dunlap, I. R. U.S. Patent 3,055,496 (1962).
30. Sonnichsen, H. M., and Bartell, C. U. S. Patent 3,245,863 (1964).
31. Hechtman, J. F., and Greenman, E. G. U. S. Patent 3,066,043 (1962).

# Chapter 20

# Hospital and First Aid Products

**Donatas Satas**
*Satas & Associates*
*Warwick, Rhode Island*

Pressure-sensitive tapes and other products are widely used for various hospital, first aid, athletic protection, and similar applications. Tapes are the most important of these products and are used for various holding purposes, such as keeping a dressing in place, or for strapping and restrictive purposes. Hospital tapes is the oldest pressure-sensitive adhesive application and the industry has grown from this use.

The beginning of the pressure-sensitive industry could be counted from the introduction of natural rubber into the adhesive formulations. In 1899, zinc oxide containing adhesive mass was introduced into hospital tape by Johnson and Johnson[1] Early pressure-sensitive tapes did not age well; they lost tackiness because the rosin used as a tackifier was easily oxidized. In 1934, Hercules Powder Company introduced hydrogenated rosin, which substantially improved the aging resistance. Introduction of dehydrogenated rosin was a further improvement. A shortage of natural rubber during World War II caused an evaluation of other polymers for pressure-sensitive adhesive applications. Polyisobutylene was adopted for hospital tapes compounded with factice, ester resin, and polybutene.[2] In the 1960's, polyacrylates and, to a lesser degree, polyvinyl ethers were introduced for hospital tapes. Polyacrylate adhesives became the most important for these applications.

Early adhesive coatings were heavy because the stiff, coated fabric backing required a heavy adhesive to provide sufficient adhesion and because these tapes were usually manufactured by calendering. Calendering is suitable for heavier adhesive coating application.

## TAPE USES

Coated or uncoated cotton fabric was used for early hospital tapes. These backings are still used for the strapping tapes which require higher tensile strength. Synthetic fiber fabrics, nonwoven fabrics and films were later introduced as backings for hospital tapes.

The proliferation of different backings and different adhesives for the hospital tapes was the result of the need to provide best suited products for various applications, rather than offering a single multipurpose tape. Many improvements in the hospital tape field were achieved because of this trend. The most important improvement was the development of tapes with a lower degree of skin irritation.

Tapes are used for many different applications in the hospital and health areas, but basically they perform one of two functions. It is used to pull something in place, restricting movement such as in various strapping applications, or it is used to hold something in place, such as a wound dressing.

In the first category, the tapes are used for various therapeutic strapping, semirigid or elastic support of soft tissues. The adhesive tape is used for treatment of sprains, ruptured muscles or tendons, and separated joints. The tape support of soft tissues is analogous to the splint support of fractioned bones. The tape provides compression and restricts the movement. Higher tensile strength fabrics and elastic fabric-backed tapes are used for these applications. Similarly, the tapes are used for compressing venous circulation in the legs and strapping fractured ribs, in conjunction with plaster casts.

Tapes are also used to prevent injuries in athletic competition by strapping the most vulnerable joints. The tape restrains the movement and prevents the muscle or tendon from being stretched beyond its limit. It also helps to absorb the impact. Wrapping a boxer's hands, which is a very elaborate procedure, or strapping a football player's back and ribs are good examples of this application.

Tape is also used to close the edges of minor wounds and skin cuts eliminating sutures.[3] Precut tape strips mounted on release paper and plastic bandages are used for this purpose.

Holding a wound dressing in place is the earliest and most common application of tapes. Lighter weight, lower strength, nonocclusive, less skin irritating tapes were developed primarily for this application. Nonwoven fabric, lightweight synthetic fabric, and polymeric film backings are used. Transparent and flesh-colored backings are employed for cosmetic reasons.

In addition to holding dressings, tapes are used to retain catheters, to secure the leads of electrocardiogram apparatus, and to secure and stabilize intravenous needles. Precut tapes on release paper are supplied for the operating room usage. These can be sterilized prior to use.

Pressure-sensitive adhesive coatings have been incorporated into unit dressings, eliminating the need of tape. Some nonadherent dressings and sanitary napkins are provided with two-faced or transfer tape strips pre-applied to the bandage and protected by a strip of release paper. The best known unit dressing is the finger bandage used to cover minor skin cuts. Vinyl film is used as a backing. It is perforated to decrease the occlusivity and to allow faster drying of the pad.

Pressure-sensitive products for the care of feet are well known. They either carry a medicament for corn removal or other purposes, or provide cushioning to decrease the pressure exerted by the shoe.

## SKIN IRRITATION

There are several causes of skin irritation. The tape construction, backing, and adhesive contribute to the degree of irritation.

Mechanical irritation may be caused by applying a stiff tape over the skin which is frequently flexed. The skin is irritated by the continuous pull by the adhesive. Usually, however, the mechanical irritation is caused by the tape removal. It can be temporary, caused by vasodilatation in which no damage of the skin occurs, or it can be of a longer duration caused by stripping the outer layer of skin. This latter case is the most common cause of skin irritation by pressure-sensitive adhesive tape. Various types of skin irritation have been reviewed by Seymour.[4]

Chemical irritation is caused by the diffusion of chemical irritants, present in either the adhesive or the backing, into the skin tissues.

Allergic irritation is caused by some tape ingredient to which an individual is allergic. This type of irritation is not as common as generally claimed. It has been shown that accelerators and antioxidants in natural rubber can cause allergic contact dermatitis.[5] The allergic irritation is easily distinguishable from the mechanical one; it spreads beyond the area over which the tape was applied, while the mechanical irritation is confined to the skin that was covered by the tape. Polyacrylates and other single component adhesives, which do not require compounding to achieve the pressure-sensitive properties, are less likely to cause allergic skin reactions. These polymers do not require antioxidants, do not have proteinaceous impurities, such as found in natural rubber, but consist of a single compound, thus decreasing the chance of allergic irritation.

Tapes can interfere with the normal physiological activity of the skin. Occlusive tapes can cause accumulation of fluid resulting in maceration, a softening of the skin. Skin, weakened by maceration, can split inside on tape removal. Overhydrated skin is also easier penetrated by chemical irritants and therefore is more susceptible to irritation of this type.

Sometimes the area beneath an occlusive tape can become completely anhydrous, because of the blockage of the openings of sweat ducts and other reasons.[1] The dehydrated tape-skin interface becomes suitable for the growth and multiplication of bacterial flora, resulting in infection.

## NONOCCLUSIVE TAPES

The use of occlusive tapes aggravates the skin irritation. Many products have been designed to overcome this problem. The approach is to provide a lighter product consisting of a porous backing and a prous, or at least a lighter adhesive coating. These tapes decrease the level of skin irritation considerably and also minimize the incidence of bacterial infection.[6]

The early work in the area of nonocclusive dressings was done by T. J. Smith and Nephew, Ltd. They prepared porous adhesive mass by blowing air[7] and used porous polyvinyl chloride films.[4]

A porous tape is prepared by applying the adhesive mass over a release paper, partially drying it, transferring the adhesive mass to a nonwoven fabric backing and then completing the drying.[8] The adhesive shrinks, leaving a porous, thin coating over a lightweight, nonwoven fabric. This process is used to produce an important product (Micropore by 3M Co.). The decrease of skin irritation by this tape has been discussed by Golden[9] and its properties have been analyzed by Kaelble and Hamm.[10] Similarly, porous adhesive coating was prepared by drying a coating on release paper, by a fast application of heat and forming fine bubbles in the adhesive mass. This coating, when transferred to a porous nonwoven fabric, is permeable to moisture.[11]

Permeable adhesive tapes have been prepared by perforation through the adhesive without perforating the backing.[12] The process is used for another important product (Dermicel tape by Johnson and Johnson). Porous adhesive mass is also applied by spraying,[13] by printing a noncontinuous pattern over a porous backing, by rapid chilling,[14] by foaming,[15] by scraping the adhesive mass, and by other methods. Tapes and other products have been perforated to achieve an overall porosity[16] and to improve tearing.[17]

Polyacrylic adhesives with hydrophilic groups can absorb moisture and allow its passage. It is believed that a nonocclusive adhesive may not necessarily be porous. It can also transmit moisture at a sufficiently high rate by absorption and diffusion.[1]

Porous polyvinyl chloride films have been used for dressings. Microporous films developed for battery separators have been adopted for this purpose.[18] These films allow the transmission of water vapor, but the pores are sufficiently small to provide a waterproof protection over the wound.

## OTHER PRODUCTS

Hospital and first aid applications require many other tapes and products with pressure-sensitive adhesives. Adhesive coated foam is used for holding electrocardiogram chest contacts. Foam tapes are used for cushioning and for compression. Soft, thick, napped cotton fabric-backed tape is also used for cushioning purposes. General-purpose tapes are used to affix labels, to identify medicine bottles, to secure charts and for many other household applications. A tape which changes color when exposed to sterilization conditions and therefore identifies sterilized supplies has a pressure-sensitive adhesive coating for ease of application. The same tape is also useful to hold the bundles together and as a writing surface to identify the contents. Sometimes an occlusive dressing is required to seal the topically applied medication. Nonporous plastic tape is used for such applications.

Transparent surgical drapes are available with a pressure-sensitive adhesive coating. The function of the adhesive is to hold the drape in an intimate contact with the skin. The incision is made through the drape. The drapes are used for abdominal, orthopedic and other surgeries.

## ADHESIVES

Natural rubber-based adhesives were the most commonly used for hospital tapes. Polyisoprene, polyisobutylene, and other synthetic elastomers have been used to replace the natural rubber. Some typical formulations are given below.

| Solvent-based Adhesive | |
|---|---|
| Smoked sheet | 100 |
| Zinc oxide | 50 |
| Stabilized rosin | 75 |
| Lanolin | 10 |
| Antioxidant | 2 |

| Calenderable adhesive | |
|---|---|
| Smoked sheet | 100 |
| Zinc oxide | 100 |
| Starch | 40 |
| Factice | 15 |
| Stabilized rosin | 100 |
| Lanolin | 10 |
| Antioxidant | 3 |

Synthetic Polymer-based Adhesive

| | |
|---|---|
| Polyisobutylene (80,000) | 50 |
| Factice | 29 |
| Glyceryl ester of hydrogenated rosin | 24 |
| Liquid polyisobutylene | 35 |
| Lanolin | 20 |
| Beeswax | 9 |
| Titania | 14 |
| Hydrated alumina | 62 |
| Antioxidant | 1.5 |

Vulcanized oil (factice) is used to reduce the cold flow of the adhesive mass. It is especially helpful in synthetic polymer formulations.

These adhesives have been largely replaced by acrylic copolymers such as a copolymer of 2-ethylhexyl acrylate/ethyl acrylate/acrylic acid or adhesives of similar compositions.

The characteristics of a good hospital adhesive are:

1. Absence of skin-irritating ingredients.
2. Good water resistance, so that the bond is maintained when skin perspires.
3. Sufficiently high cohesive strength to allow a clean removal of the adhesive from skin.
4. Preferably high water vapor transmission rate to decrease the occlusivity of the tape.
5. Rheological properties to accommodate the skin movement without lossing the bond and without excessive skin irritation mechanically.

The adhesive for application over the skin must be fairly soft of a low modulus of elasticity at low elongation. Adhesives of high modulus of elasticity debond easier on continuous flexing and also are more liable to cause mechanical skin irritation. Figure 20-1 shows the difference between the behavior of a good hospital adhesive based on acrylic copolymer and acrylic adhesives suitable for other applications.

The dotted line represents a cohesive failure and the solid line an adhesive one. Curve A represents the behavior of a higher modulus adhesive of good creep resistance such as used for many industrial tape applications. This adhesive, if used for hospital applications, would give a tape which will disbond easily on continuous flexing of the skin. Curve C represents a good adhesive for hospital tape application. At a low peel rate, the adhesive shows a cohesive failure. This assures that the adhesive will fail only when its maximum elongation is reached and will not easily disbond

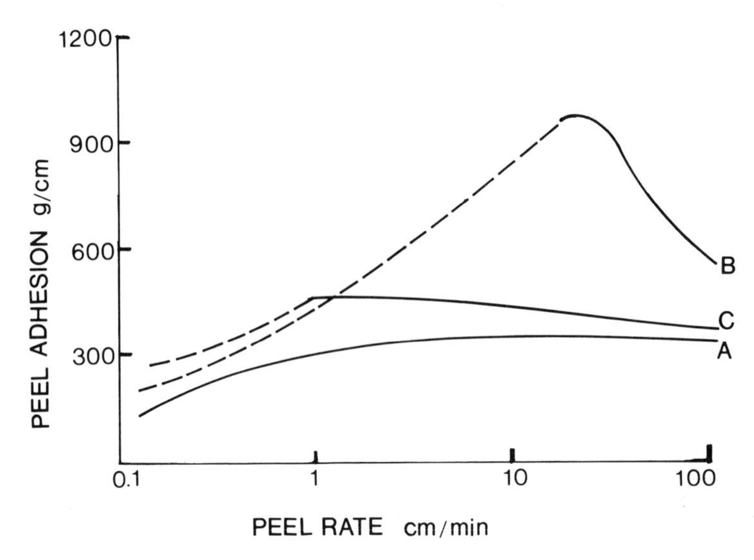


Fig. 20-1. Dependence of peel adhesion on peel rate for some polyacrylic adhesives. $A$ = High cohesive strength industrial adhesive. $C$ = Hospital tape adhesive. $B$ = Low cohesive strength adhesive.

when in use. At higher peel rates, it shows an adhesive failure and therefore will come off cleanly when removed. Curve B represents a soft adhesive which would split on removal and leave a residue.

## REFERENCES


1. *Professional Uses of Tape*, 3rd ed. Johnson and Johnson, New Brunswick, N. J., 1972.
2. Kemp, M. H. *Adhesives Age* **4**(12): 22–25 (December 1961).
3. *Adhesives Age* **6**(8): 20–21 (August 1963).
4. Seymour, D. E. *J. Pharm. and Pharmacol.* **9**:802–809(1957).
5. Fisher, A. A. *Cutis*, **1**(7):345–354 (1965).
6. Marples, R. R., and Kligman, A. M. *Arch. Derm.* **99**:107–110 (January 1969).
7. Anderson, E., Stephenson, A. R., and Lucas, A. British Patent 713,838 (1951) (assigned to T. J. Smith and Nephew, Ltd.).
8. Copeland, F. S. U.S. Patent 3,121,021 (1964) (assigned to 3M Co).
9. Golden, T. *Am. J. Surg.* **100**: 789–796 (1960).
10. Kaelble, D. H. and Hamm, F. H. *Adhesives Age* **11**(6): 30–35; **11**(7): 28–35 (1968).
11. French Patent 1,490,130 (1967) (assigned to Beiersdorf AG).
12. Blackford, B. B. U.S. Patent 3,161,554 (1964) (assigned to Johnson and Johnson).
13. Satas, D. U.S. Patent 3,364,063 (1968) (assigned to Kendall Co.).
14. Salditt, F. Canadian Patent 602,084 (1960) (assigned to Johnson and Johnson).
15. Anthony, I. B. British Patent 799,424 (1958) (assigned to Johnson and Johnson).
16. Scharr, C. H. U.S. Patent 3,073,304 (1963) (assigned to Kendall Co).
17. Blackford, B. B. U.S. Patent 3,085,024 (1963) (assigned to Johnson and Johnson).
18. Scales, Towers, and Goodman. *Brit. Med. J.* **2**: 962 (1956).

# Chapter 21

# Packaging Tapes

**Keiji Fukuzawa**
*Consulting Chemical Engineer*
*Tokyo, Japan*

and

**Donatas Satas**
*Satas & Associates*
*Warwick, Rhode Island*

Packaging application requires a large variety of tapes, encompassing most of the tape types manufactured. Packaging applications include tapes used for closing the packages, protecting the labels, sealing the packages from moisture, and strapping and bundling of loose parts. The surface protection during shipment and handling of polished metal and plastic surfaces by applying a pressure-sensitive adhesive-coated saturated paper is basically a packaging application.

The general-purpose basic packaging tape varies in its construction from country to country. Oriented polypropylene film tape has the dominant position in the United States, although cellophane, acetate film and masking tapes are also widely used for general-purpose applications. Unplasticized poly(vinyl chloride) tape is favored in Europe and the kraft paper tape is widely used in Japan. These differences in the choice reflect the backing prices, industry preference, consumer acceptance, and other factors. In Taiwan, plasticized vinyl film tapes are very popular, and they have the dominant position in packaging uses.

Some of the tapes used for packaging applications are listed in Table 21-1 along with some of their physical properties.

## Table 21-1. Tapes used for packaging applications.

| BACKING | TENSILE STRENGTH (kg/cm Width) | THICKNESS (μm) |
|---|---|---|
| Film tapes | | |
| Biaxially-oriented polypropylene | 4.5–6.5 | 50–57 |
| | 8 | 62 |
| | 9–20 | 82–95 |
| Polypropylene strapping tapes | 27–29 | 110–115 |
| | 31 | 125 |
| Polyester, box sealing | 3.9–5 | 52–57 |
| Polyester, label protection | 3.9–4.5 | 50 |
| Polyester, carton sealing, heavy-duty | 7–9 | 75–82 |
| Polyester, shrink packaging | 3.5–5 | 62 |
| Vinyl, rigid, label protection | 5 | 62 |
| Vinyl, rigid, carton sealing | 5.4 | 62 |
| Vinyl, rigid, general packaging | 3.6–3.9 | 60–67 |
| Vinyl,rigid, heavy duty | 7.9 | 98 |
| Vinyl, plasticized, colored | 3.2 | 150 |
| Cellophane film | 5.4–5.7 | 57–68 |
| Acetate film | 3.9–4.5 | 59–62 |
| Polyethylene film | 3.6–12 | 60–200 |
| REinforced | | |
| Polyester film, glass filament | 36–80 | 150–240 |
| Polyester film, polyester filament | 34 | 190 |
| Polyester film, rayon filament | 43 | 240 |
| Polypropylene film, glass filament | 36 | 140 |
| Paper, glass filament | 54–57 | 275–310 |
| Paper, reinforced | 21–60 | 250–300 |
| Paper tapes | | |
| Kraft paper | 6 | 120 |
| Kraft paper, carton sealing | 5.7–6.0 | 150–180 |
| Kraft paper, carton sealing, government specs | 8.6 | 210 |
| Saturated paper, crepe, general-purpose | 3.8–4.1 | 150–175 |
| Saturated paper, crepe, produce | 4.1 | 250 |
| Saturated paper, crepe, bandoleering | 3–3.9 | 185–200 |
| Saturated paper, flatback, general-purpose | 3.2–3.9 | 175 |
| Saturated paper, flatback, heavy-duty | 7 | 300 |
| Saturated paper, flatback, box sealing | 9 | 175–200 |
| Saturated paper, rope stock, medium strength | 3.8 | 125 |
| Saturated paper, rope stock high strength | 7–8 | 175–200 |
| Fabric tapes | | |
| Vinyl coated cotton sheeting | 6.8–8 | 275 |
| Polyethylene coated cloth | 7–8 | 275–300 |
| Vinyl coated cloth | 9 | 375 |
| Coated cloth, premium | 9 | 275 |
| Coated cloth, extra strong | 9.8 | 350 |
| Polyethylene-coated nylon | 2.7 | 250 |
| Foil tapes | | |
| Aluminum foil, sealing | 3.5–5.4 | 90–125 |

**Table 21-2. Physical properties of polypropylene films.**

| PROPERTY | FILM TYPE A | FILM TYPE B | FILM TYPE C |
|---|---|---|---|
| Thickness ($\mu$m) | 12.5–32 | 30 | 40 |
| Tensile Strength (kg/cm²) | MD 2110 | 1550 | 500 |
| | TD 2110 | 2950 | 4000 |
| Elongation (%) | MD 85 | 170 | 150 |
| | TD | 50 | 30 |
| Tensile Modulus (kg/cm²) | MD 24,600 | 21,100 | 15,000 |
| | TD 24,600 | 42,200 | 35,000 |
| Shrinkage 120°C (%) | MD 2 | 3 | |
| | TD 2 | 3 | |

Film A: Hercules B500, balanced, biaxially-oriented polypropylene homopolymer film.
Film B: Hercules T500, unbalanced, biaxially-oriented polypropylene film.
Film C: Toray Industries (Japan) easy tear polypropylene film.

## POLYPROPYLENE TAPES

Biaxially-oriented 25,30,40,50 and 60$\mu$m thick polypropylene films are used for the pressure-sensitive tapes. For general-purpose applications, 30 $\mu$m thick film is used most commonly, although some 25 $\mu$m film is also used.

Hercules, Inc., one of the important polypropylene film manufacturers, has a balanced biaxially-oriented film manufactured by the bubble process. This process is preferred for lighter gauge films. Heavier gauge (25–75 $\mu$m) films are best manufactured by the tentering process that yields unbalanced biaxially-oriented films. Easily tearable, oriented polypropylene film has been developed for use as tape backing. Tape with this backing can be torn by hand across the width. Physical properties of these polypropylene films are shown in Table 21-2.

Polypropylene tapes can be constructed by applying natural rubber-resin adhesive as shown in Fig. 21-1. This construction requires a release coating and therefore the film should be corona discharge treated on both sides. More polar acrylic adhesives release well from polypropylene sur-

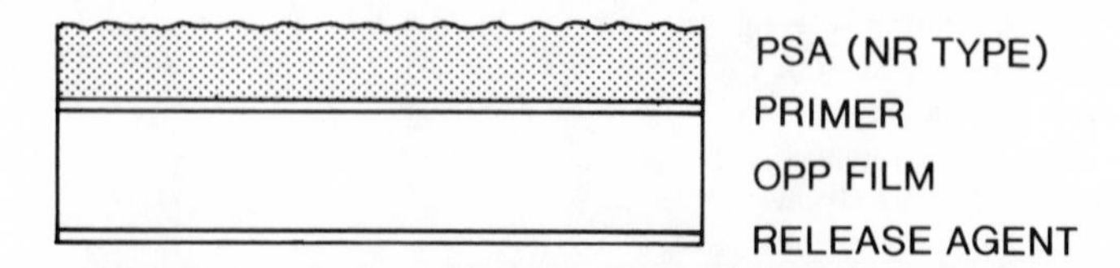


Fig. 21-1. Polypropylene film tape with a release coating.

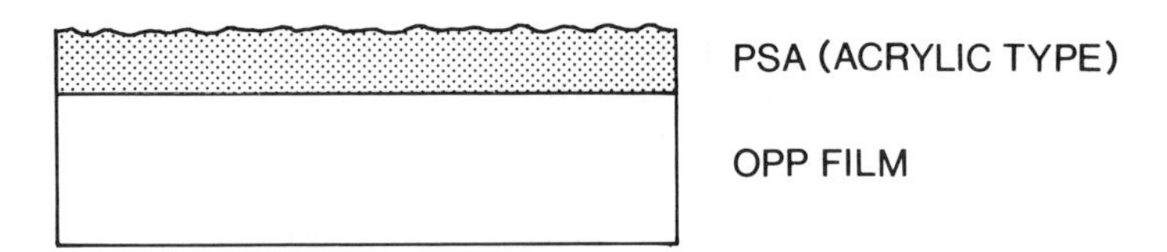


Fig. 21-2. Polypropylene film tape without a release coating.

faces. One side corona discharge treated film is used for such tapes (Fig. 21-2). Aqueous acrylic emulsion adhesives are used for polypropylene tape. Hot melt block copolymer-based adhesives make it possible to produce such tapes very economically. Generally, a release coating is required for hot melt polypropylene tapes, although a tape without a release coating is also made.

The uses of polypropylene tapes range from general-purpose packaging tape to carton closure tapes, label protective tapes, to high strength strapping tapes for bundling and strapping of heavy materials. A polypropylene fiber tape with a fused outer surface has been proposed for strapping applications.[1] An interesting application is the tear strip for corrugated cartons and other packages. A narrow strip (0.5 cm) of tape is applied over the corrugated sheet for that purpose. The application requires long rolls of tape and large spirally-wound spools are used for this application.

## UNPLASTICIZED VINYL FILM TAPES

Rigid vinyl film tapes have a dominant position as a packaging and general-purpose tape in Europe. The backing used is almost exclusively a longitudinally-oriented film. Pressure-sensitive tape products made using such films are described by Nachtsheim and Meyer.[2] An improvement in the clarity and tearability of the film is claimed. Screnock and Forse[3] describe biaxially-oriented film for a pressure-sensitive tape backing.

The vinyl films used for tape backings are usually manufactured by calendering. These films are discussed in detail by Reip.[4] The comparison of physical properties between rigid PVC and cellophane films is shown in Table 21-3.

The dependence of tensile strength and elongation on temperature is shown in Fig. 21-3 and the solvent resistance is shown in Table 21-4.

Release coating for the film made from emulsion resin is not required. The film from the suspension type polymer might require a release coating to reduce the unwind force. A commonly used release agent is 0.15% solution of Wax V (by American Hoechst Corp.) in gasoline applied to give a coating of 0.5 $g/m^2$.

## Table 21-3. Comparison of mechanical properties between cellophane and oriented PVC[a].

| PROPERTY | CELLOPHANE | EMULSION PVC LONGITUDINALLY-ORIENTED | EMULSION PVC BIAXIALLY-ORIENTED |
|---|---|---|---|
| Thickness (μm) | 40 | 40 | 40 |
| Tensile strength ($kg/cm^2$) MD | 1200 | 1500 | 1000 |
| Tensile strength ($kg/cm^2$) TD | 800 | 500 | 775 |
| Elongation at break (%) MD | 22 | 60 | 70 |
| Elongation at break (%) TD | 50 | 200 | 135 |
| Shrinkage @75°C (%) | None | ≤ 5 | ≤ 5 |
| Tear resistance | Very low | High | High |
| Tear propagation[b] (cm = g/cm) MD | 15 | 35 | 50 |
| Tear propagation[b] (cm = g/cm) TD | 20 | 150 | 70 |

[a] Data from *Adhesives Age*, **15**(3): 17–23 (1972).
[b] ASTM D689.

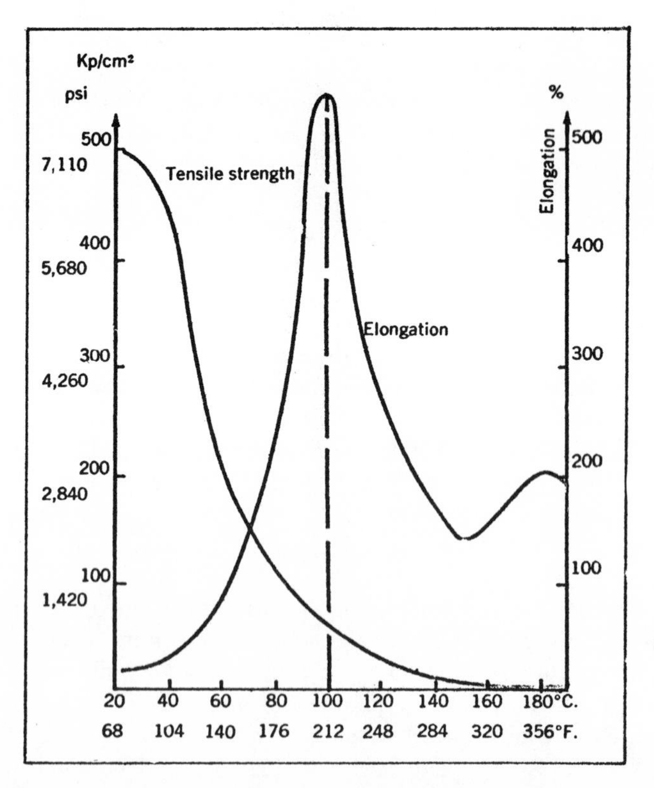


Fig. 21-3. Tensile strength and elongation of vinyl film backing as a function of temperature.

**Table 21-4. Solvent resistance at 20°C[a] (MD oriented rigid emulsion PVC film, 40 μm thick).**

| SOLVENT | TIME TO RESIST DIMENSIONAL CHANGE (sec) | BOILING POINT (°C) | RELATIVE EVAPORATION RATE |
|---|---|---|---|
| Tetrahydrofuran | 20 | 66 | 2.6 |
| Acetone | 30 | 56 | 2.1 |
| Methyl ethyl ketone | 35 | 80 | 3.3 |
| MEK/Toluol, 1 : 1 | 60 | — | 4.2 |
| Ethyl acetate | 150 | 76 | 2.8 |
| Methyl isobutyl ketone | 600 | 116 | 7.5 |
| Butyl acetate | 1700 | 121 | 12.5 |
| Toluol | 2300 | 110 | 6.1 |

[a] Data from *Adhesives Age*, **15**(3): 17–23 (1972).

## KRAFT PAPER TAPES

Kraft paper tapes are constructed by extrusion coating the paper with 18–20 g/m$^2$ polyethylene coating, corona discharge treating the surface and applying 0.6–0.8 g/m$^2$ silicone release coating. Clupak-type kraft paper is used with an elongation of 7–10% instead of 4% for the regular kraft paper. The higher elongation at break gives a much tougher tape. The properties of Clupak kraft paper used for the tape backings are shown in Table 21-5.

A pressure-sensitive adhesive coating of 35–42 g/m$^2$ is applied over the paper. The adhesive is usually natural rubber-resin type coated from solution, but hot melt adhesives are also used for this product (Fig. 21-4). The silicone release coating used has an easy release and the overlap over the backing has a tendency to flag. A kraft paper tape of the composition shown in Fig. 21-5 overcomes some of these problems, but the manufacturing costs of this tape are higher. The kraft paper tape is used as carton

**Table 21-5. The physical properties of Clupak kraft paper.**

| PROPERTY | CLUPAK KRAFT PAPER |
|---|---|
| Thickness (mm) | 0.119 |
| Weight (g/m$^2$) | 91 |
| Tensile strength MD (kg/15 mm width) | 9.5 |
| Elongation MD (%) | 7.3 |
| Tear strength CD (g) | 128 |
| Peel adhesion (g/50 mm width) | 44–64 |

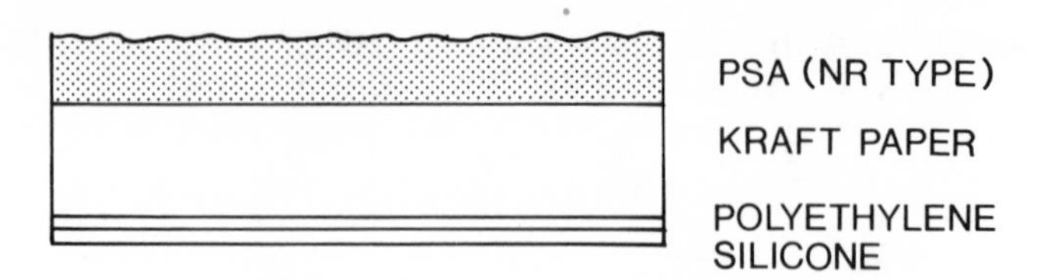


Fig. 21-4. Kraft paper tape with a silicone release coating.

sealing tape and also as general-purpose packaging tape. In Japan, this type of tape has the major position among the packaging tapes.

Kraft paper tape is also useful as tamperproof tape. The internal bond of unsaturated paper is sufficiently low so that the tape cannot be removed without destroying the backing. The construction of such product is discussed by Dabroski.[5] Various release coatings that can be used are also covered.

## CLOTH TAPES

Historically, the cloth industrial tapes have developed from the hospital tapes which were produced by calendering. Calendering is still an important method in manufacturing these tapes. Polyethylene coated fabric tapes can be manufactured by calendering in applying polyethylene and adhesive mass in the same calendering pass on a three-roll calender. A schematic diagram of this process is shown in Chapter 26.

Polyethylene coated cloth tapes can also be manufactured by applying polyethylene coating by extrusion over a low cost primed rayon or cotton fabric. In some cases, a release coating is used over polyethylene. The composition of this tape is shown in Fig. 21-6. Polyvinyl chloride coated fabric tapes are also used for packaging applications.

Various adhesives are used for cloth tapes: natural rubber, SBR, reclaimed natural rubber, block copolymer-resin systems are most commonly used.

Packaging applications of cloth tapes are mainly sealing and waterproofing. A smaller amount of high strength glass fiber cloth tape is used for bundling and strapping applications.

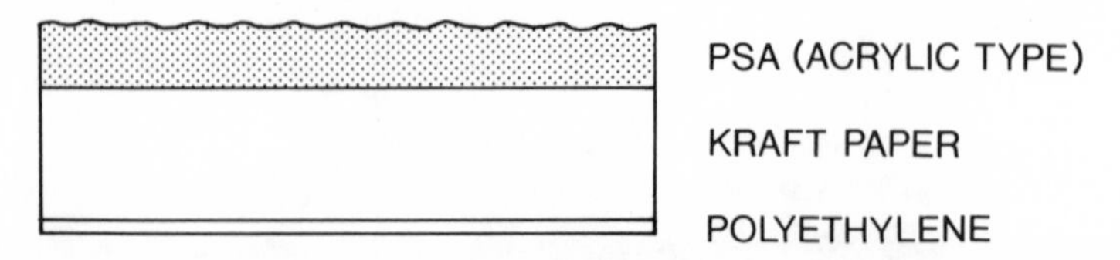


Fig. 21-5. Acrylic adhesive kraft paper tape.

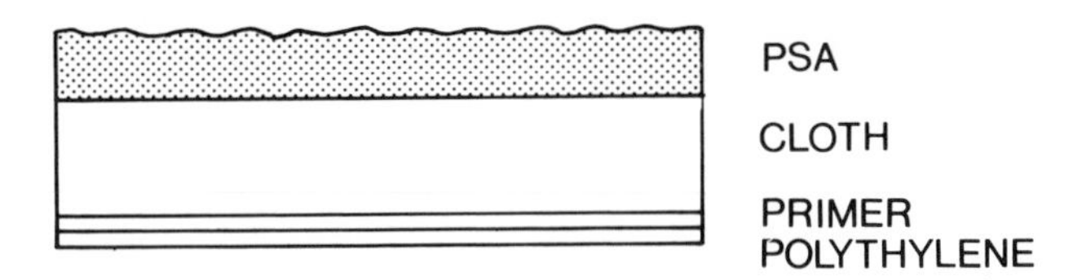


Fig. 21-6. Polyethylene coated cloth tape.

## SATURATED PAPER TAPES

Saturated paper tape is a very important product and it clearly was the most important construction for many applications, especially masking, but was also heavily used in packaging. The advent of film-backed tapes displaced the saturated paper tapes from many applications. Saturated paper tapes are widely used for many packaging applications including freezer tapes, holding produce, carton sealing, labeling, and packaging refrigerators. Bandoleering tape is used to hold electronic components, staples, nails, etc. Some of this is needed temporarily during manufacturing; some become a permanent part of the packaging such as supplying nails for a gun application. Saturated paper tapes are often used for this purpose.

Surface protection is a packaging related use. Pressure-sensitive saturated paper and sometimes film tapes are used to protect large polished metal or plastic sheeting surfaces from damage during shipping. Low tack pressure-sensitive adhesive, sometimes only lightly compounded natural rubber latex, is used. The adhesive should not leavy any residue that might require later cleaning. Typical flatback protective tape properties are shown below.

| | |
|---|---|
| Thickness | 0.16 mm |
| Tensile strength | 3.75 kg/cm width |
| Elongation | 3% |
| Quicktack | 10 g/cm width |
| Peel adhesion | 90 g/cm width |
| Unwind | 110 g/cm width |

Crepe and flat back paper is used for these tapes. The saturating process is described in a separate chapter.

## JAPANESE PAPER TAPE

Japanese paper is made of Manila hemp. Its properties are shown in Table 21-6 and the tape construction in Fig. 21-7. The paper is saturated with

**Table 21-6. The physical properties of Japanese paper.**

| PROPERTY | JAPANESE PAPER |
|---|---|
| Thickness (mm) | 0.059–0.060 |
| Weight ($g/m^2$) | 31.0–31.9 |
| Specific gravity ($g/cm^3$) | 0.52–0.56 |
| Tensile strength MD (kg/15 mm) | 5.10–5.97 |
| Tensile strength CD (kg/15 mm) | 0.96–1.17 |
| Elongation (%) MD | 2.4–3.1 |
| Elongation (%) CD | 2.9–3.5 |
| Tear strength (g) | 48–60 |

shellac or with synthetic resins. It is used for light packaging, masking, and surgical applications. It is a fairly popular tape in Japan.

## REINFORCED TAPES

Filament reinforced tapes are the most popular for bundling, strapping and other applications requiring high tensile strength. These tapes are also reasonably bulky and can take considerable abuse. The first filament reinforced tape employed was made with acetate tow. Later, glass fiber filaments and polyester fiber filaments replaced acetate. (Such tapes are discussed in References (6) and (7).)

There are also fabric reinforced tapes available, starting with low count cotton fabric and ending with glass fiber or polyester fiber high tensile strength fabrics.

## OTHER FILM TAPES

Cellophane tape was the first film tape introduced. It is still used as a general household and office tape, including some packaging applications. Cellophane film 36 $\mu m$ thick is most commonly used for tape backings.

Cellulose acetate film tape has replaced cellophane tape in many applications. The tape has a lower tear and tensile strength than other film tapes and its main application is for office mending and general household uses.

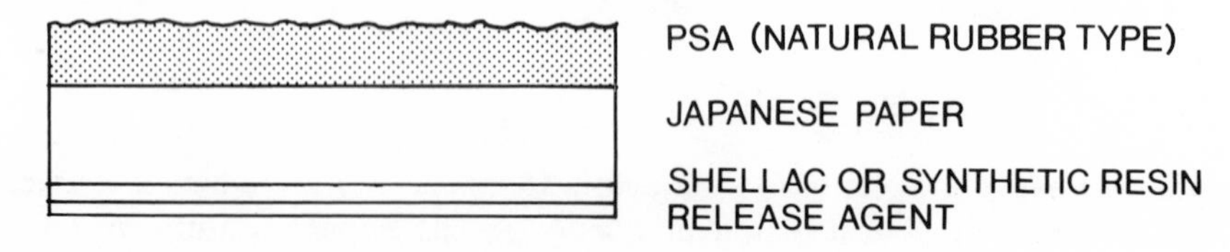


Fig. 21-7. Japanese paper tape.

A matte finish and the acceptance of ink by the film surface are the attractive features of the acetate tape. Films 38 and 44 $\mu$m thick are used for tape backings.

Polyester film tapes are also popular for various packaging applications but have not reached the acceptance polypropylene and vinyl films did.

Protective film tapes are also used along with saturated paper tapes for covering of polished metal, acrylic sheet, plastic laminate, and other surfaces. Products 50–100 $\mu$m thick are sold.

## TAPE DISPENSING

Tape dispensing for various packaging applications is very important and there are over 150 various types of tape dispensers starting from hand

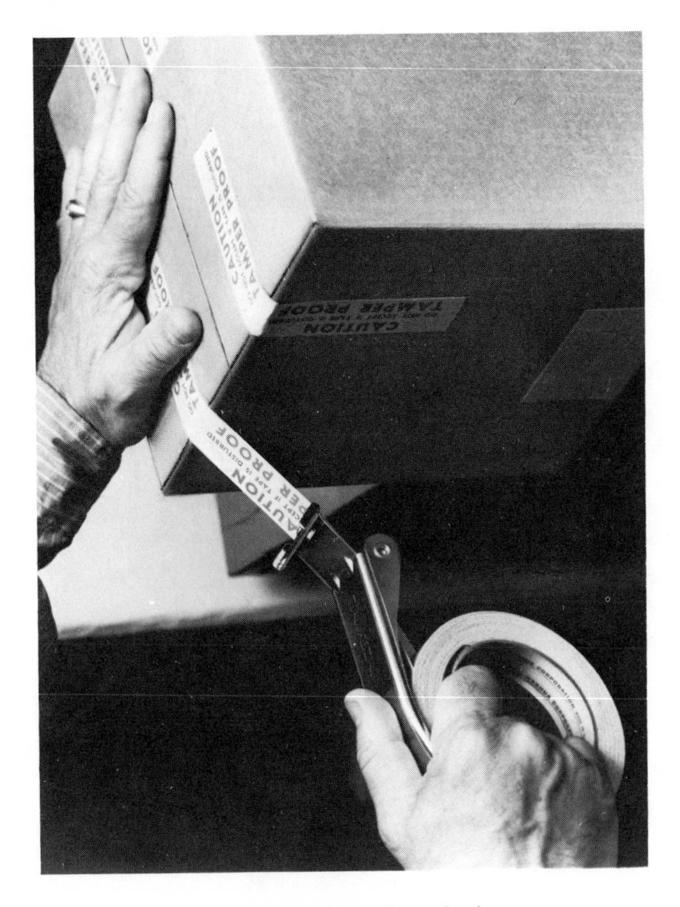


Fig. 21-8. Hand applicator of a packaging tape.

applicators to powered and automated systems. In manual application, the product is placed in position by hand and the tape is applied by hand. In semiautomatic application, the product is placed in position manually and the tape is applied by a powered mechanism. In the completely automatic system, the product is moved and the tape is applied mechanically. The required speed of application determines the type of the dispensing equipment: 10 strips per minute can be applied manually, but faster application requires automation. Simple hand applicators are shown in Fig. 21-8 and 21-9.

## TAPE TESTING

Packaging applications require some special testing specifically designed to simulate conditions to which the tape is exposed in actual use. The corruga-

Fig. 21-9. Hand applicator of a packaging tape.

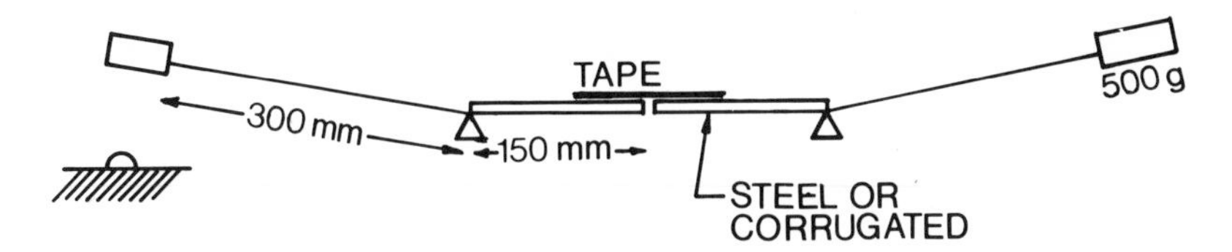


Fig. 21-10. A tape tester simulating the forces exerted on a carton closure tape.

ted box closing application has an important place among the packaging applications. The tape is subjected to continuous shear and low angle peel forces. Formulation of an adhesive that performs well has been a difficult problem. If the adhesive mass is of low cohesive strength, it fails in shear; if the shear resistance is improved, it has a tendency to fail adhesively, popping away from the corrugated surface.

The tape performance can be tested by hanging a loaded corrugated container over a bar so that the force is exerted over the taped cover. A special testing device has been used in which the tape is exposed to shear and peel forces approximating the actual condition. A schematic diagram of such a tester is shown in Fig. 21-10.

## REFERENCES

1. Morgan, B. D. U.S. Patent 3,335,021 (1967) (assigned to Morgan Adhesives Co.).
2. Nachtsheim, H. G., and Meyer, E. J., U.S. Patent 3,089,786 (1963) (assigned to Minnesota Mining and Manufacturing Co.).
3. Screnock, J. J., and Forese, J. L. U.S. Patent 3,152,004 (1964) (assigned to Mystik Tape).
4. Reip, H. *Adhesives Age* **15**(3):17–23 (1972).
5. Dabroski, W. C. U.S. Patent 3,285,771 (1966) (assigned to Johnson and Johnson).
6. Bemmels, C. W. U.S. Patent 3,073,734 (1963) (assigned to Johnson and Johnson).
7. Hauser, W. R., and Brown, R. C. U.S. Patent 3,179,552 (1965) (assigned to Minnesota Mining and Manufacturing Co.).

# Chapter 22

# Automotive Applications

**Whiteford D. Grimes**
*Stauffer Chemical Company*
*Westport, Connecticut*

Pressure-sensitive usage in the automotive industry surely began with masking tape in its traditional role as an aid to manufacturing, protecting trim areas during painting. This logical use that made it easier to offer assembled units of painted metal, glass, and bright trim is typical of the thousands of applications for adhesive-coated products that save time and labor, while providing improved utility to a product. From this basic use, pressure-sensitive materials have grown with the industry to become essential engineering materials in a wide range of applications.

In general, uses of pressure-sensitive adhesives in the automotive industry encompass every known characteristic of such materials. The reasons for their use, likewise, run the gamut of justifications described in industry trade literature and sales promotions. In short, pressure-sensitive adhesives are used for temporary benefits and to add permanence; they are used to save money, to increase functionality and to achieve a higher profit per unit of sale.

Other chapters of this book deal with the mechanics of adhesion and adhesive chemistry. In this section, the focus will be on applications, materials, performance and benefits.

The function of pressure-sensitive adhesives in all automotive applications is to fasten one material to another. In the paint masking example, the adhesive fastens a protective paper to trim and glass areas, but retains its characteristic of easy removal so that the paper can be discarded when the paint has dried. In a more sophisticated use, such as the application of body side moldings, an adhesive is used to attach a shaped vinyl extrusion to a plastic foam which is then affixed permanently to the painted side of the vehicle with yet another layer of adhesive. These two examples illustrate the complexity of pressure-sensitive uses for the automobile; they bring to mind questions of adhesion, durability, chemistry, mechanics, and even manpower utilization.

Every pressure-sensitive application must be assessed for its own merits and objectives, but of basic concern to all uses is the surface to which an adhesive is to be applied. Surface characteristics are so varied on an automobile that some understanding of principal surface materials is essential to any discussion of applications for this industry. A brief review of general material characteristics follows, but the reader is encouraged to study a specific surface in depth for a thorough understanding of its impact on a pressure-sensitive application of interest.

## PAINT

By far, the most common adhesive application surface is the exterior paint used on the vehicle body. But "paint" is not a singular term and, with the rapid development of environmental regulations coupled to inflation in costs of petroleum-based ingredients, new paint systems and materials are constantly being introduced into automotive manufacturing plants. Each paint system differs from the others in its effects on an adhesive material. This variation in properties results primarily from the chemistry of the paint formula, although the painting process can also be a factor. Paints consist of resins, pigments, modifiers, and solvents. Each is merely a general classification of a whole range of materials with its own chemical identity. The chemistry of paints, therefore, is very complex and the potential for quality differences is quite high. The following illustration may serve to develop some understanding of the effects of paint type on pressure-sensitive properties.

**Pressure-sensitive Woodgrain Adhesion (oz/in.)***

| | Days at Room Temperature | | | | | | | |
|---|---|---|---|---|---|---|---|---|
| | 1 | 3 | 7 | 10 | 14 | 17 | 21 | 28 |
| Standard Lacquer | 53 | 55 | 116 | 99 | 102 | 95 | 99 | 88 |
| Nonaqueous Dispersion | 52 | 56 | 106 | 108 | 98 | 95 | 96 | 95 |
| Water Base Enamel | 40 | 50 | 67 | 72 | 86 | 85 | 104 | 102 |

* Data courtesy of Stauffer Chemical Co.

The data show adhesion at room temperature of a single pressure-sensitive adhesive film on three different paint systems used in the automotive industry. The character of adhesion on the water base paint is decidedly different from that on the other two. Since all factors are apparently the same except for the paint, it can be assumed that some characteristics of the water emulsion are delaying the buildup of adhesion. Inasmuch as the ultimate level of adhesion for all systems is approximately the same, the performance of the product on the car may not be affected. The data illustrated were determined under laboratory conditions. That is, room temperature was controlled at 21°C; paints were formulated under close supervision; test panels were properly prepared before painting, including priming, curing, light sanding, and cleaning; application was made by controlled standard practices. Actual application on a vehicle assembly line would not be so controlled. Many normal variances of a manufacturing operation could exaggerate the differences of the data. For example, a typical application in a cold weather climate where cars are stored outside immediately after finishing could result in adhesion dramatically different from that pictured, even to the point where a bond may not exist.

## BRIGHT METALS

Many surfaces, both interior and exterior, on a car are either aluminum, chromium, or stainless steel. Pressure-sensitive application to each again varies with the type of surface and the consistency within each type. Metal treatment is an inexact science and where such operations as anodizing and plating are involved, wide differences in surface character are common between production lots. These differences result in differences in adhesion levels and aging characteristics. Even laboratory applications on anodized and plated surfaces often vary widely. Generally, these differences can be traced to the quality of the metal surface, but the frequency of "inexplicable" data on such surfaces raises doubts as to their suitability for laboratory evaluation.

Stainless steel, on the other hand, is a standard surface for adhesive characterization. This material is readily available, easily cleaned, and makes no apparent chemical contribution of its own to the total adhesion structure in a specific adhesive evaluation. In fact, the accepted method of characterizing a pressure-sensitive adhesive (not a particular application but only the adhesive itself), is to coat the adhesive on a "nonreactive material" such as a polyester film and then to apply it to a stainless steel test specimen. Thus, no chemical interaction between film and adhesive, or stainless steel and adhesive will be a factor in the characterization, and a true reading of adhesion, heat aging, humidity resistance, etc., will be obtained.

## PLASTICS

Plastic materials are increasing rapidly in use as structural and surface materials for automotive parts. ABS (acrylonitrile-butadiene-styrene), vinyl (polyvinyl chloride), glass-reinforced polyesters, polycarbonates, and polyolefins are a few of the plastic resin materials in wide use on today's automobiles. There are others and, surely, more to come. These materials are growing in popularity because of their ease of fabrication, durability, low cost, low density (potential for vehicle weight reduction), chemical resistance, and strength characteristics. Each plastic material has its own identity in relation to pressure-sensitive adhesives. Some have a strong affinity for adhesive bonding, while others totally reject conventional adhesive mechanisms. Fortunately, most plastic structural parts are painted and, although their chemistry may have made it difficult for the painting process, the paint itself may be a more consistent surface for pressure-sensitive adhesion. For direct application to plastics, specific materials must be matched for optimum performance and, in some cases, fabrication with adhesives may either be impractical or require materials and techniques not easily obtained.

Considering all surfaces, it is important to understand that the character of the material to which an adhesive product is to be applied may have an overriding impact on the suitability of that product to the application. Performance on one surface by no means implies equal performance on another. Additionally, one should keep in mind that any automotive surface must be properly prepared for adhesive application, e.g., cleaned, lightly sanded, and conditioned to a temperature within the application range for the particular adhesive being used.

## APPLICATIONS

In the following pages, specific pressure-sensitive applications for the automotive market are reviewed. This is not an all-inclusive list, but only illustrations of uses, provided for reader interest in hopes of furthering understanding of pressure-sensitive adhesives and the versatility of such materials in fulfilling market needs. Commodity uses, such as masking for painting and tape bindings for electrical harnesses, are not discussed here.

The most visible pressure-sensitive applications on automobiles are those that fulfill aesthetic objectives, e.g., striping, decals, exterior tops, and woodgrains. Less recognized applications hold down wiring, seal joints and seams from moisture, provide instructions, affix trim, eliminate rattles, and compensate for loose tolerances. The most complex of these items is probably an exterior woodgrain film because of its stringent requirements for both design and performance. An in-depth explanation of this product is presented in the following section to facilitate understanding of all pressure-sensitive materials and their use on automotive products.

## EXTERIOR WOODGRAIN FILMS

The use of wood as a decorative trim on automobiles is almost as old as the industry itself. Early in their history of infatuation with the automobile, Americans realized the practicality of combining a utility vehicle with an active family lifestyle and came up with the station wagon. The wooden body was a natural reminder of country living and became the standard for luxury and social recreation. For more than 40 years, real wooden bodies were the top of the line, but eventually the cost of fabrication and problems with durability sounded the death knell of this classic reminder of another day.

Today "wood body" station wagons are simulated by applications of woodgrain decals. Through gravure engraving and subsurface printing, beautiful woodgrains can be reproduced in exact detail, providing durable, realistic woodgrains at minimal cost.

There are two types of films used to produce exterior quality woodgrain decals, cast vinyl and calendered vinyl. The film type is derived from the process used for production. *Casting* is the process of forming a film from a liquid compound by spreading it to a desired thickness over a support web or carrier that has suitable release characteristics. An external energy source is used to fuse, cure, or dry the film while still on its carrier. *Calendering* is the process of forming a film from a melted compound by passing it successively through narrow passages formed between a series of hot milling rolls until it reaches a desired thickness and width, whereupon it is cooled sufficiently to set film properties.

Cast vinyl films are characterized by exceptional dimensional stability which results from the carrier nature of the process. Since the film is supported through all of its processing, there are no residual stresses to cause shrinkage in the web. This characteristic is quite important to woodgrain decals because they are supplied to the automotive assembly line in precut shapes that must fit within carefully defined tolerances. Cast vinyl films are produced only from dispersion vinyl resins whose high initial costs are compounded by excessive materials handling requirements and thus entail a substantial cost penalty.

Calendered films, on the other hand, are produced from bulk handling suspension resins, the commodity class of PVC resins in the United States. Fortunately, calendered films can be made dimensionally stable by additional processing and are quite suitable for woodgrain products. A further advantage of the calendered product is the versatility that evolves when a product is a commodity. This versatility in use has resulted in a mature compounding technology, not available to cast products. Proven compound additives and select suspension resins provide the wherewithal to achieve exceptional resistance to the degrading effects of ultraviolet light and heat. Calendered vinyl films used for exterior automotive woodgrains have an unparalleled record of performance in actual use, in contrast with the uncertain performance of cast films which frequently whiten and deteriorate because of limited cast film technology.

With either film, the ultimate performance against the effects of sunlight and humidity is dependent not only on the compound used to form the film, but also on each ingredient used in formulating the gravure inks that print the woodgrain beneath the clear film surface. A properly engineered woodgrain vinyl product should withstand controlled exposures in Florida and Arizona for up to 36 months without exhibiting a noticeable color change. This lack of change upon exposure to the most hostile environments for plastics and finishes in the United States is testimony to the engineering properties of vinyl and the maintenance of control over quality for each of dozens of ingredients from resin to ink. Such an exposure exceeds the expected life span of an automobile.

## Design

The woodgrain appearance of automotive films results from a gravure print beneath a clear vinyl film. The development of this print is an intricate process of design and engraving that begins with actual pieces of wood. Segments of wood or flitches exhibiting interesting grain characteristics, are photographed under various lighting conditions to isolate particular elements of the wood, such as tick and grain. From a series of such photographs, and with the use of airbrush toning techniques, the designer assembles composite photographic panels that are, in fact, a redesign of nature's work. Photographs of such composites are then used by the designer to respond to particular market assignments.

For an automotive use, design and dimension requirements vary with the specific vehicle model. This is in consideration of the size of the decal and the placement of design on the vehicle for visual appeal. Thus, when a design need arises, it is necessary to select first the grain type (walnut, cherry, birch, etc.) and then from photographic structures of that grain, an appropriate design as a starting point. Using photographs of the individual elements of the composite, the designer selects those characteristics that best suit his need, and by cut and paste techniques, restructures the woodgrain to fit the particular job assignment. To do this effectively, the designer must understand both the market and the process through which the woodgrain will ultimately find use.

The printing process requires that each of the elements of the composite design be applied to the print surface individually and successively, so that only after all are printed is the wood design finally produced. This separation is necessary to print a single ink composition (color) on the surface and to dry it before another can be printed. Thus, each of the elements of design is used to produce a gravure print cylinder through a process called photoengraving. The photoengraving process transfers the design to copper-plated steel cylinders. This involves exposure of each design separation to a light-sensitized emulsion paper which is subsequently applied to the cylinder and serves as a resist in chemical etching of the copper. The design is etched in the form of tiny cells of varying depth, several thousand per square inch, which serve as inkwells in the printing process. The engraved cylinder is plated with chromium to provide a durable wear surface and thus consistent design reproduction.

In printing, the design is applied to the clear vinyl film so that the woodlike effect is visible only through the film. In this way, the protective properties of the vinyl formulation are provided for the aesthetic effects of the inks and design. The print may be made direct to the film in a reverse order, i.e., grain, tone and ground coat, or in the more natural order, i.e., ground, tone and grain, on a release carrier for subsequent transfer to the film under heat and pressure.

In a manner similar to the transfer of printing, the woodgrain film is made pressure-sensitive by first coating an adhesive on a release paper and laminating this combination to the ink side of the printed film. The finished laminate (film, print, adhesive, and release carrier) has only to be die cut to specification dimensions to become a decal, ready for application on a vehicle assembly line.

## Adhesive

The adhesives used for woodgrain decals are selected to provide a balance of properties suitable for the film/ink combination, application techniques and performance as an exterior automotive component. The use establishes the controlling criteria. Specifications issued by the original equipment manufacturer's engineering departments spell out requirements for adhesion levels and time of achievement, resistance to environmental conditions as represented by laboratory testing, actual exposure to hostile climates and application techniques. Although the adhesive is a critical part of such evaluations, it is the total package of film, ink and adhesive, and the interaction of all three with the vehicle surface, that is under test. Experience has shown that only highly cohesive acrylic adhesives thus far have been able to satisfy the demanding range of properties necessary for automotive use.

### Performance Criteria

Each of the automobile manufacturers has its own requirements for qualifying a woodgrain product for use on its vehicles. Although these specifications vary somewhat in detail, the principal concern of all is the same: application on the assembly line and durability in use.

### Application

Woodgrain decals made for application to fenders, door panels, and tailgates are quite large, often exceeding 10 $ft^2$. Yet they must be permanently and expertly applied to a station wagon in a matter of minutes as the vehicle travels down the assembly line. To ease this difficult task, a wet application method is used. The protective release paper is separated from the adhesive-coated film and the entire area of adhesive is wet down with a mild detergent solution (about 1%) in water. This can be done with a sponge or by immersion, whatever is more convenient. The solution prevents the adhesive from sticking to itself, or to the vehicle before it is in its proper position. The wet decal is positioned by hand on the station wagon, sometimes with the aid of "key" notches in the die cut part. A hard rubber squeegee is then used to force out the solution from under the decal, thus effecting the pressure-sensitive bond. Difficult areas, such as door handle and gas cap recesses, or sharply contoured body areas may require the use of a heat gun with the squeegee to soften the film enough to conform for adhesion. The application process is one place where the differences between cast and calendered films become apparent. The cast film is a little softer and conforms more readily, but it is also weaker and tends to tear more easily during such operations. Calendered films may require more heat for conformance, but tearing is seldom a problem. Application personnel must become familiar with the characteristics of the film they are working with before proficiency is established. Both have been used in high-speed assembly areas for years and both have strong advocates, depending on experience and familiarity.

From a qualification standpoint, the automotive manufacturer is concerned with detergent concentration, heating techniques, conformance to diecut specifications and worker satisfaction. A product difficult to apply will not be long in use.

### Product Durability

The criteria for product performance in use center about adhesive properties and resistance to weather and road hazards. Evaluations of product are made in the laboratory and at test sites located in Florida and Arizona. Each automotive manufacturer establishes testing procedures and performance levels according to its own desires. Tests are made on steel panels painted in accord with plant paint procedures and using current paint formulae. Although there are significant differences in particular criteria, the overall testing programs are nearly the same for all, with the following concerns being typical.

Adhesion: Two lb/in. of width, one hr after application and $3\frac{1}{2}$ lb, or more after 72 hr.

Environmental Cycle: Exposure to alternate conditions of heat and humidity over a two-week period. A typical cycle is 17 hr at −30°C, 72 hr. at 80°C, and 24 hr. at 38°C, and 100% R.H., then repeat for the second week.

Gravelometer: Subfreezing exposure and impingement of a specific gravel mixture at road velocities.

Weather Resistance: Exposure in standard weather-o-meter cabinets (controlled temperature and humidity cycles) with prescribed ultraviolet arc lamps. 1000–1500 hr. of exposure are generally required.

Hostile Environment: Exposure at independent test sites in Florida and Arizona for approximately two years at a prescribed angle to the sun.

Other typical controlled tests include heat aging, abrasion, humidity exposure, gasoline resistance, and cleanability.

In most of the tests, the principal evaluation is that of appearance. An acceptable product is not expected to exhibit significant color change, crazing, blistering, or other indications of film degradation.

## STRIPING FILMS

Unlike woodgrains, the use of striping products is of recent vintage. Several major developments can claim responsibility for this popular practice of decorating automobiles and other vehicles with striping and decals of every description: (1) the trend to sport models that developed in the early 1970's reflecting a youth-oriented society; (2) the high cost of retooling automotive assembly lines that has extended the life of body styles to five years or more and necessitated other means of providing product differentiation; (3) the energy situation that is forcing weight reduction and resultant similarity in all car models as manufacturers struggle to meet Government-imposed limitations on miles per gallon of gasoline; and (4) the development of durable vinyl and fluoroacrylic alloy films, and high-performance adhesives that make vehicle striping easy, distinctive, and lasting.

Whereas the history of striping differs greatly from that of woodgrains, the technologies of the two are quite similar. Both cast and calendered films are widely used; printing is an important part of the striping process; and design is the key to achieving desired appearance results. Solid color stripes may be made by film pigmentation, surface printing by gravure and screen process techniques, and subsurface printing by the gravure process. Multicolored products, such as the large hood decals of birds and insects, are generally printed by screen processing on clear films. An important part of striping and decal programs is thermal die cutting that enables the film processor to provide complex designs of precise dimensions with clean edges that hold fast to the application surface.

Performance criteria for decals and striping are much like those for woodgrain films, except where screen printing is involved, surface durability cannot be expected to match that of the subsurface printed materials. Application techniques are similar, but wet application is only practical for very large decals. Most narrow striping products, or those requiring a specific arrangement in relation to other striping, require a premask application paper to support the striping and hold dimensions until adhesion is effected.

## INTERIOR WOODGRAINS

As with exterior woodgrains, woodlike pressure-sensitive decals on the interior of the automobile had their beginning in real wood. Many fine cars, until only recently, had instrument panels fashioned from exotic woods, finished with many coats of highly polished lacquer. Now woodgrain prints exhibit the same grain character as the woods they replaced, but are considerably less costly, and highlight not only instrument panels, but also door surfaces and seat frames. Some of these decorative products are merely wood prints transferred to vacuum-formed ABS, while others are pressure-sensitive vinyls or pressure-sensitive formable laminates of ABS to vinyl. Companion moldings on doors are made by laminating the printed vinyl to extrusion moldings that are electronically sealed to the door trim vinyl material.

Performance and application criteria are similar to that of exterior products, with emphasis on fabricating characteristics for extrusion and vacuum forming and less concern for ultraviolet resistance. Because the products are used on the interior, vinyl materials must also be formulated to resist chemical exudation that could form a cloudy film on windows in hot weather. Pressure-sensitive adhesives are again primarily acrylic, and applications are frequently made on bright metal surfaces, such as anodized aluminum.

## INFORMATIONAL DECALS

A modern automobile at manufacture may have as many as 20 or more pressure-sensitive decals providing use information, identifying manufacturers, and listing specifications. Examples include such things as tire pressure labels, jack operating instructions, trade names and product type, catalytic converter performance, and EPA conformance warranties. Various film and adhesive materials go into the manufacture of such decals. The most common films are vinyls because of their versatility, ease of printing and durability, but paper, polyester and acrylic films are also used. Adhesives are usually acrylic, but for unusual surfaces such as the polyolefins, or where destructibility of the decal is required (as with an EPA conformance label) to demonstrate that a label was provided, rubber-based adhesives may also be used.

Automotive manufacturers have complete specifications for every label and decal used on their vehicles. Each specification also provides test method information and application criteria.

## EXTERIOR AUTOMOTIVE TOPS

With the growing popularity of air conditioning for cars that developed in the early 1960's came the near obsolescence of the once-popular convertible. Rather than let the appeal of a convertible become a thing of the past, the automotive designers came up with a new body style, popularly known as the "hardtop convertible." This new design was merely a two-door coupe without door posts and a roof line designed to preserve the convertible look. It was not too long before the vinyl industry capitalized on the popularity of this new body style by developing an exterior coated fabric for application to the roof that made the hardtop convertible all the more realistic. By the mid-1970's, more than 50% of all American-made automobiles came off the assembly line with vinyl tops.

Application of vinyl tops was achieved by the spraying of contact cements on both the painted roof of the car and the prefabricated vinyl-coated fabric. This was, and continues to be, a complicated means of application that promises uncertain results at best. In effect, the quality of the vinyl top became the quality of the spray application, with a poor spray resulting in delamination, rusting and a short life. In 1977, Stauffer Chemical Co., with a unique combination of pressure-sensitive and coated fabric product capability, conceived the idea of an exterior top fabric with a factory-applied pressure-sensitive adhesive. The result was the product "Starlan" exterior vinyl top (patent pending), that provided improved adhesion, added corrosion resistance, and application security, while eliminating spraying operations and attendant environmental health concerns. This application of pressure-sensitive adhesives is a classic example of the way these products can save labor, improve product performance, and solve problems of manufacturing.

## ADHESIVE SYSTEM TAPES

The advent of adhesive systems in the form of double-coated and transfer tapes has been of considerable value to the automotive industry. With a variety of adhesives and product structures, these materials successfully save labor, reduce weight, solve manufacturing problems, and improve product performance. The body side molding tape mentioned in the beginning of this chapter is a good example of such a product. This tape, developed and supplied by the 3M Co., is a double-coated tape that uses an acrylic adhesive in concert with a tough neoprene foam to provide a secure bond for the trim part to the body paint (or in some cases, a woodgrain decal). The foam absorbs flex and temperature shocks and compensates for surface irregularities, while the adhesive on either side of the foam holds the assembly together. The success of such products has eliminated the drilling of holes and attaching of clips for trim applications, both sources of rust, and makes possible the use of lighter and more durable plastic trim materials.

Other automotive uses of adhesive system tapes include the application of foam underlay for vinyl tops, the holding in place of foam gaskets around light assemblies, and the sealing of welded seams at the roof line with adhesive-coated closed cell foams.

The automotive industry must continue to search for new ways of completing assemblies in order to meet market demands for performance, energy demands for weight reduction, environmental demands for worker and community health and economy demands for competitive security. Pressure-sensitive adhesives, used in imaginative ways with new and traditional engineering materials, are playing a major role in this effort. The list of uses grows longer every day because adhesive and application technologies continue to expand. The limitation of such growth is not yet chemistry or processing, but rather the creativity of those in both the automotive and adhesive industries in addressing the problems presented by a continually changing automotive market.

# Chapter 23

# Plastic Tape Pipeline Coatings

**G. M. Harris**
*The Tapecoat Company*
*Evanston, Illinois*

## HISTORY

Plastic tapes were introduced to the pipeline industry as pipeline coatings more than 25 years ago. Since then, the worldwide consumption of plastic tape pipeline coating has grown to the point that plastic tapes are used yearly on more than 50% of all new underground pipeline projects. The rate of growth in the use of plastic tape coatings was slow but steady through the first 10 to 15 years, but more recently the rate of growth of plastic tapes as pipeline coating has accelerated rapidly.

The yearly world consumption of plastic tape pipeline coatings is shown graphically in Fig. 23-1. This bar graph shows the amount of plastic tape used as pipeline coatings at the three decade intervals of 1956, 1966, and 1976. It is apparent by comparing the amounts of plastic tape used in 1966 and 1976 that within that 10-year period, plastic tape pipeline coatings experienced a major breakthrough in acceptance and consumption in the pipeline coating field. In 1976, the annual use of plastic tape coatings reached 100 million $m^2$ from the annual consumption in 1966 of 35 million $m^2$. It is forecast that the use of plastic tape pipeline coating will continue to grow, perhaps at a lower rate than that experienced between 1966 and 1976, so that by 1980, annual use will be approaching 150 million $m^2$. Of course, this assumes that pipeline construction will continue at expected rates.

In 1970, which appears to be the turning point, the annual usage of plastic tape coatings was greatly accelerated; this coincides with the 20th anniversary of the first use of plastic tapes on underground pipelines.

There are two factors accountable for this use and acceptance breakthrough which occurred in 1970. First, by that time, plastic tape pipeline coatings had established a long-term continuous record of high level in-service performance under a wide spectrum of operating and environmental conditions. Second, by this time, the plastic tape manufacturers, working with the most recent advances in plastics technology, had developed improved products.

## BASIC COMPOSITION

Although there is a wide range of plastic pipeline coating tapes commercially available made by several different manufacturers, all of the products have several common properties:

1. Their basic construction is identical.
2. They all employ a backing made from a thermoplastic polymeric material and an adhesive based on elastomeric material.
3. They all require the use of a suitable primer to enhance their bonding qualities.
4. They are all supplied in long strips wound in roll form onto a hard hollow core.

The thermoplastic polymers used as the backings are:

- Polyethylene
- Polyvinyl chloride
- Polyethylene copolymer
- Polybutylene

The elastomers used as the basic ingredient in the adhesives are:

- Butyl rubber
- Ethylene propylene rubber
- Natural rubber
- Other synthetic rubbers

These elastomers are compounded with materials such as hydrocarbon resins, polyterpene, coal tar, and asphalt in order to attain the desired bonding properties. In some cases, the bond of the adhesive to the backing is so great that the adhesive must be faced with an easy release type inter-

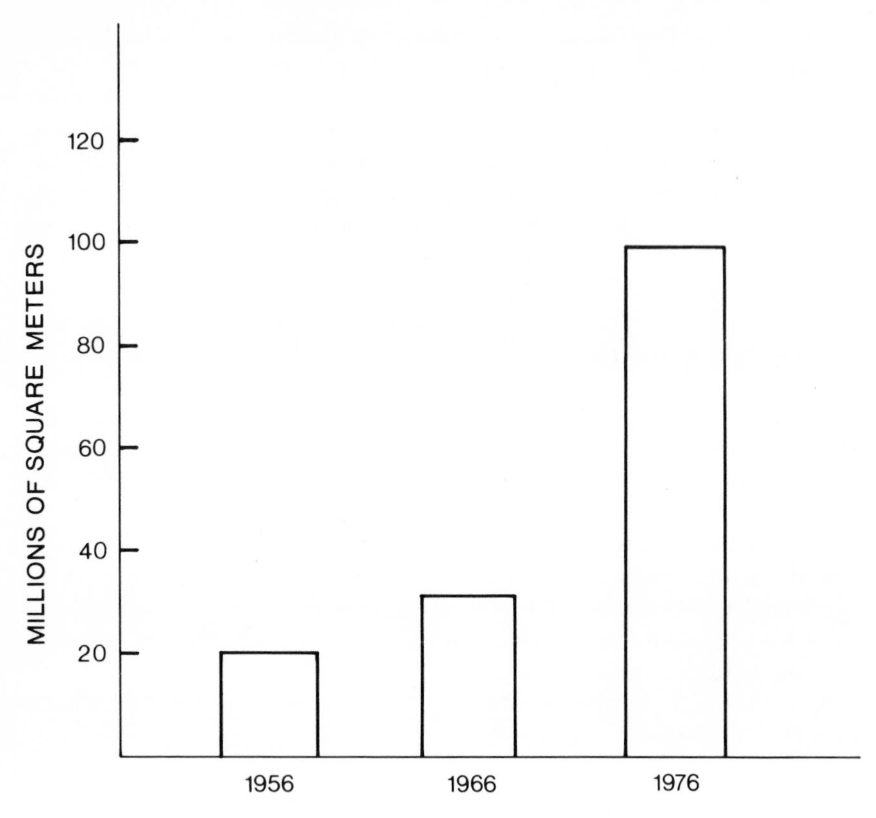


Fig. 23-1. Annual consumption of plastic tape pipeline coatings.

liner so that the tapes can be unwound from the rolls. In other cases, an interliner is not required.

The interliner may be a plastic or a coated paper. Generally, tapes with interliners lend themselves to manual application while the tapes without interliners are machine-applied. There are exceptions, notably the Alyeska Pipeline, where tapes with interliners were applied by power-driven machines.

## TYPES OF TAPE COATINGS

Taken from the standpoint of end-use, there are two types or classifications of tape coatings. These are joint wrap tape and long line tapes.

The joint wrap tapes are most often supplied in narrow width, short length rolls and are applied manually. They are used to coat short lengths of pipe and oddly shaped piping fixtures. Examples of objects that are protected by joint wrap tapes are:

- Welded field joints or plant coated pipe
- Pipe fittings such as tees, valves, elbows, and risers
- Tube turns and other sharp bends
- Tanks

Since joint wrap tapes are applied by hand and to shapes that are not perfectly cylindrical, they must be highly conformable and easy to stretch. This property is built into the joint wrap by using thinner backings and thicker adhesives than tapes used for long line purposes. They also usually have a greater overall thickness. The dimensions of a typical joint wrap tape are backing 0.125 to 0.250 mm, adhesive 0.625 to 1.125 mm giving an overall thickness of 0.875 to 1.25 mm. Joint wrap tapes have been used successfully on underground and underwater pipelines for more than 30 years. The earliest use and acceptance of plastic tape coatings was on small sections of pipe, such as welded field joints, where the tape was applied by hand. It could be said that the early success of joint wrap plastic tapes was the forerunner to the extensive use and broad acceptance of plastic tapes as pipeline coatings that exist today.

## LONG LINE TAPE COATINGS

Plastic tapes used as the only coating on pipeline projects, not just as joint wraps, are supplied in long lengths (up to 800 ft) and wide widths (up to 18 in.). They are applied by power-driven equipment both over-the-ditch, and at coating plants. They can be used on any size pipe. Many thousands of miles of 1400 mm pipe have been coated with plastic tapes outside of the United States. In 1980, a 2250 mm water pipe project in Colorado was coated with a plastic tape plant coating system.

Long line plastic tape coatings are tougher and have much greater tensile strength than joint wrap tapes and can only be applied smoothly, tightly and with high comformability by machines designed for tape application.

Tapes used as the corrosion protective coatings on pipelines usually have backings and adhesives of close to the same thickness—but most often the backing is thicker. Typically, the backing will make up 55 to 60% of the total tape thickness. Until recently, a tape with 0.50 mm total thickness was the most commonly used, but now the tendency is towards use of a tape with 0.625 mm overall thickness and greater in some special cases.

Plastic tapes for corrosion protection have always been used with an outerwrap to afford mechanical protection to the plastic tape. Outerwrap materials such as kraft paper, saturated asbestos felts, and plastic films have been widely used for this purpose in the past, but now almost 100% of the time, the outerwrap is a second tape consisting of a plastic backing and a suitable adhesive. These double-layer plastic tape systems have all the properties required to meet the mechanical and corrosion mitigation needs of a pipeline coating. Plastic tape pipeline coatings are most often applied over-the-ditch but the consumption of plastic tapes as plant coatings has increased significantly since 1978.

The application of plastic tapes over-the-ditch is well established in terms of specifications, procedures and equipment. The first over-the-ditch experience using power-driven equipment with plastic tape coatings occurred in 1953.

A typical over-the-ditch tape coating coating specification includes the following elements:

- Cleaning by use of wire brushes
- Priming
- Application of the corrosion protective tape (inner wrap)
- Application of the mechanical protective tape (outer wrap)
- Lowering in coated pipe directly from the coating machine
- Backfill

One piece of line travel equipment travels on top of the pipe and is capable of cleaning the pipe, priming the cleaned pipe surface, and applying the two-layer tape system in one efficient simultaneous operation.

A typical construction spread for the application of tape over-the-ditch will show a savings of 60% in equipment and labor over that necessary for the over-the-ditch application of hot applied enamel-type coatings. It is also possible to attain a higher daily coating rate with tape over-the-ditch than any other type of coating. Tape is applied to a large diameter pipe at the average rate of 2 miles per day. These rates can be achieved over ambient temperature ranges of $-30$ to $+50$°C. No other pipeline coating can be applied in the field over this broad temperature range.

In the segment of the underground pipeline industry where coatings are specified for over-the-ditch application, plastic tape coatings are used almost without exception.

Plant coating systems have been developed using both types of plastic outerwrap, i.e., plastic film only and plastic tapes with an adhesive (Fig. 23-2). Since 1976, the use of plant coating systems using film outerwraps have been discontinued.

A discussion of plastic tape plant coating systems using the adhesive

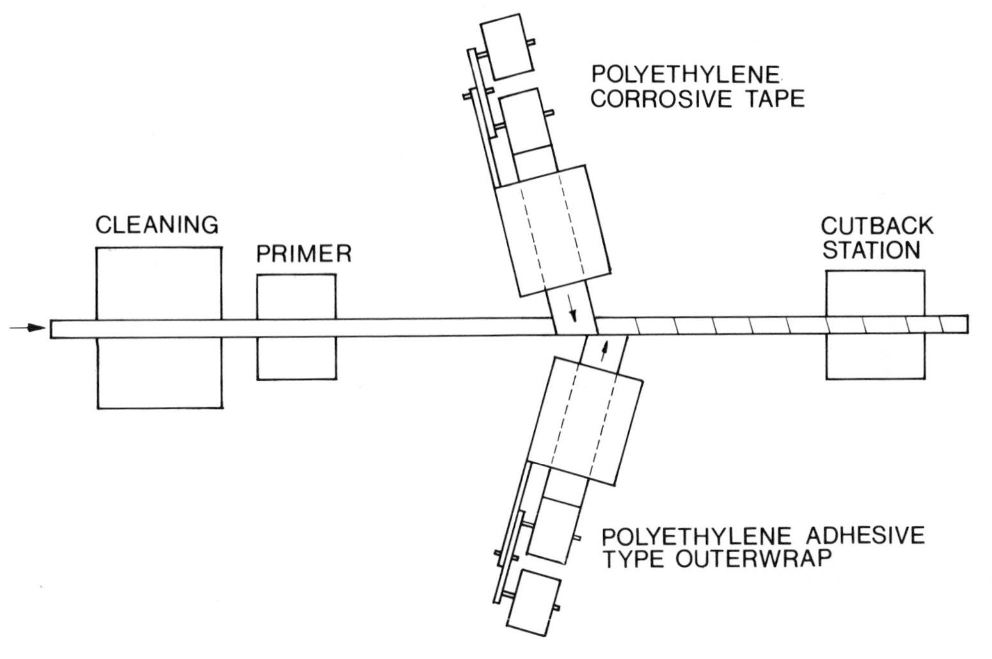


Fig. 23-2. A schematic diagram of multi-layer polyethylene tape plant coating process.

type plastic outerwrap will serve to discuss the progress made by plastic tapes as plant coatings.

The multilayer polyethylene tape plant coating process consists of:

1. The pipe to be coated is placed on racks where the pipe is dried if necessary and then fed into the cleaning station. The cleaning station may be either a high-speed wire brush device or grit or sand-blasting equipment.
2. Next, the cleaned pipe is fed into the priming station of the coating machine where a thin layer of primer is applied.
3. Then the primed pipe passes the tape wrapping station where the polyethylene pipeline coating tape is applied with tension from the tape dispensing spindles.
4. The primed and tape coated pipe then passes the outerwrap station where the outerwrap, which is a polyethylene film with an adhesive, is applied. The adhesive on the outerwrap forms a continuous bond to the tape previously applied next to the pipe surface.
5. The final step in the process is to cut back the coating at the end of each separate joint of pipe as the coated joints of pipe leave the coating machine and proceed to the outgoing racks.

Fig. 23-3. Application of polyethylene plastic tape coating to large diameter pipe by line travel equipment under winter conditions in Canada. (*Courtesy The Kendall Co.*)

## POLYETHYLENE TAPE PLANT COATING SYSTEMS

The multilayer polyethylene processes and systems offer important advantages in the area of product performance. These product advantages are:

1. The corrosion protection layer is completely protected during outdoor storage. The outerwrap used as part of the system completely covers and shields the corrosion protective layer from ultraviolet rays, cold, heat, and moisture. With all other types of plant coatings, including epoxy and extruded or sintered polyethylene, the corrosion protective layer during outdoor storage is exposed directly to deterioration from ultraviolet rays, cold, heat, and moisture. Any deterioration of the corrosion protective layer will reduce the level of corrosion protection after the coated pipe is installed underground.
2. It remains flexible in cold weather. The polyethylene-based tape and outerwrap remain ductile and flexible down to −40°C. Other plant coatings become brittle at temperatures considerably above −40°C.
3. It resists bending damage because of good ductility and flexibility throughout a wide temperature range.

4. It can be concrete coated. Pipe coated with the multilayer polyethylene tape coating systems can be coated with concrete by all of the commonly used concrete coating techniques.
5. Coating is tough and resists mechanical damage during handling and transportation. The polyethylene outerwraps absorb the mechanical forces which cause damage and prevent damage to the primary corrosion protection layers. With all other plant coating systems, all of the mechanical forces which cause coating damage act directly on the primary corrosion protective coating, resulting in damage to the primary corrosion coating.
6. Field joints can be coated and repairs can be made easily. Polyethylene tape is used to make field joints and repairs when required.

## SPECIFICATION

The specification for the application of joint wrap and long line plastic tape are identical in concept and principle but vary in that joint wrap application involves mostly manual steps while the long line tapes involve

Fig. 23-4. Application of hand wrap plastic tape coating to 16-inch gas pipeline. (*Courtesy The Tapecoat Company*)

power-driven mechanical means. A typical application specification includes the following:

1. Removal of moisture and foreign material
2. Removal of corrosion products and mil scale
3. Application of primer
4. Application of tape
5. Application of outerwrap

With joint wrap tapes, the outerwrap may be omitted but it is usually substituted by using two layers of the joint wrap tape.

The specifications of the various plastic tape vary as would be expected

**Table 23-1. Joint wrap specifications.**

| PROPERTY | REQUIREMENT | TEST METHOD |
|---|---|---|
| Thickness | 20 mils ± 5%, min. Type I and II<br>10 mils ± 5%, min. Type III | ASTM D1000 |
| Water vapor transmission (perms) | 0.25, max. | ASTM E96, method BW |
| Dielectric breakdown (V/mil) | 400 | ASTM D1000<br>ASTM D149 |
| Insulation resistance (megohms) | 500,000, min. | ASTM D1000<br>ASTM D257 |
| Adhesion (oz/in. width, with primer) | 20, min. | ASTM D1000, method A |
| Cathodic disbondment | 2 in. diameter (average of 5 test samples) | ASTM G8 method A, 30 days |

from manufacturer but they all have a strong similarity. The American Water Works Association (AWWA) Standard C209 includes a specification for joint wrap tapes. This specification given below will serve as a good example of a specification for joint wrap tapes.

A proposed standard for long line plastic tapes is in process for approval as an AWWA standard. The specification for the tape, outerwrap and total system to be included in this proposed AWWA standard serves well as a typical specification for plastic tape pipeline coatings (Table 23-2).

**Table 23-2. Physical properties of tape coating for long line pipe (inner layer, outer layer, total system).**

PHYSICAL PROPERTIES OF INNER LAYER TAPE

| PROPERTY | REQUIREMENT | TEST METHOD |
|---|---|---|
| Thickness | 15 mils min | Section 2.1.4 |
| Thickness deviation | ±10%, not to exceed 5 mils | Section 4.4.2 |
| Width deviation | ±5% of width or 1/4 in. whichever is smaller | Section 4.4.1 |
| Adhesion (oz/in. width, with primer) | 200 min | ASTM, D1000 |
| Water absorption (24 hr) | 0.2% max | ASTM D570 |
| Water vapor transmission | 0.2 max, perms/ft$^2$/hr/in. HG | ASTM E96, method B |
| Dielectric strength (volts) | 6,000 min, single thickness | ASTM D149 |
| Insulation resistance | 500, 000 megohms, min | ASTM D1000 |
| Tensile strength (lb/in. width) | 20 min | ASTM D1000 |
| Elongation (%) | 100 min | ASTM D1000 |
| Thickness | 25 mils min | Section 2.1.4 |
| Thickness deviation | ±10% not to exceed 5 mils | Section 4.4.2 |
| Width deviation | ±5% of width or 1/4 in. whichever is smaller | Section 4.4.1 |
| Adhesion to inner layer tape (oz/in. width) | 20 min | ASTM D1000 |
| Tensile strength (lb/in. width) | 40 min | ASTM D1000 |
| Elongation (%) | 100 min | ASTM D1000 |
| Thickness (mil) | 40 min | ASTM D1000 |
| Dielectric strength (volts) | 12,000 min/two layers | ASTM D141 |
| Cathodic disbonding | 2 in. diameter (average of 5 samples) | ASTM G8 method A-30 days |
| Impact resistance (in.-lbs) | 25 in.-lbs min | ASTM G14 |
| Penetration resistance | 20%, max penetration | ASTM G17 |

## IN-SERVICE PERFORMANCE

A short review of the in-service record which plastic tapes have compiled since they were introduced as a means of mitigating corrosion on pipelines will establish their credentials.

The performance of plastic tape coatings has been monitored continuously by the "Methods for Evaluating In-Service Field Performance" given in NACE Standard RP-01-69, Section 5, Table 6. Two of the methods given in this table are measuring leakage conductance of pipeline coatings and the rate of change in current required for cathodic protection. These methods have been employed as recommended by making a comparison of initial measurements with subsequent periodic measurements.

Table 23-3 compares the year-to-year leakage conductance of a polyethylene pipeline coating which was applied to 48 km of 150 mm pipe in 1957. The latest leakage conductance measurements were made by a corrosion consultant in July 1978.

The leakage conductance of the polyethylene tape coating was determined to be 845 micromhos/m$^2$ (78 micromhos/ft$^2$). This is a low level of leakage conductance after 23 years and confirms that plastic tape coatings retain the integrity of their physical, chemical, dielectric, and moisture resistance properties.

Table 23-4 compares the current requirements during the period 1958–1974 for a polyethylene plastic tape coating on 48 km of 200 mm pipe. The current requirements for this coated pipeline changed from 7.5 $\mu A/m^2$ (0.7 $\mu A/ft^2$) initially to 15.8 $\mu A/m^2$ (1.6 $\mu A/ft^2$) after 16 years of service. This means that for the entire pipeline, the protective cathodic protection levels in excess of $-1.0$ volt as measured by a $Cu/CuSO_4$ reference electrode were maintained by 0.61 A in 1974. The change in total current requirements

**Table 23-3. Coating conductance performance of a polyethylene tape and coal tar coatings after 16 and 12 years underground in comparable environments.**

| AGED UNDERGROUND (YEARS) | COATING CONDUCTANCE | |
|---|---|---|
| | COAL TAR ENAMEL (MICROMHOS/m$^2$) | POLYETHYLENE TAPE (MICROMHOS/m$^2$) |
| 1 | 80.2 | — |
| 2 | 178.2 | 45.9 |
| 3 | 467.0 | 70 |
| 4 | 258.2 | 76.5 |
| 5 | 903.8 | 78.3 |
| 6 | 1248.2 | 70 |
| 7 | 1635.0 | 106.4 |
| 8 | 1108.3 | 98.3 |
| 9 | 1291.2 | 108.9 |
| 10 | 1527.9 | 188.1 |
| 11 | 2388.7 | 239.0 |
| 12 | 2238.1 | 212.4 |
| 16 | — | 520 |
| 19 | — | 700 |
| 23 | — | 845 |

**Table 23-4. Current requirements of polyethylene tap pipeline.**

| 48.3 km OF 203 mm DIAMETER PIPE (CONSTRUCTION: 1958) | | |
|---|---|---|
| YEAR | CURRENT REQUIREMENTS ($\mu A/m^2$) | PIPE-TO-SOIL POTENTIAL (MILLIVOLTS) |
| 1958 | 7.5 | 1270–2000 |
| 1963 | 11.4 | 1275–1875 |
| 1964 | 10.5 | 1350–1700 |
| 1965 | 12.6 | 1200–1675 |
| 1968 | 13.6 | 1175–1600 |
| 1969 | 14.8 | 1125–1600 |
| 1970 | 16.1 | 1050–1325 |
| 1971 | 18.1 | [a]950–1650 |
| 1972 | 14.9 | [a]950–1650 |
| 1974 | 16.9 | [a]1000–1650 |

[a] Low P/S potential reading was taken at point of bond to a crossing pipeline.

increased only 0.4 A in 16 years. This level of performance once again attests to the enduring stability of polyethylene tape coatings on pipelines in the underground environment.

It is generally accepted in the pipeline coating industry that plastic tape coatings exhibit a high level of corrosion protection in conjunction with cathodic protection.

The worldwide use of plastic tape coatings since the mid-1950's has proven that these coatings provide a high level of corrosion protection as measured against cathodic protection current requirements. In general, plastic tapes have shown better corrosion protection qualities than other types of coatings. Table 23-5 illustrates this point by making the comparison of a tape-coated section and a coal tar-coated section of a 500 mm

**Table 23-5. Current requirements on a product pipeline coated with a polyethylene tape coating system and coal tar enamel (construction: 1971).**

| | |
|---|---|
| Pipe diameter | 500 mm (20 in.) |
| Total length of line | 234 km |
| Length of polyethylene tape section | 159 km |
| Length of coal tar enamel section | 76 km |
| Current Requirements: 1974 | |
| Polyethylene Tape | Coal Tar Enamel |
| 31.2 $\mu A/m^2$ | 132.3 $\mu A/m^2$ |

products pipeline built in 1971. In this case, the coal tar enamel coating requires greater than four times the amount of cathodic protective current to achieve acceptable protective pipe to soil potential levels than the plastic tape coating. Many other similar case histories of the performance of plastic tape coating are cited in the literature.

It is anticipated that the use of plastic tape pipeline coatings will continue to grow in the future because plastic tapes satisfy the engineering parameters of good performance and good economics.

## REFERENCES

1. NACE Standard RP-01-69, *Recommended Practice—Control of External Corrosion on Underground or Submerged Metallic Piping Systems* (NACE August (1969).
2. *May* 1976 *Coating Conductance Test*, The Kendall Company (1976), unpublished.
3. Benson, T. D. *PG & E's Polyethylene Wrap* (Gas Industries, Vol. **16**, No. 11 (September 1972).
4. Smith, T. E. *Arctic Construction Experience on the Trans-Alaska Pipeline* (Paper 15 (1976) Interpipe Convention).
5. Mange, E. A. O. *Leach Primer Offers Solution to Stress-Corrosion Cracking* (International Pipe Line Industry, Vol. **45**, No. 3 (September 1976).
6. Fesslar, R. R. *Studies Reveal Causes of Stress-Corrosion Cracking* (Pipe Line Industry March 1976 pp. 37–39).
7. Harris, G. M. *A Review of Recent Developments and the Performance of Tape Coatings for Underground Pipelines* (Materials Protection, September 1979).

# Chapter 24

# Graphic Art Applications

**Norman DeBastiani**
*Chartpak, A Times Mirror Company*
*Leeds, Massachusetts*

Pressure-sensitive products are currently used extensively in many segments of the diverse graphics industry. These products feature repetitive uniformity, ease of application, versatility, and reproducibility. They are now constant elements used in the preparation and composition of commercial art, advertising, charts, graphs, visual aids, etc., minimizing tedious manual drawing and typesetting. The utilization of graphic art pressure-sensitive products has become a significant cost saving factor.

Four general groups of products will be discussed:

1. Tapes
   a. Stripping tape
   b. Projectable tape
   c. Colored matte and gloss tape
   d. Crepe tape
   e. Fluorescent tape
   f. Printed graphic tapes
2. Films
   a. Color films
   b. Shading films
   c. Pattern films
   d. Masking films
   e. Blockout films
   f. Drafting applique film

3. Die Cut Products
   a. Die cut lettering
   b. Die cut symbols
   c. Printed circuit electronic die cut symbols
4. Dry Transfer Lettering

The fabrication and manufacture of these products require the implementation of numerous specialized and technical skills, processes and disciplines.

This chapter describes the general physical properties, construction and specialized end-uses of these products.

## GRAPHIC TAPES

The many different styles, kinds, and types of pressure-sensitive tapes used in the graphic arts industry are designed to eliminate long hours of repetitive hand work. They are excellent aids for printers, publishers, designers, industrial engineers, architects, commercial artists, and chart, graph and map makers.

### Litho Blockout Tape

Litho blockout tape, otherwise known as stripping tape, is designed for blocking out work when setting up negatives for printing forms. It is also used for opaqueing and edging negatives in photographic labs. Stripping involves the application of a light-blocking pressure-sensitive tape to a negative for the purpose of blocking out unwanted graphics. The light-blocking tape must photograph clear when producing a positive.

The first produced blockout tape was an opaque black paper which was applied over the unwanted graphics. The development of a transparent red tape was a significant improvement because the user could see through the tape and see the graphics he wanted blocked out.

The most common backing materials used in its construction are cellophane, acetate, polyester, and polyvinyl chloride films. The light-blocking compounds are added into the resin matrix and extruded into a solid red film. This involves the selection of the correct light-blocking ingredients, (pigments, dyes, absorbers, etc.) that will block out the radiation frequencies to which orthochromatic photo film is sensitive.

The red wavelength radiation emitted by the red blockout film does not affect the orthochromatic photo film and after processing, the blocked out area will result in a clear window in the positive produced. The clear window can now be used by the artist to insert new graphics without

**Table 24-1. Physical properties of litho blockout tapes.**

| PSTC# 1 | PSTC# 1 | PSTC# 6 | PSTC# 7 | PSTC# 31 | PSTC# 33 |
|---|---|---|---|---|---|
| INITIAL | 24 HRS. | QUICK STICK | SHEAR | ELONGATION | THICKNESS |
| g/2 5mm. | g/2 5mm. | cm. | hrs. | % | mm. |
| 600 | 800 | 2.5 | 22 | 40 | 0.063 |

remaking the entire artwork. This positive is now converted to a second-generation positive which will then be used to generate a printing plate.

The pressure-sensitive adhesive applied to the red film is usually a clear, medium tack, high shear system demonstrating a good balance of adhesive and cohesive properties (Table 24-1).

The adhesive is applied smoothly to the red film. Figure 24-1 illustrates a typical litho blockout tape construction.

The most common widths used range from 1/32 to 3 in.

## Projectable Tapes

Projectable pressure-sensitive tapes are constructed similarly to stripping tapes. The adhesive properties are usually very similar to those used for blockout tapes as illustrated in Table 24-1. Diacetate or triacetate backing is preferred, because of its superior clarity, cuttability, and heat resistance; PVC, cellophane, and polyester films are also sometimes used.

The backing is colored or tinted by coating the color uniformly on one or both sizes of the backing film. Another method is to laminate two clear films together with a colored adhesive. The film is then release coated on one side and pressure-sensitive adhesive is applied on the other side, again similarly to stripping tape.

These projectable transparent colored tapes are used for preparing

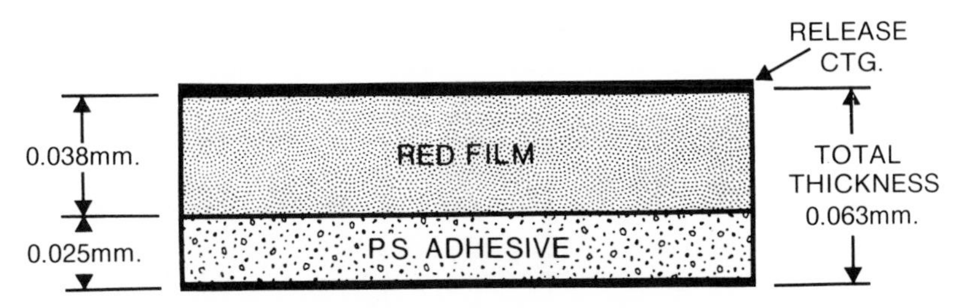


Fig. 24-1. Construction of litho blockout tapes.

overhead transparencies and overlays. There are 11 or more colors available ranging in width from 1/32 to 1 in.

## Colored Matte and Gloss Tape

A very popular family of graphic tapes are colored matte and gloss tapes. These tapes are designed to be used in preparing visual presentations such as bar charts, graphs, posters, signs, maps, etc. Their brilliant colors add a professional appearance to all artwork.

Gloss and matte tapes are constructed exactly the same way, except for their surface finish. Matte tapes are used where nonreflective colors are needed, as in photographic reproduction or in presentations where glare from surrounding light might make viewing difficult.

The physical requirements for colored matte and gloss tapes are brilliant opaque colors, easy tearability, good tensile strength, and permanent adhesion with good cohesive strength. Typical physical properties are illustrated in Table 24-2.

The backing is usually a laminate construction consisting of paper and film. The laminating adhesive is pigmented to produce the desired colored backing. The laminating adhesive must be permanent to prevent the film and paper from delaminating. The surface of the film is either matte or gloss to produce the desired finish while the color is common to both matte and gloss tapes. The film is generally diacetate or polyester. The paper is usually hemp fiber paper. Hemp fiber paper possesses a high tensile strength in the grain direction combined with easy tearability in the cross-grain direction. This is an important property to the artist end-user who insists on easy tearability when preparing charts.

A typical example of chart tape construction is illustrated in Fig. 24-2. The pressure-sensitive adhesive must be permanent, exhibiting a good balance between adhesive and cohesive properties that will yield good creep resistance and adhesion to a wide variety of receiving surfaces.

**Table 24-2. Physical properties of colored matte and gloss tape.**

| PSTC# 1 | PSTC# 6 | PSTC# 7 | PSTC# 31 | PSTC# 33 |
|---|---|---|---|---|
| INTIAL | QUICK STICK | SHEAR | ELONGATION | THICKNESS |
| g/2 5mm. | cm. | hrs. | % | mm. |
| 1200 | 2.5 | 24+ | 1-4 | 0.135 |

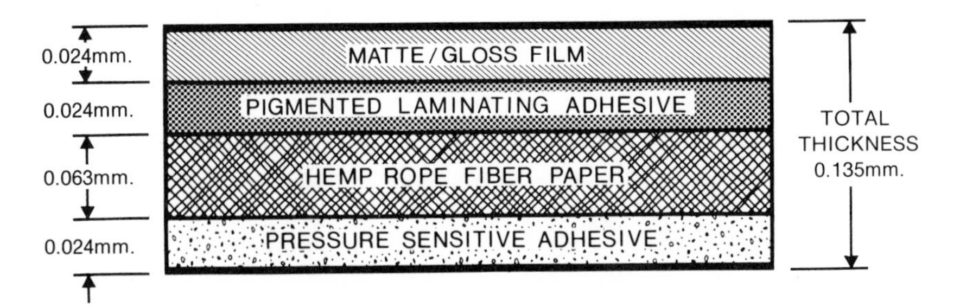


Fig. 24-2. Construction of colored matte/gloss tapes.

### Colored Crepe Tapes

Colored crepe tapes are used by artists and draftsmen where tape is required to make curves down to one inch or less of arc. These tapes are extremely flexible and curve easily with little or no distortion. Uses include graphs, topographical maps, cross-sectional drawings, etc.

The backing is colored prior to the creping process. The crepe in the backing will allow the tape to curve without stress. When combined with a permanent pressure-sensitive adhesive, the crepe tape will stay firmly in place without any stress buildup.

### Fluorescent Tapes

These tapes are constructed similarly to the colored crepe tapes. The backing is usually a stretched-out crepe paper. These tapes are used where there is a need for eye-catching emphasis to a chart or drawing. They can also be used with black light for extra emphasis.

### Printed Graphic Tapes

These tapes are printed with various graphics and are used as:

- Decorative border tapes
- Pattern tapes
- Plant and office layout tapes
- Statistical and symbol tapes
- Mechanical production tapes
- Register and crop mark tapes

These tapes are of the same construction; only the graphics vary. The most common construction is a two-ply lamination of the same film. The first film is printed with the desired graphics and the second film is laminated over the printing to protect the graphics. Figure 24-3 illustrates a typical two-ply construction.

Diacetate film is the most popular because of its superior clarity, stability, and cuttability. Cellophane, PVC and polyester films are also used. These tapes are available with a gloss or matte finish. They are also available in a variety of colors, but black is the most popular.

These tapes are also produced on opaque white backing in a variety of colored graphics.

Graphic tapes offer the artist, drafting and layout composition departments an opportunity to simplify their operations. The tedious repetitive pen and ink drawings of lines, patterns, symbols, shading, etc., can be eliminated. One simply applies the desired printed pattern to the subject drawings as one would apply any tape. Figure 24-4 illustrates several examples of patterns available as printed tapes.

In addition to black, printed graphic tapes are offered in transparent red, blue, green, orange, yellow, and brown patterns. They are matched to complement the projectable tapes and films used to prepare overhead transparencies.

## Drafting Tape

Drafting tape is a natural colored crepe tape with a low tack removal adhesive. This tape is designed to hold down drawings, blueprints, tracings, etc., onto a drafting table but can be easily removed without damage to the drawings. Table 24-3 illustrates the physical properties of a typical drafting tape.

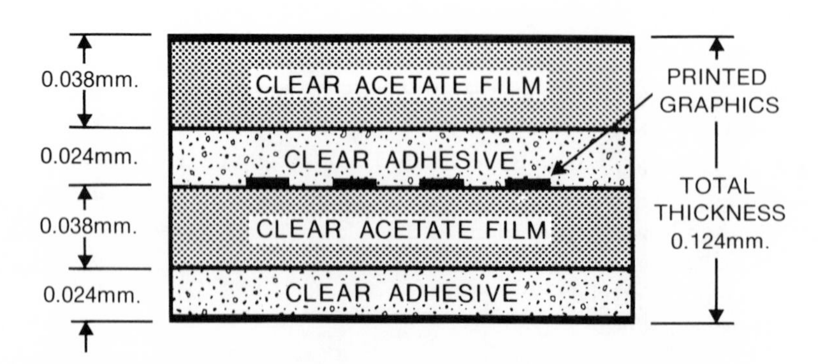


Fig. 24-3. Construction of a two-ply printed graphic tapes.

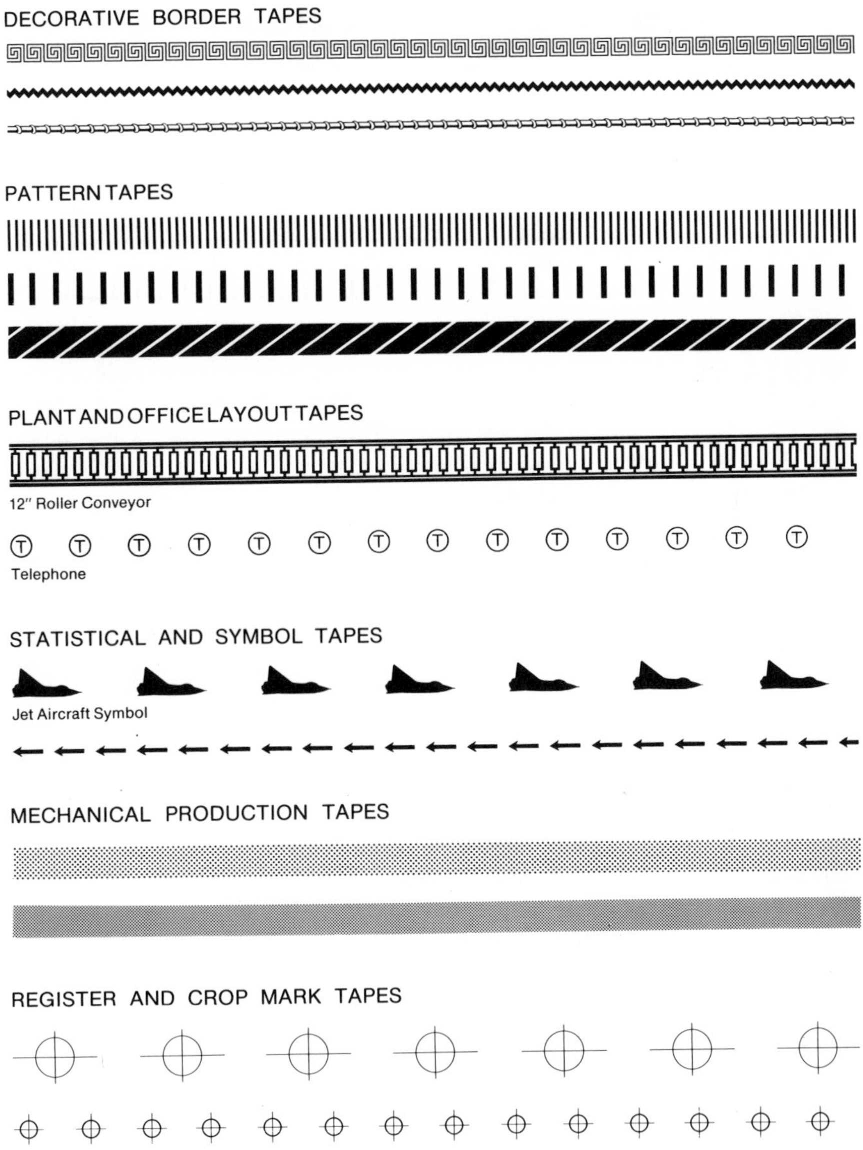

DECORATIVE BORDER TAPES

PATTERN TAPES

PLANT AND OFFICE LAYOUT TAPES

12" Roller Conveyor

Telephone

STATISTICAL AND SYMBOL TAPES

Jet Aircraft Symbol

MECHANICAL PRODUCTION TAPES

REGISTER AND CROP MARK TAPES

Fig. 24-4. Examples of printed pattern tapes.

**Table 24-3. Physical properties of drafting tape.**

| PSTC# 1 | PSTC# 1 | PSTC# 33 | PSTC# 31 | PSTC# 31 |
|---|---|---|---|---|
| INTIAL | 24 HRS. | THICKNESS | ELONGATION | TENSILE |
| g/2 5mm. | g/2 5mm. | mm. | % | g/2 5mm. |
| 400 | 500 | 0.16 | 9 | 500 |

### Masking Tape

Masking tape is usually the general-purpose tape that can be purchased anywhere. It is used in the normal bundling, protecting, and holding applications around an art department.

### Black Photo Tape

Black photo tape is an opaque dense black crepe tape designed for optimum photographic masking and sealing of light-sensitive materials such as films, plates, etc. This tape has an excellent long-term holding strength. It is resistant to moisture, will not creep, and easy to use.

## GRAPHIC FILMS

In the graphic arts industry, there is an equal demand for assorted pressure-sensitive graphic films as there is for tape. These films are also designed for the purpose of eliminating many hours of repetitive hand drawing for artists, architects, engineers, or draftsmen.

### Color Film

Color films are produced by applying a carefully controlled solid color to one side of the film. Cellulose diacetate or triacetate film is preferred because it can be cut easier than polyester film with an art knife. Both matte and gloss finish are available.

The most popular films use a repositionable adhesive in its construction. The films can be applied, cut out and removed from an art drawing without damage to the rest of the work. Ordinary pressure-sensitive adhesive would tear the surface or lift the paper fibers and damage the art drawing. The repositionable adhesive must perform on a variety of re-

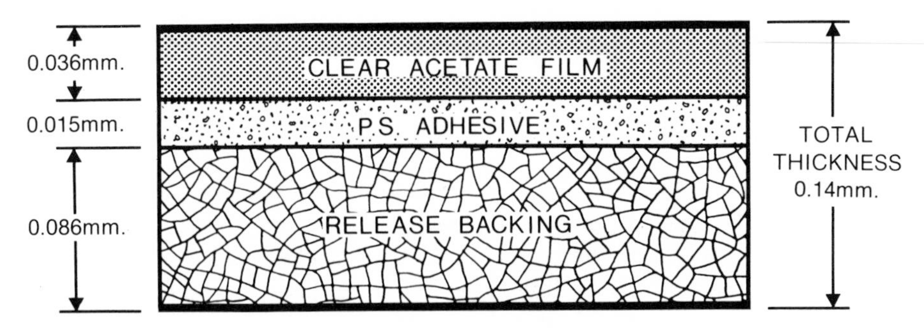


Fig. 24-5. Typical construction of color films.

ceiving surfaces such as vellum and drafting film, gloss polyester and acetate film, art board, and assorted inexpensive art papers.

The adhesive system must be clear so as not to interfere with the color of the film. For example, an adhesive with a tendency to yellow on aging causes a blue color to appear green.

The final step in the construction of color film is the lamination of film and adhesive to a suitable paper release backing. The release backing will protect the film from damage and at the same time aid the artist in its use. The release properties of backing paper have to be matched with the adhesive properties for optimum performance. Should the adhesion between the adhesive and backing be too light, the sandwich construction will have a tendency to tunnel. Should the adhesion be too strong, the film will have a tendency to stretch and tear when it is removed from the release backing, making it very difficult to use.

Figure 24-5 shows a typical color film construction and Table 24-4 identifies typical adhesive properties on various receiving surfaces.

**Table 24-4. Physical properties of color film.**

| RECEIVING SURFACE | PSTC# 1 INITIAL | PSTC# 1 24 HRS. |
|---|---|---|
| | g/2 5mm. | g/2 5mm. |
| STAINLESS STEEL | 200 | 250 |
| VELLUM | 35 | 50 |
| DRAFTING FILM | 175 | 200 |
| POLYESTER FILM | 220 | 275 |
| ART BOARD | 75 | 125 |
| BOND PAPER | 35 | 100 |

Color films are provided in over 200 opaque and transparent colors. There are also many special color effects that can be provided.

## Projectable Color Film

Projectable color films are constructed basically the same as opaque color films. The difference is that the colors must be clear and transparent. These films are usually used in the preparation of overhead projector slide mounts and transparencies. The adhesive system must also be clear and must adhere to polyester or acetate slide mounts without air pockets or lumps that distort the colors or interfere with projection.

## Shading and Pattern Film

Shading and pattern films are basically constructed the same as color films i.e., on gloss or matte acetate and also on polyester films. An assortment of topographic, architectural, thematic, grids, textures, line tints, screens, graduated tones, and patterns are printed on them. They are used by artists to tone and shade their line artwork. Just like the graphic tapes, they function as time savers for the end-user. A few examples may be seen in Fig. 24-6.

The black printing should be of sufficient density to yield good photo reproduction. Sometimes the image is reverse printed and laminated to the adhesive side of the film. This technique will protect the printing from scratches and abrasion while being used. However, with the advent of new ink technology, these patterns can now be surface printed to produce films with excellent scratch and abrasion resistance.

## Masking Film

Masking film consists of a red or amber coating applied to a clear, dimensionally stable polyester film. The red or amber coating is usually applied from a cellulose nitrate resin solution that has been pigmented and compounded to blockout orthochromatic light. The amber film provides better "see-through" when used over dark and complex artwork, but it will not blockout as well as the red film.

The artist can use masking film to create color separations and overlays rapidly, to cut intricate lines, and to add dropout marks.

Masking films have a unique adhesive system which permits readhesion to itself, to its own polyester backing or to other receiving surfaces, while demonstrating very easy removability.

The adhesive is the first coating applied to the polyester backing. It is

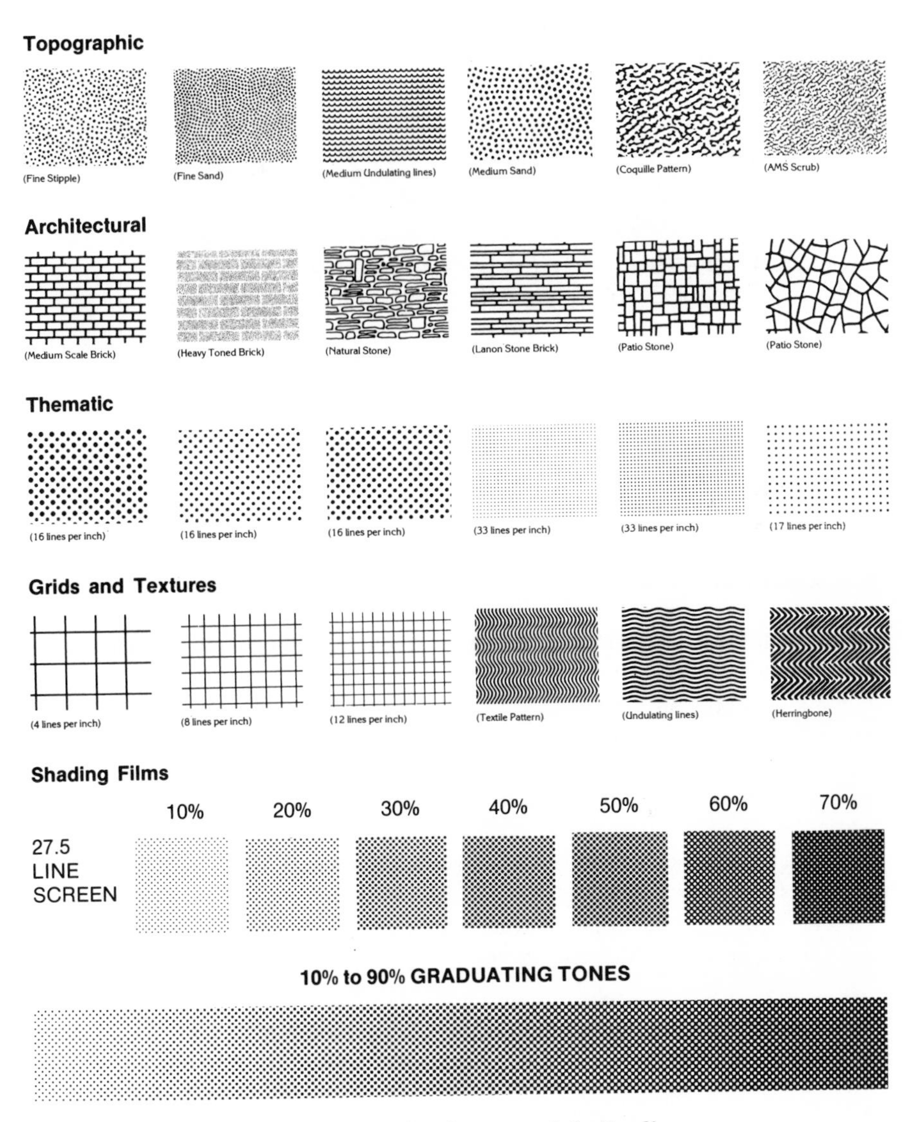


Fig. 24-6. Examples of pattern and shading films.

applied very thin. The second coating is the red or amber coating which is applied to a thickness of 1.5 mil. It usually requires two coating passes to achieve that coating thickness.

After the cellulose nitrate coating has dried, the adhesive will transfer and become attached to it. Some manufacturers offer products with several

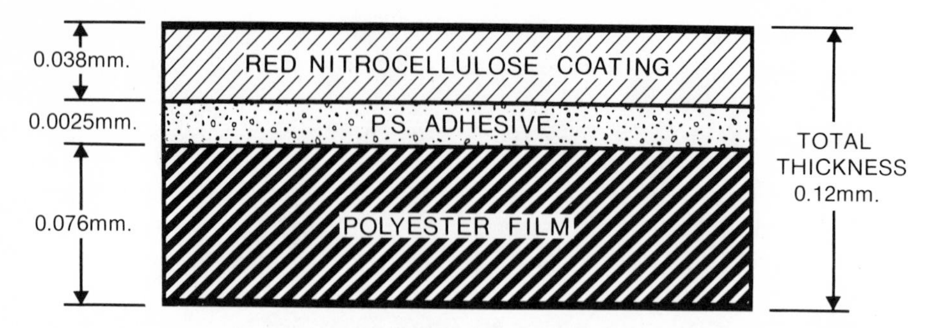


Fig. 24-7. Construction of a masking film.

adhesive levels. The most popular level average is 8–15 g/cm. Figure 24-7 illustrates a typical masking film construction.

The film is offered on a 3 or 5 mil polyester base. The 3 mil base is the most popular.

## Blockout Film

Blockout films are constructed differently than masking film. They consist of a red acetate film coated with a low tack adhesive and laminated to either an opaque or translucent silicone coated release backing. The construction is very similar to color films, except that the red acetate film is pigmented like stripping tape but laminated to a release liner.

Use of blockout films also differs from masking film in that it is mostly used to blockout negatives and positives rather than to separate colors. It is used mostly in conjunction with litho blockout tapes where large areas of negative need to be blocked out.

The adhesive used for blockout films must be repositionable and must apply free from air bubble entrapment.

## Drafting Applique Film

Drafting applique film is primarily designed for "scissor drafting" applications. These films have a specially formulated matte coating that will accept many common marking and printing systems i.e., india ink, drafting pencil, typewriter, ball-point pen, felt marker, xerography, and offset printing. Images or lettering that are repetitive on drawings can be imaged in multiples on drafting applique film via xerography or standard printing methods, and cut out and applied to the subject drawing. It is a major time-saver for the end-user.

Drafting applique film is a laminate construction consisting of a matte-coated polyester top film and either a polyester or paper silicone-coated release liner. The adhesive is a permanent, nonremovable, nonyellowing system that cannot be separated from the drawing once it is applied. For applications where removability or repositionability are desired, a removable adhesive system is available. The back side of the release film is coated with a suitable friction coat to assist drafting applique film through the feed mechanism of the various plain paper copiers available today. Figure 24-8 illustrates a typical drafting applique film construction.

## DIE CUT PRESSURE-SENSITIVE PRODUCTS

### Die Cut Lettering

Die cut lettering is produced on a 4 mil vinyl stock and is backed with an adhesive which is suitable for both indoor or outdoor applications. These items are designed to assist the architect, building contractor, maintenance engineer, autobody painter, hobbyist, or homemaker who has specific lettering or sign requirements. They can be applied to painted or sealed wood, plastic, and metal receiving surfaces. The adhesive is specially formulated to resist heat and cold, and it is weatherproof.

The adhesive-coated vinyl is laminated to a suitable paper release liner. The lettering is die cut on the release liner which also serves as a spacing system.

### Die Cut Symbols

Die cut symbols are produced similarly to the die cut lettering. They are used for logos, insignias, etc.

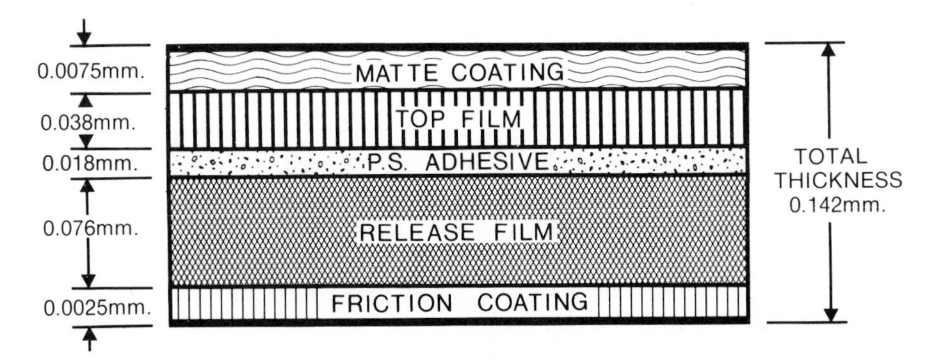


Fig. 24-8. Construction of a drafting applique film.

There is a second family of die cut symbols produced for the visual communication industry. These products are die cut from a variety of materials such as colored tapes, crepe tapes, films, etc. They are used indoor for visual communication on such things as charts, maps, and drafting films.

Electronic die cut symbols, cut from black tape, are used by electronic engineers, designers and draftsmen in the design of printed circuits. They require precision tolerances and an agressive adhesive system. They can be applied quickly and accurately to a drawing, thus eliminating the need of repetitive drawings of subject symbols.

## DRY TRANSFER LETTERING

Dry transfer lettering is a unique and unusual product. It consists of a plastic carrier film, usually styrene, polyethylene, or polyester that has been coated with a pigmented coating formed in the shape of letters or indicia and overcoated with a unique adhesive.

A silk screen printing process is used to apply the pigmented coating and form the indicia. The adhesive is formulated so it can be flood-coated evenly over the indicia and dry to a nontacky surface.

The adhesive will become activated when the indicia is burnished or rubbed from the opposite side through the carrier film. This will cause the indicia to adhere to the receiving surface before the carrier film is lifted away and positioned for the next transfer.

The original product was invented in England some 25 to 30 years ago. The inventor capitalized on the fact that a cellulose nitrate lacquer will not adhere to a styrene or polyethylene film.

The pigmented coating or ink must silk-screen print indicia with good edge definition and density. The indicia must transfer cleanly and easily without distortion and cracking. The indicia must adhere to a wide variety of receiving surfaces, i.e., art board, bond paper, vellum, drafting film, polyester film, acetate film, glass, etc. Transfer errors should be correctable by rubbing the indicia off with a rubber eraser or by lifting them off with a piece of a pressure-sensitive tape.

Dry transfer lettering is widely used in place of typesetting offering a real savings when creating compositions, layouts, and other artwork prepared for generating printing plates. There are hundreds of letter styles, sizes and colors available, making this product very convenient to use. Custom sheets of any design, color or multiple of colors are also manufactured. These custom sheets represent a fast way to apply frequently used logos, technical symbols, trademarks, etc., eliminating the need for the artist to draw each one. Dry transfer lettering is also used in drafting to take advantage of the convenience it offers.

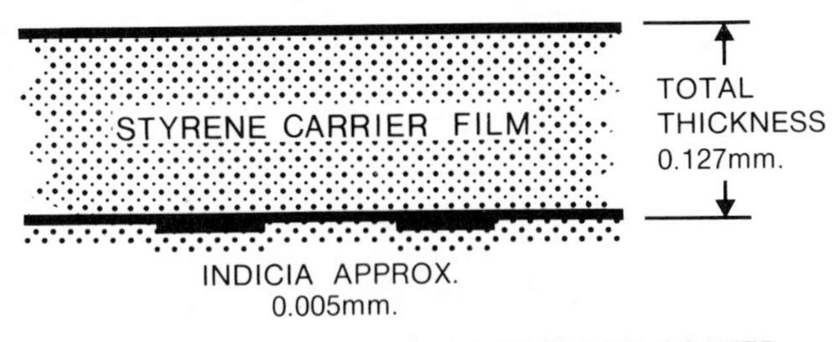


Fig. 24-9. Construction of dry transfer lettering.

Figure 24-9 illustrates a typical dry transfer construction. The carrier film thickness will vary from 3 to 6 mils, depending on the manufacturers preference. The thickness of the indicia is normally about 0.2 mil and the adhesive about 0.1 mil.

Most manufacturers flood coat the adhesive over the indicia, covering the entire carrier film. This method requires that the adhesive be formulated so that there will be no legging or stringiness at the indicia and adhesive interface when transfer is taking place. A second requirement is that the adhesive be dry enough to prevent the sheets from blocking or pre-transferring when stored in piles for a long period of time. The adhesive must also display five or more years longevity to ensure consistent indicia transfer.

# Chapter 25

# Labels and Decals

**Jim Komerska**

*Paper Manufacturers Company*
*Southampton, Pennsylvania*

One of the widest applications for pressure-sensitive adhesives is in labels and decals.[1] They are generally information carriers which are affixed to some item and are intended to be informative, instructional, decorative, or some combination of these functions. Methods of label attachment vary from mechanical (such as riveting or screwing a sign in place) to adhesive systems.[2] Liquid adhesives can be employed from sophisticated systems such as epoxy, and cyanoacrylate to the more conventional dextrins, animal glues, polyvinyl acetates, etc.

Usually, however, labels and decals are furnished to the end-user already coated with one of the following adhesives:

1. Remoistenable: Usually a dextrine or animal derived system, the animal glue being used where strong bonds are called for.
2. Instantaneous Heat Seal: Heat and pressure must be applied simultaneously to make the bond while the label is held in contact with the surface being labeled.
3. Delayed Action Heat Seal: An adhesive specifically formulated to become pressure-sensitive with exposure to heat and remain tacky for a prolonged period so that label application requires only pressure to form a permanent bond with the substrate being labeled.[3,4,5]
4. Pressure-Sensitive: An adhesive formulated so that it is permanently tacky and instantly adheres to whatever surface the label is designed for.[6,7,8] This system normally requires a liner (also called a release sheet) which covers the adhesive until the label is ready to be fastened

in place. At that time, the release sheet is removed and the tacky adhesive surface of the label is exposed. When the label is mated to the surface, the pressure-sensitive mass forms an instant bond with that surface.

In all instances, the fastening role of the adhesive is secondary to the informative or decorative purpose of the label. Over the past several years, pressure-sensitive adhesives have gained widespread preference over other label fastening adhesive systems primarily because of the following reasons:

1. Only slight pressure is needed; no messy gumming or other adhesive application equipment or after-drying step is required.
2. No activation step is needed, such as with pregummed labels where water activation is required. Nor is a heating step necessary to use, such as for heat seal adhesives, where care must be taken to assure adequate bond formation in the few seconds available prior to loss of adhesive tack properties, as the adhesive reverts back to a nontacky state.
3. Reliability of adhesive performance is ensured once the label or decal has been positioned.
4. The variety of pressure-sensitive adhesive types is at your fingertips: permanent, removable by peeling, and even water washable. The user has a choice of selection not readily available with other methods of fastening. In most cases, and after various periods of elapsed time, the whole label can be removed (via the correct adhesive choice) without any noticeable deleterious effect to the previously adhered substrate.
   a. Permanent: Conventional attempts at removal usually result in destruction of the label.[9]
   b. Removable: Removal is accomplished merely by careful peeling away from the surface.[10]
   c. Cold Temperature: For frozen food, the label is applied at subfreezing temperatures.[11]
   d. Specialty: Water-soluble pressure-sensitives, silicone rubber-based pressure-sensitives[12] and others are possible.

## ADHESIVES COMPOSITION

Pressure-sensitive label adhesives as they will be dealt with here can somewhat arbitrarily be further categorized as follows:

1. Rubber Based. Compounded adhesive with the elastomer system comprised singly or with any combination of natural rubber (or synthetic natural rubber), styrene-butadiene rubber, butyl rubber, nitrile rubber or a

thermoplastic elastomer. To the elastomer backbone polymer(s) are added the necessary tackifiers (resins), antioxidants (if called for, vulcanizing), the crosslinking agents with their required accelerators, activators, etc. Fillers such as clays and whiting (calcium carbonate) are added whenever possible to reduce adhesive unit cost. Opacification of any adhesive is via addition of at least 5 phr rutile titanium dioxide. Zinc oxide may also be used for opacification but is not as cost-effective as titanium dioxide. The compounded rubber family was the original pressure-sensitive type available and remains a workhorse for most applications. See Table 25-1 for a typical formula.

2. Single Component. Generally, these are acrylic-based copolymer or terpolymer systems[13] with the monomers chosen to confer specific desired performance properties with respect to tack, adhesion, creep resistance, heat stability, adhesive bleed, etc. This family is rapidly growing in areas requiring long service life (greater than five years). A specific adhesive may or may not be designed to crosslink additionally in service for added creep resistance.[14]

## METHODS OF APPLICATION

### Hot Melt

The adhesive is applied as 100% solids generally by slot die extrusion, roll or gravure coater. The adhesive mass may be applied directly to the facing,

**Table 25-1. Example of rubber-based pressure-sensitive adhesive.**

| | PARTS BY WEIGHT |
|---|---|
| No. 1 pale crepe rubber | 75 |
| Styrene-butadiene polymer (23% styrene) | 25 |
| $\beta$-pinene polyterpene resin (S.P. 115°) | 50 |
| Pentaerythritol ester of hydrogenated resin (S.P. 104°) | 25 |
| Calcium carbonate pigment (2.5 $\mu$m average) | 25 |
| 2, 2′-methylenebis-(4-methyl-6-*tert*-butyl phenol) antioxidant | 2 |
| Total | 202 |

The above formula may be applied by calendering at 100% solids or solvent coating by dispersing admixed elastomers and fillers and antioxidants into 300 parts *n*-heptane simultaneously with the tackifying resins.

**Table 25-2. Example of hot melt pressure-sensitive label adhesive (Non Removable)**

| | PARTS BY WEIGHT |
|---|---|
| Styrene/isoprene block copolymer | 100 |
| Dipentene derived polyterpene resin (S.P. 115°) | 67 |
| Antioxidant (sterically hindered phenol) | 1 |
| Total | 168 |
| Viscosity Characteristics: | |
| Brookfield viscosity at 350°F = 20,700 cps | |
| Brookfield viscosity at 390°F = 8600 cps | |

or may first be coated onto the release liner or a transfer liner prior to mating with the final facing material. A typical nonremovable hot melt formula is given in Table 25-2.

### Calendered

While a common method of adhesive application in the past to cloth backings, this method is seldom used for label and decal production today.

### Solvent Coating

The adhesive is dissolved in appropriate solvent(s) generally not exceeding 45% solids on a weight basis, and is then cast either directly onto the facing, or first onto the release liner and then later the adhesive/liner combination is laminated to the desired facing.

### Water Based

The adhesive is dispersed (latex form) in water and applied with appropriate means in the same fashion as in solvent coating, either directly to the final facing material or by transfer coating and subsequent lamination.

## RELEASE LINERS

The rapid acceptance of pressure-sensitive adhesives for mechanical (or machine applied) labeling over the past 20 years is directly attributable to the availability during this period of new release liners with evenly controlled release properties.[15] Basically, automatic labellers all use practically the same technique.[16,17,18]

A pressure-sensitive label in a roll is released from a protective backing by being pulled around a knife edge (release liner against the knife edge) so that the label lifts away from the release paper. A vacuum head picks up the label on the printed side at a precise time, triggered by an electric eye or mechanical trip, cuts the vacuum and a positive air blast propels the label onto the piece to be labeled. Other machines pick up with a vacuum head and apply the label to the article being labeled direct with mechanical pressure. Others use one of these techniques plus a rolling, wrapping action to apply to circular objects, etc. Figs. 25-1 through 25-3 show the automatic label application machines and Fig. 25-4 shows a hand labeler.

As one can see from the description of application methods, the release characteristics of the liner are key to successful uninterrupted machine performance.

Label adherence to the liner must allow intermediate processing such as die cutting, waste stripping, and printing without premature label lift or separation from the liner. Finally, when label application is called for, the label must come away from the liner as dictated by the application method.

Because a variety of adhesives exist in the marketplace (i.e., removable vs. permanent), no single liner will work satisfactorily with all adhesives. Different adhesives and facings both affect release from a given liner. The end result is that to satisfy mechanical label application of various different

Fig. 25-1. Automatic labeler suitable for short runs and a wide range of container sizes. (*Courtesy NJM, Inc.*)

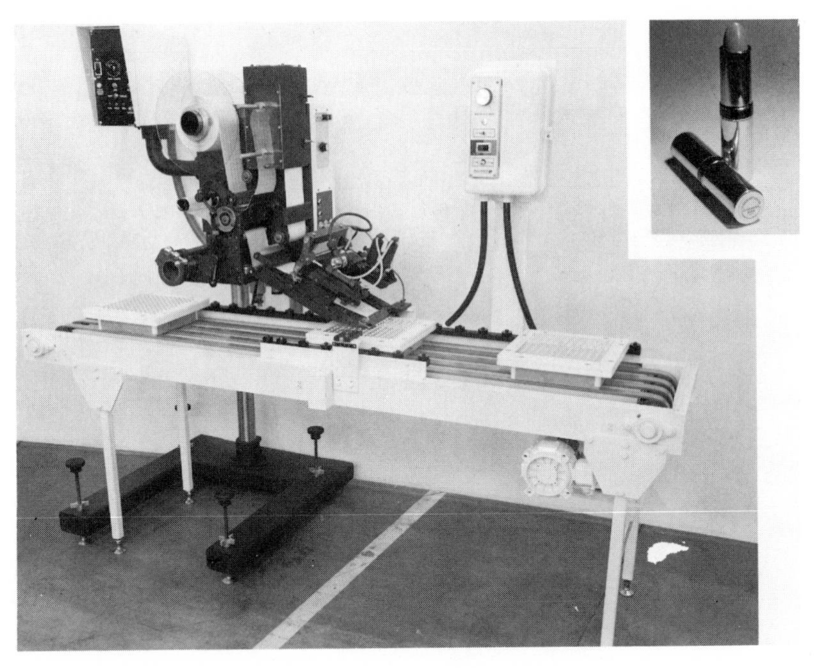

Fig. 25-2. Pressure-sensitive labeler for applying small labels to the bottom of lipstick cases at speeds up to 300/min. (*Courtesy NJM, Inc.*)

label sizes, adhesives, etc., a whole host of release liners has evolved. Release liner construction is discussed in some detail since correct liner performance is the key to a successful pressure-sensitive labelling operation.

## LINER BACKINGS

### Paper

Usually a bleached or unbleached high internal strength kraft, super-calendered for exceptional smoothness is used. Basis weights start at about 49 g/m$^2$ and go up as high as 147 g/m$^2$. For optimum smoothness, the kraft is extrusion coated many times with 10–16 g/m$^2$ (6–10 lb/3000 ft$^2$) polyethylene prior to release coating.

### Film Liners

Usually biaxially-oriented polyester, polypropylene, or coextrusions of film combinations which exhibit good smoothness and uniform thickness. Most

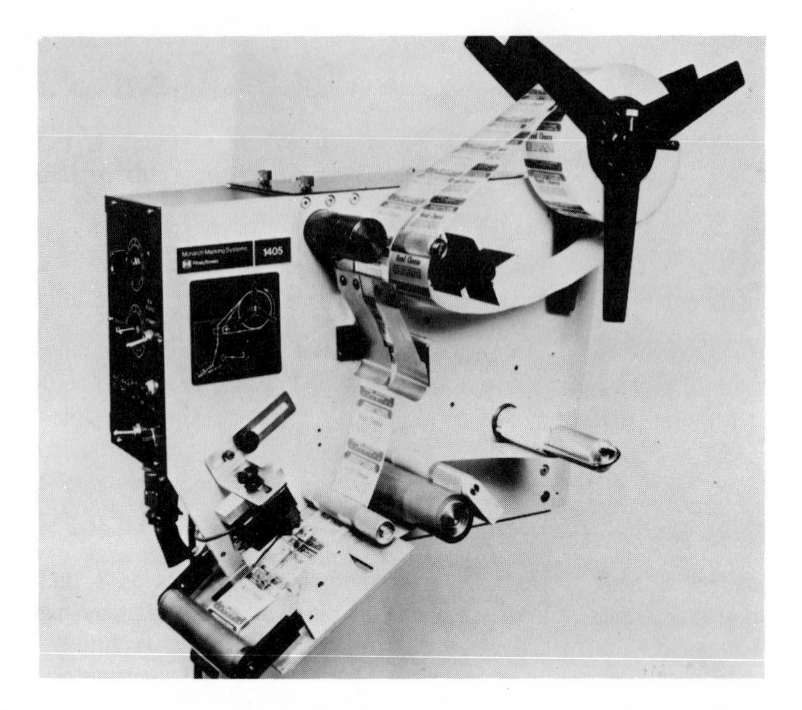


Fig. 25-3. A pressure-sensitive label applicator to apply labels to products moving on a conveyor. (*Courtesy Monarch Marking Systems*)

constructions used are in the thickness range of approximately 0.0025–0.0038 cm.

### Others

Specialty sheet constructions can be release coated, such as vulcanized fiber sheet, or extruded thermoplastic sheet, but these are not important for the mechanical high-speed labeling market.

## RELEASE COATINGS

All commonly used release coatings are silicone-based. Silicone release coatings are commonly applied from solvent solution, but aqueous application is gaining ground along with 100% solids application. Release coatings are normally applied at coat weights from 0.16 to 0.81 g/m$^2$ (0.1 to 0.5 lb/3000 ft.). Because of the low coat weights, desired application methods from solvent or water are those best suited for controlled low add-on, such as gravure, or better yet offset gravure, wire-wound rod, and for water-based, the air doctor. In most instances, some sort of post-coating application smoothing is employed prior to solvent or water drying.

## Curing

This is the critical step in release liner manufacture. All silicone release coatings must be completely cured to form an insoluble three-dimensional polymeric network if they are to perform as intended without causing a pressure-sensitive adhesive to lose tack. The polymeric structure of an incompletely crosslinked release coating shows an affinity for a pressure-sensitive adhesive and there occurs some mutual polymer interaction (intermingling or partial solvation into each other) with the disastrous result that the pressure-sensitive adhesive loses all of its quickstick or tack properties and cannot adhere to anything. Any reputable liner manufacturer has very rigid quality control standards to test for and guard against incomplete release coat cure.

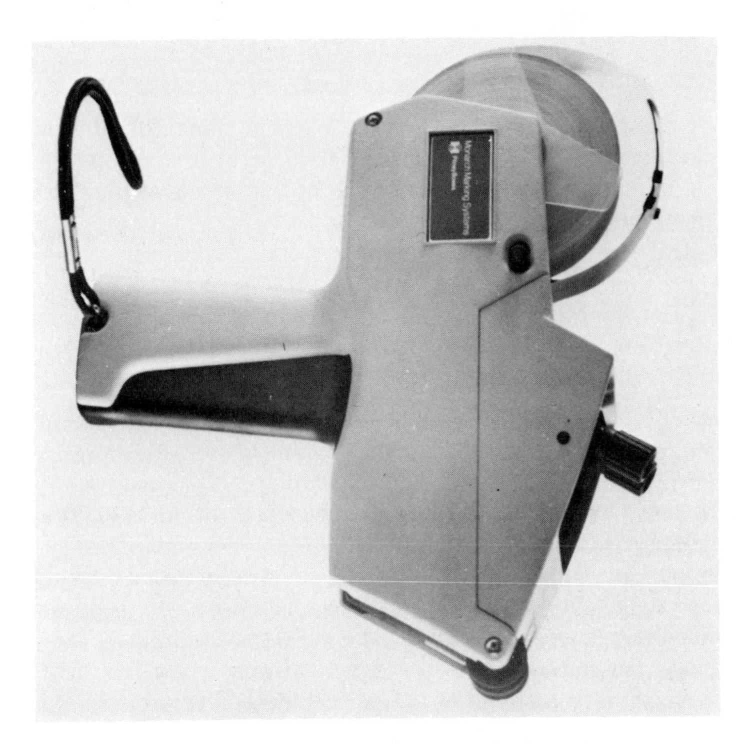

Fig. 25-4. A hand held labeller. (*Courtesy Monarch Marking Systems*)

## RELEASE LEVELS

Release coatings are formulated by liner manufacturers to have controlled levels of release depending on the adhesive with which it is to be used. Usually, the terminology for levels of release is related to how easily a pressure-sensitive adhesive label can be removed from its liner. A generally accepted method of testing release level is called the Keil test, which has been adopted in various forms by most users and manufacturers. One common set of test conditions is to peel a 1 in. wide strip of label from the liner at a 180° peel angle at the rate of 30 cm/min.

The best release papers are said to have "full" or "easy" release, and Keil values for these papers by the above test conditions usually run under 25 g. There is also a higher rate Keil test used employing a 2 in. strip peeled at the rate of 1500 cm/min. (600 in./min.). Using this peel rate, a liner with "full" release will run between 30–90 g peel.

Intermediate release liners are said to have "moderate" release and these liners usually have Keil values of 75–150 g peel.

At the extreme end are the "low" or "tight" release liners which have the poorest release characteristics, and against a "standard" adhesive would yield Keil values in excess of 200 g peel. Regardless of the specific test employed, it is important that there exists commonly agreed upon terminology relative to liner release levels. Without standardized terminology within an organization, confusion can arise between such groups as purchasing, manufacturing and sales, with the end result of purchasing procuring the wrong liner.

An ideal liner should have a very easy release to deliver a label off that liner when that label is coated with a very agressive, high quick stick adhesive. On the other hand, it should have a sufficiently tight release to hold a low tack label during die cutting, printing, and waste stripping until the time of label application.

## DELAYED ACTION ADHESIVES

An important type of label adhesive for pharmaceutical, cosmetic, and food packaging use is the delayed action system. Delayed action adhesives (DAA) represent a very important group because of cost advantages over conventional pressure-sensitives since DAA labels do not require a release liner prior to or during application. The main disadvantage of DAA is that they remain pressure-sensitive for periods of only a few months to a few years at most after activation. After this time interval, they lose their pressure-sensitive character and label failure can be induced. A typical DAA formulation is given in Table 25-3.

**Table 25-3. Example of delayed action adhesive formula.**

| | PARTS BY WEIGHT |
|---|---|
| Dicyclohexyl phthalate (m.p. 58–65°) | 58 |
| Pentaerythritol tetrabenzoate (m.p. 94.8°) | 7 |
| Glycerol tribenzoate (m.p. 71°) | 43 |
| Vinylpyrrolidone/ethyl acrylate copolymer emulsion (40% solids by weight, vinylpyrrolidone/ethyl acrylate ratio of 60/40 respectively by weight) | 50 |
| Total | 158 |

DAA are not inherently pressure-sensitive systems such as conventional rubber-based compounded adhesives or polymeric engineered acrylic multipolymers. Rubber-resin blends and the acrylics are applied as pressure-sensitives and remain so until outside influences such as ultraviolet light, oxidation, or ozone degradation, or combinations thereof, chemically alter the systems to reduce their quickstick properties gradually. Bond strengths (or peel strengths) of conventional pressure-sensitive adhesives usually remain high, unless chain scission from oxidation results, in which case the adhesive becomes extremely tacky and will show creep under stress. DAA systems, on the other hand, as applied by the manufacturer and as supplied to the printer/converter, are not pressure-sensitive or tacky to the touch, and require no release liner. Just prior to final label application on the line, the label is momentarily heated and the label becomes pressure-sensitive, tacky to the touch.

Tackiness is achieved because the DAA coating consists of a discrete mixture of at least two different solid (at room temperature) plasticizers held to the label with a binder vehicle. When the label is heated, the plasticizers form a melt which has a freezing point lower than their individual freezing (melting) points. This is a eutectic mixture. The materials which have this property also adhere well to the surfaces desired, but show the undesirable characteristic of having limited solubility in one another and hence separate in time by recrystallization. The end result of the recrystallization is that over a period of time, DAA's gradually show a weakening of the bond formed and after complete recrystallization, bond separation from the adherend can be induced or may occur with flexing or shock. Recrystallization is retarded by either very low or very high temperatures. Either subfreezing temperatures, or temperatures in excess of 65°C, will stabilize the systems. Temperatures in the range of 38–45°C accelerate crystallite formation (with tack loss) in most commercial systems. The adhesive loss mechanism of a delayed action adhesive with age is a physical phenomenon and is reversible with reheating, though this is not a practical solution for products already labeled and packaged. It is worthwhile, however, to know and be able to distinguish the method of adhesive loss and label failure between conventional pressure-sensitive adhesives and delayed action adhesives. From a practical standpoint, bond loss and possible bond failure can occur more rapidly with DAA systems under relatively mild adverse (above room temperature) product storage conditions.

One very important and desirable characteristic of DAA systems is a fast bond development rate after label application. It is typical of DAA labels to develop permanent bonds (label tearing) to the adherend within 24 hr after application, if not sooner. By definition, all DAA labels are permanent in drug label use and cannot be removed without label destruction, within their serviceable lifetime.

## LABELS

### Pricing

For cans, boxes, bottles and flexible packages of nonperishable foods, practically all labeling uses pressure-sensitive adhesives. Affixing is accomplished by means of portable or hand labelers, with the pricing imprinted at the time of labeling. Label stocks are generally paper with an English finish (EF), or coated-one-side (C1S) litho stocks and are 97 $g/m^2$ (60 $lb/3000\ ft^2$) or lighter. The labels are coated with a permanent adhesive and often are prescored to aid in self-destruction. Thus, tampering with price marking (price label switching) is discouraged. Many retail label stocks, including some pricing labels, are furnished to the printers in fluorescent colors, especially red, orange, yellow, and green, to command instant customer attention.

Price labeling for nonfood soft goods and hard goods, where pressure-sensitives are used, employs the same label types with similar application equipment.

Labeling of most perishables, such as frozen foods, fresh meat, poultry, cheese, and luncheon meats employs either a specialty pressure-sensitive cold temperature label or a delayed action adhesive. The specialty cold temperature pressure-sensitives are usually block copolymer-based adhesives, and are applied individually by handheld labelers. Delayed action adhesive labels, on the other hand, are usually used in conjunction with a packaging operation. The label is price marked, activated, cut, and applied as part of the weighing step. Major manufacturers of such equipment are Toledo Scale Div. and Hobart Corp. Millions of packages of poultry, meat, cheese, and other perishables are processed and price marked on a daily basis automatically using DAA labels. Water removable adhesives are used on dishes, housewares, fabric materials, submersible small appliances, etc. Wetting with water is all that is required to remove the label. This adhesive is also repulpable so that trim from label stocks can be sold back to mills for reuse without gumming up mill equipment. These stocks can be tailored to dissolve in cold water, hot water, or alkaline detergent solution, so that the labels, such as beer bottle labels, can be resistant to removal by cold water when being chilled, but can be removed with a detergent solution when bottle is washed for reuse. A typical water-removable pressure-sensitive adhesive might be a polyvinyl methyl ether maleic anhydride co-polymer. Such adhesives are normally coated out of solvents, such as acetone or methyl ethyl ketone. They are not easily coatable out of water, since they tend to be aqueous system thickeners and produce thixotropic mixes.

### Machine Labelers for Pressure-Sensitives

These are some of the companies making pressure-sensitive label-applying machines:

- Avery
- Excelsior Tape
- Label Aire
- Dennison Eastman
- Kimbal Systems
- Monarch Marking Systems
- New Jersey Machine
- W. H. Brady Co.
- NJM, Inc.

This list comprises only some of the machine manufacturers and is not intended to be all-inclusive.

### Office/Inventory Labels

This is a market segment that covers the whole spectrum of label types. Certain needs will call for permanent adhesives such as "ACCEPT" and "REJECT" labels on raw materials after incoming quality inspections. Removable adhesives are called for in some inventory tasks and filing operations. Most office/inventory label applications call for a paper facing with a lightweight liner sheet. Labels are supplied in both roll and sheet form, depending upon the volume of use. For specific customer wording, the job printer can buy individual 8-1/2 × 11 in. sheets of pre-die cut labels for either offset or letterpress printing. Paper facing sheet finishes common are English finish (EF), coated-one-side litho (C1S) or cast coated, with label stock weights of 81–98 g/m$^2$ (50–60 lb/3000 ft$^2$). Office label paper face sheets must accept ball-point pen, pencil, typewritter ribbon inks or transfer tape. These labels should also accept stamp pad inks and offset ink.

## Tape Cassette and Reel Labels

These products usually have a face sheet made of high strength paper stock, either saturated or rope stock, so that a label with a tight, strong peel and semiremovable adhesive can be removed easily without tearing the label facing. The adhesive must adhere to several different surfaces such as plastic and metal and be cleanly peelable from these surfaces. Moreover, while the cassette label must adhere tightly, it should be removable and transferable, if needed, to another cassette.

## Transparent Labels

These products are intended for office purposes and can be used as identification and information on fiche, microfilm jackets, and engineering drawings. They must also be copyable. They can also be used in any other areas where a clear label is desired. The surface of the face sheet must be capable of accepting typewriter or toner where copy is made through a copying machine. In addition, adhesive normally required is permanent, but easily repositionable initially, nonyellowing, nonoozing, or truly removable but nonyellowing and nonoozing.

## Labelon* Adhesive Labels

A specialty label construction is discussed here because of its extensive office use. Labelon labels can be imaged or marked merely with stylus pressure, type impact, or ball-point pen pressure. The Labelon label marks as the result of deformation and/or flow of an opaque blushed wax layer to expose a contrasting dark undercoating so that the imparted mark appears dark against a light or white background. The label usually is supplied in roll form from a cellophane tape-type dispenser, without a liner and with a conventional pressure-sensitive adhesive on the backside. A disadvantage of

* Trademark by Labelon Corp., Canandaigua, New York

the Labelon label tape is in its inherent property of continuing to accept new image marks for the lifetime of the label simply by exposure to new pressure marking situations. The Labelon tape label is best suited for applications that preclude further sharp impact situations once initial imaging has been accomplished.

## Address and Shipping Labels

The address label market is another huge market segment that is extremely cost-conscious. Much of the market employs gummed as well as instantaneous heat seal or hot melt means of label fixation. Paper stocks for address labels, also called EDP or computer printout labels, are usually in the 81–98 $g/m^2$ weight range (50–60 $lb/3000\ ft^2$). The finish is EF or C1S litho with typically 32 $g/m^2$ conventional pressure-sensitive adhesive. Liner stock can vary from 48 to 81 $g/m^2$ (30 to 50 $lb/3000\ ft^2$) usually of easy or full release. While most of the pressure-sensitive label stock for address label use was produced via solvent coating in the past, there is a pronounced shift in this item to coating via hot melt techniques with the concomitant cost savings.

EDP labels, also called tab labels by some manufacturers, are usually supplied to the end-user die cut, waste stripped and fan-folded in cartons. These labels must accept clear computer printouts or ink jet addressing. EDP labels have the following properties and must meet these requirements:

1. The printing must not smudge (smudgeproof label stock is necessary, especially for ink jet printing).
2. Adhesive usually is a semipeelable, so that the label will stay put without danger of coming off, but can be removed (with care) if need be.
3. The composite product (bleached kraft liner with smudgeproof face sheet) must be capable of processing through a high-speed computer printing unit without the carrier sheet tearing, especially at the pin feed holes. It must not jam the computer by labels coming off or fan-fold performations breaking. There cannot be any adhesive bleed at the edges of cut labels since this would jam the feed mechanism of the computer.
4. Most computer labels are made in a fan-fold configuration being perforated at the fold. A stack of fan-folded labels is fed into the computer where the labels unfold as they go through printing and refold and stack once again as they leave the printing unit.

## Duplicate-Copy Address Labels

They are used where a duplicate label is needed when the original label is printed out. The face sheet must not smudge after printing, and the release sheet may have a built-in encapsulated imaging system, which causes release sheet to image when the face sheet is printed out.

The duplicate-copy system allows the original label to be removed and placed (with its pressure-sensitive adhesive intact) onto a return mail item. The release liner in this instance is itself adhered to the item with a hot melt, another pressure-sensitive adhesive layer (permanent) or with a conventional wet gum or glue.[19]

## Packing List Labels

Clear plastic film envelopes for packing lists, etc., with pressure-sensitive adhesive on one side have gained wide popularity. The clear plastic (usually polyolefin) envelope is so constructed that shipping documents can be inserted only prior to envelope placement onto the carton or skid. Once in place with its pressure-sensitive adhesive, the envelope contents can be removed only by destructive entry into the envelope.

## Bottle Labels

One of the largest label markets for pressure-sensitive and delayed action adhesives lies in bottle, jar and vial labeling for the consumer, cosmetic, drug and pharmaceutical industry. Labeling both over-the-counter and ethical drugs and pharmaceuticals has in the past been dominated by delayed action adhesives, but it is being challenged by pressure-sensitives. Pressure-sensitive labels are making inroads in areas such as clear labels for small vials. DAA's are not available on clear film. Besides, they tend to be milky or tan in color and are not transparent. Because of the necessary activation temperatures, DAA's are usually supplied on paper stocks; most films might distort during activation.

The transparent printed pressure-sensitive labels, when correctly applied, tend to disappear and give the appearance of the printing being directly on the vial or bottle. Hence, these new labels do their job best by appearing not to be there. Delayed action adhesives, however, are still widely used. They form the necessary (required by the FDA) permanent label destruct bond to the variety of container body materials used in the drug and cosmetic industry. DAA systems develop excellent paper tearing bonds to plain glass, silicone-treated glass, polystyrene, high-density polyethylene, polypropylene, Barex, polyvinyl chloride, and other bottle plastics.

Mechanical label application with DAA labels is even faster than with pressure-sensitive labels since it is not necessary to strip the label from a release liner. Activation can be done in fractions of a second.

Inroads, however, are being made by pressure-sensitives in a second new area—container labels for nonperishable foods and hardware items, bumping out conventional wet glue labeling.

For the pharmaceutical, cosmetic, ethical drug and other applications, paper labels conventionally employ either C1S litho or cast coated stocks in the 81–98 $g/m^2$ range. DAA coatings are applied at approximately 25 $g/m^2$ (15 $lb/3000\ ft^2$), whereas pressure-sensitive labels usually have coat weights of 32 $g/m^2$ (20 $lb/3000\ ft^2$). Liners for the clear vial pharmaceutical labels are usually release-coated polyester film so as to give the smoothest release surface, and are furnished free from entrapped air which shows as undesirable blemishes when it occurs. These superclear label stocks are usually made by casting the clear acrylic adhesive from solvent onto the release-coated film liner, removing the adhesive solvent and carefully laminating the film facing to the clear dry adhesive mass to avoid air entrapment. In label application to a vial or bottle, it is likewise tricky to avoid air bubble entrapment during label laydown onto the container surface. Special techniques are often needed.

Pressure-sensitive adhesives for paper-backed labels are of the conventional permanent type, whereas the DAA coated label calls for the long shelf-life product.

All of the above labels are usually supplied to the end-user in roll form. Most labels for pharmaceuticals are printed by either offset lithography or gravure, with some high-quality flexographic work being used also. Many cosmetic and consumer product labels are also printed by all three methods. Many labels are printed in four or more colors including any varnish or lacquer overcoating for label protection.

## Identification—Graphic Arts

The primary use in this area is of pressure-sensitive numbers, letters and symbols. In many instances, the user makes a do-it-yourself sign. The facing materials are paper, plastic coated cloth, films such as polyvinyl chloride or Tedlar and various other film laminates. The combinations may be embossed, debossed or sometimes metallized. All are supplied on release liners of varying thicknesses and weights. All types of pressure-sensitive adhesives are used, depending on the length and severity of service called for and the quality of materials used.

## Identification—Wire Markers

This is a specialty use of pressure-sensitive labels where the end-user, usually an electrician or electronics technician, applies either a numbered or color-coded adhesive label strip to a terminal and a wire going to that terminal. Wire markers are extensively used in computers, aircraft, ships, factory machinery controls, etc. One DC-10 airplane may contain over one million individual wire markers. Individual wire marker strips are commonly 6.3 × 19.5 mm or 6.3 × 39 mm long. The marker facing can be vinyl-coated fabric, foil, polyester vinyl film, polyvinyl fluoride film, spun bonded nylon, polyimide film, or laminations of the above. Pressure-sensitive adhesives for wire markers are formulated to resist wire coating plasticizers as well as withstand exposure to various oils,[20] jet fuels and in some cases, high temperatures. These adhesives comprise conventional rubber-based masses compounded for oil resistance, single component acrylics, and even silicone-based pressure-sensitives for high (300°C) temperature service. Inks for the wire marker legends can be surface printed offset (wet or dry) or trap printed. Wet offset printing is conventional offset lithography employing a planar lithographic plate. The image areas to be printed are rendered lipophilic to accept an oil-based ink, whereas the nonimage areas are rendered hydrophilic to wet out with a water-based fountain solution and resist ink uptake. Dry offset printing utilizes a raised printing plate where only the area to be printed is raised and comes in contact with the inking rolls. Dry offset printing does not lend itself to reproduction of very small type or halftones. Trap printing is simply printing the information, design, etc. on the underside of a transparent film so that the printing is not easily marred, damaged or removed. Letterpress printing is also done. In all cases, the printed legend (or copu) must be extremely durable to environmental factors for the expected life of the wire marker. Added protection for surface printed legends can be attained with an overlaxquer or other clear protective overcoating.

## Library Labels

These labels require a durable facing of archival quality. Bleached krafts, free from added acidic components are the best, along with coated spunbonded polyolefin sheet (Tyvek). The pressure-sensitive adhesive used should be well aging, such as an acrylic system and of low edge bleed properties.

## Signs

Pressure-sensitive signs are really nothing more than large labels. Pressure-sensitive signs are found in great variety from the eye-catching retroreflective WARNING or CAUTION signs to the lowly bumper sticker. They range in size from a postage stamp (really a label if it is this small) to a few square feet.

Sign facing materials can be paper, vinyl, polyester, polycarbonate, polyvinyl fluoride film, foil, spun bonded polyolefin or various combinations. Printing can be either surface or trapped. Printing methods can be wet or dry offset, letterpress or silk screen. Laminated structures can be "printed" by masking off the nonimage or print area and then subjecting the composite to chemical milling or acid bath etching[21] to remove a top layer and reveal a contrasting sublayer.

The sign background can be plain or overall coated after printing the legend or indicia. The background can also be vacuum metallized, retroreflective, or even phosphorescent,[22] as for EXIT signs.

Most signs are prepared in sheet form with the last step in manufacture being the lamination of the pressure-sensitive adhesive backing, just prior to final die cutting. The pressure-sensitive is applied using a two-sided supported (or unsupported) adhesive from roll form, usually supplied with a single differential release liner. Many manufacturers may elect to go through a liner change prior to final die cutting to supply the finished sign with a high quality liner.

The pressure-sensitive adhesive for a particular sign is chosen depending on the serviceability desired and whether it will be used indoor or outdoor. Signs to be adhered to rough surfaces such as brick walls may employ a heavy adhesive of 0.12 mm thickness cushioned by foamed plastic 1.58 mm thick for conformability on application. The foam plastic layer is sandwiched between the sign facing and the pressure-sensitive adhesive.

## NAMEPLATES

Nameplates are distinguished from signs in that a nameplate usually carries more information than a sign including, say, a serial number for a specific item. Nameplates are usually designed for long service life under adverse conditions, on such items as motors, small and large appliances, etc. Some dashboard components such as heater/air conditioner control back panels and speedometer panels are laminates fastened to the dashboard assembly with an acrylic-based pressure-sensitive adhesive for long (greater than five years) service life.

For many nameplates, the facing material can be foil, polycarbonate film, polyester film, vinyl, with surface or reverse (trap) printing by offset or silk screen. The polyester and other films are often coated and metallized. The metallized films are most popular of the configurations substituting for aluminum foils. Brush finished metallized films give the appearance and wearability of brushed metal films. Often the facing is embossed or textured.

Again like the manufacture of signs, the pressure-sensitive adhesive is applied to the nameplate as one of the last steps prior to the final die cutting or trimming. Most nameplates are furnished as individual pieces, and are usually placed into final position by hand.

## PROTECTIVE AND DECORATIVE SHEETS

This is a fast growing market segment and while they cannot be called label stocks, protective and decorative sheets are used many times in conjunction with labels. An example is an overlay protecting clear film which goes over nameplates or other labels.

These can be vinyl, polyester, biaxially-oriented polypropylene, or other clear film with a clear pressure-sensitive adhesive. Overlamination must be done carefully to prevent air inclusion which appears as a blemish or defect.

## DECALS

Decals fall into several categories which are listed below, but for most purposes are intended for decoration as well as for protection. Decal facing materials include the same facings as for the nameplates listed above, except that the variety of laminate combinations are greater. Metallized polyester films in thicknesses ranging from 0.038–0.25 mm are widely used along with brush finished metallized films in the same thickness ranges. Different pressure-sensitive adhesives are laminated to the decals depending upon the end-use.

Typical decal uses are:

- Football helmet insignia
- Appliance decorative trim
- Manufacturer's logo or insignia on appliances
- Automotive dealer decal
- Auto component decorative decal (seat belt buckle)
- Sports equipment decorative decal (other than helmet)
- Emblem type decal
- Toy decoration decal
- Housewares decoration
- Furniture decal
- Promotional advertisement decal

Some of the decals offered are quite fancy and beautiful with combinations of debossing and metallizing.

## EMBOSSABLE-IMAGIBLE TAPES

The face sheet of embossable tapes is made of a special vinyl film incorporating a specific metallic hydride. When the face sheet is creased, such as occurs during embossing letter formation, a distinct white character appears, with the contrasting tape background color imparted by a pigmented coating on the facing underside.

The adhesive which is coated on the back of the tape is designed to adhere to many surfaces such as wood, metal, plastic, glass, film, etc. The label is faced with release-coated polyethylene film. There are currently at least two manufacturers of imagible pressure-sensitive tapes that are designed to compete with embossable tapes. Kalograph (sold by W. H. Brady Co.) uses an ultraviolet light source to image its tape one digit at a time just as an embossing tape tool operates. Kalograph tape is offered on a variety of colored backgrounds. Kroy Tape Type 80 (Kroy Industries) images a clear tape, also digit by digit.

The imagible tapes have an aesthetic advantage over the plain, stark lettering attainable with embossing tape, and also have more flexibility in type fonts from which to select.

## REFERENCES


1. *Packaging Digest*, pp. 31–37, December 1979.
2. *Modern Packaging*, pp. 42–43, November 1979.
3. Holt, F. W., Jr., U.S. Patent 2,746,885 (May 15, 1956).
4. Sirota, J. U.S. Patent 3,082,108 (March 19, 1963).
5. Cochrone, W. D., *et. al.* U.S. Patent 3,154,428 (October 27, 1964).
6. Tierney, H. J. U.S. Patent 2,319,959 (May 25, 1943).
7. Drew, R. G. U.S. Patent Re 19,128 (April 3, 1934).
8. Weymon, H.P., *et. al.* U.S. Patent 3,413,246 (November 26, 1968).
9. Huber, E. W. U.S. Patent 2,845,728 (August 5, 1958).
10. Crozier, R. N., *et. al.* U.S. Patent 2,816,655 (December 17, 1957).
11. Tamburro, A. J. U.S. Patent 3,149,997 (September 22, 1964).
12. Currie, C. C., and Keil, J. W. U.S. Patent 2,814,601 (November 26, 1957).
13. Ulrich, E. W. U.S. Patent 2,884,126 (April 28, 1959).
14. Weidner, C. L., *et. al.* U.S. Patent 3,102,102 (August 21, 1963).
15. Dickard, L. R. U.S. Patent 2,985,554 (May 23, 1961).
16. Western Electric Co., U.S. Patent 2,920,780 (January 12, 1960).
17. 3M Co., U.S. Patent 3,107,193 (October 15, 1963).
18. Kleen Stik Products, U.S. Patent 3,093,528 (January 11, 1963).
19. Kelly, G. E. U.S. Patent 2,563,340 (August 7, 1951).
20. Ulrich, E. W. U.S. Patent 2,973,286 (February 28, 1961).
21. Schilling, R. W. U.S. Patent 3,128,202 (April 7, 1964).
22. Striker, A. M. U.S. Patent 3,075,853 (January 29, 1963).

# Chapter 26

# Coating

**Donatas Satas**
*Satas & Associates*
*Warwick, Rhode Island*

The behavior of pressure-sensitive adhesive solutions and dispersions in coating operation depends on the rheological properties of the fluid. The dependence of viscosity and shear stress on the shear rate determines the behavior of these adhesives in coating. We distinguish Newtonian, dilatant, pseudoplastic, and thixotropic behavior of the fluids.

The shear stress and shear rate are related by

$$\tau = kD^n, \tag{1}$$

where

$$\begin{aligned} \tau &= \text{shear stress} \\ D &= \text{shear rate} \\ k \text{ and } n &= \text{characteristic constants.} \end{aligned}$$

In the case of fluids which follow Newtonian behavior, the constant $n$ is equal to 1 and the shear rate and shear stress are related by a simple first-order equation.

*Dilatancy* is the isothermal reversible increase of viscosity with increasing shear rate, showing no measurable time dependence. This may or may not be accompanied by a change of volume. The term dilatancy has a dual meaning: shear rate thickening or volume increase with shear. We are concerned with the first meaning of this term. In Equation (1), the constant $n$ is greater than 1 for dilatant behavior. Figures 26-1 and 26-2 show various modes of shear stress and viscosity response to changing shear rate.

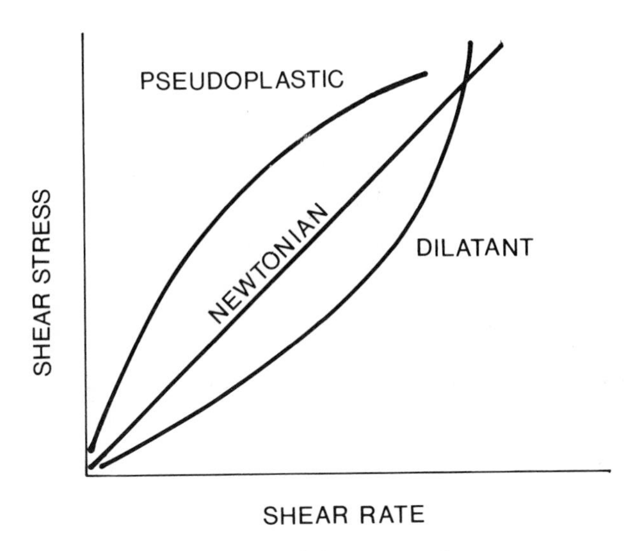


Fig. 26-1. The effect of shear rate on shear stress.

The application of force can cause orientation and alignment of polymer particles, causing an increase in viscosity. It also can cause breaking of the intermolecular bonds, causing a decrease in viscosity. If the making of bonds predominates, there is dilatant behavior. If the breaking of bonds is more prevalent, the behavior is thixotropic. Dilatant solutions are known to exhibit a climb up the shaft as shown in Fig. 26-3.

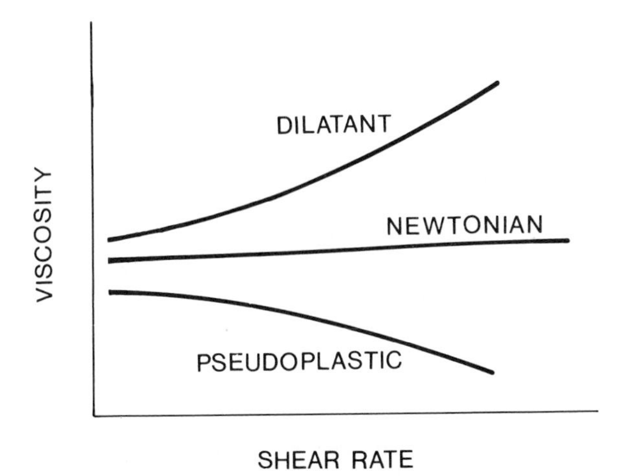


Fig. 26-2. The effect of shear rate on viscosity.

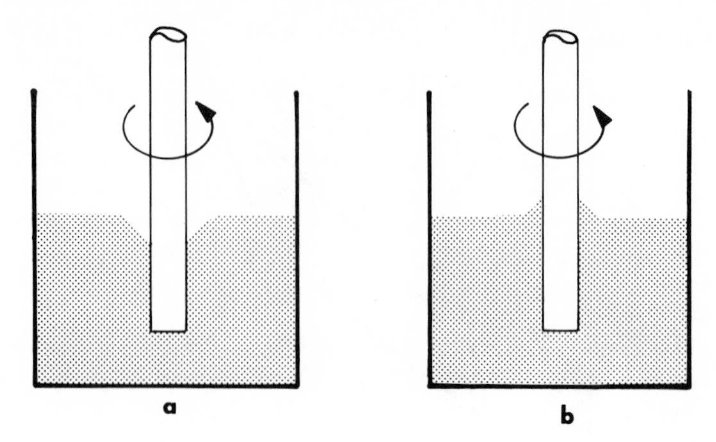


Fig. 26-3. Fluid behavior during mixing. $a$ = Newtonian; $b$ = dilatant.

Dilatant behavior has been observed with both solutions and latexes. In the case of latexes, this behavior is not restricted to any particular particle size, but can occur over a wide range of particle size from 1–700 $\mu$m.

In addition to the effect of rate of shear, the polymer solutions and latexes can exhibit the effect dependent on the duration of shear. Thixotropy is the decrease of viscosity as a result of mixing over a period of time. Rheopectic behavior is opposite to the thixotropic one. Figure 26-4 shows these modes of behavior graphically.

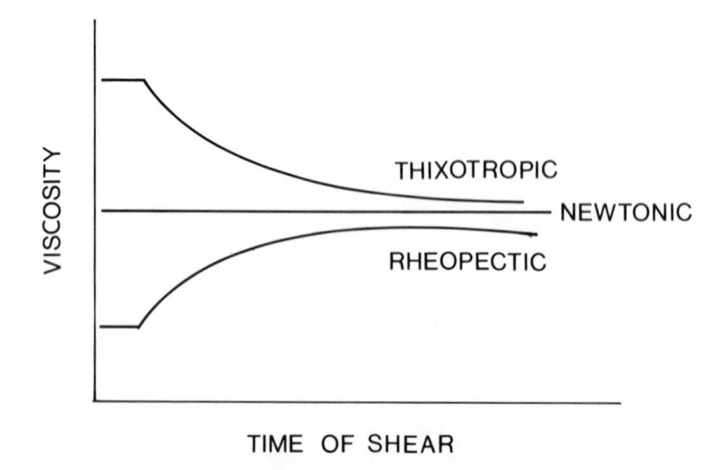


Fig. 26-4. The effect of time of shear on viscosity.

## MACHINEABILITY

Deviation from a Newtonian response to shear rate can cause problems in the coating operation. Especially dilatancy can interfere with the deposition of a smooth and uniform adhesive coating. In the knife-over-roll coating, dilatancy might require application of excessive tension, causing distortions of the web and under extreme conditions even knife deflection. Severe dilatancy may cause "chatter," leaving void spots in places on the web. In reverse roll coating, dilatant solutions or latexes will require more power and will cause more splitting.

Various coating imperfections for vinyl plastisol have been discussed by Lazor[1] and are shown in Fig. 26-5. Similar behavior is also observed with pressure-sensitive adhesives, except that the deviations from Newtonian behavior are not as common or as severe.

Pressure-sensitive adhesive latexes must be thickened for the best handling on coating equipment. The types of thickeners used have an effect on the rheological properties of the latex. Cellulosic thickeners, poly(vinyl alcohol) and aqueous solutions of alkali soluble acrylic emulsions are widely used for thickening latexes. Sanderson and Gehman[2] discuss the use of acrylic thickeners that impart various flow characteristics to the adhesive latex. For the gravure coating applications, a blend of Acrysol ASE-60, which gives a short, buttery consistency, and Acrysol ASE-95, which gives a long, leggy consistency, are recommended. The exact blend depends on the cell depth. For the reverse roll coating applications, ASE-60 is recommended. For the air knife coating, various types are suitable.

Figure 26-6 shows the effect of shear rate on the viscosity of formulated pressure-sensitive acrylic latexes.[3] The solutions of pressure-sensitive adhe-

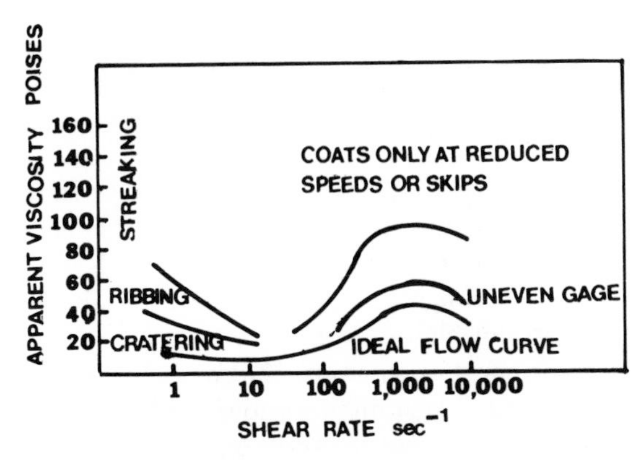


Fig. 26-5. The coatability of plastisols.[1]

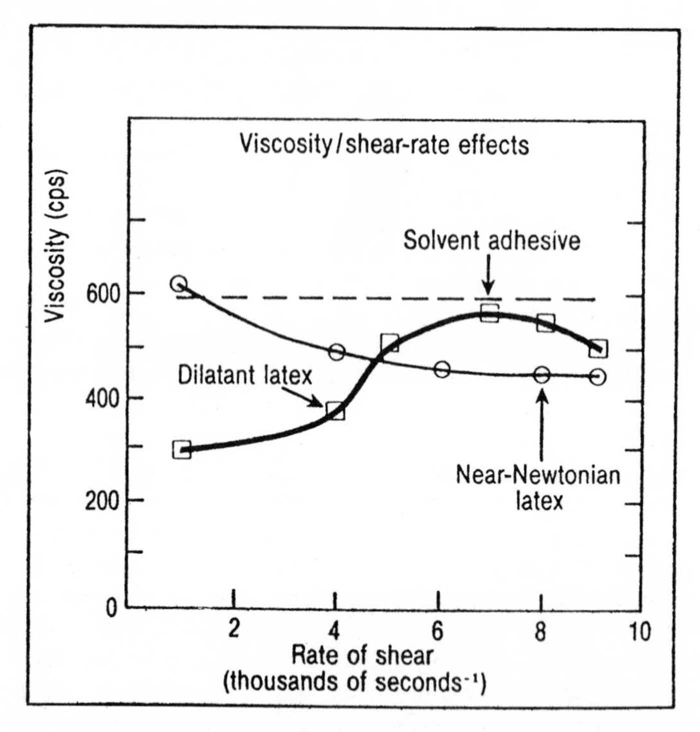


Fig. 26-6. The effect of rate of shear on viscosity of pressure-sensitive adhesive latexes.[3]

sives are usually shear insensitive, but compounded latexes can show dilatant behavior, although the decrease of viscosity with increasing shear rate is more common.

The adhesive is exposed to high shear rate during the coating operation. Reverse roll coater develops shear rates in the range of 100–1000 $sec^{-1}$, and high-speed knife-over-roll coating in the range of 1000–10,000 $sec^{-1}$. The viscosities, however, are usually measured at low shear rates. Such viscosity data might fail to predict the deviations from ideal behavior at high shear rates. Table 26-1 shows the shear rate range for various viscometers.[1] It is obvious that one viscometer is not sufficient to characterize the adhesive coating completely. The Brookfield viscometer is suitable for viscosity measurements at low shear rates, while Severs extrusion rheometer is often used for high shear rate.

In addition to the effect on viscosity, the compounding of latexes might be required to improve the wettability of low energy, difficult to coat surfaces, such as silicone-coated paper. Addition of surface active agents de-

**Table 26-1. Shear rate range for various viscometers.**

| VISCOMETER | SHEAR RATE ($sec^{-1}$) |
|---|---|
| Brookfield model HAT | 0.6–24 |
| Sun Chemical viscometer | 24–480 |
| Severs extrusion rheometer | 50–1000 |
| Instron extrusion rheometer | 500–46,000 |

creases the surface tension sufficiently. The choice of these additives must not affect the adhesive properties.

Latex adhesives often exhibit plastic behavior: they do not flow without application of a definite shear stress. This might result in poor leveling. Scratch marks imparted under shear may remain in the coating when the shear application is removed. Latex can be formulated to improve its leveling properties. Practical problems related to the conversion from solvent-based to latex pressure-sensitive adhesives have been discussed by Bafford.[4]

## COATING METHODS

Coating of adhesive solutions and latexes can be carried out by several methods. In this process of coating, two different operations should be distinguished: application of the coating and its metering. In some cases, an additional leveling operation is also used to obtain the coating of required surface smoothness.

The most commonly used methods for coating of pressure-sensitive adhesives are reverse roll and knife-over-roll. In addition, gravure, wire-wound rod, floating knife, air knife and other methods are also used. Pressure-sensitive adhesive coatings after drying are usually in the thickness range of 0.04–0.06 mm but rarely below 0.025 mm. Some of these methods are not suitable for applying coatings that thick. Manufacturing of pressure-sensitive tapes and labels also requires application of very thin coatings for release, priming and other purposes. Gravure and wire-wound rod methods are especially suitable for the application of low coating weights.

Table 26-2 summarizes the main characteristics of some of the coating methods of interest for pressure-sensitive adhesive products which are coated as solutions or emulsions. Various coatings methods and equipment are described by Booth[5] and by Weiss.[6]

**Table 26-2. Summary of operating characteristics of various coating methods.**

| COATING METHOD | VISCOSITY (cps) | MAXIMUM SPEED (m/min) | COATING WEIGHT[a] RANGE ($g/m^2$) |
|---|---|---|---|
| Knife-over-roll | 5000–15,000 | 150 | 20–100 |
| Trailing blade | 100–20,000 | 1500 | 3–15 |
| Wire-wound rod | 10–200 | 200 | 3–25 |
| Air doctor | 10–1000 | 750 | 3–20 |
| Reverse roll: | | | |
| Nip fed | 1000–6000 | 300 | 10–200 |
| Pan fed | 200–6000 | 100 | 10–200 |
| Gravure roll | 100–2000 | 700 | 3–20 |
| Slot orifice | 500–20,000 | 200 | 20–100 |

[a] Dry coating weight, assuming 25–50% solids, is given.

The viscosity range given above represents normal operating range. Higher viscosities have been run on most of the coaters.

Maximum speed attainable is not a very important characteristic for pressure-sensitive coaters and such coaters are rarely run at maximum speed. Pressure-sensitive adhesive coatings are reasonably heavy and the maximum attainable speed depends on the drying rate rather than the mechanical capability of the coating equipment. The raw material costs of pressure-sensitive products are high as compared to the labor costs. This decreases the importance of high-speed operation and increases the value of any waste that might be incurred. The waste level usually increases with increasing processing speed.

## KNIFE AND BLADE COATERS

The most direct and simplest to operate coating method is knife-over-roll. The coating thickness is determined by the gap between the knife and the roll. The main disadvantage of this method is that the coating thickness is affected by the variations in the backing thickness. Streaking can also be a problem if any undispersed particles or lumps are present in the adhesive. The knife-over-roll coating method can be used over a wide viscosity range, although the viscosities of 5000–15,000 cps are most common. Adhesives of lower than 5000 cps viscosity are difficult to keep in the puddle without excessive leakage around the dams. Adhesives of viscosity as high as 40,000 cps have been coated by this method, but skips, poor leveling and excessive variation in coating thickness might appear at high viscosities.

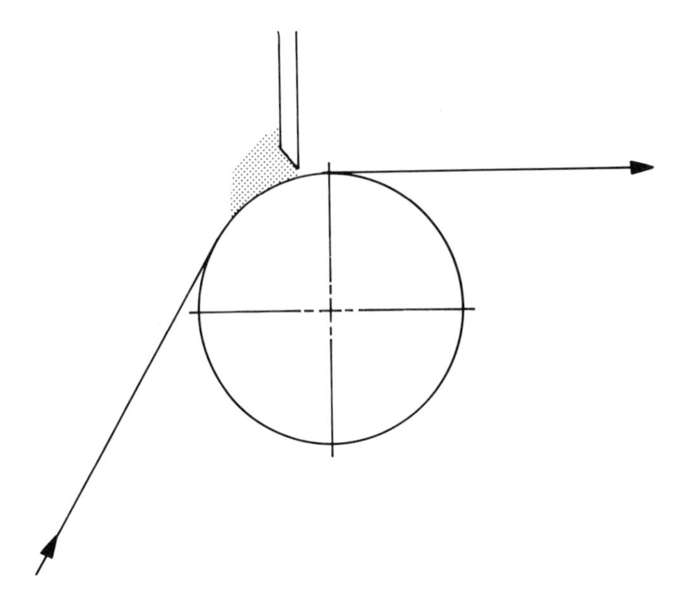

Fig. 26-7. A schematic diagram of knife-over-roll coater.

The dry adhesive coating thickness below 0.025 mm is difficult to deposit with sufficient accuracy by this method. Coatings thicker than 0.1 mm are best deposited in the multiple pass operation: the drying of heavy coatings is difficult.

The coating knife is mounted over the chrome plated steel roll slightly off center (Fig. 26-7). The adhesive is placed in a trough, or as a free puddle restricted by side dams only. The adhesive puddle is replenished by pumping, gravity flow directly from an adhesive drum, or manually by a scoop. A photograph of a knife-over-roll coating station is shown in Fig. 26-8.

The roll and the knife edge must be machined as true as possible. The total run out as small as 0.0125 mm is possible, usually 0.02–0.05 mm is sufficient. The knife can be so designed that the gap is adjusted across the entire width by warping the blade.

There are several blade designs used for pressure-sensitive adhesive coating as shown in Fig. 26-9. The bevelled knife (a) is the simplest to manufacture. It is bevelled to about an angle of 10° and a land of 1.5 mm. The shear forces exerted on the coating in this type of knife design are high. Sometimes this knife is used in the opposite position with the straight leading edge. This arrangement exerts the maximum horizontal shear force on the coating, but minimizes the vertical component of such force responsible for driving the coating into the substrate.

Fig. 26-8. Knife-over-roll-coater. (*Courtesy Liberty Machine Co., Inc.*).

The rounded edge knife (b) and (c) exerts less shear. Hook knife (d) is a modified rounded edge knife with a groove on the trailing edge. The groove is intended for the accumulation of small particles and is supposed to minimize the streaks. This knife design is probably the most popular for pressure-sensitive adhesive coating. Bull nose (e) knife has the largest radius and exerts less shear than the other knife designs. This design is popular with a major pressure-sensitive adhesive product manufacturer. A sharp blade (f) is useful for priming and similar applications.

For pressure-sensitive adhesive coating, the knife coater is usually used as shown in Fig. 26-7. Instead of a knife-over-roll arrangement where the deposit thickness is determined by the gap setting, a knife over rubber roll method is quite different (Fig. 26-10a) despite a similar superficial appearance. The deposit is controlled by the pressure of knife against the roll,

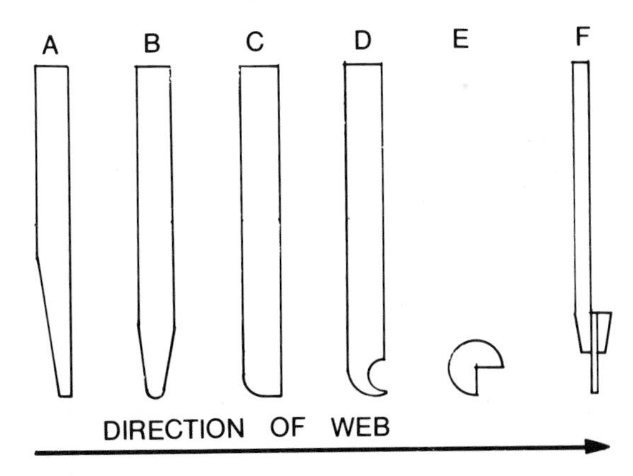


Fig. 26-9. Various types of blades used in knife-over-roll coating. $A$ = beveled; $B, C$ = rounded edge; $D$ = hook; $E$ = bull nose; $F$ = spanishing knife.

knife angle and some other variables such as adhesive viscosity, web tension and coating speed. This method is often used for fabric coating. If the strike through the open weave is a problem, double wrap around the backup roll can be used, as shown in Fig. 26-10b.

There are many other knife coating arrangements. Knife-over-blanket coating is shown in Fig. 26-10c. The method is similar to knife-over-rubber-roll coating. The driven blanket allows the use of less strong substrates.

Floating knife coating is shown in Fig. 26-10d. The knife is pressed against the web which is supported on both sides of the knife by either rolls or nonrotating supports, such as channel. The method is suitable for back-

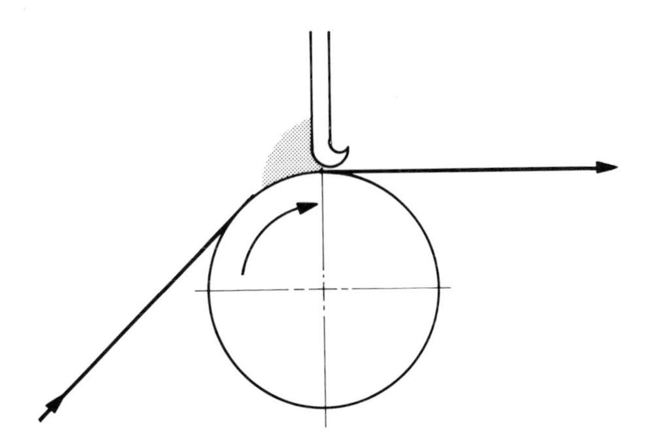

Fig. 26-10a. Knife over a rubber roll coater.

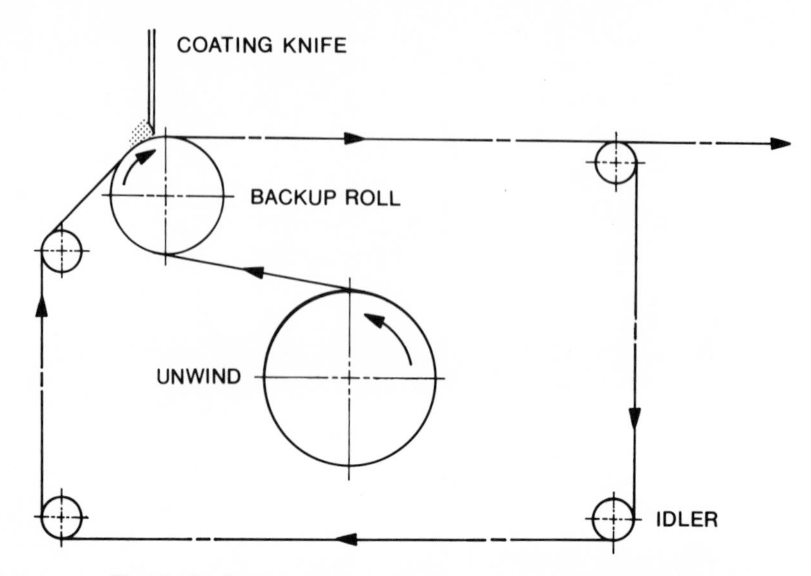


Fig. 26-10b. Coating with a double wrap to prevent a strike-through.

coating of textile materials and also has been used for coating pressure-sensitive latex adhesives.

Inverted knife coating is shown in Fig. 26-10e. It is a floating knife arrangement in the inverted position. The strike through is decreased as compared to floating knife. The coating is controlled by the knife position, web tension and the rheological properties of coating dispersion.

Two or more floating knife coaters can be used in tandem. Such a sequential knife coater allows deposit of different coatings one after another.

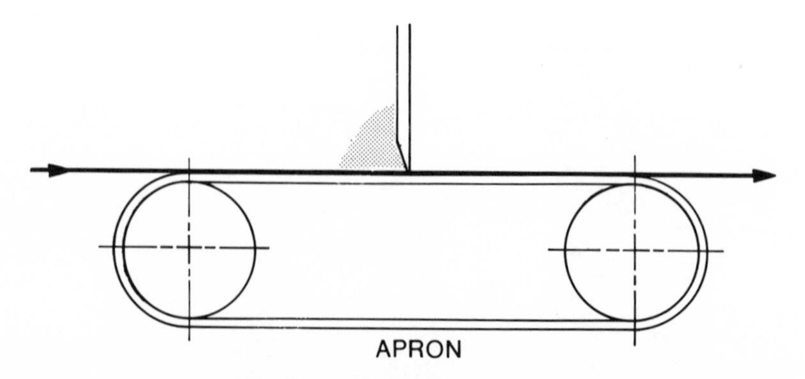


Fig. 26-10c. Knife-over-blanket coater.

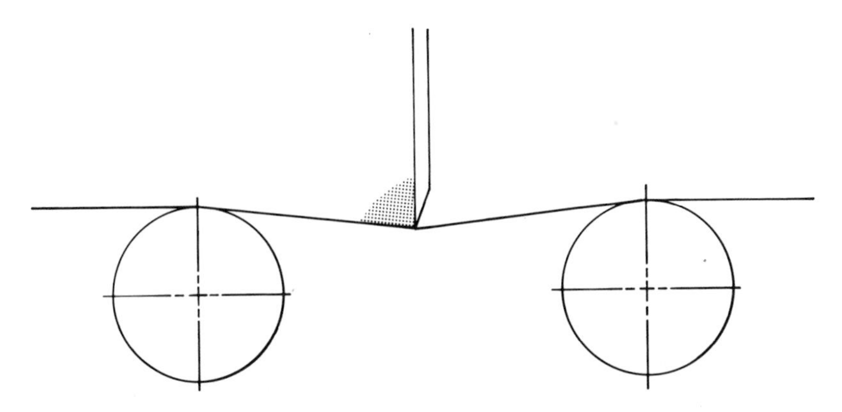

Fig. 26-10d. Floating knife coater.

Blade coaters are similar to the knife coaters, except that a flexible blade is used instead of a rigid knife. The blade is wiping the surface of the web against a backup roll. The coating can be in a puddle in front of the blade as in case of the trailing blade coater shown in Fig. 26-11. The coating can also be preapplied by a roll and the blade is used for metering purposes only. The blade can be used in a inverted position and in several other combinations. The trailing blade coating is important in high-speed paper coating. In pressure-sensitive product applications, the method is applicable for release coating applications and other lightweight coatings. The increasing use of water-based pressure-sensitive adhesives might increase the interest in blade coating for adhesive application as well. A recent development of an extended blade coater with an inflatable pressure tube near the blade tip provides an accurate control of blade pressure and allows the deposition of heavier coatings.

Blade coaters develop high shear forces. The shear rates of $10^6$ $sec^{-1}$ are reported for a blade coater at 900 m/min coating speed.[7]

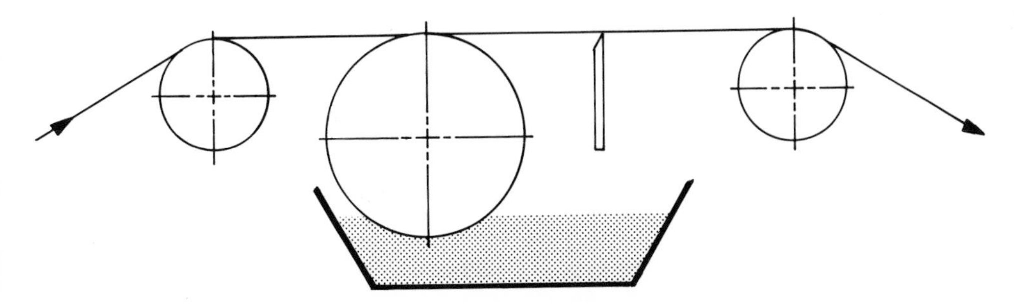

Fig. 26-10e. Inverted knife coater.

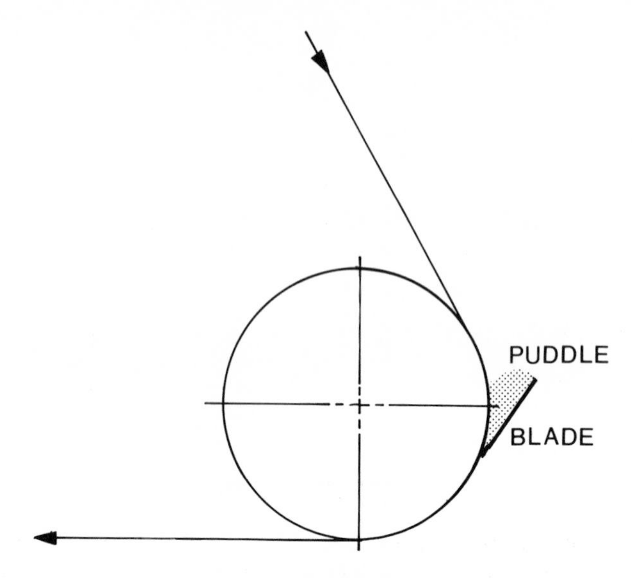


Fig. 26-11. Trailing blade coater.

## ANALYSIS OF KNIFE AND BLADE COATING

Knife coating is a high shear method. Its coating mechanism is complex and has not been studied that much. The process is empirical and any mathematical model requires many simplifying assumptions. Freeston[8] has published a theoretical analysis of knife coating over porous substrates. A knife coating model as shown in Fig. 26-12 has been analyzed. The analysis is based on a simplified geometric model of the knife-coating-substrate interface and on several other simplifying assumptions.

Freeston derived the equation

$$p_m = \frac{4\mu U}{l\theta^2} \tag{2}$$

where

$p_m$ = the difference betweem the total pressure and the hydrostatic pressure
U = web velocity
$\mu$ = viscosity
$l$ = knife width
$\theta$ = knife angle

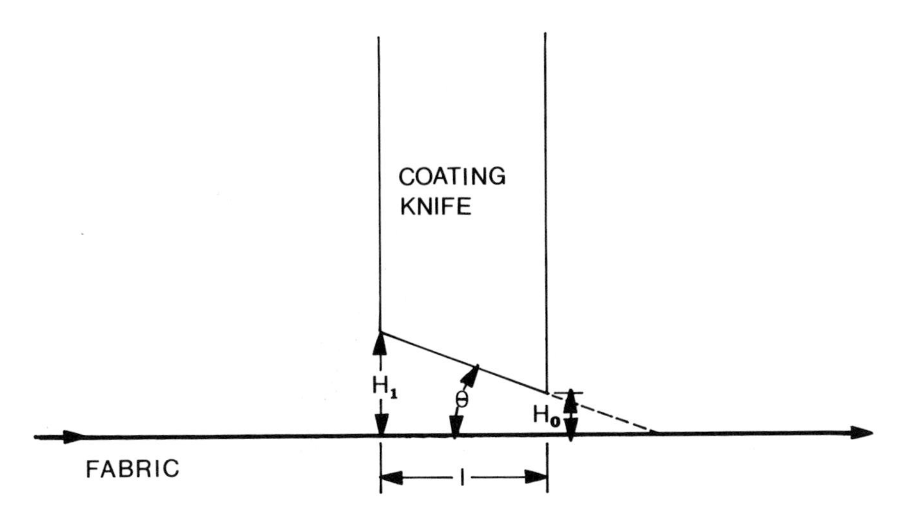


Fig. 26-12. The geometry of a coating knife.

The expression shows that the higher the web velocity, the thinner the knife, and the smaller the knife angle, the greater is the dynamic pressure generated under the knife. Increasing the coating viscosity increases the pressure, but the viscosity is a complex variable and the relation is not as simple as shown in Equation (2).

The penetration of the coating under the knife into the porous substrate is also discussed by Freeston.[8] He concludes that the depth of penetration is mainly determined by the pressure developed under the knife. The depth of penetration is independent of web velocity, but increases with an increasing degree of coating pseudoplasticity. The penetration is increased by decreasing the angle between the blade and the substrate and by increasing the pore size of the substrate.

Hwang[9] has presented a detailed analysis of hydrodynamic forces involved in a coating operation employing a rigid blade and assuming a Newtonian fluid. Two different analyses are presented: a simplified one employing five variables and a complete one employing fifteen variables. In the simplified version, the coating thickness is related to gap width, liquid viscosity, density, surface tension and the web speed. The relationship is shown in Equation (3). A schematic diagram of the coating geometry is shown in Fig. 26-13.

$$T = \frac{h}{2} + \frac{1}{12\mu}\left(\frac{\sigma}{2h^2} + \rho g_x\right)\frac{h^3}{u_o} \tag{3}$$

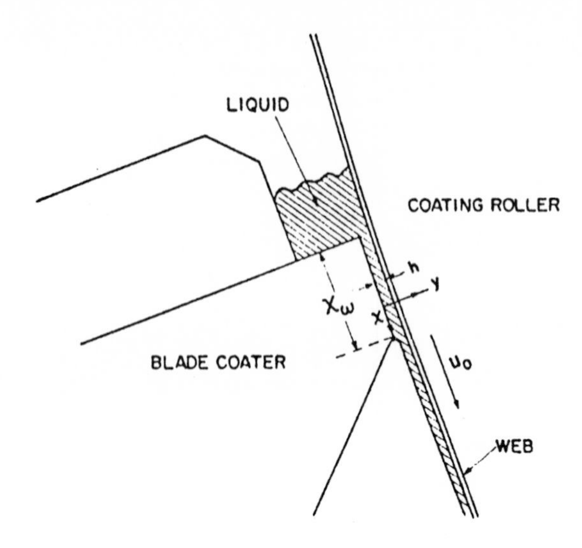


Fig. 26-13. The geometry of a blade coater.

where

$T$ = coated liquid film thickness
$h$ = coater gap setting
$\sigma$ = surface tension of the liquid
$\mu$ = viscosity
$\rho$ = liquid density
$g_x$ = gravity constant in the direction of $x$ coordinate
$u_o$ = web speed

The results calculated by Equation (3) are compared to experimental data in Fig. 26-14.

Particles in the adhesive can clog the gap between the knife and the substrate, causing streaks. Therefore, it is desirable to coat from the largest possible gap, so that these particles would be able to pass under the knife. Usually, the liquid film thickness $T$ is about half of the coating gap $h$. An accurate coater optimization leading to a minimum film thickness for a given gap thickness requires a complete analysis employing 15 variables. The most important variable, however, is the gap width $h$; the other variables can affect only 10% change in the film thickness. The second most important variable is the blade thickness.

Blade coating is also analyzed by Middleman[10] for both rigid and flexible blades. The analysis of the flexible blade coating becomes more

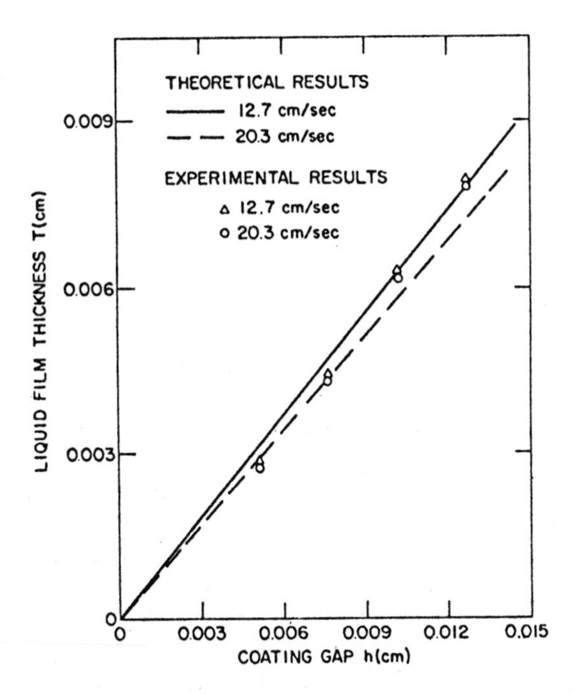


Fig. 26-14. Comparison of theoretical calculations based on Equation (3) with experimental results.

complex. The case of Newtonian coating solution is discussed and a nonlinear behavior of a fluid following the power law is also considered as well as the effect of viscoelasticity in general.

## AIR KNIFE COATING

Air knife coating has been developed in the 1930's for paper coating application. The method is suitable for high-speed coating of printing paper, label paper and diazo paper. The air knife has not been employed much for pressure-sensitive adhesive coating. With the increasing use of pressure-sensitive adhesive latexes, this method becomes of interest. It is most suitable for low coating weight applications: up to 20 $g/m^2$ per pass.

Air knife coating is a metering system which removes excess coating applied to paper. Unlike knife-over-roll or blade coating methods, which coat to a constant total thickness, the air knife coating follows the contours of the paper: it coats to a constant coating thickness, similarly to the roll coaters. This difference is illustrated in Fig. 26-15a and b.

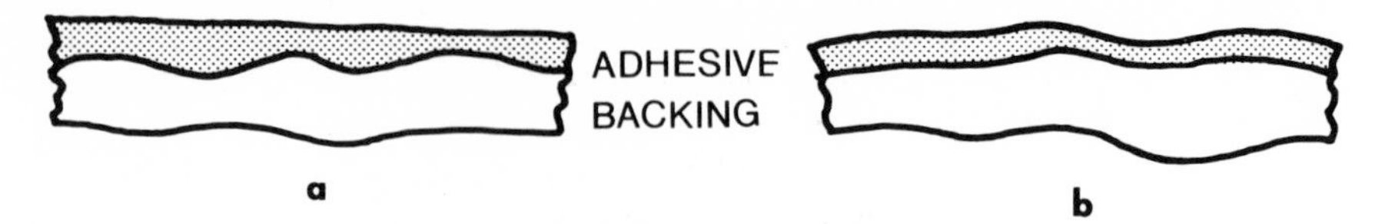


Fig. 26-15. Difference in the nature of coating. $a$ = Coating to constant total thickness. $b$ = Coating to constant coating thickness.

The air knife can be used in three different ways. In metering a highly filled latex coating on absorbent backing, the air knife removes the outer, less viscous layer. The absorbent material removes the water from the coating and the layers closer to the web are higher in solids and viscosity and are more difficult to remove by shearing. The shearing is done by the air stream forced through a gap between the blades as shown in Fig. 26-16. This coating mechanism is similar to the formation of a filter cake on a filter screen. In case of nonabsorbent webs or less filled lower viscosity coatings, the air knife removes a portion of the coating. The air velocities used are lower in this case. Air knife and brush air doctor can be used for finishing purposes only. A light stream of air helps to level the coating which has been applied by some method.

A schematic diagram of an air knife coater is shown in Fig. 26-17. The excess coating is removed by the air knife and is recirculated. The aqueous coatings are inclined to foam and various designs of the vessel receiving the excess coating are available, facilitating the deaeration of the coating. Judicious use of defoaming agents can also help. The coating can be applied by

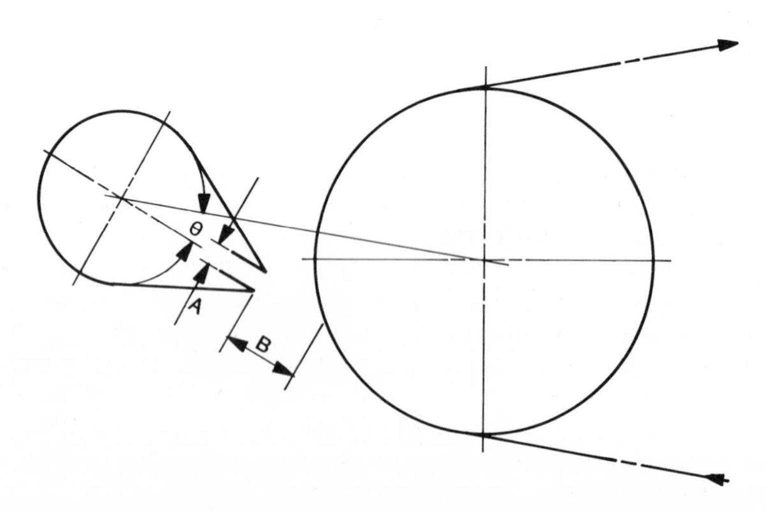


Fig. 26-16. A schematic diagram of air knife coater head.

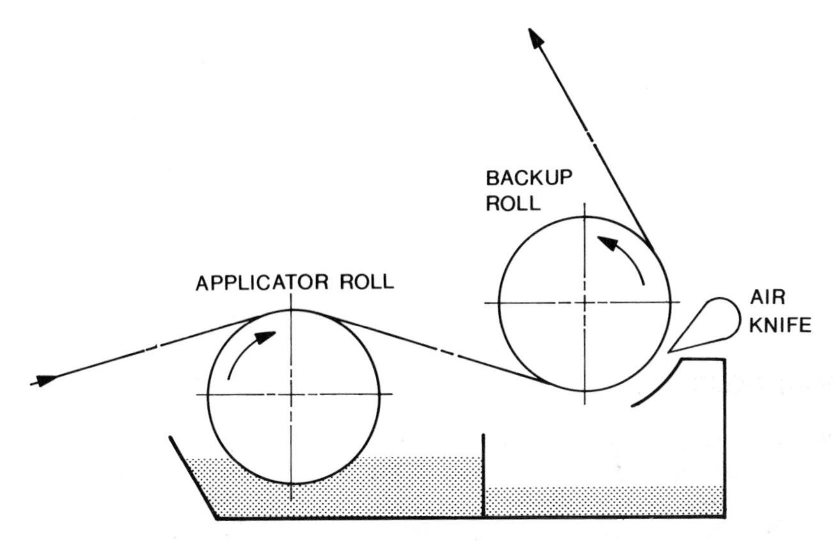


Fig. 26-17. A schematic diagram on an air knife coater.

various methods: single roll as shown in Fig. 26-17, multiple roll system, fountain or some other method.

The air knife as shown in Fig. 26-16 consists of a plenum chamber and a set of blades which can be adjusted to regulate the gap A. The main variables are the air velocity, distance between the lip and the drum B, lip gap A, angle of attack ($\theta$). The range of these variables is given below.

| | |
|---|---|
| Lip gap | 0.25–0.75 mm |
| Lip distance | 3.5–5.5 mm |
| Angle of attack | 5–12 degrees |
| Air velocity | 30–140 m/min |
| Air pressure* | 0.1–0.6 atm |
| Air volume | 2–10 $m^3$/hr/cm |
| Coating weight | 3–20 $g/m^2$ |
| Coating speed | 10–750 m/min |

* The air pressure range is given for coating of absorbent materials, such as paper. The air pressure for nonabsorbent materials is 5–10% of that and the air velocity and air volume are correspondingly lower.

The effect of various process variables is discussed by Taylor.[11]

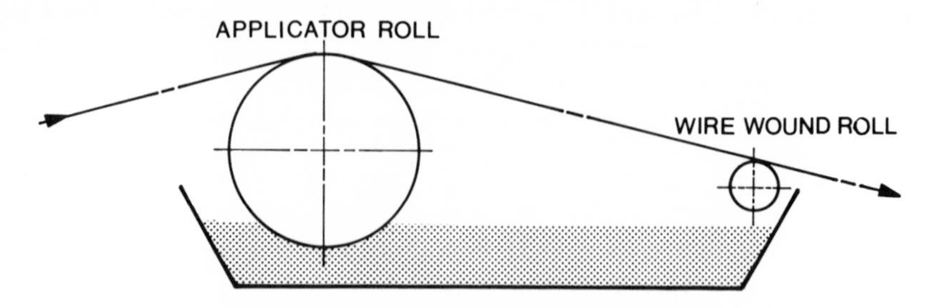


Fig. 26-18. A schematic diagram of a bar coater.

## BAR COATERS

Bar coaters are widely used for metering and smoothing applications. Lightweight pressure-sensitive adhesive coatings are applied by bars over films. Coating bars are often used for release or prime coating application. The coating is applied by some other means, usually a kiss roll on the bottom side of the web. The bar is used to remove the excess of the coating or to smooth it out as shown in Fig. 26-18.

Wire-wound bar or Mayer rod is the most popular version of bar coaters used for metering purposes. Wire is wound around a small diameter rod. The rod wipes of the excess coating except what passes between the wires. The amount of the coating remaining depends on the wire diameter: the larger the diameter, the more coating remains on the substrate. Wire diameters 0.07–0.8 mm and heavier are used for wrapping the rod. To minimize clogging, the thickest possible wire should be used.

Figure 26-19 shows schematically the wire-wound rod and the gaps between windings. The area of one gap is equal to $2r^2 - \pi r^2/2 = 0.429r^2$. Ideally, this would give a coating thickness of $0.2145r$, if the gap area were the only factor that determines the coating weight. Unfortunately, the amount of coating deposited depends on variables such as web tension, wrap angle, rheological properties of the coating, web speed, bar rotating

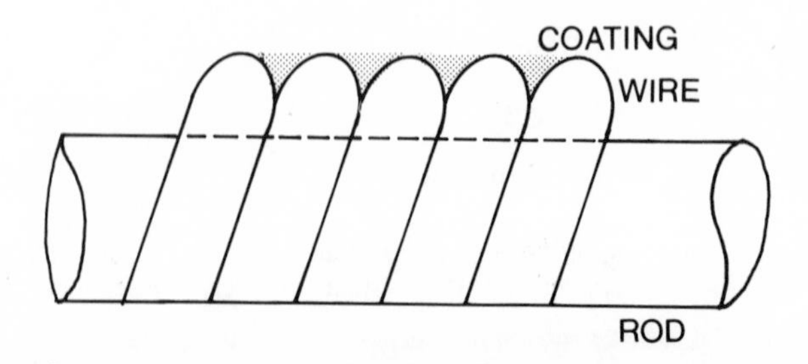


Fig. 26-19. A schematic diagram of a wire-wound-bar.

speed and absorbency of the substrate. The amount of coating, therefore, must be determined experimentally. The type of wire used gives only a general idea of the deposit that could be expected.

The rod is continuously rotated in either direction by a separate drive in order to distribute the wear. The rod should be installed in such a way that the rotation tightens the wire to the rod.

Various holder designs are used. The holder keeps the rod in place and allows it to rotate without excessive friction. The holder could consist of simple clamps holding the ends of the rod, or it could consist of a magnet holding the rod sufficiently firmly but allowing minimum resistance to rotation. Several other designs are available. A fairly elaborate wire-wound rod assembly is shown in Fig. 26-20.

Most adhesive applications require smooth coatings. Some uses, such as metallized polyester film that is applied over the window glass for solar radiation control, require optically perfect adhesive coating. Polishing bars are used to smooth out the adhesive coatings. A common coating imperfection is ridges in the machine direction formed by splitting of the coating in

Fig. 26-20. Wire-wound-bar coater assembly. (*Courtesy American Tool and Machine Company.*)

roll coating machines. Mushel[20] describes a fast rotating smoothing bar which improves on this condition not by completely eliminating these striae but by decreasing their amplitude. The flow of the coating completes the leveling before it dries. Polishing bars can also help to remove the air bubbles, which might be introduced into the coating during mixing or during the coating operation itself.

The polishing bar consists of a smooth rod or tube that is brought into contact with the adhesive surface. It may be stationary or it may be equipped with a drive in order to rotate the bar, usually in the direction opposite to the web movement. The polishing bar smoothes the surface irregularities by simple leveling, or it may remove a substantial amount of the coating and redeposit it. In most cases, the mode of operation is between these two extremes.

## REVERSE ROLL COATERS

Reverse roll is the most versatile coating method and the most widely used for pressure-sensitive adhesive coating applications. It is preferred over the knife-over-roll coating arrangement for many pressure-sensitive applications. The normal viscosity range is 1000–6000 cps, but it can handle much higher viscosities, reportedly up to 40,000 cps. Coating speeds up to 100 m/min are often used, but speeds up to 300 m/min are claimed to be possible. Most of the time, the speed is determined by the rate of drying.

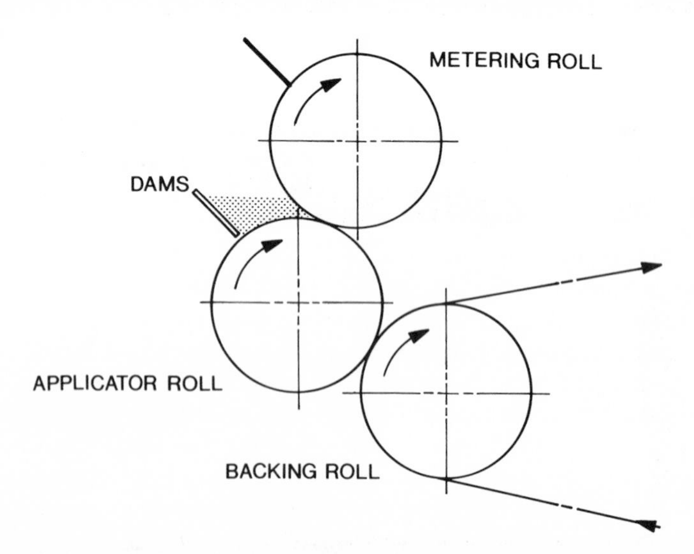


Fig. 26-21. A schematic diagram of a three-roll nip-fed reverse roll coater.

Reverse roll coater is somewhat more difficult to operate than a knife-over-roll coater, although the basics of the operation are relatively simple. Figure 26-21 shows a three-roll nip fed reverse roll coater. The applicator roll carries the coating to the web. The amount of the coating on the applicator roll is determined by the gap set between the applicator and metering rolls. These are chilled iron rolls made very accurately with a total indicator readout as low as 0.0025 mm. The metering roll is running very slowly and some operators prefer to hold it still. The applicator roll rotates 2–3 times faster than the backing roll carrying the web and wipes the coating onto the web. The differential speeds eliminate scratches and other imperfections in the coating. The amount of coating deposited on the web is determined by the gap between the metering and applicator rolls and by the speed ratio between the applicator and backing rolls. Typical reverse roll operating conditions for coating of a pressure-sensitive acrylic latex are given below.[12]

| | |
|---|---|
| Applicator roll speed | 32 rpm |
| Metering roll speed | 3.5 rpm |
| Metering gap | 0.0076 mm |
| Backing roll speed | 10 rpm |
| Coating weight (wet) | 71 $g/m^2$ |

The reverse roll coater deposits a premetered uniform coating thickness which is independent of backing thickness variations.

Any imperfections in the backing are less likely to cause tearing in the reverse roll coater than on a knife coater. Weak substrates are handled easier on a reverse roll coater. For example, the second pass coating of a two-faced film tape is difficult to run on a knife-over-roll coater because of the frequent tearing, but this product can be easily handled on a reverse roll coater.

Another popular arrangement of a reverse roll coater is a pan fed machine as shown in Fig. 26-22. A pan fed coater is suitable for lower viscosity coatings which cannot be contained between the dams in the nip.

Various other arrangements and modifications of reverse roll coaters are available. Die feeding has been proposed as an alternate method to nip and pan fed coaters.[13] A four-roll coater as shown in Fig. 26-23 can be run at much higher speeds,[14] because the roll in the pan can rotate at a slow speed, preventing foaming and splashing of the coating. Reverse roll coaters are extremely versatile machines and can be converted to many different coating arrangements. A reverse roll coater that is also convertible to a knife-over-roll coater is shown in Fig. 26-24.

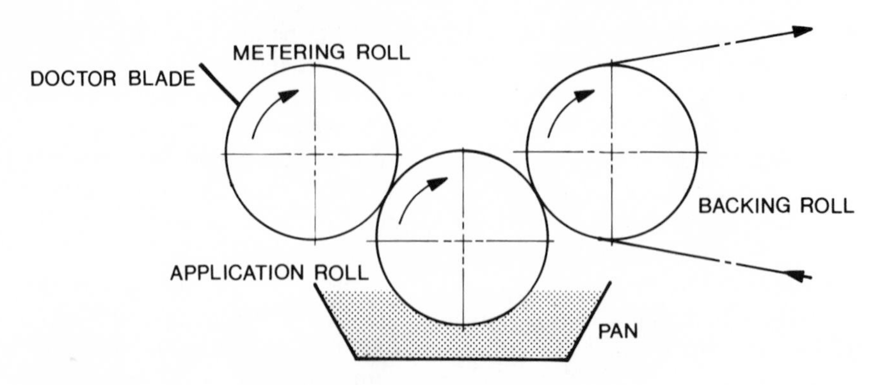


Fig. 26-22. A schematic diagram of a three-roll pan-fed reverse roll coater.

## MECHANISM OF ROLL COATING

The mechanism of roll coating involves compression of a thin wedge of a liquid and splitting it at the other side of the nip. The porosity of the backing, coating thickness, coating speed and the rheological properties of the adhesive influence the coating process. Miller and Meyers[15] have studied the film splitting mechanism and divide the flow in the nip into four regions: laminar flow, cavitation, cavity expansion and filamentation. Surface tension and viscosity resist the expansion of the cavity after its forma-

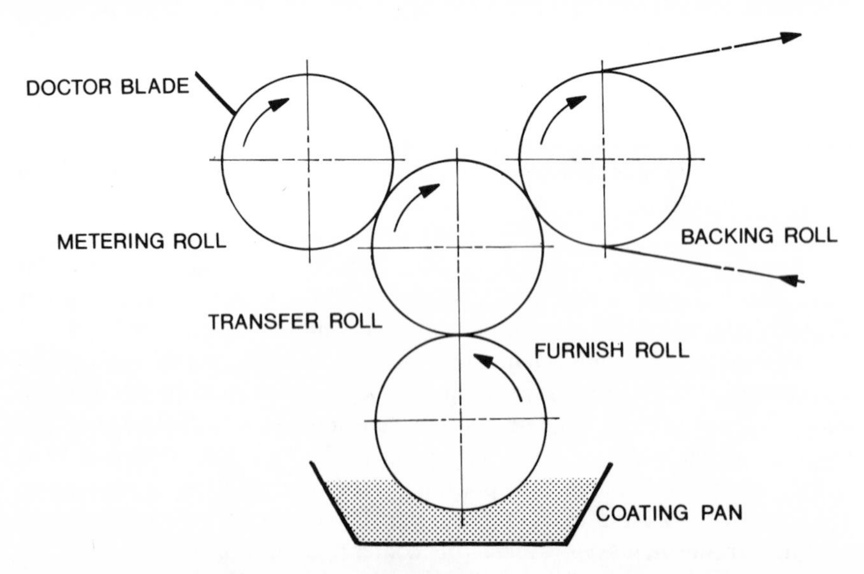


Fig. 26-23. A schematic diagram of a four-roll pan-fed reverse roll coater.

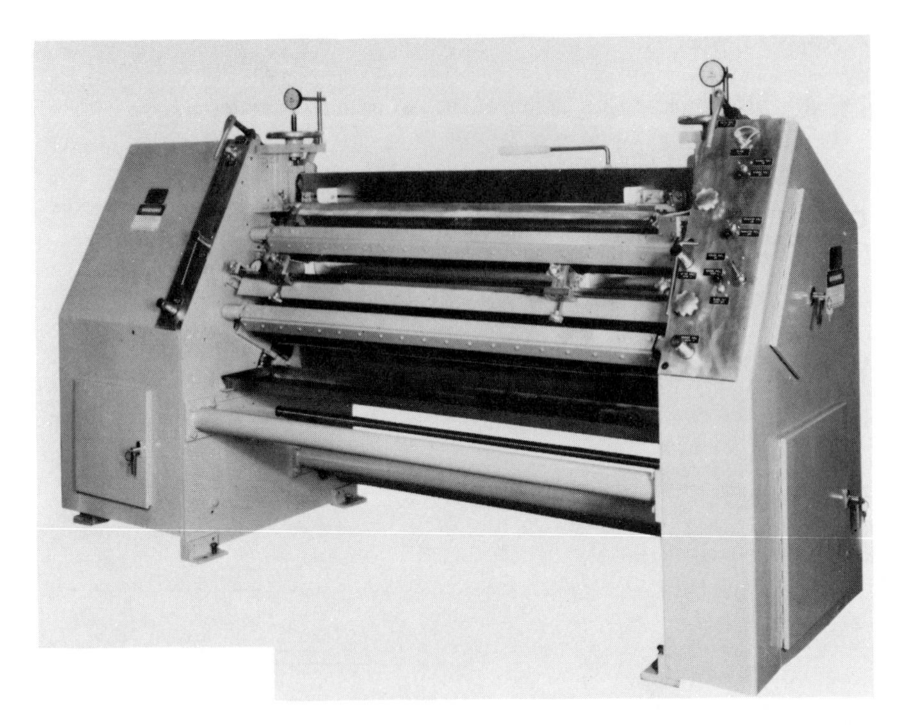

Fig. 26-24. Reverse roll coater and knife-over-roll coater combination. (*Courtesy American Tool and Machine Co.*)

tion. Air present in the cavities provides PV work to the system during expansion. Surface tension is the most important factor in the cavity expansion. Various regions in a roll nip and the pressure profile created are shown in Fig. 26-25. Minimum clearance $h_o$ and clearance at the point of cavitation $h_1$ together provide an important geometrical parameter.

The roll coating was also analyzed by Middleman.[10] Coating thickness is related to geometric, rheological and operating variables. The analysis is carried out, assuming Newtonian behavior of the coating fluid and also a case in which the coating fluid follows a power law.

The dependence of wet film thickness was investigated experimentally by Schneider[17] in a two-roll setup. Roll speed had no effect on the wet film thickness; the wet film was directly proportional to the roll clearance. The increase in viscosity caused a slight increase in the film thickness. The surface tension had no effect on the film thickness. A simple mathematical expression

$$t = K(c/2) \tag{4}$$

is proposed, where

$t$ = film thickness on either of the two equal diameter rolls rotating at equal speeds
$c$ = clearance between rolls
$K$ = factor including the effect of all the other variables. The maximum variation of $K$ is $1 \leq K \leq 4/3$. For conditions in which the inlet bank wets the top roll surface at about 10°, $1.192 \leq K \leq 4/3$.

## OTHER ROLL COATERS

Many other roll coating arrangements are used for pressure-sensitive adhesive coating.

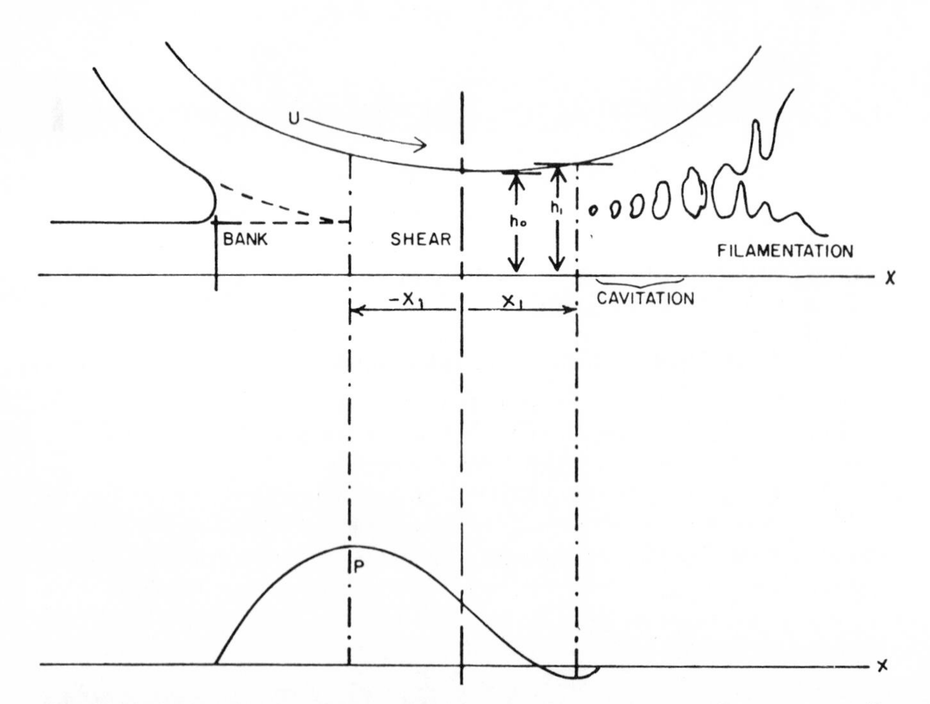


Fig. 26-25. Regions in a roll nip and the pressure profile created according to Meyers.[16] *(Reprinted by permission of John Wiley and Sons, Inc.)*

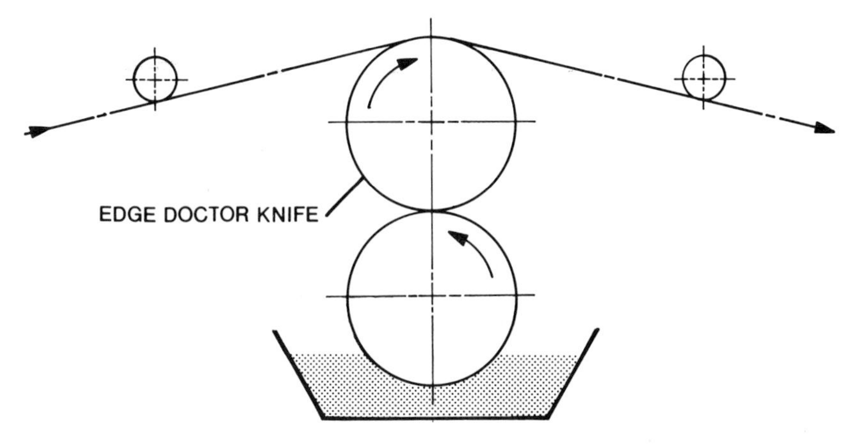


Fig. 26-26. A schematic diagram of a two-roll kiss coater.

Kiss roll coating is often used to apply the coating for subsequent metering by a wire-wound rod, air knife or other means. It is also used without additional metering where the coating weight control is not that important. One-, two-, or three-roll arrangements are employed. Figure 26-26 shows a two-roll kiss coater.

A pressure roll coater is shown in Fig. 26-27. It consists of rubber and steel rolls. The deposit is controlled by the pressure applied. The coating is picked up at the nip by the web. Film splitting in the nip can cause a stippled or ridged pattern in the coating. The severity of this coating disturbance depends on the rheological properties of the coating. Fast relaxing, low viscosity coatings are less likely to leave a pattern. The resiliency of the rubber roll has an effect on the ridged pattern. The pressure type coaters are mainly used in the paper industry and are applicable for prime or release coating of pressure-sensitive products.

Roll-over-roll coater is shown in Fig. 26-28. The coating principle is identical to that of the knife-over-roll. In this respect, this coater does not belong in the category of roll coaters where the coating is generally premetered in some way.

The bottom roll can be either rubber-covered or steel. When the

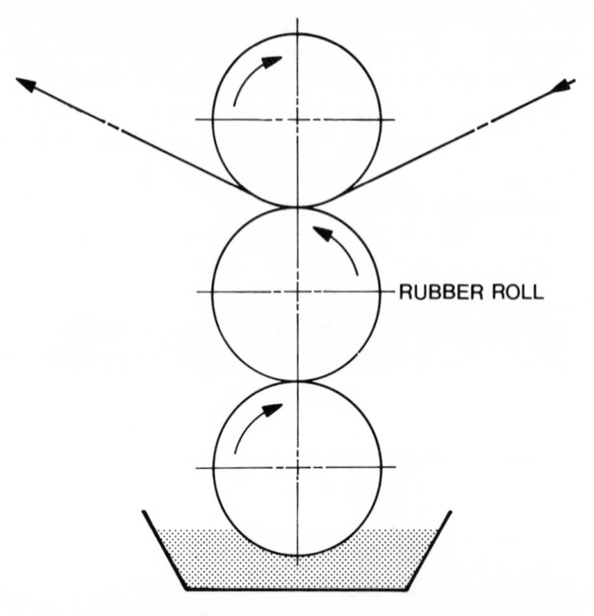


Fig. 26-27. A schematic diagram of a three-roll pressure coater.

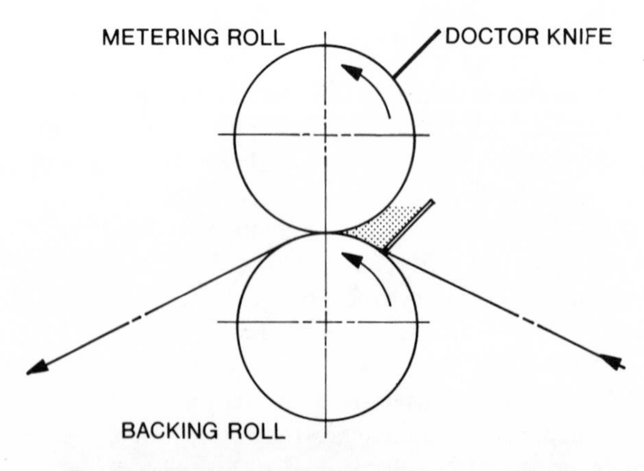


Fig. 26-28. A schematic diagram of roll-over-roll coater.

rubber-covered roll is used, the coating is controlled by the nip pressure. When the steel roll is used, the deposit is controlled by a gap setting in a manner identical to knife-over-roll coating. This coating method produces less scratches than knife-over-roll process. It has been used to apply pressure-sensitive adhesive latexes.[14] The same coater can be used in a horizontal arrangement or it can be pan fed by a kiss roll. The roll-over-roll coater is also known as a levelon coater.

Microtransfer coater[18] is suitable for an accurate application of very light coatings, as low as 0.4 g/m$^2$. It is useful in applying 100% solid silicone release coatings. A schematic diagram of such a coater is shown in Fig. 26-29. Premetering is accomplished in a conventional way as in a reverse roll coater. The coating is contained in a bank between two accurately machined-chilled iron rolls and is transferred to a rubber-covered roll which rotates 10–20 times faster than the chilled rolls in the premetering section. The coating is transferred onto the web in the next nip between the rubber applicator roll and the backing roll.

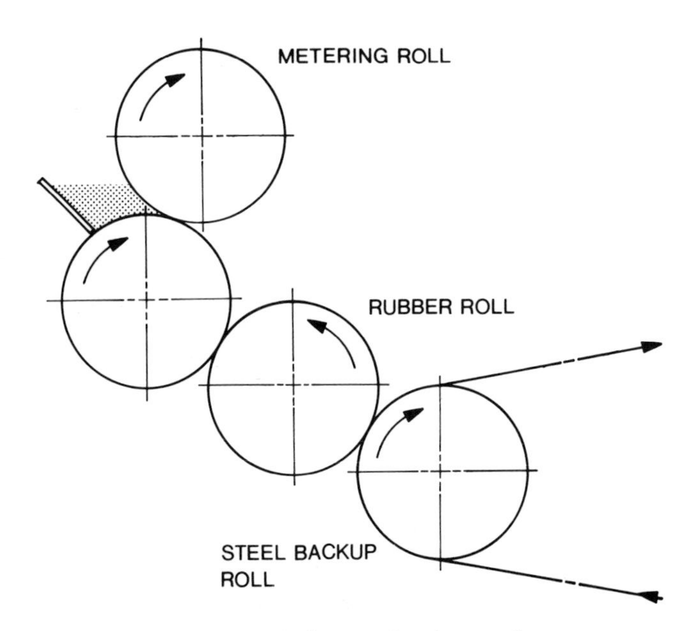


Fig. 26-29. A schematic diagram of a microtransfer coater.

## GRAVURE COATING

Gravure coating is a highly reproducible method, mainly for low coating weight application. The coating is picked up by an engraved roll and the amount of coating deposited is mainly determined by the depth of engraving. There are two variations of the gravure coating, direct and offset. In the direct gravure, the adhesive is transferred from the engraving to the backing. In the offset gravure the coating from the engraved roll is transferred to another roll and then to the backing. A better flow and less patterning are obtained by offset coating.

Several types of engravings are used for coating applications. Quandrangular cell engraving is used for direct gravure application of lighter coating weights. Hexagonal cell engraving is also used for the same application. Less often, trihelical engraving is used for heavier coatings and pyramid cell for offset coating. For very low weight coatings, smooth or just sandblasted rolls are used.

The applicator rolls are engraved by using the appropriate knurled tool and then are chrome-plated. For additional wear resistance, the rolls can be covered with a ceramic coating.

Gravure application is best suited for coatings up to 10 g/m$^2$, Although heavier application is possible, the coating viscosity is usually 100–2000 cps. Heavier viscosity coatings do not flow fast enough from the cells. Gravure coating is suitable for fast speed applications: two-roll coaters up to 400 m/min, three-roll coaters up to 700 m/min. In pressure-sensitive adhesive products, it is used for application of release and prime coatings.

A two-roll direct gravure coating set up is shown in Fig. 26-30. The

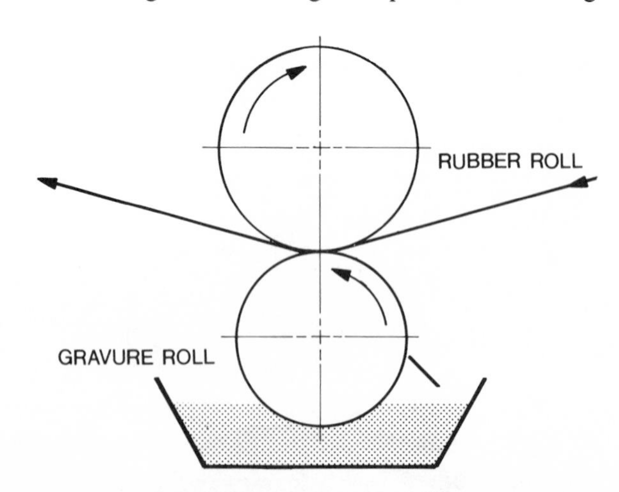


Fig. 26-30. A schematic diagram of a two-roll direct gravure coater.

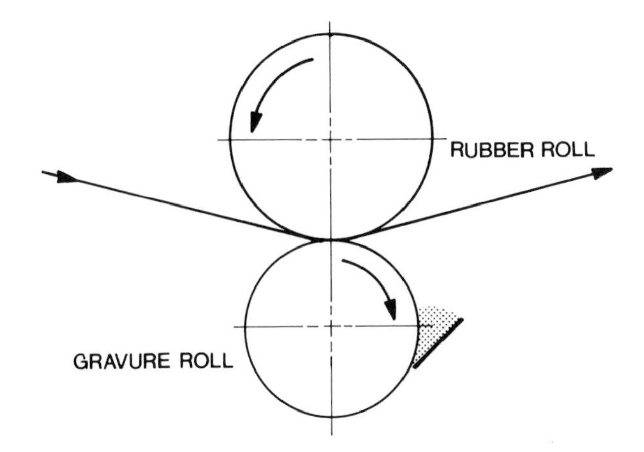


Fig. 26-31. Direct gravure coater, coating reservoir between the roll and the blade.

coating weight, in addition to the roll engraving, is also controlled by the doctor blade angle, blade pressure, coating viscosity and the nip pressure. Nip pressure of 1.5–150 kg/cm width is generally used. Rubber rolls of 90 Shore A durometer are usually used. The type of rubber is selected on the basis of required solvent resistance. The rubber roll may be undercut narrower than the width of the web if the coating is not to be applied up to the edges. This also helps to prevent exposure of rubber to solvents, which might cause swelling.

Several other arrangements of direct gravure are used. Figure 26-31 shows a method where the coating reservoir is held between the knife and the gravure roll. No pan is needed. For higher speeds, a slower rotating fountain roller is used to apply the coating onto the engraved roll as shown in Fig. 26-32.

Offset gravure is best suited for lower coating weights. Better flow out of the coating is obtained and the coating appearance is improved. Offset gravure is one of the methods in applying 100% solid silicone release coatings. A schematic diagram of the offset gravure setup is shown in Fig. 26-33.

## CALENDERING

Calendering has been used for a long time in coating of pressure-sensitive adhesives. The early hospital and electrical friction tapes were coated by calendering and many cloth and polyethylene tapes are still made by the calendering method.

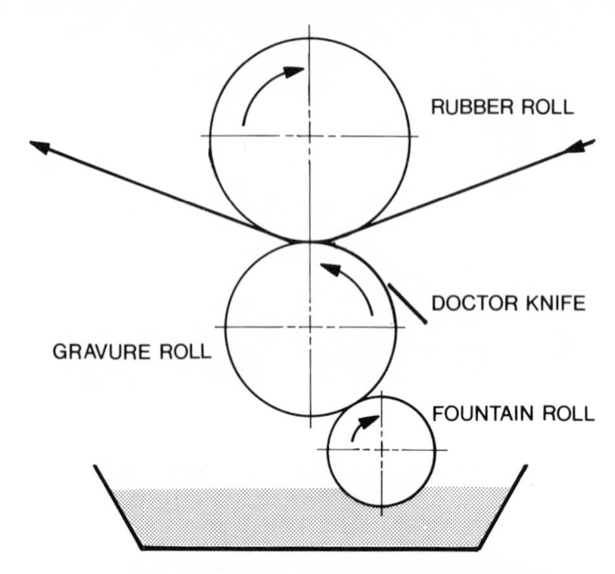


Fig. 26-32. Direct gravure coater, coating applied by a fountain roll.

In calendering, the material softened by heat is dragged through a narrow gap between the rolls, producing a sheet of polymeric material. This sheet is then pulled away from the roll either alone or together with the backing. The main difference between the calendering and roll coating is that the adhesive sheet is removed cleanly from the roll in calendering, while in roll coating splitting of the adhesive takes place.

Three roll calenders are most commonly used for pressure-sensitive adhesives. A typical arrangement is shown in Fig. 26-34. The adhesive is compounded in a Banbury type mixer, prewarmed on a two-roll mill and fed into the calender. The typical roll temperatures could be 127°C for the slow rotating top roll, 38°C for the center roll and 93°C for the bottom roll. Adhesive can also be calendered on a two-roll calender as shown in Fig. 26-35.

Polyethylene pressure-sensitive tapes used for corrosion protection of oil and gas lines are manufactured by calendering. Polyethylene and adhesive are calendered simultaneously as shown in Fig. 26-36. Similarly, a tape with a supporting scrim in the center can be made as shown in Fig. 26-37.

The adhesive is stripped off the calender roll at a fairly high force, despite a high roll polish. Therefore, the backing must have a sufficiently high tensile strength. The use of low tensile strength backings requires special processing. Doctor's knives have been used to separate the adhesive from the roll in case of fragile backings.

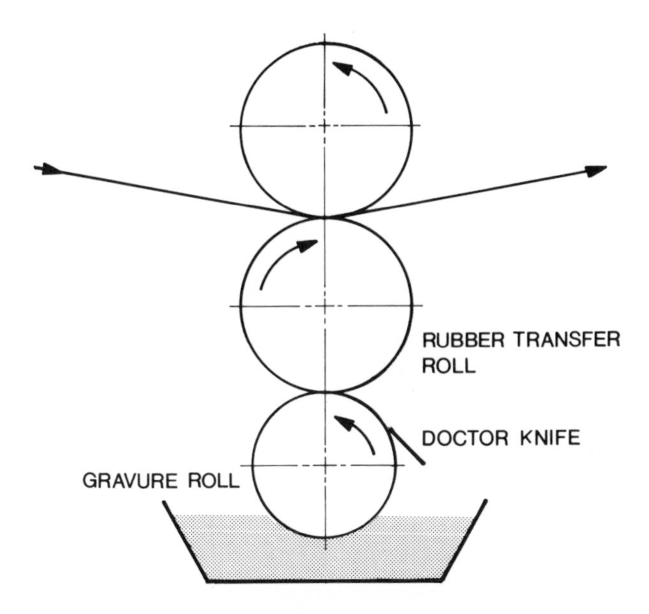


Fig. 26-33. A schematic diagram of an offset gravure coater.

Calendering is suitable for application of heavy adhesive coatings. While coatings as thin as 0.05 mm could be produced by calendering, the uniformity of the coatings below 0.1 mm might not be acceptable. The process is susceptible to weight variation across the width because of roll bending. The weight variations because of bending are compensated by crowning of the rolls. This decreases the gap in the center and compensates for bending when coating. Roll crossing and roll bending are continuous methods to compensate for the increase of the gap in the calender center. Roll crossing is accomplished by moving the bearings horizontally, thus increasing the gap at the roll ends. Roll bending is accomplished by applying hydraulic pressure at the bearings to counteract the shear forces causing the bending.

Calendering equipment is expensive and it is best suited for large volume manufacturing of products which require a heavy adhesive coating. Calendering can convey a soft material at high rates at a much lower mechanical energy input then extrusion. The largest volume product made by calendering is the corrosion protective tape.

The analysis of calendering process requires simplifying assumptions regarding polymer behavior. Middleman[10] presents a mathematical analysis of several cases of calendering process assuming Newtonian and power

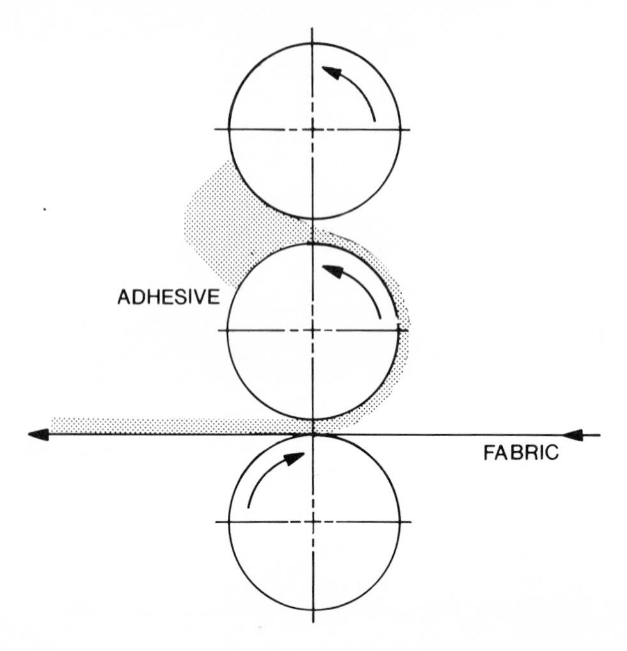


Fig. 26-34. A schematic diagram of a three-roll pressure-sensitive adhesive tape calender.

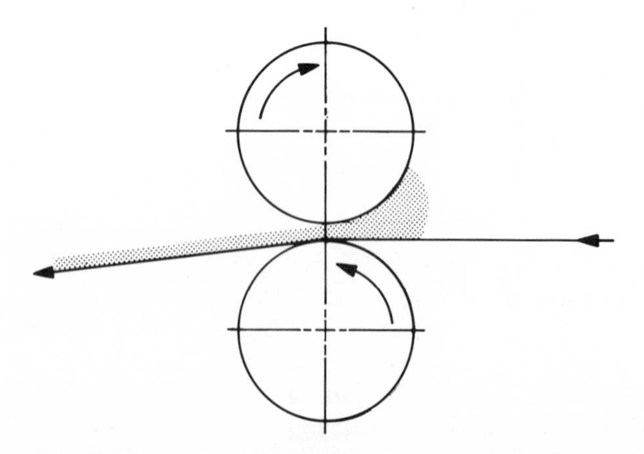

Fig. 26-35. A schematic diagram of a two-roll adhesive calender.

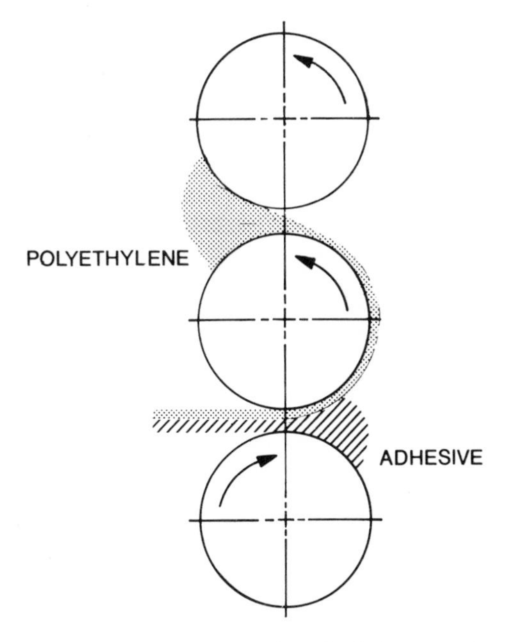


Fig. 26-36. Simultaneous polyethylene and adhesive calendering to produce polyethylene tape.

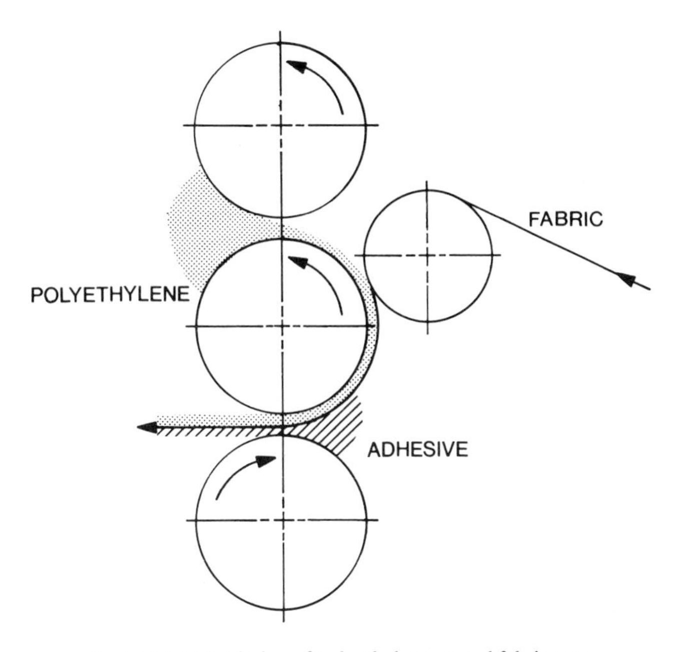


Fig. 26-37. Calendering of polyethylene-coated fabric tape.

law models of polymer behavior during calendering. Calendering analysis by the finite element method is reviewed by Tadmor and Gogos.[19]

## REFERENCES


1. Lazor, T. J. *Modern Plastics* **42**: 149–152, 154, 157, 182 (May 1965).
2. Sanderson, F. T., and Gehman, D. R. *Pressure Sensitive Adhesive Council, Technical Meeting on Water Based Systems*, Chicago, June 13–14, 1979, pp. 91–108.
3. Sanderson, F. T. *Adhesive Age* **21**(12): 31–36 (1978).
4. Bafford, R. A. *Adhesive Age* **22**(12): 21–24 (1979).
5. Booth, G. L. *Coating Equipment and Processes*. New York: Lockwood Publishing Co., Inc., 1970.
6. Weiss, H. L. *Coating and Laminating Machines*. Converting Technology Co., Milwaukee, Wisconsin, 1977.
7. Follette, W. J., and Fowells, E. W. *TAPPI* **43**(11) (1960).
8. Freeston, W. D., Jr., *Coated Fabrics Technology*, pp. 25–41. Technomic Publishing Co., 1973.
9. Hwang, S. S. *Chem. Eng. Sci.* **34**: 181–189 (1979).
10. Middleman, S. *Fundamentals of Polymer Processing*. New York: McGraw-Hill Book Co., 1977.
11. Taylor, P. C. *Paper Technology and Industry* **19**(9): 305 (1978).
12. Grant, O. W., *et al. Pressure-Sensitive Tape Council, Technical Meeting on Water Based Systems*, Chicago, June 13–14, 1979 pp. 165–182.
13. Zink, S. C. *Pressure-Sensitive Tape Council, Technical Meeting on Water Based Systems*, Chicago, June 21–22, 1978, pp. 129–137.
14. Grant, O. W. *Pressure-Sensitive Tape Council, Technical Meeting on Water Based Systems*, Chicago, June 21–22, 1978, pp. 113–128.
15. Miller, J. C., and Meyers, R. R. *Trans. Soc. Rheology* **II**: 77–93 (1958).
16. Myers, R. R. *J. Polym. Sci.* Part C, **35**: 3–21 (1971).
17. Schneider, G. B. *Trans. Soc. Rheology* **VI**: 209–221 (1962).
18. Kosta, G. U. S. Patent 4,029,833 (1977) (assigned to Midland Ross Corp.).
19. Tadmor, Z. and Gogos, C. G. *Principles of Polymer Processing*, New York: John Wiley and Sons, 1979.
20. Mushel, L. A. *Package Printing*, June 1979.

# Chapter 27

# Drying

**Donatas Satas**
*Satas & Associates*
*Warwick, Rhode Island*

Pressure-sensitive adhesive, when coated onto a substrate from either a solution or a latex, requires removal of the vehicle by drying. The rate of drying determines the production speed. It is usually the slowest step in the coating process. Drying also has a great effect on the quality of the coating.

During the drying process, heat and mass transfer take place simultaneously. Heat is transferred to the coating and backing. The vehicle moves to the surface as a liquid or vapor where it evaporates and is removed with the exhaust. The overall drying rate depends on the rate of these steps, or more exactly, on the rate of the slowest step.

The study of adhesive drying can be approached by studying various steps in the drying process or by studying the external conditions of the drying process.

A study based on the internal drying mechanism requires identification of the slowest step, which determines the drying rate. The drying rate is increased by finding a means to improve the rate of the slowest step. The study of the drying process by external conditions requires observation of the effect of temperature in various zones, web speed, coating thickness, solvent volatility, air velocity, and other process variables. While the first method gives a better insight into the drying process and might suggest less obvious solutions how to improve the drying rate or the coating quality, a study based on external factors is simpler and can offer more immediate improvements based on the experimental observations, even if the nature of the problem is not understood.

## MASS TRANSFER MECHANISM

Heat is introduced into the adhesive from one or both sides and vapor is removed from the top. In the case of a coating over a porous web, some vapor can also escape through the bottom.

The mass movement can take place via several mechanisms.

1. Diffusion in a homogeneous adhesive mass
2. Capillary flow, if capillaries are developed during drying
3. Flow caused by shrinkage and pressure gradients
4. Gravity flow
5. Flow by vaporization-condensation sequence

It is believed that, in the case of pressure-sensitive adhesive solvent-based coatings, the mass transfer mechanism is diffusion. In the case of aqueous emulsions, some capillary flow could be expected. The driving force for the diffusional flow is the concentration gradient: movement of solvent or water molecules towards the lower concentration, i.e., the surface from which the solvent is evaporating. Capillary movement takes place in the direction of the decreasing capillary diameter.

The drying rate varies throughout the drying process. Initially, the concentration of the liquid to be evaporated is high and the movement to the surface is sufficiently rapid to keep the surface at a saturated condition. In this stage, called constant rate drying, the rate of mass transfer is determined by the rate of heat transfer into the adhesive. An increase of the drying rate can be realized by increasing the heat transfer rate.

At the conditions of dynamic equilibrium between the heat transfer into the adhesive film and the vapor removal from the surface, the drying rate is expressed by the following equation.

$$\frac{dw}{d\theta} = \frac{hA\Delta t}{\lambda} = k_g\, A\Delta p, \tag{1}$$

where

$\frac{dw}{d\theta}$ = drying rate

$k_g$ = mass transfer coefficient
$h$ = heat transfer coefficient
$A$ = area for heat transfer and evaporation
$\lambda$ = latent heat of evaporation at $t_s$

$\Delta t = t - t_s$, where $t$ = gas temperature
and $t_s$ = temperature of evaporation surface
$\Delta p = p_s - p$, where $p_s$ = vapor pressure of liquid at the surface temperature $t_s$ and $p$ = partial pressure of liquid in the gas.

As soon as the liquid concentration at the surface starts decreasing below saturation, the drying enters into the falling rate period. The rate of solvent removal decreases with drying time in this period. The initial portion of the falling rate period is sometimes called the warming up period, because the adhesive surface temperature increases during this period. This is a period of short duration. Afterwards, the drying rate becomes dependent entirely upon the transport of the liquid through the adhesive film. The drying rate determining step becomes the mass transfer to the surface, rather than the heat transfer into the adhesive.

It is believed that in the case of pressure-sensitive adhesives cast from solution, the movement of the liquid takes place by diffusion through the adhesive film. The adhesive surface is dry and impervious to the vapor, except to the molecular diffusion. If the temperature is raised above the solvent boiling point, the vapor forms a blister and escapes by blowing a hole in the adhesive film. This results in distortion of the adhesive surface. Therefore, the temperature of the adhesive film should not be raised above the solvent boiling point until the solvent concentration becomes quite low.

The general diffusion equation for the rate of flow is

$$\frac{\partial w}{A \partial \theta} = -D \frac{\partial c}{\partial x}, \tag{2}$$

where

$w$ = the liquid flow by diffusion
$\theta$ = time
$c$ = concentration of liquid in weight per unit volume
$x$ = half thickness of the solid layer when the drying takes place from both surfaces
$D$ = diffusivity of the liquid
$A$ = area of mass transfer.

The general differential equation for variation in moisture content is given by

$$\frac{\partial c}{\partial \theta} = D\left(\frac{\partial^2 c}{\partial x^2}\right). \tag{3}$$

If we let $T$ = moisture content subject to diffusion per unit weight of dry solid, and $\rho$ = density of the dry solid, then, when shrinkage is negligible, Equation (2) becomes

$$\frac{\partial w}{A\partial\theta} = -D\rho\frac{\partial T}{\partial x}, \tag{4}$$

and Equation (3) becomes

$$\frac{\partial T}{\partial\theta} = D\left(\frac{\partial^2 T}{\partial x^2}\right) \tag{5}$$

Some of the solutions of Equation (5) for various boundary conditions were solved by several authors and are given by Hougen, McCauley, and Marshall.[1]

Sherwood[2] has solved the diffusion equation for the falling rate period assuming an uniform initial liquid distribution.

$$\frac{w - w_e}{w_c - w_e} = \frac{8}{\pi^2}\left[\exp(-D\theta[\pi/2x]^2) + \frac{1}{9}\exp(-9D\theta[\pi/2x]^2) + \frac{1}{25}\exp(-25D\theta[\pi/2x]^2) + \cdots\right] \tag{6}$$

where

$w$ = average moisture content at any time 0
$w_c$ = average moisture content at the beginning of the falling rate period
$w_e$ = average moisture content in equilibrium with the environment.

For long drying times, Equation 6 simplifies to

$$\frac{w - w_e}{w_c - w_e} = \frac{8}{\pi^2}\exp(-D\theta[\pi/2x]^2) \tag{7}$$

It can be differentiated to

$$\frac{dw}{d\theta} = \frac{-\pi^2 D}{4x^2}(w - w_e). \tag{8}$$

Equation (8) shows that the drying rate is directly proportional to the liquid content ($w - w_e$) and the liquid diffusivity $D$ and varies as the square of the material thickness.

It is desirable to keep the coating thickness as low as possible. Better overall production rates are obtained by applying the coating in several stations with short drying ovens between stations, rather than by a single coating station and a longer oven. Obviously, the equipment costs are much higher in the first case.

The diffusion coefficient can be estimated from the following expression which is based on the Stokes-Einstein equation

$$D = \frac{T}{\mu F} \tag{9}$$

where

$T$ = absolute temperature
$\mu$ = viscosity of the solution
$F$ = a constant dependent upon the molecular volume of the solute.

Small, compact molecules have a higher diffusion coefficient and would give higher drying rates in the falling rate period.

The diffusion equations fit the experimental data reasonably well. Hougen, McCauley, and Marshall[1] point out that the validity of the diffusion equations is not established by apparent reliability in calculating the overall moisture content of the solid during drying. Fairly good matches of experimental data to theoretical predictions can be achieved, even if the drying process does not follow the theoretical equations. The true criterion is in the accurate prediction of moisture distribution throughout the thickness. Such data are very difficult to obtain in thick slabs; they are practically impossible to obtain in thin adhesive films.

The deviation of the theoretical diffusion equations from the experimental results is due to the assumption that the diffusion coefficient is constant. In fact, it is affected by many variables. It decreases with decreasing liquid concentration, with decreasing temperature, with increasing pressure and with increasing density. Van Arsdell[3] has elaborated on the diffusion calculations, assuming a variable diffusion coefficient in the falling rate period. Hansen[4] points out that the diffusion coefficient for solvents in polymers is highly dependent upon the solvent concentration. This dependence is exponential.

In drying latex films, the mass transfer mechanism, at least during the part of the falling rate period, is capillary movement of water to the surface.

The equation below describes the drying rate in such a case.

$$\frac{dw}{d\theta} = \frac{h(t - t_s)(w - w_e)}{\rho \lambda x(w_c - w_e)} \tag{10}$$

## DRYING CURVES

The drying characteristics of a product are represented graphically in one of several ways. The most common ways are to plot solvent content in the adhesive as a function of time, drying rate as a function of drying time, or drying rate as a function of solvent content in the adhesive. The curves for a typical pressure-sensitive adhesive coating containing 25% solids, solvent (heptane-toluene blend, 80/20), coating weight (105 g/m$^2$), drying temperature (66°C) and heat transfer by convection, are shown in Figs. 27-1 through 27-3.

Figure 27-1 shows the solvent remaining in the adhesive vs. the drying time. Drying rate variation with changing solvent content and with time is not clear from this curve. The changes of the drying rate are better illustrated by plotting the drying rate against time (Fig. 27-2). Various drying periods are clearly shown. At the beginning of drying, immediately after the coating head and before the web enters the oven, rapid evaporation takes

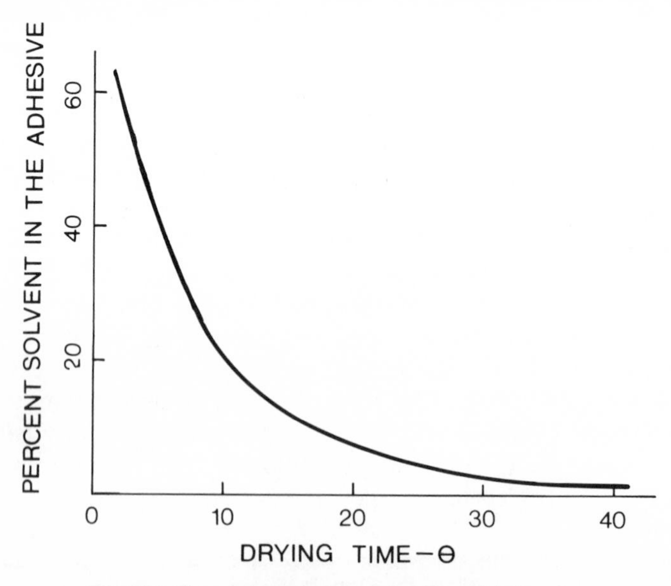


Fig. 27-1. Remaining solvent as a function of drying time.

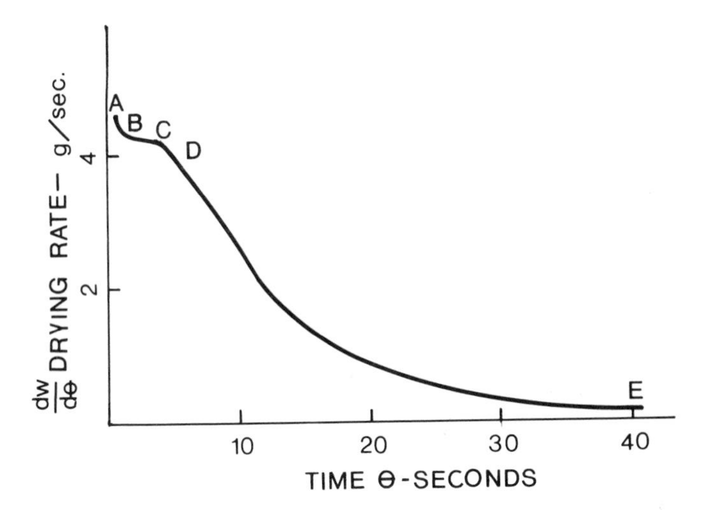


Fig. 27-2. Drying rate as a function of drying time.

place, accompanied by lowering of the film surface temperature. This cooling down period is shown by the A-B segment of the curve. The B-C segment represents the constant rate drying period, which is usually short for solvent-based adhesive coatings. Obviously, it is advantageous to increase the length of this period because the evaporation rate drops very rapidly at the end of constant rate drying period. Segment C-D represents the transition from the constant to the falling rate period. The evaporation takes place from an incompletely saturated surface and the temperature of the adhesive film starts rising. This period sometimes is called the warming-up period. Segment D-E represents the falling rate drying period which is the longest period in the drying of a typical solvent-based pressure-sensitive adhesive. It can be accelerated by increasing the temperature. The increase is limited by bubble formation in the adhesive mass which is caused by evaporation of the residual solvent.

It is advocated by some that an addition of a higher boiling solvent, which will leave the coating last, elevates the coating temperature and increases the drying rate. It should be also kept in mind that bulky molecules have lower diffusion coefficients and therefore diffuse at a slower rate, counteracting the effect of the increased temperature. It has been observed that addition of a liquid plasticizer to the adhesive formulation increases the drying rate. Apparently, the liquid plasticizer provides a path of lower resistance for the diffusion of solvent molecules.

The drying data are sometimes represented by plotting the drying rate vs. the solvent content in the adhesive (Fig. 27-3).

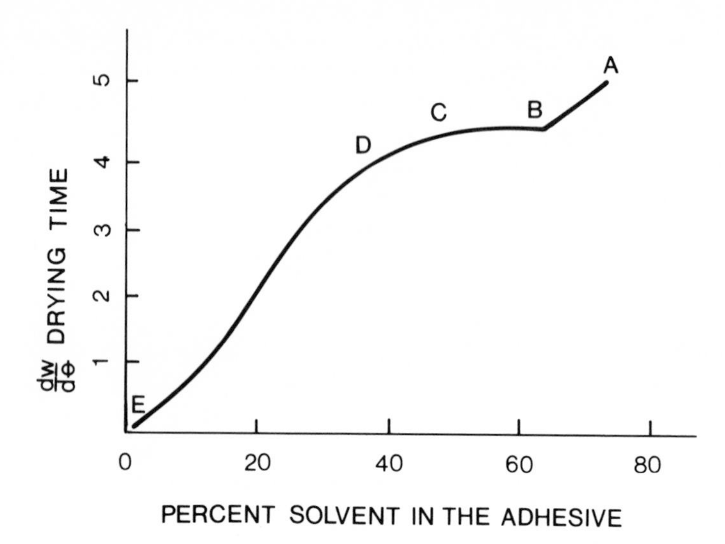


Fig. 27-3. Drying rate as a function of remaining solvent.

## SOLVENT RETENTION

The solvent movement during the falling rate period is diffusion controlled. The driving force, the concentration gradient, continuously diminishes as the adhesive film is drying and the diffusion coefficient decreases as the solvent concentration decreases. Therefore, it is not surprising that solvents are retained in polymer films, even after drying for a long period of time. Adhesives containing large amounts of retained solvent might appear dry on casual inspection. An adhesive film containing as much as 2–8% solvent might appear dry.

Hansen[4] suggests the applicability of free volume theory to explain and generalize the solvent diffusion and retention processes. Free volume refers to the presence of unoccupied spaces or holes in the polymer structure. These holes not only allow the movement of polymer chain segments, but also accommodate the solvent molecules. The solvent diffusion takes place by jumping of solvent molecule from hole to hole. Bulkier solvent molecules require larger holes, and therefore bulky molecules diffuse at a slower rate. Molar volume is the only factor that affects solvent diffusion and retention. Steric hindrance is very important in determining the mobility of solvent molecules. Short straight chain molecules diffuse easier, while molecules with side groups diffuse slower and are easier entrapped in the polymer. For example, 2-nitropropane is retained more than 1-nitropropane and 1-

nitropropane more than nitroethane. *N*-butyl acetate, being linear, is retained less than other solvents of similar molar volume, but is branched.

Newman, Nunn, and Oliver[5] studied solvent retention and they also came to conclusion that the level of retention cannot be explained by solvent volatility or by polymer-solvent interactions. It can be explained in terms of size and shape of the solvent molecule. The size and branching increase the level of retention. Planar molecules, such as toluene, are released easier than nonplanar ones, such as cyclohexane.

Suzuki *et al.*[6] have reported a gas chromatographic method for the estimation of residual solvents in adhesives. The amount of residual solvent in common well-dried adhesive tapes was found to be 0.01–0.1%.

The effect of residual solvents on natural rubber-based pressure-sensitive adhesives has been reported by Toyama, Ito, Suzuki, and Moriguchi[7] and has been reviewed later by Toyama and Ito.[8] Using *n*-hexane with a small amount of a higher boiling solvent such as *n*-nonane, cyclooctane, or *n*-hexanol, the peel force and probe tack values show a maximum at 0.02% residual solvent as shown in Fig. 27-4. It was also

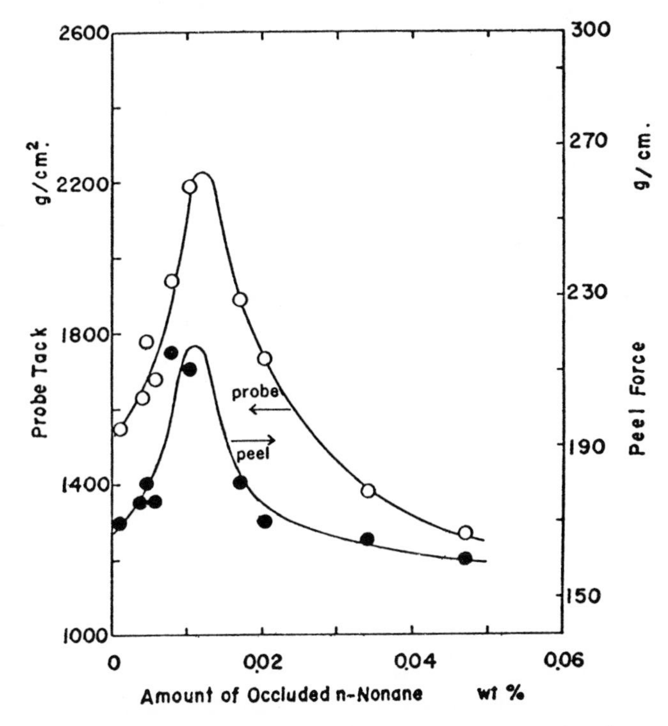


Fig. 27-4. Effect of the residual solvent on the pressure-sensitive properties of a natural rubber based adhesive. Data by Toyama *et al.*[7]

reported[8,9] that the peel force of pressure-sensitive tapes depends on the solvent used. This is attributed to the different solubility of various components used. The dispersion of resin in rubber-based adhesive is homogeneous if a good solvent system is used. If the solubility of one component is poor, it can precipitate during drying and give a heterogeneous adhesive of decreased peel adhesion.

Engel and Fitzwater[10] report that the adhesion of methyl methacrylate/lauryl methacrylate (90/10) copolymer films depend on the solvent composition used for casting of these films. The polymer chains are extended in good solvents. The probability of functional groups contacting the substrate is higher than in case of poor solvents where the polymer chains are coiled.

## DRYING OF LATEXES

Various mechanisms of the film formation from aqueous polymer dispersions have been proposed. Dillon, Matheson, and Bradford[11] have suggested that the coalescence of latex particles is caused by the surface tension forces similar to the sintering process. Brown[12] has pointed out that in addition to surface tension, a capillary force furnishes the main energy for coalescence. The capillary force is a result of the surface tension of water in the capillaries between the particles. Vanderhoff *et al.*[13] have modified the proposed sintering mechanism to wet sintering. Dry sintering mechanism is inconsistent with the fact that water evaporation and film formation take place simultaneously. In dry sintering, the film formation would start only after water has evaporated. Sheetz[14] reviews the other theories and proposes a diffusion mechanism. According to this theory, at the early stages of latex coating drying, capillarity, wet sintering and diffusion are important. In the later stages, diffusion of water through the polymer particles to the film surface becomes the dominant mass transfer mechanism.

Sheetz points out that the rate of water evaporation would become very slow at the end of drying if the water were supplied to the surface by capillary forces as proposed by Brown. Experimental data show that the rate of water evaporation becomes slower at about 70% solids, but not as slow as would be expected from the capillary theory. Furthermore, the drying rate depends on the type of polymer used, which also suggests a diffusional process.

Adhesive coatings from latexes can be dried at higher temperatures and at higher heat transfer rates than the solvent-based coatings, although it is advisable to keep the temperature in the early stages of drying below 100°C. The fast decrease of drying times with increasing temperature for various thicknesses of acrylic latex coatings is shown in Fig. 27-5.[15] Many latex

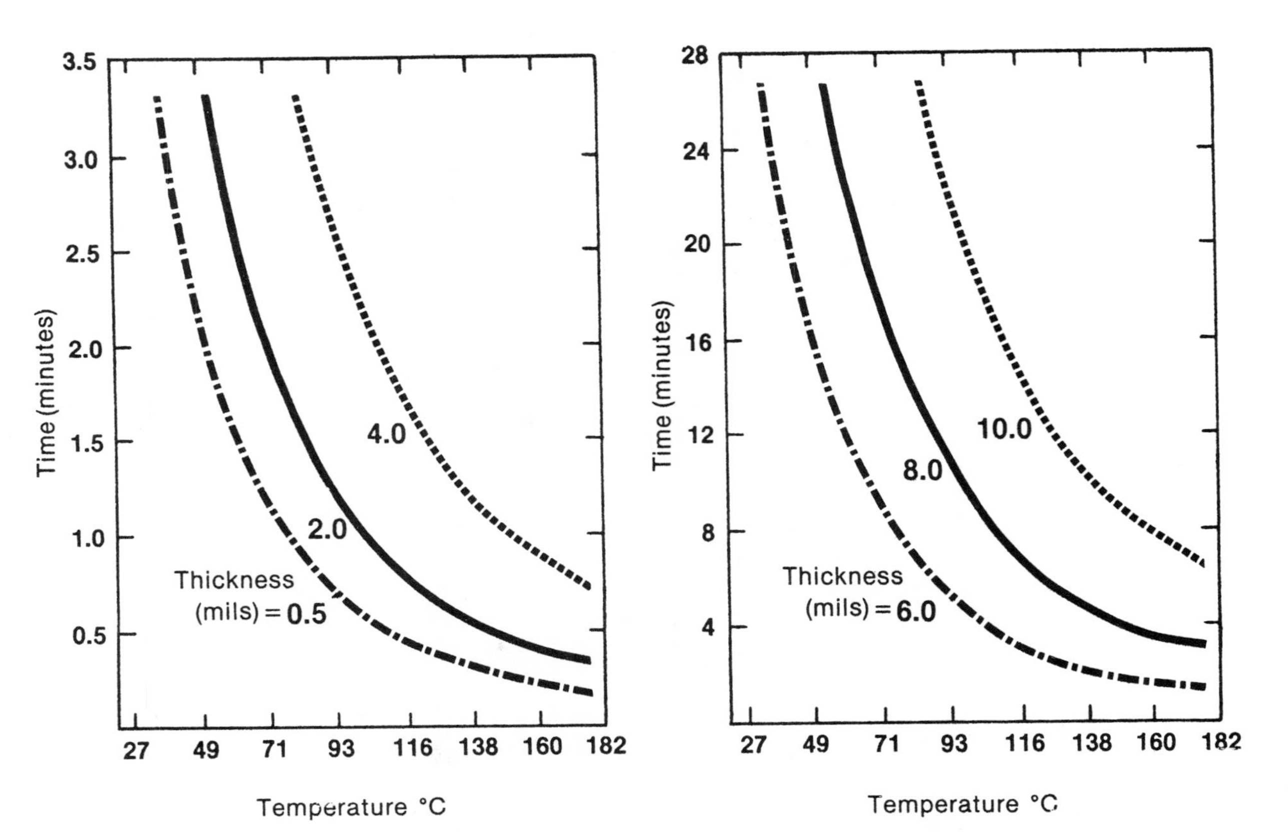


Fig. 27-5. Drying time versus oven temperature for several film thicknesses of an acrylic polymer. (*Courtesy Goodrich Chemical Co.*[15]).

adhesives are curable at elevated temperatures and the last drying zones can be utilized to crosslink the polymer.

Organic solvents are sometimes added to the latexes to improve the coatability and to help the coalescence of latex particles, although this is rarely required for pressure-sensitive soft polymers. The coalescence of latex particles is improved if the cosolvent is last to leave the polymer coating. The concept of Critical Relative Humidity (CRH) has been developed to describe the simultaneous evaporation of water and cosolvent.[16] CRH is defined as the humidity at which water and cosolvent evaporate at the same relative rate. At humidities above CRH, the evaporation of water is retarded and the cosolvent leaves the coating at a higher rate.

## HEAT TRANSFER

During the constant rate drying period, when the surface of the adhesive coating is saturated with solvent or water, the rate of drying depends on the heat input into the coating. There are limits on the optimum drying rate in the initial constant rate period. Increase of the rate of solvent removal from the surface beyond the rate at which the solvent can be delivered to the surface is self-defeating. A polymer film is formed on the surface, which is poorly permeable to the solvent, and the drying enters into the falling rate period sooner and with a larger amount of trapped solvent in the film than in the case of slower drying during the constant rate period.

In circulating air ovens, heat is transferred almost solely by convection; in infrared ovens, heat transfer is by radiation, conduction (within the adhesive film), and convection.

In the case of a turbulent air stream passing over the adhesive coating, the heat must be transferred through a boundary layer of stagnant air, which offers resistance to the heat flow. The thickness of the boundary layer determines the heat flow, and it is a good measure of the resistance to the heat transfer.

Most of the air convection ovens operate well in the turbulent flow region at high Reynolds numbers $[LV(\rho/\mu)]$. At Reynolds numbers above 500,000, the Colburn relation can be used.[17]

$$\frac{h}{C_p V \rho}\left(\frac{C_p \mu}{k}\right)^{2/3} = \frac{0.036}{[LV(\rho/\mu)]^{0.2}}, \tag{11}$$

where

$h$ = heat transfer coefficient
$C_p$ = specific heat at constant pressure

$V$ = average velocity
$\rho$ = density
$\mu$ = viscosity
$k$ = thermal conductivity
$L$ = heated length

For air, Equation (11) can be simplified to

$$h_m = 0.0128G^{0.8}, \tag{12}$$

where

$h_m$ = heat transfer coefficient based on length mean $\Delta t$
$G$ = mass flow ($V\rho$)

In this case, heat transfer is increased by the 0.8 power of the air velocity.

Continuous increase of air velocity to improve the drying rate creates several problems: power consumption is increased and the high velocity air may disturb the coating.

At Reynolds numbers below 80,000, laminar boundary layer determines the heat transfer, and the Pohlhausen equations become valid.[18]

$$\frac{h_x}{C_p V_1 \rho_1}\left(\frac{C_p \mu}{k}\right)^{2/3} = \frac{1}{3}\left(\frac{x V_1 \rho_1}{\mu}\right)^{-1/2} \tag{13}$$

$$\frac{h_m}{C_p V \rho}\left(\frac{C_p \mu}{k}\right)^{2/3} = \frac{2}{3}\left(\frac{L V \rho}{\mu}\right)^{-1/2} \tag{14}$$

where

$h_x$ = local heat transfer coefficient based on local $\Delta t$

For further reading, McAdams[19] is recommended. The Pohlhausen equations can be simplified to:

$$h = n\left(\frac{V}{L}\right)^{1/2} \tag{15}$$

where

$n$ = a constant dependent on the properties of the air

Heat transfer is increased as the square root of air velocity and as the reciprocal square root of $L$, which depends on the discharge nozzle spacing. The closer the nozzles are spaced together, the smaller is $L$.

Gardner[20] has designed a drier to operate at a sufficiently low Reynolds number to obtain laminar flow at the coating surface. Closely spaced nozzles keep the boundary layer low. High heat transfer rates are obtained.

## INFRARED DRYING

Infrared drying in combination with convective drying is used quite often for pressure-sensitive adhesives. Infrared ovens require less capital investment than convection ovens, but they are not necessarily less costly to operate. The presence of hot elements near the coated web present some safety problems.

Infrared heating units are either electric or gas-fired. Electric units emit infrared radiation of a definite wavelength corresponding to the temperature at a given line voltage. Quartz radiators emit infrared energy in the wavelength region of 2.5–3.5 $\mu$m. Gas infrared units allow variation of the temperature of the refractive surface, depending on the gas input into the combustion chamber.

It is possible to expose the adhesive to a high-density heat flux by means of infrared radiators. A typical infrared unit is capable of a radiation output of 25,000–45,000 kcal/m$^2$-hr. Depending on the adhesive mass properties, either the energy is absorbed and raises the temperature of the coating, or it is transmitted and heats up the substrate or the oven belt, or it is reflected to the surroundings. The adhesive absorbs the energy of a particular wavelength and reflects the rest. In order to avoid losses because of reflection, the wavelength of the radiation source must match the wavelength that can be easily absorbed by the coating. Infrared sources emitting 1–8 $\mu$m wavelength radiation are generally used. For aqueous coatings, 3–6 $\mu$m wavelength is the best; for post-curing the adhesive surface, 6–8 $\mu$m wavelength radiation is most efficient.

Infrared drying does not require heating of large volume of air, as required in convection ovens. It can cause less surface hardening. Depending on the nature of coating and the characteristics of emitter, the energy could be transmitted through the film and heating could be started at the adhesive-substrate interface. This way, drying takes place from the bottom of the coating and can considerably improve the drying rate.

## COATING IMPERFECTIONS

Coating imperfections can be caused in the coating process or during the drying of the adhesive. Blisters are usually formed during the drying and can be due to the following reasons:

1. Trapping the liquid underneath the adhesive film. It evaporates when exposed to a higher temperature and forms blow holes or blisters.
2. Introduction of air during mixing.
3. Introduction of air at the coating head. Reverse roll coaters and gravure coaters can introduce air into the adhesive bank.
4. Disturbing of the coating by high velocity air.
5. Poor wetting of the substrate by the adhesive coating.
6. Water in the substrate, quite common in case of paper substrates.

The most common cause of blisters in the adhesive mass is the trapping of solvent underneath the solid surface coating. Coating of impervious webs is especially susceptible to the formation of blisters. This can be remedied by lowering the rate of evaporation in the early zones in order to delay the formation of surface skin. A decrease in temperature, air flow or line speed helps in this case.

Presence of a high boiling diluent is especially conducive to the formation of blisters. The extreme case is the presence of water in the solvent solution. Water, unless it forms an azeotrope, would tend to stay behind. It does not diffuse easily through the polymer film and later causes the blow holes in the coating, when subjected to a temperature above its boiling point. Smaller quantities of trapped solvent, especially water, might be insufficient to form blow holes, but sufficient to form a foamy layer within the adhesive coating, usually close to the substrate. This coating is weak and fails easier in shear than in the rest of the adhesive.

## CURLING

Drying often causes the coated web to curl. The cause of curling is two-sidedness, and therefore, nonuniform dimensional changes when exposed to temperature changes or to changes in humidity, in case of material that absorbs moisture. Paper tapes are especially subject to curl. Shrinkage of the coating on drying causes the material to curl towards the coating. Wetting of the backside of a paper tape causes it to curl away from the coating because of the swelling and expansion of paper fibers on the wetted surface. In an anisotropic material, the presence rather that the absence of curling is more natural. The absence of curling requires that the surface stresses were either absent or equal on both sides of the web. To eliminate curl by trying to equalize the stresses is a very difficult task. It is much easier to attempt to eliminate the stresses by relaxation. The stresses that develop in a polymeric coating relax with time and the material loses its tendency to curl if wound up in a roll or kept flat as a sheet. Some materials have a tendency to accept the curl of a roll after some period of time.

In the case of uncoated paper, the difference in fiber orientation between top and wire sides is the most important cause of curl. Some contribution to curl is also made by a nonuniform distribution of fillers and sizes. Moistening one side of the paper sheet results in a permanent tendency to curl toward the previously moistened side.

Spitz and Blickensderfer[21] have proposed that a property called curl tendency is more fundamental than curl. In the case of coated and uncoated papers, this property can be measured and quantitatively defined.

During the coating of paper tapes or label stock, if the direct coating method is used, paper loses moisture. When exposed to normal room humidity, paper tends to pick up moisture from the air in order to come back to the equilibrium moisture content. This often results in curling. Moisture lost during processing is often added at the end in order to minimize later curling. Coated products might be stored in a high humidity atmosphere, restrained in the flat state. Or, moisture might be added at the end of coating before the product is rolled up. Moisture may be added by steaming, water spray or roll coaters.

Pagendarm[22] describes the apparatus for this purpose, consisting of a steam chamber. Problems of curl in extrusion-coated papers have been discussed by Bates and Marsella.[23]

## EQUIPMENT

Web drying equipment can be classified according to the method of energy transfer to the coating:

- Convection dryers—air dryers
- Radiant energy dryers—infrared dryers
- Contact or conduction dryers—hot cylinders
- Microwave dryers

The last two categories are not used for the pressure-sensitive coatings. Contact drying is not suitable because the heat must be applied alternatively from both sides, and microwave drying is an expensive process.

The dryers can also be classified according to the method of web conveyance:

- Conveyor dryers
- Arch dryers
- Floater dryers
- Airfoil dryers

Solvent-based pressure-sensitive coatings are best dried in zoned ovens. The solvent evaporation rate is very fast immediately after the coating application, and often an unheated short section, equipped with an air exhaust only, is provided before the web enters the oven. It is common to construct the ovens with at least three different temperature zones. The first zone is usually kept at a lower temperature (60–70°C) in order to extend the constant rate drying period. The temperature in the second zone is higher. The temperature in the third zone is maintained at a still higher level to remove the remaining solvent and to cure the adhesive if required. Ovens with a higher number of zones or with only two zones are also used.

Drying of latexes requires less careful temperature adjustment. Higher heat input can be used without entrapment of the vehicle and higher temperatures are used throughout the oven.

The paper industry has been the leader in the development of new drying methods. The large size of the industry and the requirements to remove huge quantities of water at high speeds are the factors that placed the requirements of the paper industry in the forefront. Unfortunately, the requirements of the paper industry and pressure-sensitive product drying are not quite the same. In the paper industry, the requirement is to remove the water at the fastest possible speed from an open web. The drying rate is largely dependent on the heat input rate. Drying of pressure-sensitive adhesives is a process dominated by the diffusion of the vehicle through the polymeric coating, and heat input rate is much lower.

## CONVECTION DRYERS

The air flow in a convection oven consists of the air input from the heater and makeup air and discharge. Part of the air is recirculated. A schematic diagram of the air flow is shown in Fig. 27-6. In drying solvent-based coatings, the recirculation is limited by the allowable concentration of flammable vapor. Without instrumentation, the concentration of flammable vapor must be kept below 25% of the lower explosive limit (L.E.L.). With a proper vapor detection system, the concentration of the flammable vapor can be increased to 40% of the L.E.L., which varies with the solvent used, but generally is about 1.1—1.3 volume percent for the commonly used solvents. Most of the solvent evaporates in the first zone and therefore less recycling is possible in that zone. To conserve the energy, cascading can be employed. Cascading is reusing the exhaust gases of one zone and using it as the supply air for another zone. Cascading can be used effectively in pressure-sensitive adhesive drying, because the air in the zones toward the end of drying does not have much vapor and can be gainfully employed in the earlier zones as the supply air. Heat recovery units to extract the heat

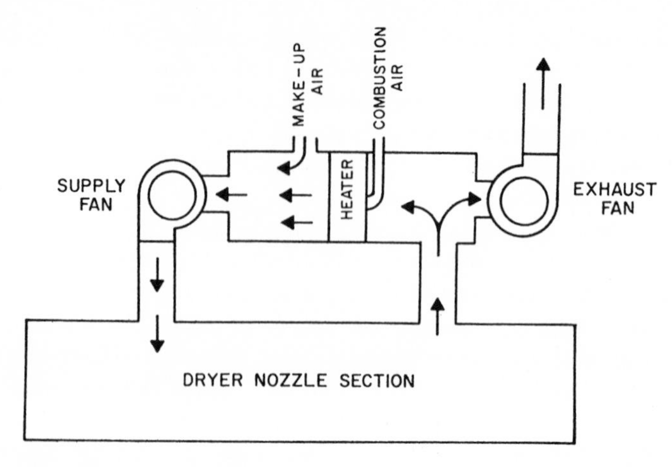


Fig. 27-6. Schematic diagram of the air flow in a convection oven. (*Courtesy Overly, Inc.*[24])

from the exhaust and to utilize it for preheating of makeup air are also used. Lawton[24] considers various means to conserve energy.

Air recirculation in the ovens used to dry water-based adhesive coatings is much higher, resulting in lower overall energy requirements, even though the heat of evaporation of water is higher than that of solvents. In this case, the extent of air recirculation is not limited by safety considerations, but by the effect of increasing vapor concentration on the drying rate. In convection ovens, most of the energy is used not for liquid evaporation but for heating large volumes of air.

Air was circulated at a low velocity in the early convection dryers. The heat transfer rates were considerably lower than in the case of impingement dryers in which the air is discharged through the restricted openings at high velocity. An impingement dryer has a 3–4 times higher heat transfer rate than that of a parallel flow tunnel dryer. Gardner[25] gives some heat transfer coefficients for comparison purposes of various drying oven constructions.

| | |
|---|---|
| Quiet air | 10 kcal/(hr)($m^2$)(°C) |
| 15 mph wind | 29 kcal/(hr)($m^2$)(°C) |
| Early dryers | 50–120 kcal/(hr)($m^2$)(°C) |
| High velocity air dryer | 170–370 kcal/(hr)($m^2$)(°C) |

The following water drying rates have been achieved in various dryers.

| | |
|---|---|
| Tunnel dryer | 7–15 kg/($m^2$)(hr) |
| Impingement dryer | 35–65 kg/($m^2$)(hr) |

| | |
|---|---|
| Air foil dryer | 20–50 kg/(m$^2$)(hr) |
| Infrared dryer | 50–140 kg/(m$^2$)(hr) |
| High velocity air dryer* | 140–150 kg/(m$^2$)(hr) |
| Maximum rate achieved on high velocity air dryer** | 290 kg/(m$^2$)(hr) |

The air velocities in an air impingement tunnel are not over 2000–2500 cm/sec, while in high velocity dryers the air velocities of 5000–10,000 cm/sec are developed. The air velocities are measured at the discharge from a nozzle or a slot. The heat transfer coefficient depends upon the air velocity at the point of impact with the drying surface. Therefore, other factors, such as the distance between the nozzle and the coated surface and the air volume discharged, become important in determining the oven's effectiveness. The effect of air velocity on the evaporation rates of solvent and water-solvent blends were studied experimentally by Eaton and Willeboordse.[26]

There is some disagreement whether round or slotted nozzles are more effective in a convection dryer. Gardner[25] shows that the slotted orifices provide 35–50% higher heat transfer rate based on purely geometric considerations. Round orifices are expected to produce more turbulence that could be helpful to improve the heat transfer rates. Sections of a slot orifice drying oven are shown in Fig. 27-7. A typical dryer might have the nozzles mounted 8 cm from the surface, 15–25 cm apart. The width of the slots could be 5 mm. Such dryer would operate in either turbulent or transitional flow region at $80{,}000 < Re < 500{,}000$.

Gardner[20] has proposed a dryer that operates in the region of laminar air flow. A typical dryer of this type might have the nozzles spaced 2–2.5 cm apart, 3–12 mm away from the surface, and a slot width less than 0.6 mm. The air velocity is 5000–10,000 cm/sec at the orifice. Claims of a considerable increase in heat transfer rates are made, and the weaker jets cause less disturbance of the coated surface.

The dryers are classified according to the method of conveying the web and according to the design of air nozzles. The web can be simply pulled in a tunnel dryer over idlers in a straight pass. An improvement over the straight pass is the arrangement of the idlers in an arch, so that the web wraps each of the idlers 1–5°. The arch helps to prevent the edge curl. Part of the tension force applied to the web is translated into the cross-machine directional force that keeps the edges from curling. The idlers can be driven

* Hot air only at a discharge rate of 100 kg/(m$^2$)(hr).

** Thin paper web was in contact with a high pressure steam cylinder shrouded with a high velocity air dryer.[25]

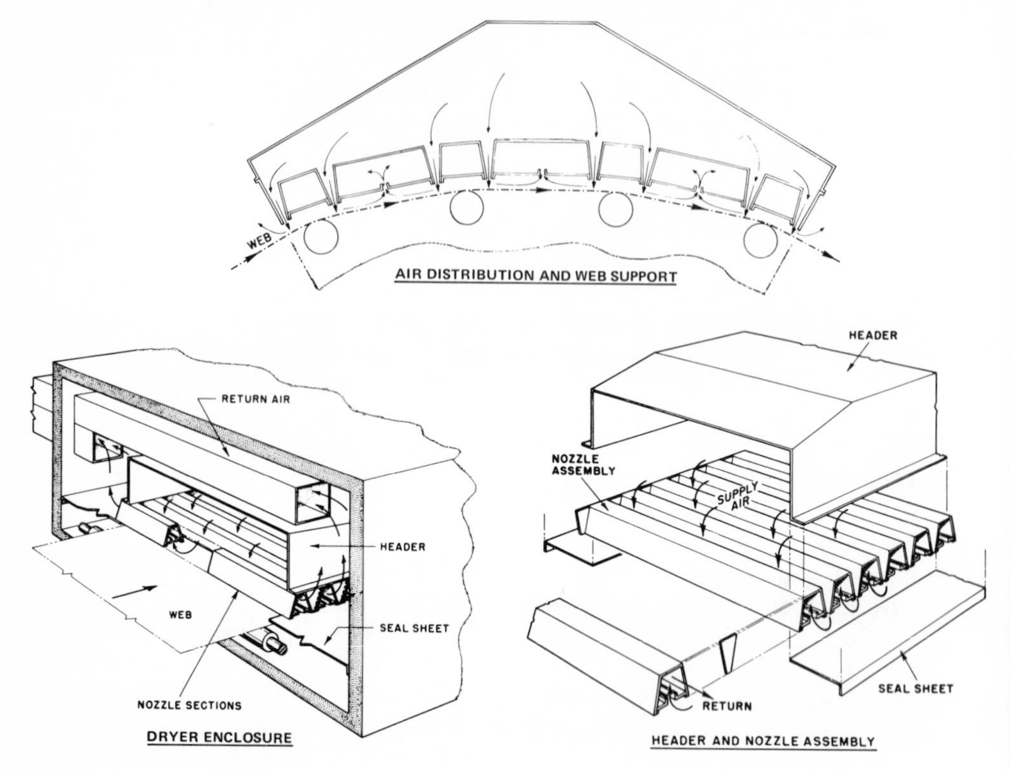


Fig. 27-7. Air distribution in a slotted nozzle convection oven. (*Courtesy Black Clawson Co.*)

to decrease the drag and the tension on the web. A typical arch oven is shown in Fig. 27-8.

Conveyor type dryers allow to transport the web at very low tension through the oven and are widely used for pressure-sensitive products. In this dryer the coated material is placed on the belt or directly on the conveyor bars. Various types of dryers used for coated paper are reviewed by Marsh and Jepson.[27]

## FLOATER DRYERS

The floater dryer was introduced in the early 1960's mainly to dry the offset printed webs. Floater oven has air nozzles directed from both sides of the web as shown in Fig. 27-9. The advantages of this dryer are that no additional web support is required and that the web is heated from both sides

Fig. 27-8. Arch oven. (*Courtesy Faustel, Inc.*)

simultaneously. The disadvantages are the requirement to balance the jets on both sides and to maintain a high air flow rate in order to have the oven operating properly. A coating line with floater ovens is shown in Fig. 27-10.

Fig. 27-9. Floater oven, nozzle, and web positioning. (*Courtesy Pagendarm KG*).

Fig. 27-10. Pressure-sensitive adhesive coating line employing floater ovens. (*Courtesy Pagendarm KG.*)

Further development of the floater oven is the airfoil oven. The airfoil oven is based on Bernoulli's theorem that has to do with the mechanical energy balance in flowing fluids. It states that an increase in velocity in a

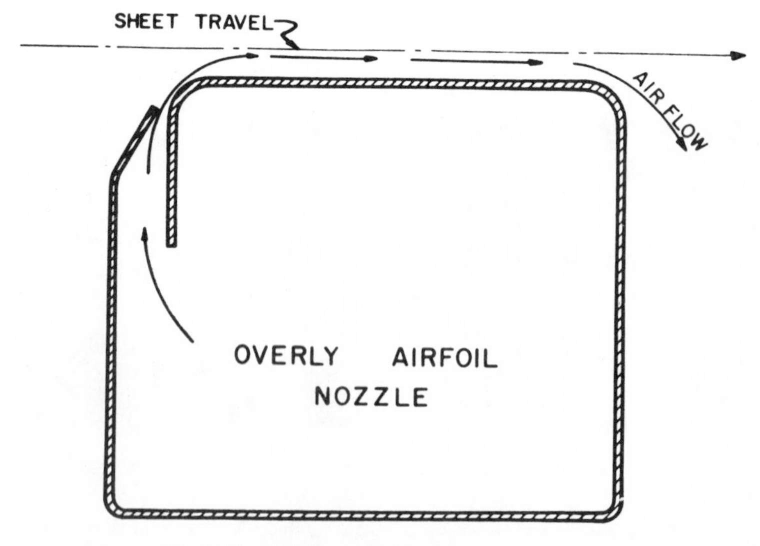


Fig. 27-11. Airfoil nozzle. (*Courtesy Overly, Inc.*)

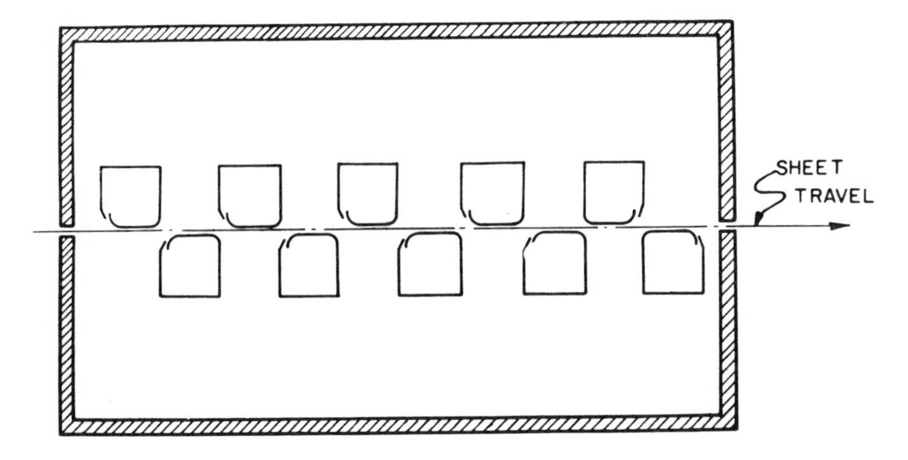


Fig. 27-12. A schematic diagram of an airfoil floater dryer. (*Courtesy Overly, Inc.*)

fluid stream causes a corresponding decrease in pressure in order to keep the amount of mechanical energy constant. This principle is utilized to provide lift to an airplane wing. It also has been utilized in airfoil drying ovens to float the web. The air is discharged through a nozzle as shown in

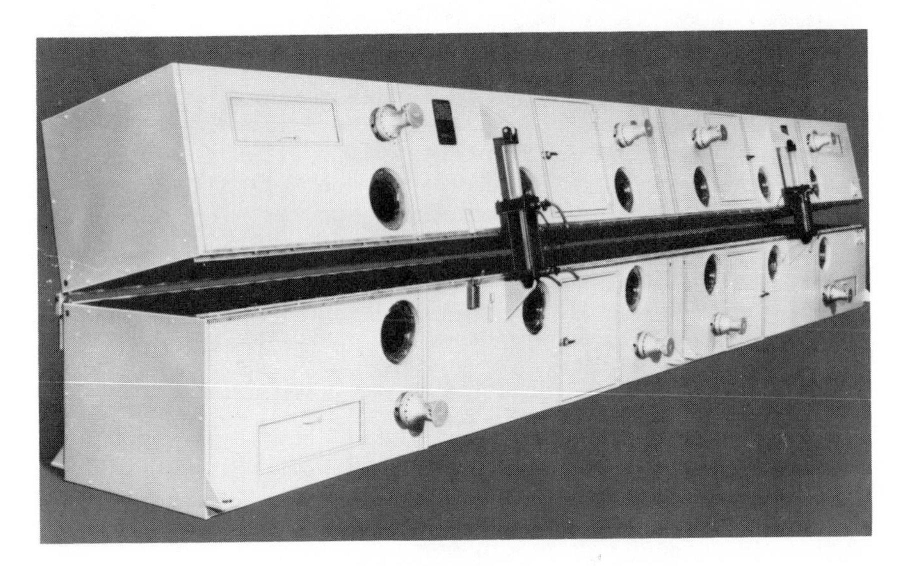

Fig. 27-13. A section of an airfoil floater oven. (*Courtesy Overly, Inc.*)

Fig. 27-11 and flows parallel to the web. At the point of *vena contracta*, near the discharge of the nozzle, the air pressure is lower and the air velocity is higher. This causes an attraction of the sheet to the nozzle, but does not allow the web to touch it, because of the air cushion. A schematic diagram of such an oven is shown in Fig. 27-12, and a photograph of one section is shown in Fig. 27-13. An airfoil dryer has a greater freedom in the adjustment of the air flow. The dryer can be used with the bottom jets only.

A pilot airfoil dryer was constructed in 1969 mainly for drying starch-coated paper. The oven has gained acceptance in the industry since that time for a variety of applications. The construction of the airfoil dryer is discussed by Overly and Pagel.[28] The application of airfoil dryer for drying of coated paper has been described by Richardson and Lawton[29] and by Lawton.[30] Airfoil drying was compared to aircap drying in a pilot plant study by Johns.[31]

In both types of floater dryers, the air has a dual function: it is the heat transfer medium and it provides mechanical support for the web. This might cause some limitations in the control of the air flow.

## REFERENCES


1. Hougen, O. A., McCauley, H. J., and Marshall, W. R. *Trans. Am. Inst. Chem. Engrs.* **36:** 183–202 (1940).
2. Sherwood, T. K. *Ind. Eng. Chem.* **21:** 12 (1929).
3. Van Arsdell, W. B. *Trans. Am. Inst. Chem. Engrs.* **43:** 13–24 (1947).
4. Hansen, C. M. *Official Digest* **37**(480)**:** 57–77 (1965).
5. Newman, D. J., Nunn, C. J., and Oliver, J. K. *J. of Paint Technology* **47**(609)**:** 70–78(1975).
6. Suzuki, M., Tsuge, S., and Takeuchi, T. *Anal. Chem.* **42:** 1705(1970).
7. Toyama, M., Ito, T., Suzuki, M., and Moriguchi, H. *Preprints of the Ninth Conference on Adhesion*, p. 13, Tokyo, 1971.
8. Toyama, M., and Ito, T. *Polymer-Plast. Technol. Eng.* **2**(2)**:** 161–229 (1973).
9. Toyama, M., Ito, T., Hino, K., and Kono, T. *Preprints of the Tenth Conference on Adhesion, Osaka*, p. 41, 1972.
10. Engel, J. J., Jr., and Fitzwater, R. N. *Adhesion and Cohesion* p. 89, (P. Weiss, ed.) Amsterdam: Elsevier Publishing Co.,1962.
11. Dillon, R. E., Matheson, L. A., and Bradford, E. B. *J. Colloid. Sci.* **6:** 108(1951).
12. Brown, G. L. *J. Polym. Sci.* **22:** 423–434 (1956).
13. Vanderhoff, J. W., Tarkowski, H. L., Jenkins, M. C., and Bradford, E. B. *J. of Macromol. Chem.* **1:** 361 (1966).
14. Sheetz, D. P. *J. Appl. Polymer Sci.* **9:** 3759–3773(1965).
15. Hycar Acrylic Latexes. Bulletin L-17. B. F. Goodrich Co.
16. Dillion, P. W. *J. Coating Technol.* **19:** 634 (1977).
17. Colburn, A. P. *Trans. Am. Inst. Chem. Engs.* **29:** 174–210 (1933).
18. Pohlhausen, E. *Z. angew. Math. u. Mech.* **1:** 115 (1921).
19. McAdams, W. H. *Heat Transmission.* New York: McGraw-Hill Book Co., 1954.
20. Gardner, T. A. *TAPPI*, **43**(9)**:** 796–800 (1960).
21. Spitz, D. A., and Blickensderfer, P. S. *TAPPI* **46**(11)**:** 676–689 (1963).

22. Pagendarm, R. *Wochenblatt fuer Papierfabrikation*, **17**: 670–673 (1979).
23. Bates, A., and Marsella, L. J. *TAPPI* **47**(7): 133A–135A, 168A–170A (1964).
24. Lawton, D. W. Paper given at TAPPI Coating and Graphic Arts Conference, 1975.
25. Gardner, T. A. *Pulp Paper Mag. Canada* **62**(6): T327–T332 (1961).
26. Eaton, R. F., and Willeboordse, F. G. Evaporation Behavior of Organic Cosolvents in Water Borne Coating Formulations. Union Carbide Corporation, R&D Department, Bound Brook N. J.
27. Marsh, G. R., and Jepson, M. D. *TAPPI Monograph Series No. 28*, 93–103 (1965).
28. Overly, W. F., and Pagel, K. J. U.S. Patent 3,629,952 (1971) (assigned to Overly, Inc.).
29. Richardson, C. A., and Lawton, D. W. *TAPPI* **56**(4): 86–89 (1973).
30. Lawton, D. W. *TAPPI* **57**(6): 105–107 (1974).
31. Johns, R. E. *TAPPI*, **61**(2): 41–44 (1978).

# Chapter 28

# Hot Melt Application

**Christopher Watson**

*Kleenstik Laminating Div.*
*CCL Industries, Inc., Ajax, Ontario*

and

**Donatas Satas**

*Satas & Associates*
*Warwick, Rhode Island*

Hot melt pressure-sensitive adhesives have become important products. Improved application equipment, allowing the use of higher viscosity hot melts, new elastomers suitable for formulation of improved hot melt adhesives and economical factors which make solvent-based adhesives more expensive to use are the factors contributing to the growth of hot melt pressure-sensitive adhesives.

## PRODUCTS

Hot melt technology has made extensive penetration in some areas of pressure-sensitive product manufacturing. The hot melts were most successful in large volume, lower performance products. The main growth has been in the label and floor tile areas. Acceptance of hot melts for tapes is growing: biaxially-oriented polypropylene tapes, saturated paper masking tapes and other tape products are made by hot melt technology.

The advantages of hot melt application are obvious. Drying is not required, thus eliminating space occupied by expensive ovens needed for processing of solvent-based or aqueous coatings. The energy requirements are much lower. Hot melt coating technology appears to be quite suitable for small-scale operations, and this has opened a possibility for many label converters to produce their own stock. Small machines resembling laboratory equipment are used for production purposes. Because of high coating rates, coaters can supply sufficient stock for a small converter. Such a coater that handles a 16 in. web is shown in Fig. 28-1.

Typical coating weight for label stock is 18–27 g/m$^2$ (12–18 lb/3000 ft$^2$). There are some applications where a higher coating weight is needed, such as applications on difficult to adhere surfaces: corrugated packaging and automobile tires. These high coat weights accentuate the differences of hot melt adhesives in die cutting. Hot melts are softer materials and do not fracture under the high shear of a knife edge, like some of the solvent-based adhesives. Because of this, it is important that the depth of the cut be controlled so that the knife edge cuts through at least 80% of the adhesive mass. Otherwise, label lifting may occur or the waste matrix could break.

## ADHESIVES

Early hot melt pressure-sensitive adhesives were based on ethylene-vinyl acetate (EVA) copolymers. These adhesives suffered from low cohesive strength, especially at elevated temperatures. EVA copolymer is compounded with tackifying resins. Wax is not used for pressure-sensitive adhesive formulations, although it is a regular ingredient in EVA-based hot

Fig. 28.1. Narrow width hot melt coater. (*Courtesy Acumeter Laboratories, Inc.*)

melt adhesive and coating formulations. Wax is not always compatible with the system and has a tendency to bloom to the surface, affecting tack and adhesion. High acetate content (40%), high melt index EVA polymers are generally used for pressure-sensitive adhesives. This results in an adhesive of low shear strength. Some improvement in this property can be achieved by addition of low melt index (high molecular weight) polymer. Generally, a mixture of tackifying resins is preferred over a single one. This helps to minimize the loss of tack on aging. Introduction of synthetic hydrocarbon-based tackifying resins helped to improve the thermal stability of hot melt pressure-sensitive adhesives. Litz[1] has discussed the formulation of EVA based pressure-sensitive hot melt adhesives. A typical generalized formula is given below.

| | |
|---|---|
| EVA copolymers | 35–50% |
| Plasticizer | 0–20% |
| Tackifying resin | 30–50% |
| Filler | 0–5% |
| Antioxidant | 0.5% |

EVA copolymer, when compounded, might give good tack and peel strength initially, but these might drastically decrease on aging. This is usually caused by incompatibility of tackifying resins with EVA copolymer. Incompatibility can be revealed by microscopic observation as evidenced by the presence of a discontinuous phase.[9] The incompatibility could also be

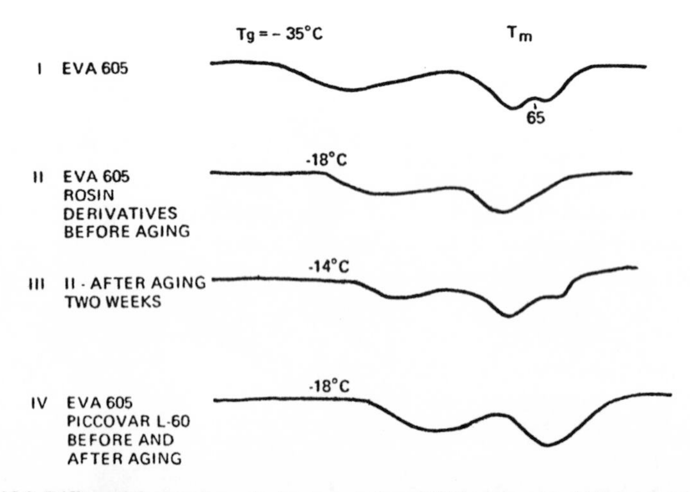


Fig. 28.2. Differential scanning calorimeter graphs of EVA resin with different tackifiers.[2]

detected by differential scanning calorimetric measurements as described by Rifi.[2] Endothermic transitions due to crystalline sites in the EVA copolymer are disrupted by compounding with a tackifying resin. These can reappear later on aging, indicating incompatibility as shown in Fig. 28-2.

The introduction of block copolymers increased the use of hot melt pressure-sensitive adhesives considerably. These polymers and their use are described in Chapter 11. The block copolymers secured for themselves the most important place in hot melt pressure-sensitive adhesives. They have replaced EVA copolymers in most applications and contributed to the acceptance of the hot melt pressure-sensitive products. These adhesives are used not only for labels, but for carpet laying tapes, masking tapes and packaging tapes on various film backings.[3] The carpet laying tapes require a heavy adhesive coating of about 100 g/m$^2$. For coatings that are heavy and difficult to dry, the hot melt technique is especially suitable when manufacturing these tapes.

Block copolymers are used with other polymers in compounding pressure-sensitive adhesives. Russell[8] describes the use of block copolymers with EVA as shown in the formulation below.

| | |
|---|---|
| Ethylene-vinyl acetate copolymer (Elvax 40) | 12.75 parts |
| Block copolymer (Kraton 1107) | 12.75 parts |
| Hindered phenol antioxidant (Ionox 330) | 0.5 part |
| Ethylene-vinyl acetate copolymer (Elvax EP-3643) | 26 parts |
| Tackifying resin (Piccovar L-60) | 19.15 parts |
| Tackifying resin (Nevindene R-6) | 19.15 parts |

The adhesive was used for vinyl-asbestos floor tiles. This is an important application for hot melt adhesives. In some cases, fillers might be used to decrease the cost, because of the heavy adhesive deposit required.

Amorphous polypropylene is a by-product of polypropylene manufacturing, and therefore is available at a low cost. It is a soft, tacky polymer that can be compounded to hot melt pressure-sensitive adhesives for applications which do not require a high shear resistance. The low cost of such pressure-sensitive adhesives makes them attractive for such applications as floor tiles and various laminating uses. A typical formulation is shown below.

| | |
|---|---|
| Atactic polypropylene (16–20,000 m.w.) | 78 |
| Ethylene-vinyl acetate copolymer | 5 |
| Polyterpene tackifying resin | 17 |

Ethylene-vinyl acetate copolymer is used to improve the shear resistance. Other polymers such as block copolymers, crystalline polypropylene might be used for the same purpose. A formulation using block copolymers has been described by Park.[10]

| | |
|---|---|
| Amorphous polypropylene (viscosity is 3000 cps at 190°C) | 50 |
| Rosin ester (Foral 105) | 15 |
| Block copolymer (Kraton 1107) | 14.75 |
| Polybutylene (Indopol H-1900) | 10 |
| Polyisobutylene (Vistanex LM-MS) | 10 |
| Crystalline polypropylene (inherent viscosity = 1.7) | 1 |
| Antioxidant (Ethyl 702) | 0.25 |

Other polymers have been also used for hot melt pressure-sensitive adhesives. Styrene-isobutylene copolymers have been described by Shenfeld *et al.*[11] and by Scardiglia and Hokama,[12] polybutadiene by Jurrens.[13] Hinterwaldner[4] has discussed graft copolymers for application in hot melt pressure-sensitive adhesives. Post-cured block copolymers and 100% solids coatings which are cured after their application offer many interesting possibilities for the future. The hot melt equipment and technology is suitable for handling these novel systems.

## EQUIPMENT SELECTION

There are many process and product variables that must be considered in selecting the most suitable hot melt application equipment. Some of these variables are listed below.

1. *Coating types.* Is the equipment to be used exclusively with pressure-sensitive hot melts or should it be capable of handling heat seal, barrier, or protective coatings? It is important to know the viscosity range of the adhesives or coatings to be used and the degree of versatility required. Lower viscosity coatings allow the construction of a coater with wider tolerances and less structural integrity. A coater designed for high viscosity hot melts might not handle the low viscosity coatings properly, unless it is specifically designed for it.

2. *Web types.* What kind of webs will be run on the machine? This is an important factor in determining the web handling system required. Heavy board stock requires larger diameter idler rolls; lightweight extensible films require the ability to run at low tensions; foils require special graduated preheats. If the substrate to be coated is expected to have baggy edges or thickness bands, additional web handling devices might be needed to compensate for these mechanical imperfections.

3. *Structures to be produced.* How versatile should the equipment be in order to accommodate various structures (laminates, two-side coating, three-web lamination)? Such requirements are best incorporated initially, or at least provisions should be made for future incorporation. The decision must be reached whether it is more economical to provide two coating heads in line for two-side coating, or the product is best made in two passes. The horsepower requirements should be established for current and possible future needs in order to avoid future replacements.

4. *The size of production run.* For short runs, the change over time might be more significant than the machine speed. As the run length decreases, the start-up waste increases. These considerations will determine whether the emphasis should be placed on the machine speed or on the ease of change. The length of run, the supply roll size, and the machine speed will determine the splicing devices most suitable for the operation.

5. *Contamination.* The importance of cross-contamination of hot melts should be established. Pigmented coatings, wax containing coatings before a hot melt pressure-sensitive run, or incompatible blends might place an important emphasis on ease of cleanup.

How much versatility should be incorporated into the coater is always a difficult decision. The decision to forego versatility should be an active one, not one decided by default. If the products planned are expected to generate sufficient volume to warrant the acquisition of a second coater, a choice might be made to have the first coater versatile, but not necessarily the fastest or most economical to run. The second coater would then be purchased for a well-established product mix and will not require additional versatility.

There are basically two types of hot melt coaters suitable for hot melt pressure-sensitive coating: slot orifice and roll coater. Slot orifice coater can handle hot melt viscosities of 400–200,000 cps. The early roll coaters designed for coating waxes were not capable of handling high viscosity hot melts. The new generation of roll coaters can easily handle coatings up to 200,000 cps viscosity.

In addition to the above coaters, an extruder has been designed for pressure-sensitive adhesive application. It is not a hot melt by strict definition: it can handle polymers of 250,000–1.5 million cps at 177°C. Such viscosities are outside of what is normally classified as a hot melt.

## SLOT ORIFICE COATERS

The principle of the slot orifice coater is simple. The coating flows through a slot forming a thin sheet of the coating material which is deposited on the web. One type of coater is the curtain coater, where the slot orifice is positioned above the web and the coating is allowed to flow by gravity onto the web surface. Curtain coaters generally handle viscosities of 200–10,000 cps of hot melt, solvent, or water-based coating.

In a fountain-type slot orifice coater, the coating head is located underneath the web and the coating material is moved upwards by employing a pump. The coating can be metered by controlling the flow and the web speed, or it can be applied in excess and then metered by a separate device.

There are many variations of coaters using the slot orifice head. They are discussed in some detail by Booth.[14] The first fountain system for hot melts was the Genpac coater, capable of handling viscosities of 1000–20,000 cps and depositing coating weights of 4–30 g/m$^2$.

A schematic diagram of a fountain type slot orifice coating head is shown in Fig. 28-3. The melted coating is supplied to a pump which moves the melt into the coating head. The coating head is a long cavity the width of the machine with a slot facing the web of material to be coated. The melt fills this cavity with a slot facing the web of material to be coated. The melt

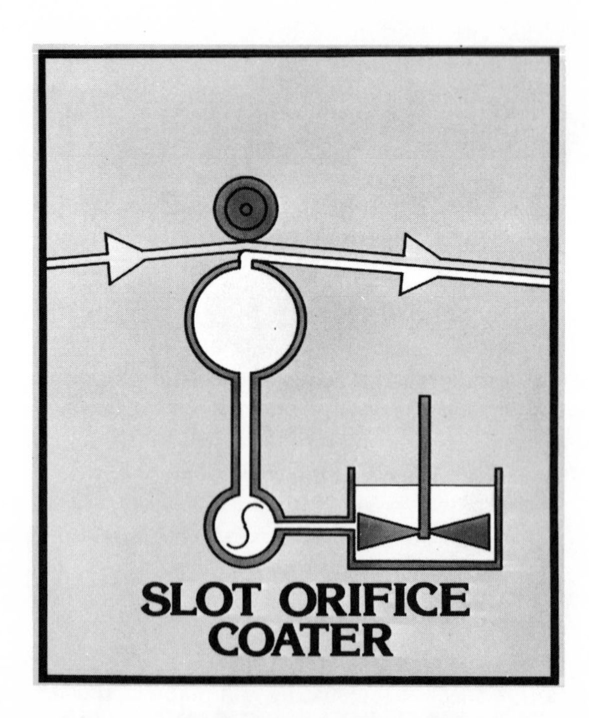


Fig. 28.3. A schematic diagram of a slot orifice coaters.

fills this cavity and is then pushed out the slot. At this point, the melt contacts the moving web and the coating is applied. Because the web completely covers the slot in the coating head, a uniform pressure across the width is obtained. This provides a constant coating weight across the web. Some machines add the backup roller over the die to control the contact of the web with the die. The hot melt delivery is tied to the machine speed through a coupled pump drive or a constant pressure is maintained by a bypass arrangement.

The slot orifice coater has established itself as a good coater for pressure-sensitive adhesives and it is the most popular coater in the United States. Its basic operating concept is simple, and it offers a potential capability to handle the highest viscosity hot melts. The machine relies on a uniform flow of melt out of the die and therefore its operation can be affected by factors such as viscosity changes, because of temperature variations, or the effect of oxidation on the hot melt. The condition of the web to be coated is important. Baggy edges, excessive lint, or nonuniform thickness may affect the coating uniformity. The slot orifice is difficult to use with low viscosity hot melts. Low viscosity increases the possibility of pressure variations and therefore the variation of the coating weight.

Park coater (Bolton-Emerson) has found the most wide acceptance for pressure-sensitive adhesive application. The die is used not only to distribute the coating evenly, but also to smooth out the adhesive coating. Coating weight tolerances are claimed to be $\pm$ 5% and the coating weight can be varied over a wide range[5] of approximately 0.5–600 g/m$^2$. A photograph of this coater is shown in Fig. 28-4.

As discussed previously, narrow width slot orifice coaters have found an application for low volume manufacturing of label stock (Fig. 28-1).

## ROLL COATERS

Roll coating and calendering are basically similar processes in which the coating material is pulled through a narrow gap between the two rolls in such manner as to form the adhesive into a continuous sheet. The main difference between coating and calendering is in the area of separation of the coating from the roll. In the case of calendering, the coating material remains on the substrate and pulls away cleanly from the roll. In the case of a coating operation, the adhesive splits between the web and the roll. The rheological properties of the coating and the surface tensions involved determine the behavior of the adhesive. Adhesives of high viscosity and elasticity are applied by calendering. Adhesives of lower viscosity are coated.

The coating weight is controlled by the gap between rolls, roll speed, web speed, and other variables. The operation of a hot melt roll coater is

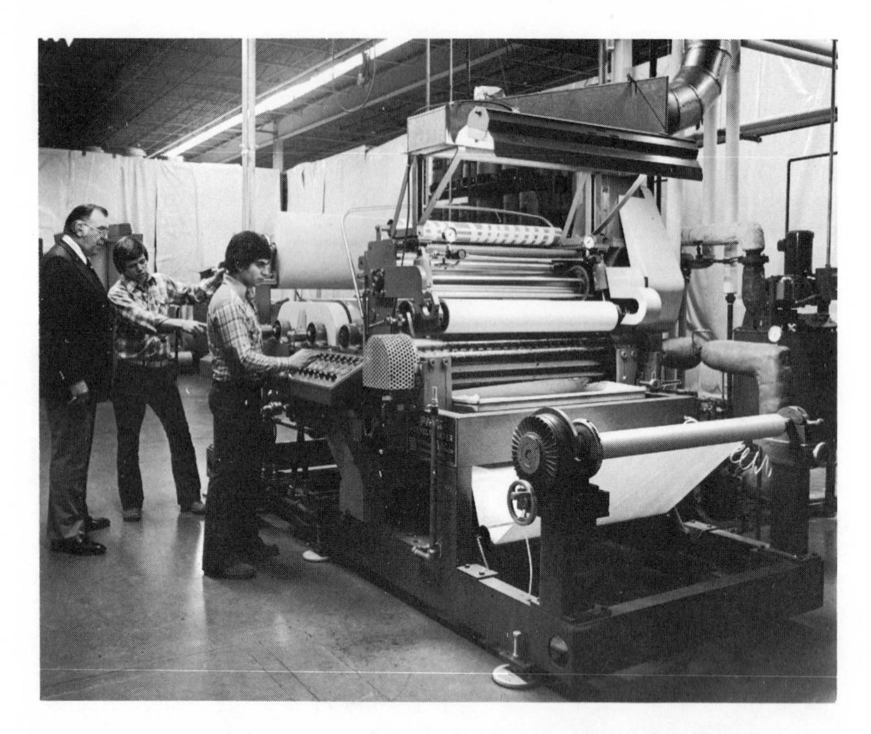

Fig. 28.4. Park coater, slot orifice type. (*Courtesy Bolton-Emerson, Inc.*)

similar to the operation of a three-roll reverse roll coater. Figure 28-5 shows a schematic diagram of a hot melt roll coater. The melt in the coating pan wets out the lower surface of rolls A and B. As they rotate, they bring the melt up to the nip produced by rolls A and C. At this point, the melt contacts the moving web. Now the coating can be cooled, laminated or sent directly to rewind. In order to handle the relatively high melt viscosities, rolls A and B rotate with a surface speed much slower than the web speed. It is common practice to keep roll B stationary so that it acts as a metering roll, wiping the surface of roll A of excess adhesive melt. Rotating rolls A and B slowly also reduces the hydraulic pressure in the nip.

High viscosity hot melts ($> 100{,}000$ cps) can cause roll bending because of the high hydraulic pressures involved. This is corrected by roll crowning or by axis crossing. Crossing the axis increases the gap at the ends, keeping it the same in the center. This counteracts the effect of roll bending. The coating variation of $\pm 5\%$ across the width is normal.[6] The maximum limit of the coating weight is 300 $g/m^2$.

A control is available over the pressure by pushing the rolls A and B

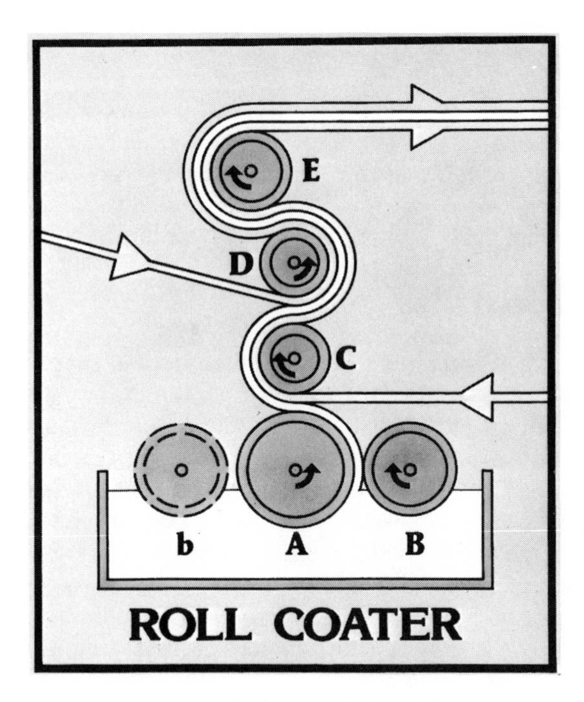


Fig. 28.5. A schematic diagram of a hot melt roll coater.

together. The rolls can also be pushed with a high pressure against a fixed gap between the rolls. The speed of the rolls is controlled independently of the web speed. Another point of control is the pressure exerted on roll C against roll A and the independent speed control of these rolls.

The roll coater can handle a wide range of viscosities and it is more suitable for low viscosity coatings than the slot orifice coaters. The upper viscosity limit is 200,000 cps.

Since the coating film is preformed before it is applied to the web, the roll coating offers the advantage of being able to handle heat-sensitive webs such as polyethylene or poly(vinyl chloride) films. The webs can be run at a lower tension than in the slot orifice coater.

Where thorough cleanup between different melts is important, the roll coater offers better accessibility because of its open design.

The roll coater is susceptible to film splitting when the coating adheres to both the web and the roller surface. Differential speeds or reverse roll application minimizes coating splitting, but splitting is a characteristic of the coating operation.

The open design of the coater and the fact that the coating is preformed before it is applied to the web make this design more conducive to the oxidation of the hot melt adhesive.

Figures 28-6 and 28-7 show hot melt roll coating machines manufactured by Maschinenfabrik Max Kroenert of Hamburg, Germany. Figure 28-6 shows the PAK 600 machine, which is a universal wax and hot melt coater, and Fig. 28-7 shows the PAK 400, a smaller model specifically designed for the application of pressure-sensitive adhesives.

## EXTRUSION COATERS

Extrusion coating of pressure-sensitive adhesives has been introduced to handle high viscosity adhesives in a large continuous coating operation. One of the products made by this method is saturated paper tape. The normal hot melt pressure-sensitive adhesive viscosities range in the area of 5000–50,000 cps at 177°C. The extrusion coating handles adhesives of viscosity 250,000–1.5 million cps at 177°C and reportedly could process the adhesives in the range of 20,000–2 million cps at 177°C.[7]

The method allows mixing and coating pressure-sensitive adhesives in one pass with a total residence time of 2–3 min. The adhesive is exposed to high temperature for only a short period of time, and the problem of thermal degradation is minimized. The substrate used does not need to have thermal stability; poly(vinyl chloride) and polypropylene films have been coated by this method.

Fig. 28.6. A universal hot melt roll coater, PAK 400, Maschinenfabrik Max Kroenert. (*Photo courtesy Pacon Machines Corp.*)

Fig. 28.7. Pressure-sensitive adhesive hot melt coater, PAK 600, Maschinenfabrik Max Kroenert. (*Photo courtesy Pacon Machines Corp.*)

Standard plastic extrusion equipment is not suitable for handling of pressure-sensitive adhesives. The adhesive sticks to the flight of the screw. A special single screw extruder has been developed.[15] It is equipped with an auxiliary wiper screw in the feed section which continuously clears the flights of the feed screw of any material that might be sticking to it. In order to obtain proper mixing of all ingredients, a large length-to-diameter ratio is required (40 : 1). The adhesive, after passing through the extruder, is properly mixed and it is then extruded through a die onto the substrate. This method is suitable for a large volume operation: a 88.9 mm extruder mixes 270 kg/hr, and a 152 mm one mixes 900 kg/hr.

## COATER OPERATION

The operation of a hot melt pressure-sensitive adhesive coater presents some unique problems. These might fall into the area of web handling, machine start-up, maintaining coating uniformity, and control of the hot melt supply.

Control of the web, as it moves from the unwind to the coating head and on to the rewind, is important in all coating operations. In hot melt coating, higher tensions are generally used at the coating head, especially in case of slot orifice coating. These tensions must be isolated, and the best way to do so is to employ driven nip rolls. Also, because of the high viscosity of the coating, the web at the coating head must be held flat in order to obtain good adhesive transfer. Skewable rolls can help to pull out soft edges. If these rolls are not set properly, tension wrinkles might be formed. Tension wrinkles, when looking in the direction of web travel, will point to the side of high tension. With this as an indicator, a proper tension balance can be set up.

If the tension in the coating head can be isolated, then the unwind tension should be set at a minimum level required for the web to track straight and to give a sufficiently firm roll. Several methods of winding can be used: center winding, surface winding and surface-center winding, which is the combination of both methods.

In the case of center winding, force is applied on the shaft. It is transmitted through a clutch, core and layers of tape already wound on the core. If a simple slip clutch is used, the torque applied to the roll remains constant. As the roll diameter grows, the tension on the web decreases, since the torque is a product of web tension and roll radius. Thus, winding at a constant torque results in continuously softer wind as the roll size increases.

It is possible to program the winding tension in various ways to obtain better wound rolls. A simple way is to increase torque continuously as the roll size increases. This minimizes the drop-off in web tension. The roll can be evenly wound at constant tension with proper programming.

The web carries an air layer. At high winding speeds, air can get entrapped in the roll-forming balloons, providing a low friction surface so that a shift of lower levels can take place. This can result in gapping or telescoping of the rolls and cause problems later in slitting and processing of the material. Top riding rolls may be employed to smooth out the rewind roll. They help to remove the air layer and are especially useful in winding films with gauge bands, winding materials at low tensions and other problem cases.

In surface winding, the force is produced by one or two winding drums which are in contact with the outer surface of the roll being rewound. The roll is wound at a constant tension which does not depend on the roll diameter.

Surface-center winding combines the two methods discussed. The web is wrapped around a driven winding drum and a torque is applied to the core.

Winding becomes especially important in the final slitting operation or in cases where the roll produced might be sliced without another rewinding. Slitting and rewinding techniques have been reviewed by Rienau.[16,17]

The biggest concern with the start-up of a cold machine is that the coating weight changes as the coating head and idler rolls slowly come up to temperature. Even with preheating of the machine, some changes in coating weight might take place several hours later. The greatest change occurs during the first hour of operation. The performance of the machine should be watched carefully and appropriate corrections made. The machine is predictable as to the amount of change and the time frame in which the changes happen. With this information, start-up waste can be greatly reduced by anticipating the corrections needed.

Besides the coating waste, the bond between the adhesive and the substrate or the release liner might vary. Because the temperature of the whole machine is lower, the bond takes a longer time to develop.

Hot melt pressure-sensitive adhesives are heat-sensitive materials and are subject to thermal degrading. Therefore, care must be taken to restrict the adhesive's exposure to oxygen. The exposure to oxygen is more detrimental than the thermal history. A nitrogen blanket is recommended over all adhesive reservoirs.

If the adhesives are purchased, they have to be melted just prior to use. There is a number of commercial units available for quickly melting the adhesive. Drum melters which use 55-gallon steel drums and pump the adhesive out as it melts are available. This system minimizes the exposure of the molten adhesive to air, and it melts only what is needed for the operation. Another type is a drum unloader that requires a smaller container, approximately 100 lb. The adhesive is pushed out of the container into the melting section where it is cut up and melted. This system has a very high capacity (up to 1500 lb/hr) but does not bring all the adhesive up to temperature. A holding tank is required. Another method utilizes hot rotating rolls between which pieces of adhesive are dropped and melted. A storage tank of molten adhesive is also needed.

The molten adhesive might contain some foreign material and should be filtered. For most applications, a 100-mesh screen is sufficient. Once the adhesive is melted and filtered, it must be brought to application temperature. This can be done when pumped through the supply hoses, which can be electrically heated, or in a holding tank. The viscosity is very sensitive to the temperature changes and it is important to maintain a uniform temperature in order to maintain a constant coating weight.

If the adhesive is made in-house, it can be pumped directly from the compounding equipment or from an intermediate storage tank.

There are many factors that affect the quality of the hot melt at the coating head. Temperature uniformity of the adhesive is important, but so is the temperature profile of the coating head itself. If it has a poor heat distribution, hot and cold spots will develop, giving heavy coating weights in the cool areas and low coating weights in the hot areas. Heat also causes the expansion of coating head and rollers. This expansion is not uniform and the coating weight profile will change. Therefore, it is important to recheck the coating profile after the machine has been run for some time.

The temperature of the coating head has a great effect on how well adhesive is transferred from the steel of the coating head to the moving web. If the temperature is too low, the adhesive might not flow properly because of the excessively high viscosity. It also might not wet the surface well enough, causing fine skips in the coating. The appearance of these skips is very similar to the pattern created when the contact between the web and the steel surface is insufficient. The operator might be tempted to apply more pressure when the real problem is insufficient adhesive temperature.

With a porous web, such as paper or fabric, some adhesive is absorbed by the material. Some penetration helps to create a good bond, but too much penetration increases the adhesive consumption. Inadequate penetration of paper label stock requires a longer storage to improve the anchorage of the adhesive. By controlling the pressure, temperature, and speed, the properties of the finished label stock lamination can be controlled so that very little change is observed in the release levels and anchorage to the label stock as the lamination is aged.

The label stock can be manufactured by either coating the release liner and then transferring the adhesive to the face stock, or by coating the face stock directly. The most common method is to coat the release paper. The release paper has a smooth surface and usually a more uniform thickness. With this method, the penetration of the adhesive into the face stock is controlled in the laminating nip. Direct coating is used when characteristics of the release liner, such as embossing, make it difficult to transfer coat. Excessive laminating pressure can cause wrinkling and side curl. This is particularly important if the label stock is sheeted for subsequent printing.

## REFERENCES


1. Litz, R. J. *Adhesives Age* **32**(7)(1971).
2. Rifi, M. R. *Course Notes 1977, Cavalcade of Hot Melts, Their Future, Their Problems*, April 18–21, 1977, Hilton Head Island, S. C., Technical Association of the Pulp and Paper Industry, pp. 17–24.
3. De Jager, D. and Borthwick, J. B. *Course Notes, 1978 International Hot Melt Short Course*, May 8–11, 1978, Frankfurt, Germany. Technical Association of the Pulp and Paper Industry, pp. 223–226.
4. Hinterwaldner, R. *Course Notes, 1978 International Hot Melt Short Course*, May 8–11, 1978, Frankfurt, Germany. Technical Association of the Pulp and Paper Industry, pp. 247–250.
5. Houseman, G. *Course Notes, 1978 International Hot Melt Short Course*, May 8–11, 1978, Frankfurt, Germany. Technical Association of the Pulp and Paper Industry, pp. 85–87.

6. Womack, H. G., and Wallace, R. C. *Course Notes, 1978 International Hot Melt Short Course*, May 8–11, 1978, Frankfurt, Germany. Technical Association of the Pulp and Paper Industry, pp. 94–97.
7. Palermo, F. C., and Korpman, R. *Course Notes, 1978 International Hot Melt Short Course*, May 8–11, 1978, Frankfurt, Germany. Technical Association of the Pulp and Paper Industry, pp. 115–118.
8. Russell, T. E. U.S. Patent 3,630,980 (1971)(assigned to Flintkote Co.).
9. Powers, P. O. Growth and Change of Hot Melts. *Adhesion Science and Technology* (L.-H. Lee, ed). New York, London: Plenum Press, 1975.
10. Park, V. K. U. S. Patent 3,850,858 (1974)(assigned to Eastman Kodak Co.).
11. Shenfeld, R.S., Musser, F. M., and Kothari, G. U. U. S. Patent 3,664,252 (1972) (assigned to Velsicol Chemical Corp.).
12. Scardiglia, F., and Hokama, T. U. S. Patent 3,835,079 (1974) (assigned to Velsicol Chemical Corp.).
13. Jurrens, L. D. U. S. Patent 4,025,478 (1977) (assigned to Phillips Petroleum Co.).
14. Booth, G. L. *Coating Equipment and Processes*. New York: Lockwood Publishing Co., Inc., 1970.
15. Smith, D. J. U. S. Patent 3,929,323 (1975) (assigned to Egan Machinery Co.).
16. Rienau, J. H. *Paper, Film and Foil Converter* **53**(4): 61–66 (1979).
17. Rienau, J. H. *Paper, Film and Foil Converter*, **53**(5): 172–178 (1979).

# Chapter 29

# Polymerization

**Donatas Satas**

*Satas & Associates*
*Warwick, Rhode Island*

Polymerization of elastomers used in compounding pressure-sensitive adhesives is outside the sphere of interest of an adhesives technologist. Their original properties are accepted and only changes are sought. Various elastomers, including natural rubber, are broken down mechanically and compounded to obtain the desirable adhesive properties. In the case of pressure-sensitive acrylates, the polymerization process itself has become a part of the activity of an adhesives technologist. The in-house polymerization of acrylates for a captive use is common, unlike the manufacture of other synthetic elastomers. There are several reasons for this difference. The development of pressure-sensitive polyacrylates for tape and label use originated with pressure-sensitive product manufacturers. This is unique, because the other elastomers were developed by polymer manufacturers and only adopted for pressure-sensitive applications, which are usually less important than other, much larger volume applications. The growth of block copolymer use for pressure-sensitive applications was largely promoted by the polymer manufacturers.

Polymerization of acrylates by either solution or emulsion polymerization are well known processes and such techniques are well known throughout the industry. Small-scale operations are possible and indeed large volume operations have only a very slight economical edge. The equipment costs are low; equipment required for compounding of adhesives is more expensive if it also involves the mechanical breakdown.

The literature on various aspects of polymerization is voluminous. The purpose of this chapter is only to give a brief review and to emphasize the areas of specific interest to pressure-sensitive products which might not be covered elsehwere. For a more detailed and general review of vinyl polymerization, many references are available. Luskin and Meyers[1] have written a short review on the polymerization of acrylic esters and several books have been published on the subject.[2,3,4,5]

Solution and emulsion polymerization processes have their own advantages and shortcomings. Table 29-1 summarizes the most important differences between these two processes.

**Table 29-1. Differences between solution and emulsion polymerization process and products.**

| PROPERTY | TYPE OF PROCESS | |
|---|---|---|
| | SOLUTION | EMULSION |
| Molecular weight | Low | High |
| Branching | High | Low |
| Residual monomer | High | Low |
| Impurities | Low | High |
| Cost | Higher, because of solvent | Low |
| Yield per reactor space | Lower | High |
| Polymerization rate | Low | High |
| Ease of tailoring | Poorer | Good |
| Coating | Directly as is | Directly as is or coagulated and dissolved |

## SOLUTION POLYMERIZATION

Solution polymerization of pressure-sensitive acrylates is a homogeneous free radical polymerization process. The monomers are soluble in the solvent and the polymer is also usually soluble in the same solvent or solvent blend. The main advantage of the solution polymerization is that a pure polymer is obtained in a ready-to-coat condition. Unfortunately, the solvents used also act as chain transfer agents and cause a decrease in the molecular weight that has a negative effect on the physical properties of the adhesive. Ethyl acetate, cyclohexane, toluene, and *n*-heptane are commonly used solvents. The two former ones give a higher molecular weight polymer. The lower molecular weight must be compensated by cross-linking, hydrogen bonds, or other means. The lower the chain transfer constant, the higher is the molecular weight. The chain transfer constant is defined as the ratio of the rate constant for transfer with the solvent (termination) and for propagation. Impurities in the solvent increase the chain transfer.

The molecular weight also depends on the monomer concentration.

Lower molecular weight polymers are obtained at low monomer concentrations. High polymer concentrations in the solution increase branching and crosslinking due to chain transfer with a dead polymer.

Various chemical initiators are used to start the polymerization. Organic, oil-soluble peroxides are commonly used. Their main disadvantages are occasional formation of peroxide crosslinks and the dependence of the decomposition rate on the type of solvent used and other variables. Azo compounds, such as 2,2′-azobisisobutyronitrile, do not possess some of the disadvantages of peroxides and are especially useful for solution polymerization. Initiator concentrations of 0.05–2% of the monomer weight are used. A lower amount of initiator favors the higher molecular weight. The rate of the polymerization reaction is directly proportional to the monomer concentration and it is related to the square root of the initiator concentration.

One of the disadvantages of solution polymerization is the increase of viscosity with an increasing degree of conversion, increasing molecular weight and solids content. Formation of difficult to handle gels can occur easily if any crosslinking takes place.

The most common way to carry out the solution polymerization of acrylic pressure-sensitive adhesives is to start with a high monomer concentration (as high as 70%) using a minimum amount of catalyst, choosing a solvent of a low chain transfer constant and running the reaction at the reflux conditions that help to remove the heat of reaction. Remaining monomer is added as the reaction proceeds, solvent is added to regulate the viscosity and additional catalyst keeps the reaction going. The rate of reaction in solution polymerization is much lower than that in emulsion polymerization and the amount of residual monomer is higher. This might not be very important for pressure-sensitive adhesives which are coated in thin layers, favoring an easy removal of residual volatile materials.

The sequence of addition of comonomers can be varied for special reasons. It is claimed[6] that improved hot melt adhesives are obtained if consisting of a blend of several species of different glass transition temperatures. Such blends can be prepared, besides blending of different polymers, by sequential polymerization, where high $T_g$ monomers are polymerized first and softer, lower $T_g$ monomers are polymerized later by controlling their addition.

## EMULSION POLYMERIZATION MECHANISM

The mechanism for an emulsion polymerization of a water insoluble monomer has been qualitatively described by Harkins.[7] It is assumed that the polymer is soluble in its own monomers. It is generally agreed that Harkins' theory describes reasonably well the emulsion polymerization mechanism of acrylic monomers, with the exception of water-soluble monomers.

Emulsifier molecules associate to form micelles and a small amount of monomer is solubilized in the micelles. The remaining monomer is dispersed in small droplets. The locus of the polymerization reaction is within the micelles. The monomer is supplied from within the micelle. As the polymer particle grows and the monomer is used up, the polymer absorbs more monomer that diffuses from the droplets through the aqueous phase to the micelles. Monomer droplets act as a reservoir only and very little polymerization is initiated in the monomer particles. Some particle initiation may take place outside the micelles in the aqueous phase. Growing polymer-monomer particles adsorb the dissolved soap, which leads to the disappearance of micellar soap. The original monomer particles also disappear and enter the polymer particles swollen with monomer and protected by adsorbed surface active agents. Thus, during the emulsion polymerization, we have several species in the aqueous media: dissolved soap, monomer, and free radical particles. Dissolved soap is in equilibrium with the soap in micelles, the soap adsorbed on the surface of monomer particles and later, as the reaction proceeds, adsorbed on the surface of growing polymer-monomer particles.

The quantitative theory of emulsion polymerization based on the Harkins description of the mechanism has been developed by Smith and Ewart[8] and later refined by Gardon.[9]

Smith and Ewart shows that the number of particles formed during the polymerization reaction is

$$N = \chi\left(\frac{\rho}{\mu}\right)^{2/5}(aS)^{3/5}, \tag{1}$$

where

$N$ = number of particles
$\rho$ = rate of radical production
$\mu$ = rate of volume increase of a particle
$a$ = area occupied by one soap molecule
$S$ = total amount of soap present in the system
$\chi$ = a constant which varies between 0.37–0.53

The constant's value is 0.53 if an assumption is made that only micelles can capture radicals. Since it is clear that polymer particles can capture free radicals as well, the constant's value of 0.53 constitutes the upper limit. The value of 0.37 for the constant $\chi$ is obtained if an assumption is made that the number of radicals entering a particle is independent of the particle radius. This constitutes the lower limit. The emulsion particle size can be controlled by several variables. An increased amount of surfactant decreases the particle size simply by providing a larger number of micelles. The delay in monomer and surfactant addition increases the particle size as compared to adding all the surfactant at once. An increase in reaction temperature increases the particle size, as well as the presence of electrolytes. Acrylic acid and other water-soluble monomers can reduce the latex particle size.

Gershberg[10] shows that butyl acrylate emulsion polymerization does not obey the Smith-Ewart theory. Butyl acrylate is, of course, a very important monomer for pressure-sensitive acrylic adhesives. The reasons for the deviation are slow termination and probably radical transfer out of particles. Monomers which are soluble in water also deviate in their behavior from the Smith-Ewart theory because some polymerization takes place in the aqueous phase.

## EMULSIFIERS

Emulsifiers perform several important functions in emulsion polymerization as described above. Pressure-sensitive polymers are soft and have a tendency to coagulate during polymerization. The emulsifier protection is especially important during the critical stage when all the monomer has migrated to polymer/monomer particles which are quite soft and can coalesce upon impact.

A mixture of anionic and nonionic emulsifiers is generally used. Cationic surfactants are rarely used for polymerization of pressure-sensitive acrylates. Anionic surfactants stabilize the particles by a charged double layer. They improve the mechanical and storage stability. Nonionic surfactants stabilize the latex by surrounding the particles with a hydrated layer of emulsifier. They have a high tolerance for electrolytes. About 2–5 parts of surfactant per 100 parts of monomer are sufficient to carry out emulsion polymerization and to obtain a reasonably stable latex. Further stabilization of latex by addition of more surfactants, or of colloidal substances might be needed.

The emulsifier concentration has an effect on the physical properties of the polymer, especially on molecular weight. Increased emulsifier concentration increases the micelle concentration and the number of particles formed, as shown in Equation 1. Molecular weight is inversely proportional to the number of polymer particles formed.

Emulsifiers may have an effect on pressure-sensitive properties. Low molecular weight surface active materials have a tendency to accumulate at the interfaces and to affect the properties which depend on the surface condition. Adhesives containing surfactants might exhibit poor water resistance, increased water absorbency, decreased tack, and adhesion.

It has been important for the successful use of pressure-sensitive acrylic emulsions to eliminate the detrimental effect of surfactants. Several approaches have been attempted: use of fugitive surfactants, use of oligomeric and polymerizable surfactants, and carrying out the polymerization in the absence of micelle-generating additives.

Fugitive emulsifiers are simply systems which later decompose because one of the components is sufficiently volatile at the conditions employed in drying. Morpholine salt of oleic acid has been prepared *in situ* as a fugitive emulsifier for various oil in water emulsions. Ammonium salts of acrylic or maleamic acids have been used as surfactants for pressure-sensitive acrylic polymerization. While one part of the surfactant is fugitive, the other is polymerizable and is incorporated into the polymer chain. Quaternary ammonium salt of vinyl sulfonic acid was used in emulsion polymerization of acrylic esters.[11] About 5 parts of surfactant per 100 parts of monomers were used.

Polymerizable surfactants such as sodium-2-sulfoethyl methacrylate have been described by Sheetz,[12] and the improvement of the mechanical stability of paint latexes by this surfactant was studied by Mills and Yocum.[13] Use of surfactants containing many other α-unsaturated vinyl groups such as sulfoesters of α-methylene carboxylic acids, carboxy-terminated alkyl esters of α-methylene carboxylic acids, sulfoalkyl allyl ethers, acrylamide alkane sulfonates, vinyl alkyl sulfonate esters, monoalkyl ethers of oxyalkylene esters of carboxylic acids, and alkylaryloxypolyalkoxy esters have been described.[14] Unsaturated carboxylic acid partial esters, such as *p*-dodecyl-phenoxy(nonaethoxy)ethyl hydrogen maleate, have been used by Fallwell.[15] Use of 2-acrylamido-2-methyl propane sulfonic acid allowed the reduction of regular surfactants to concentrations as low as 0.1%. Use of 2-acrylamido-2-methyl propane sulfonic acid is reported to have improved peel adhesion.[16] Both 9- and 10-acrylamido stearic acids were used for polymerization of styrene-butadiene as a copolymerizable surfactant.[17,18,19]

Polymerization of styrene-butadiene was also carried out with such mild colloidal stabilizers as sodium metaphosphate, although the reaction rate was predictably slow and the particle size large.[20] *N*-vinyl-2-pyrrolidone and similar monomers have been used as colloidal stabilizers in carrying out emulsion polymerization.[21]

Oligomeric surfactants are prepared by polymerizing the surface active monomers to a low molecular weight and using them to carry out the polymerization of pressure-sensitive polymers. Pairs of copolymers in a mole ratio of 1 : 3 were used: dimethylaminoethyl methacrylate and ethyl methacrylate; 2-vinyl pyridine and ethyl methacrylate; *tert*-butylaminoethyl methacrylate and ethyl methacrylate; 4-vinyl pyridine methyl methacrylate and ethylene glycol dimethacrylate.[22] The monomers to be reacted are added to this oligomeric dispersion system and the reaction is carried out in the usual manner. About 5 parts of oligomer per 100 parts of monomer are sufficient to impart the required stability during emulsion polymerization. Acrylonitrile and acrylic acid oligomers have been used for the same purpose.[23] Their molecular weight was kept low by chain transfer agents such as *n*-octyl mercaptan.

## INITIATORS

Chemical free radical producers are most often used to initiate the emulsion polymerization. The initiators are classified into dissociative and redox systems. The former yields free radicals upon dissociation, and the latter, upon reaction between a reducing and an oxidizing agent.

Persulfate ion is the most commonly used dissociative initiator producing free radicals according to the following reaction:

$$S_2O_8^{2-} \longrightarrow 2SO_4^{-}\cdot$$

$$SO_4^{-}\cdot + H_2O \longrightarrow HSO_4^{-} + \cdot OH$$

$$2\cdot OH \longrightarrow H_2O + \tfrac{1}{2}O_2$$

Ammonium or potassium persulfate is added to the emulsion and decomposed by raising the temperature to 50°C or higher. About 0.5% of the initiator based on the monomer is sufficient to carry out the polymerization. In order to obtain the highest molecular weight possible, the initiator level should be kept low at the start of polymerization. Once the number of reaction loci is fixed, the level of persulfate has no effect on the polymerization rate.

Organic peroxides are sometimes used as secondary catalysts to promote the complete conversion of monomers in emulsion systems. They decompose according to the following reactions:

| | |
|---|---|
| Organic peroxides | $ROOR \longrightarrow 2RO\cdot$ |
| Hydroperoxides | $ROOH \longrightarrow RO\cdot + \cdot OH$ |
| Azo compounds | $RN{=}NR \longrightarrow 2RN\cdot$ |

Many various redox systems are used to initiate the polymerization of acrylic pressure-sensitive emulsions. The hydrogen peroxide/ferrous ion system is suitable for low temperature initiation. Hydrogen peroxide is added to the emulsion first and then a solution of ammonium ferrous sulfate is introduced gradually. The reaction can be initiated at temperatures as low as 5°C. Keeping the reaction temperature low helps to maximize the molecular weight. Addition of ascorbic acid helps to decrease the induction period. Chelating agents forming complexes with iron are also sometimes used and help to control the ferrous ion concentration and to decrease the polymer discoloration because of ferric ion.

Organic hydroperoxides have been used in ferrous ion systems instead of hydrogen peroxide. A typical catalyst system is shown below.

| | |
|---|---|
| *t*-butyl hydroperoxide | 1.8 g |
| Ferrous ammonium sulfate | 0.6 g |
| Ascorbic acid | 2.4 g |
| Chelating agent | 0.05 g |

Hydroperoxide polyamine redox systems eliminate the use of iron which can cause discoloration and can catalyze oxidative degradation of the polymer. The presence of ionic materials is undesirable for electrical tape applications.

The persulfate/ferrous ion system can be used for initiation of emulsion polymerization and proceeds according to the following reaction:

$$S_2O_8^{2-} + Fe \xrightarrow{2+} SO_4^{-}\cdot + SO_4^{2-} + Fe^{3+}$$

The persulfate/bisulfite system is a popular redox system for initiation at 50°C and higher temperatures. Free redicals are produced according to the reaction:

$$S_2O_8^{2-} + HSO_3^{-} \rightarrow SO_4^{2-} + SO_4^{-}\cdot + HSO_3\cdot$$

The sulfate radical is more important in the initiation reaction than the bisulfite radical. To polymerize 10 kg of monomer, 25 g of potassium persulfate dissolved in 1 liter of water and 5 g of sodium bisulfite in 500 ml of water were sufficient. Persulfate solution was added to the emulsion and the bisulfite was added gradually. The reaction was carried out at 70°C. Persulfates give a higher conversion of monomer to polymer than peroxides.

The polymerization reaction does not start immediately upon addition of the catalyst. Usually a delay of 5 to 30 min is observed. This induction period is attributed to the presence of inhibitor in the monomers and to the oxygen in the air. The length of the induction period varies depending on these variables, initiator concentration, pH and the rate of mixing. Fast mixing tends to inhibit the reaction. For commercial manufacturing of acrylic pressure-sensitive adhesives, it is not necessary to remove the inhibitors from the monomers or to carry out the reaction under an inert atmosphere.

After addition of the initiator and after a suitable induction period, a temperature rise is observed. The reaction can be carried out at approximately constant temperature by applying cooling and withholding the addition of the initiator until the temperature starts decreasing. Further addition of the initiator causes a rise again and the reaction is allowed to proceed at some control temperature. Sometimes the control temperature is chosen to be the boiling point of the lowest boiling monomer. The removal of the heat of reaction is much easier if the reaction is run at reflux conditions. Vinyl acetate containing pressure-sensitive adhesives are often polymerized at 72–73°C, the boiling point of vinyl acetate.

The monomer emulsion can be prepared separately in a disperser and introduced pre-emulsified into the reactor, or the monomer mixture can be added to the reactor containing water and emulsifiers and dispersed by mixing. A portion of the monomers can be withheld and added gradually as the reaction proceeds. This technique allows an easier control of the exotherm. Likewise, some of the surfactant can be withheld in order to control the number of particles and thus the particle size. The withheld surfactant does not participate in micelle formation and is used for protection of polymer particles only.

A small amount of the monomer remains unreacted in the polymer particles. In the case of pressure-sensitive adhesives, the small amount of residual monomer is not important. The adhesive coating is spread in a thin layer and dried at an elevated temperature in an enclosed oven where the residual monomer is removed. Therefore, it is not usually required to remove the unreacted monomer from the emulsion prior to coating. In cases where the remaining monomer might be detrimental, various techniques to strip the emulsion are used. Raising the temperature at the end of the reaction, adding a good dosage of catalyst, and sparging the emulsion with steam are the techniques used to remove the unreacted monomer.

## EQUIPMENT

Acrylic monomers are stored in storage tanks or in drums depending on the volume used. Drum storage area temperature should not exceed 32°C. Drums should be unloaded by either gravity or pumping. Use of compressed air is not safe for drum unloading. The empty drums should be washed and steamed. Remaining monomer vapor in an empty drum constitutes an explosion hazard.

Storage tanks are usually constructed from 304 SS, aluminum, phenolic-, or epoxy-lined carbon steel. Carbon steel or cast iron is also acceptable, if iron contamination is not important. Pipes valves and pumps are usually constructed from 316 SS.

Storage tanks are loaded by pumping the monomer from the tank truck or car. A vapor return system should be provided to avoid the discharge of monomer vapor to the atmosphere. The storage tank is provided with a 2 psi pressure release valve, conservation vents (to minimize water vapor intake from the atmosphere and the discharge of monomer vapor) and flame arrestors on the vents. The conservation vent might also be equipped with a dehumidifier ($CaCl_2$) to minimize the moisture entrance into the storage tank.

Either underground or aboveground storage tanks are used, depending on the proximity to other buildings, local fire and safety regulations and other reasons. Installation of above ground tanks requires a reservoir to contain the monomer in case of spillage. Analysis of the monomer in storage for inhibitor, water and iron content and for the presence of polymerized material is advocated every 4–6 weeks. No monomer should be left in pipes or pumps exposed to direct sunlight.

The acrylic monomer odor is a problem whenever the monomers are handled. Conservation vents on the storage tanks, vapor return during unloading and activated carbon filters on vents help to minimize the escape of the monomer. To destroy acrylic monomer spills, hydrated lime as a dry powder or a slurry (25%) is useful. Sodium sulfite is also helpful, but not as effective.

Monomer is pumped from the storage to the reactor, preblending tank, or a disperser, if used in the process, via a volumetric measuring device. The reactor is a jacketed, baffled vessel, constructed from 316 SS, from glass-lined steel, or other suitable material. It is equipped with a condenser, or at least a cooled vapor pipe, and a mixer. A turbine type mixer with variable speed control or at least two speeds is preferred. Faster speed is used to prepare the emulsion and slower speed is used during the polymerization. Agitation is quite important. Fast mixing can extend the induction period and cause polymer coagulation. The emulsion particles are very soft at the beginning and can easily coagulate if the shear forces are too high. The temperature is controlled by cooling and by withholding initiator addition that slows down the reaction. Figure 29-1 shows a schematic flow diagram of a batch polymerization process.

The polymerization can also be carried out in a continuous manner. The advantages gained in decreasing the size of the equipment are minor and continuous polymerization is not important for polymerization of pressure-sensitive adhesives.

The polymer emulsion is filtered, preferably by a rotating screen, and stored for further use. Stainless steel, aluminum alloy, reinforced polyester or epoxy resin-lined tanks are used for emulsion storage. The tanks should be equipped with a mechanical mixer or a pump-actuated recirculation

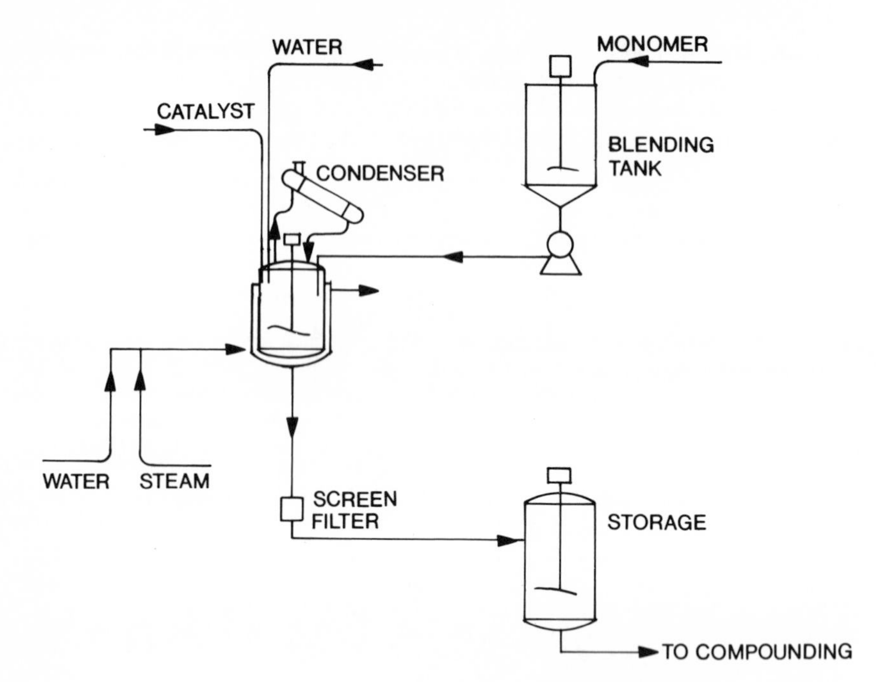


Fig. 29-1. Schematic diagram of batch polymerization process.

loop for periodical agitation. At least 15 minutes of agitation per day is required. Stainless steel storage tanks can be cleaned with 5% NaOH solution at 70°C or hotter. Aluminum- or glass-lined tanks are cleaned with soapy water by soaking for 24 hr. Emulsion is transferred by low shear pumps. Suitable centrifugal, diaphragm, or single-screw rotary pumps are employed.

In some cases, the emulsion polymer could be coagulated, washed and dissolved for solvent coating, or dried for hot melt or calendering applications.

## REFERENCES

1. Luskins, L. S., and Meyers, R. J. "Acrylic Ester Polymers" in *Encyclopedia of Polymer Science and Technology*. Vol. I, pp. 246–328. New York: John Wiley and Sons, Inc. 1964.
2. Ham, G. E. *Vinyl Polymerization*. New York: Marcel Dekker, 1969.
3. Ham, G. E. *Copolymerization*. New York: Interscience Publishers, 1964.
4. Tsuruta, T., and O'Driscoll, K. F. (eds.). *Structure and Mechanism in Vinyl Polymerization*. New York: Marcel Dekker, 1969.
5. Blackley, D. C. *Emulsion Polymerization*. New York-Toronto: John Wiley and Sons, 1975.

6. Guerin, J. D., Hutton, T. W., Miller, J. J., and Zdanowski, R. E. U. S. Patent 4,152,189 (1979) (assigned to Rohm and Haas Co.).
7. Harkins, W. D. *J. Am. Chem. Soc.* **69**: 1428 (1947).
8. Smith, W. V., and Ewart, R. H. *J. Chem. Phys*, **16**: 592 (1948).
9. Gardon, J. L. *Rub. Chem. Technol.*, **43**: 74 (1970).
10. Gershberg, D., A.I.Ch.E. *I. Chem. E. Symposium Series No.* 3, 4–15 (1965). Inst. Chem. Engrs. London.
11. Wolf, H., *et al.* German Patent 1,163,545 (1964) (assigned to Badische Anilin and Soda Fabrik AG).
12. Sheetz, D. P. U. S. Patent 2,914,499 (1959) (assigned to Dow Chemical Co.).
13. Mills, T. L., and Yocum, R. H. *J. of Paint Technol.* **39** (512): 532–535 (1967).
14. Silver, S. F., Winslow, L. E., and Zigman, A. R. U. S. Patent 3,922,464 (1975) (assigned to Minnesota Mining and Manufacturing Co.).
15. Fallwell, W. F. U. S. Patent 3,395,131 (1968) (assigned to Monsanto Co.).
16. Baatz, J. C., and Corey, A. E. U. S. Patent 4,012,560 (1977) (assigned to Monsanto Co.).
17. Greene, B. W., Sheetz, D. P., and Filer, T. D. *J. Colloid and Interface Sci.* **32**: 90 (1970).
18. Greene, B. W., and Sheetz, D. P., *J. Colloid and Interface Sci.* **32**: 96 (1970).
19. Greene, B. W., and Saunders, F. L. *J. Colloid and Interface Sci.* **33**: 393 (1970).
20. Willis, J. M. *Ind. Eng. Chem.* **41**: 2272 (1949).
21. Perry, W. M. U. S. Patent 3,166,525 (1965) (assigned to General Aniline and Film Corporation).
22. Samour, C. M. Canadian Patent 776,184 (1968) (assigned to Kendall Co.).
23. Roe, C. P. in *Polymer Colloids* (R. M. Fitch ed.) pp. 139–152, New York: Plenum Press, 1971.

# Chapter 30

# Radiation Curing

**R. Dowbenko**
*PPG Industries, Inc.*
*Allison Park, Pennsylvania*

Pressure-sensitive adhesives are made by irradiating an uncured mixture of polymerizable compounds either by ultraviolet light (UV) or by electron beam (EB) to form a polymeric film of an adhesive. For short, they are called radiation adhesives or UV (EB) adhesives. They are intended to be formed directly on the substrate on which they will be used. Many intermediate steps employed in the manufacture of conventional pressure-sensitive adhesives are eliminated. In some cases, the radiation-curable adhesives are formed on release paper. This way of preparation still retains the advantages of these adhesives over the conventional pressure-sensitive adhesives.

The interest in radiation pressure-sensitive adhesives derives from several intrinsic advantages of these materials. The most important properties of these materials involve the abatement of air pollution because no solvents (or very reduced amounts of them) are used, savings in manufacturing cost because of elimination of solvents, as well as simplification of the manufacturing process. Moreover, the performance characteristics are improved, which is an additional reason for a high degree of interest in these materials. Other special features of these UV and EB adhesives will become apparent later on in this chapter.

Preparation of radiation pressure-sensitive adhesives is very simple in concept. As mentioned, the formation of the adhesive occurs directly on the substrate to be used or on the release paper. In a most direct and simple form, a mixture of monomers is applied in a film thickness which is used with conventional adhesives and is then irradiated by either ultraviolet light or by an electron beam. The adhesive is formed directly on the substrate to be used later as desired. The operation is usually a continuous one on a web of a release paper or another substrate, or by coating and curing discrete objects on a conveyor belt, such as adhesive-backed floor tiles. This is the essential difference in manufacturing of radiation and conventional pressure-sensitive adhesives—a separate polymerization step in a reactor, as with conventional adhesives, is avoided, and the materials are handled and processed as essentially 100% solids materials. This latter fact probably constitutes one of the most important factors which is responsible for the high degree of interest accorded to radiation adhesives.

Despite the interest, only one review of this field of work is available.[1] Yet the knowledge has steadily increased over the years to the point that a broader review for those working in this field and for others who have a tangential interest seems appropriate.

The work on radiation-curable pressure-sensitive adhesives is thoroughly practical and goal-oriented. Radiation pressure-sensitive adhesives are in a way an outgrowth of radiation-cured polymers in general, and of coatings in particular. Intrinsically, they present no great theoretical chemical interest, but are thoroughly practical. For this reason, as is to be expected, all of the published information in this field originates from nonacademic organizations, mostly from the industrial laboratories of corporations with some interest in pressure-sensitive adhesives and allied fields, be it from the direct involvement in this business, as a raw material supplier, as a manufacturer of radiation equipment, or for other reasons. The available information is found in the patent literature or in the periodicals dealing with applied industrial chemistry. These two types of literature are the primary sources of information which have been used in this chapter.

## ULTRAVIOLET AND ELECTRON BEAM POLYMERIZATION

As mentioned earlier, radiation pressure-sensitive adhesives are an outgrowth of a much larger and growing endeavor, that of radiation polymerizable materials, in particular, of radiation-cured coatings.[2] The principles are the same for polymerization or crosslinking of polymers by ultraviolet light or by electron beam, and, just as in conventional solution (or emulsion) polymerization, the characteristics of the final product depend essentially on the choice of monomers used and the process conditions.

The concept of radiation pressure-sensitive adhesives can be described as follows. Whether in UV or in EB, the process depends on polymerizability of classes of compounds which are reactive when irradiated by these two types of energy, and on the formation of a polymer of a reasonably high molecular weight by a free radical initiated addition polymerization. Therefore, while there are many other factors to which attention must be paid, the problem is reduced to an efficient, high-conversion addition polymerization.

Many books have been devoted to both the theoretical and practical aspects of UV and EB polymerization.[3,4]

## FREE RADICAL POLYMERIZATION

Free radical initiated polymerization occurs when a monomer, such as 2-ethylhexyl acrylate, is subjected to the action of free radicals which are derived from a free-radical initiator, such as benzoyl peroxide or azobis(isobutyronitrile). There are several key steps in the chemical mechanism of this process, the first being the decomposition of the initiator, benzoyl peroxide, to form free radicals (1). Next, the free radical, which has been formed as a result of the decomposition of the initiator, reacts with a molecule of a monomer, such as 2-ethylhexyl acrylate or methyl methacrylate, to form an intermediate (2) which reacts further (3) with additional monomer molecules in a propagation reaction. This propagation or polymer growth reaction continues with numerous monomer molecules until the polymeric free radical encounters a molecule which has a hydrogen atom that is able to react with the polymeric radical. When that happens, a termination occurs (4), and an inactive polymer molecule is formed which theoretically does not continue to undergo polymerization. These four reactions, roughly speaking, describe the intimate steps in the free radical addition polymerization process.

$$R'-R' \longrightarrow R'\cdot + R'\cdot \tag{1}$$

$$R'\cdot + CH_2 = CHR \longrightarrow R'-CH_2-\underset{\bullet}{C}HR \tag{2}$$

$$R'-CH_2-\underset{\bullet}{C}HR + n(CH_2 = CHR) \longrightarrow R'-(CH_2-CHR)_n-CH_2-\underset{\bullet}{C}HR \tag{3}$$

$$R'-(CH_2-CHR)_n-CH_2-\underset{\bullet}{C}HR \xrightarrow{R''H} R'-(CH_2-CHR)_n-CH_2CH_2R + R''\cdot \tag{4}$$

Radiation polymerization is in principle completely analogous to the conventional addition polymerization with one important difference: a different compound is used to initiate the polymerization in UV and in EB.

In the case of UV, a photoinitiator is used. Its function is completely analogous to the free radical initiator. Both furnish free radicals which are necessary for initiation of polymerization, but with a photoinitiator, the process of radical formation is different. In contrast to the conventional initiators which possess a thermally labile bond which is cleaved to form free radical species, the photoinitiator has a bond which breaks upon absorption of radiant energy. It forms free radical species which initiate polymerization by an addition reaction to a monomer molecule and then continues by propagation and eventual termination. The photoinitiator is therefore one of the key components in a UV polymerization, and the outcome of such a polymerization is critically dependent on the choice of the photoiniator, including its chemical nature and the amount.

As mentioned earlier, both the UV and EB polymerizations occur essentially by free radical processes. In case of UV polymerization, the energy required to break the bond is associated with a wavelength of about 300 nm and is equivalent to approximately 4 eV. This light is absorbed by a photoinitiator which furnishes a reactive free radical species. No such initiator is required in EB polymerization. The passage of energetic electrons through organic media can lead to production of free radical species without using photoinitiators.[5] Although other types of reactions occur in EB polymerization (ion formation, rearrangement, crosslinking), free radical reactions are considered to be the most important in polymerizations of coatings and adhesive films.[6]

## PHOTOINITIATORS

As mentioned above, a photoinitiator is one of the important and necessary constituents in UV polymerizations of pressure-sensitive adhesives. Little or no polymerization will take place without a photoinitiator, and if it does, it will proceed at impractically slow speeds. For this reason, the activity of a photoinitiator is one of the more important properties which must be considered when choosing a photoinitiator, since the speed of polymerization of pressure-sensitive adhesives is one of the important advantages of these adhesives. In addition to the activity of a given photoinitiator, other properties must be kept in mind. Of the more important ones, the following must be mentioned: storage stability, color, cost and availability. These are self-explanatory properties and requirements, but for a successful UV polymerization, all of them must be considered.

Of the many photoinitiators that can be used in UV polymerizations, only those classes that are readily available and that have achieved commercial importance will be discussed here.

### Benzoin Ethers

These readily available compounds, shown in the accompanying formula, undergo a fragmentation on UV irradiation to form two free radical species:

$$C_6H_5-\overset{\overset{O}{\|}}{C}-\overset{\overset{OR}{|}}{CH}-C_6H_5 \xrightarrow{h\nu} C_6H_5-\overset{\overset{O}{\|}}{C}\cdot + \cdot\overset{\overset{OR}{|}}{CH}-C_6H_5$$

Of the two radical species, the benzoyl radical is the more effective radical in UV polymerization, although the substituted benzyl radical[7,8] also initiates some polymerization. All of the lower alkyl ethers, especially those of secondary alcohols, can be used, as well as benzoin itself (R = H in the formula above), but whereas the ethers have in general good storage stability, benzoin has poor storage stability and is almost never used in practical systems where storage stability is of importance.

## Benzyl Dialkyl Ketals

The well-known 2,2-dimethoxy-2-phenylacetophenone is a very efficient photoinitiator and decomposes on UV irradiation according to the following scheme:

$$Ph-\overset{\overset{O}{\parallel}}{C}-\underset{OCH_3}{\overset{OCH_3}{\overset{|}{\underset{|}{C}}}}-Ph \xrightarrow{h\nu} Ph-\overset{\overset{O}{\parallel}}{C}\cdot + \cdot\underset{OCH_3}{\overset{OCH_3}{\overset{|}{\underset{|}{C}}}}-Ph \longrightarrow CH_3\cdot + CH_3O-\overset{\overset{O}{\parallel}}{C}-Ph$$

with the dimethoxybenzyl radical further fragmenting to methyl benzoate and the methyl radical.[9] Compositions containing this photoinitiator have fast cure and good storage stability.

## Benzophenone

On irradiation, this compound undergoes a conversion to the singlet benzophenone, followed by a conversion to the triplet excited state. The triplet then abstracts a hydrogen atom from a different compound to form a benzhydryl free radical which is an active initiating species.

$$Ph-\overset{\overset{O}{\parallel}}{C}-Ph \xrightarrow{h\nu} S_1 \longrightarrow T_1 \longrightarrow Ph-\underset{\bullet}{\overset{\overset{OH}{|}}{C}}-Ph$$

The activity of benzophenone can in general be enhanced by addition of small amounts of amines.[10] Benzophenone is an inexpensive photoinitiator and can be used in mixtures with the benzoin ethers and with 2,2-dimethoxy-2-phenylacetophenone.

Michler's ketone is a substituted benzophenone which initiates polymerization in a similar manner. It can also be used together with the photoinitiators discussed above, but its main disadvantage is its dark color.

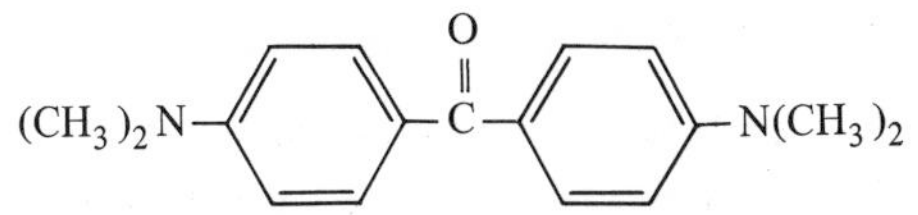

$$(CH_3)_2N-C_6H_4-C(=O)-C_6H_4-N(CH_3)_2$$

### Acetophenone Derivatives

Although acetophenone itself is not an effective photoinitiator, substituted derivatives of this compound are effective. The best known of these, 2,2-diethoxyacetophenone,[11] undergoes fragmentation

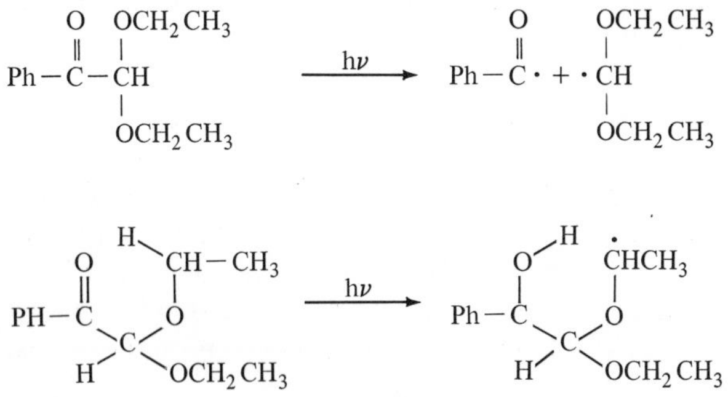

$$Ph-C(=O)-CH(OCH_2CH_3)_2 \xrightarrow{h\nu} Ph-\overset{O}{\overset{\|}{C}}\cdot + \cdot CH(OCH_2CH_3)_2$$

$$PH-C(=O)-CH(H)(O-CH(H)-CH_3)(OCH_2CH_3) \xrightarrow{h\nu} Ph-C(OH)-CH(H)(O-\dot{C}HCH_3)(OCH_2CH_3)$$

and gives good initiation of UV polymerization. The initiator has particularly good storage stability and can also be used in conjunction with other photoinitiators. Some chloro-substituted acetophenones are useful photoinitiators.

## RADIATION CURE EQUIPMENT

### UV Radiation Sources

The most useful and widely used equipment for UV curing consists essentially of a mercury vapor arc enclosed in a tubular quartz on glass envelope[13] which usually is provided with an envelope for cooling for proper operation. Although mercury arcs with as low a pressure of mercury as a few millimeters and as high as tens of atmospheres are available, the type of the mercury lamp that enjoys the widest use in UV curing of adhesives and coatings is the medium pressure mercury lamp, which operates with a pressure of 1 to 2 atm. of mercury and at about 200 watts/in. of length. Most lamps come in various lengths and can be purchased together with a housing which includes a reflector, usually of highly polished metal, which may be used directly on a conveyor line.

## Electron Beam Generators

There are currently available two types of electron beam generators, scanned and linear.[14] The former operates by energizing a tungsten filament to emit a stream of electrons which are typically accelerated through 300–500 kV in a vacuum. The beam issues through a thin metal window and is magnetically scanned through the length of the line. This type of equipment requires thick shielding and constant caution to prevent exposure to people.

The linear accelerator is highly efficient and compact, requires much less shielding and comes in small and compact units[15,16,17,18] with an effective length of as little as 1.7 m. The linear filament system is of lower energy than the scanned beam system, but it is quite adequate for curing thin coatings of organic materials.

## Measurement of UV Intensity and Dose Rate

Among the many variables that can affect the outcome of radiation curing is the intensity and the rate of deposition of radiant energy into the adhesive composition to be cured. There are some commercially available on-line devices to measure the intensity of UV radiation.[19] The system should also be periodically checked by curing a known composition and measuring its properties.

In electron beam curing, the dose rate is measured by megarads (Mrad) per second (one Mrad is energy absorbed equal to 10 joules). It can also be estimated by similar devices measuring the amount of color increase in radiation-sensitive films. Here also it is advisable to check the equipment by curing a composition with a known rate.

## Atmosphere in Radiation Curing

There are several types of compositions known, especially in the area of coatings, that can be effectively cured in air atmosphere. This is especially true of compositions polymerizing by ionic mechanisms (which will not be discussed here since such pressure-sensitive adhesives apparently are unknown). Most free radical polymerizations are more or less strongly inhibited by molecular oxygen. Propagating free radicals react with oxygen to form a peroxy radical which is useless for further propagation.

$$\cdots CH_2 \cdot + O_2 \longrightarrow \cdots CH_2-O-O \cdot$$

The effect is to produce considerably lower molecular weight species, especially on the surface where subsequently the adhesive bond is formed.

Because of the lower molecular weight, such species contribute much poorer adhesive properties. For this reason, many of the radiation-cured (both UV and EB) adhesives are made in an inert gas atmosphere, most commonly in the relatively inexpensive nitrogen. For the present, this practice must be continued until some solution is found to counteract the polymerization inhibition by oxygen.

### A Generalized Procedure for Radiation Curing

For a better appreciation of the experimental procedure, it will be instructive to give at this point an outline of a generalized experimental process for producing radiation-cured adhesives in a laboratory. A mixture of suitable monomers which have been selected is prepared together with additives as may be required (tackifying agents, UV stabilizer, chain transfer agents, etc.). The photoinitiator, at a level of about 0.5–3.0% based on the composition, is then dissolved in the mixture. An additive that deserves special mention is some kind of a viscosity control agent since most monomers are low viscosity liquids which are difficult to apply to a substrate in a continuous film of uniform thickness. The viscosity control agent may be any high molecular weight material which dissolves in the monomer mixture to give increased viscosity. The amount of such a polymer used is not critical, but mainly depends on the degree of viscosity increase which in turn depends on the method of application of the composition.

Next the liquid composition is deposited on a suitable substrate (release paper, plastic film or metal), usually by a draw bar, and the film is then subjected to UV or EB radiation by sending it into the appropriate piece of equipment on a conveyor belt and by noting the operating conditions of the equipment. The cured adhesive sample is then subjected to appropriate adhesives tests.

## POLYMERIZABLE COMPOSITIONS FOR RADIATION PRESSURE-SENSITIVE ADHESIVES

It will be useful to present at this point an overview of the types of compositions that yield pressure-sensitive adhesives by UV polymerization, and to make some general statements about them. In this way, it will be easier for the reader to follow the review of the literature which is given below.

First of all, it may be fairly accurate that the compounds that are useful in UV polymerizations to prepare pressure-sensitive adhesives are also those that are used in conventional solution or emulsion polymerization to prepare pressure-sensitive adhesives. Thus, simple acrylic and vinyl esters are most commonly used in conventional adhesives. As with the latter compounds, those having longer alkyl groups ($C_4$ and higher) give the best balance of properties in the UV products. Methacrylates are used rarely in UV polymerization because of their slow rates of polymerization. Maleates, fumarates, styrene and substituted styrenes fall in the same category. Another important factor is that UV polymerization for pressure-sensitive adhesives nearly always occurs in films of less than about 1 mil. Therefore, volatile monomers, such as vinyl acetate or ethyl acrylate, are used only rarely, because they usually evaporate faster than they can be polymerized. For this reason, volatile compounds are in general unsuitable for preparation of UV pressure-sensitive adhesives.

In addition to the simple acrylic and vinyl esters which contain only a polymerizable group and an ester group, other monomers with different functionality can be used. Thus, ethers, amides, carbamates, hydroxyacrylates, acids and other functional compounds can be used in UV pressure-sensitive adhesives where special properties are desired.

Compounds containing two or more allyl groups present a special case in that they undergo an addition with di- or trithiols as shown to form

$$CH_2=CHCH_2 \cdots CH_2CH=CH_2 + HS \cdots SH \longrightarrow [SCH_2CH_2CH_2 \cdots CH_2CH_2CH_2S]_n$$

polymers which, with the proper choice of backbones that carry the allyl and the mercaptan groups, can give useful pressure-sensitive adhesives.[12] The processing of such compositions in UV polymerization is completely analogous to polymerizations with acrylic compounds, including the necessity of a photoinitiator.

The possibilities of raw materials to prepare UV pressure-sensitive adhesives are indeed numerous and can encompass nearly every type of compound. In contrast to the solution or emulsion pressure-sensitive adhesive polymers to which additives (such as tackifiers, UV and heat stabilizers, etc.) are usually added in a separate formulating step, the composition which will form a pressure-sensitive adhesive by radiation polymerization must contain all the additives before polymerization. This is obvious since little can be done to add another component to a thin film of a pressure-sensitive adhesive once it has been formed. Therefore, all formulation to produce a complete final product must be done before the composition is polymerized by radiation.

Functional monomers of all kinds may be used in radiation pressure-sensitive adhesives. In addition to this, it is often useful, as it is in UV-cured coatings, to use compounds containing two acrylic (or vinyl) double bonds. Although this practice must be used carefully in conventional solution adhesive polymers since it would lead to unwanted gelation, in UV-cured pressure-sensitive adhesives this approach is often used to counteract the generally lower molecular weight obtained by radiation polymerization and to promote the formation of higher molecular weight polymers. In radiation-cured adhesives, the finished product is often formed in a fraction of a second, favoring low molecular weight compounds. Other components (additives) in the composition may not favor the formation of high molecular weight polymers.

From the point of view of the types of compositions reported, as well as from the conceptual viewpoint, the published information can be divided into two broad categories. The first category comprises essentially compositions that do not have pressure-sensitive adhesive properties in their own right in the uncured state. Such compositions generally contain one or more polymers dissolved in one or more monomers as well as the initiator and any special additives. They may be conveniently termed pressure-sensitive adhesive syrups.

The second broad class of radiation pressure-sensitive compositions comprises materials whose adhesive properties are to be improved by irradiation. Such compositions have some measure of adhesive properties even before irradiation. The compositions in this broad category are applied as solutions in a volatile solvent or as aqueous dispersions and are irradiated, after the evaporation of the solvent, or they may be lower molecular weight hot melt adhesives. In addition to the photoinitiator, such compositions may include reactive functional groups for attainment of properties as well as small amounts of di- and trifunctional monomers for building up the molecular weight.

## ADHESIVE SYRUPS

One of the simplest ways to prepare a pressure-sensitive adhesive is to cure a mixture of acrylic monomers. A mixture of an alkyl acrylate or a methacrylate with less than 10 carbon atoms in the alkyl group and 0.2 mole or less of a polar vinyl monomer (containing carboxyl, cyano, keto or amido groups) has been polymerized with ionizing radiation (EB, gamma rays and other sources) at a dose of about 5 Mrads to produce pressure-sensitive adhesives.[20] The peel adhesion and shear adhesion of the radiation adhesives were in most cases higher than those of conventional adhesives (750 g/cm and 10 hr vs. 600 g/cm and 8 hr). The polar monomer is said to be important for attaining good adhesive properties. Compositionally similar adhesives are prepared by EB cure at about 7 Mrads.[21]

Nearly as simple a method, both conceptually and in practice, is to cure a mixture of a polymer and a monomer.[22] Among the monomers used are simple alkyl acrylates and methacrylates as well as hydroxyalkyl, alkoxyalkyl, cyanoalkyl acrylates and methacrylates, and substituted amides and maleates and fumarates. Homopolymers of the corresponding acrylates, methacrylates and substituted acrylamides and methacrylamides were used at levels between 2–50%, and the compositions were cured at doses between 2 and 6 Mrads to produce pressure-sensitive adhesives. No adhesive properties were listed, but another report[23] describes similar compositions which are cured by both EB and UV (with conventional photosensitizers) to produce pressure-sensitive adhesives. Because of the generally excellent adhesive properties of these materials as compared with the properties of conventional solvent adhesives, it is of interest to quote them here (Tables 30-1 and 30-2).

In the adhesive cured by EB, high peel and high tack combined with high shear strength are especially noteworthy as such properties are difficult to achieve in conventional pressure-sensitive adhesives.

Other work is based on the use of compositions containing polyvinyl ethers and acrylic monomers, both mono and polyfunctional, together with a light initiator.[24] Even though the main claim of the patent encompasses compositions containing both mono and polyfunctional acrylates, one of the best adhesives prepared (peel adhesion 570 g/cm, shear time 60 hr.) used neopentyl glycol diacrylate but no monofunctional monomer. Urethane acrylates made from polycaprolactone diol, isophorone diisocyanate and hydroxyethyl acrylate are also used in this work. Very similar compositions

**Table 30-1. Performance of an EB syrup adhesive.**

**Conditions of Cure: 10 Mrads, nitrogen atmosphere containing 250–300 ppm of oxygen cured on 2 mil Mylar film**

| | | | EB ADHESIVE | CONVENTIONAL ADHESIVE |
|---|---|---|---|---|
| Peel adhesion, stainless steel (g/cm) | Initial | 20 min. | 396 | 318 |
| | | 24 hr. | 735 | 643 |
| | 7 days roll storage @ 70°C | 20 min. | 299 | 404 |
| | | 24 hr. | 648 | 579 |
| Quick stick (Grams, triplicate values) | Initial | | 780 | 480 |
| | 7 days roll storage @ 70°C | | 360 | 370 |
| Dead load strength (hr., 0.91 kg. weight) | Initial | | 9.5 | 4 |
| | 7 days roll storage @ 70°C | | 9 | 4 |

**Table 30-2. Performance of a UV syrup adhesive.**

Conditions of Cure: $N_2$ flow 1 $m^3$/min; lamp intensity 333 W/in., conveyor speed 30 m/min, 12 passes

| | | | UV ADHESIVE | CONVENTIONAL ADHESIVE |
|---|---|---|---|---|
| Peel adhesion, stainless steel (g/cm) | Initial | 20 min. | 401 | 362 |
| | | 24 hr. | 541 | 386 |
| | 7 days roll storage @ 70°C | 20 min. | 145 | 75 |
| | | 24 hr. | 240 | 132 |
| Percent shrinkage, 13 × 13 cm vinyl/steel, 7 days @ 70°C | Machine direction | | 0.5% | 1.05% |
| | Cross direction | | 0.0% | 0.55% |

from 2-ethylhexyl acrylate, acrylic acid and polyvinyl isobutyl ether are described in an earlier patent.[25]

The utility of polyvinyl alkyl ethers, especially polyvinyl ethyl ether, is shown in further work of Steuben *et al.*[26] where, in addition to conventional liquid acrylic monomers, such as cyclohexyl, 2-ethoxyethyl and isobornyl acrylates, an unusual monomer, 2-(*N*-methylcarbamyloxy)ethyl acrylate, is used to dilute compositions for UV or EB cure. Also noteworthy is an exhaustive list of photoinitiators which include, to name a few, 2,2-diethoxyacetophenone, 2,2-dimethoxy-2-phenylacetophenone, 4-bromoacetophenone, benzaldehyde, benzoin, dibenzosuberone, xanthone and benzophenone, the initiators, preferably benzophenone, to be used in combination with amine activators, such as tributylamine, methyldiethanolamine, piperidine, and others. Many adhesive properties of various compositions are given, but one of the compositions containing 62.5% polyvinyl ethyl ether, 37.5% 2-(*N*-methylcarbamyloxy)acrylate and 3 phr of benzophenone gave, after 6 sec of UV cure under the conditions given, peel adhesion of 652 g/cm and shear adhesion of more than 140 hr.

The use of 2-(*N*-methylcarbamyloxy)ethyl acrylate as a diluent monomer in UV-cured adhesives is described in another patent[27] with polyoxyethylene/polyoxypropylene copolymers. The latter is said to require at least 40% of the oxyethylene units for best adhesive properties. Comparative examples with either polyoxypropylene or polytetramethylene oxide homopolymers in place of the copolymers claimed, or with isobornyl acrylate substituted for 2-(*N*-methylcarbamyloxy)ethyl acrylate of the claims, gave poor adhesive properties. It is noteworthy that the cure of the compositions of this patent can be carried out in an inert atmosphere or in air, a result which can perhaps be accounted for by the facilitation of polymerization by the oxyalkylene copolymers by virtue of formation of peroxy ether radicals in the presence of oxygen in the air.

The use of acrylic/vinyl acetate copolymers in conjunction with monofunctional and trifunctional acrylate monomer (2-ethylhexyl acrylate and trimethylolpropane triacrylate) is exemplified in another patent[28] on UV cure, and also in a different one[29] where, in addition to the acrylic/vinyl acetate copolymer, monomers with carboxyl, cyano, and amido groups are included in EB cure at doses between 3–35 Mrads. The use of copolymers of 2-ethylhexyl acrylate and vinyl acetate for adjusting viscosity (at least 100 cps) of liquid acrylic monomers, also containing crosslinkers, is exemplified by EB cure of such compositions at 8–10 Mrads in absence of oxygen.[30]

A different approach to radiation-curable adhesive syrups is based on the so-called thiolene reaction[12] in which a di- or trithiol adds in the presence of photoinitiators a polyallyl compound to form pressure-sensitive adhesive polymers. Polythiols can be prepared conveniently by the reaction of $\beta$-mercaptopropionic acid with a polyol to form a $\beta$-mercaptopropionate, and the polyallyl compounds, the second constituent in the syrup, from an isocyanate terminated prepolymer, and allyl alcohol. Then, to form an adhesive, a diallyl compound, prepared from a mixture of polypropylene and polyethylene glycols, toluene diisocyanate and allyl alcohol, is mixed with pentaerythritol tetra($\beta$-thiopropionate) and a small amount of benzophenone. After curing with a sun lamp for 20 sec, the resulting adhesive had a peel adhesion between 187 and 464 g/cm. Another adhesive is reported to have a peel adhesion of 630 g/cm.[12]

## ADHESIVE POLYMERS FOR RADIATION CURE

The volume of work which has been recorded in the literature with respect to radiation cure of essentially polymeric materials for the purpose of improving their properties is much larger than the volume reported on adhesive syrups.

One of the earliest references in radiation cure[31] describes EB irradiation at 1–24 Mrads of polymers such as polyacrylates, polyethers and rubber-tackifier blends to obtain pressure-sensitive adhesive tapes with improved high temperature shear adhesion. Similar work is described in another patent[32] where solutions containing a photoinitiator and elastomers, such as crepe rubber, styrene/butadiene rubber, 2-ethylhexyl acrylate/vinyl acetate copolymer, polyvinyl ethyl ether, and others are irradiated in UV, after evaporation of the solvent, to improve cohesive and shear strengths of adhesives.

A different approach, although still working with essentially "nonfunctional" polymers, is exemplified by another patent[33] in which a conceptual departure is made from previous work in that solvent-free, 100% solids hot melt adhesives are irradiated to produce pressure-sensitive adhesives with substantially improved adhesive properties. Predominantly acrylic polymers, with a Williams' plasticity number of up to about 1.5 mm and a melt viscosity (at 75°C) between 500–100,000 cps, are applied as hot melts and irradiated at doses between 0.5–12 Mrads. Compositionally, most of them are 2-ethylhexyl acrylate and vinyl acetate copolymers with small amounts of *N,N*-dimethylaminoethyl methacrylate, diacetone acrylamide, *N*-vinylpyrrolidone, or acrylic acid. Polyfunctional acrylates, such as trimethylolpropane trimethacrylate, and thermal stabilizers can also be used.

Since the data are extensive,[23] it will be instructive to quote them completely for both the EB and the UV adhesives (Tables 30-3 and 30-4).

**Table 30-3. Performance of an EB hot melt adhesive.**

| | | | EB ADHESIVE | CONVENTIONAL ADHESIVE |
|---|---|---|---|---|
| Plasticity No., original (mm) | | | 1.34 | — |
| Plasticity No., cured (mm) | | | 2.55 | — |
| Weight (mg/cm²) | | | 1.6 | 1.6 |
| Peel adhesion | Initial | 20 min. | 429 | 358 |
| (g/cm) | | 24 hr. | 618 | 382 |
| | 7 days roll | 20 min. | 185 | 74 |
| | storage @ 70°C | 24 hr. | 280 | 130 |
| Shrinkage, 13 × 13 cm | Machine direction | | 0.42% | 1.05% |
| vinyl/steel, | | | | |
| 7 days @ 70°C | Cross direction | | 0.37% | 0.55% |

**Table 30-4. Performance of a UV hot melt adhesive.**

| | | | UV ADHESIVE | CONVENTIONAL ADHESIVE |
|---|---|---|---|---|
| Plasticity No., original (mm) | | | 0.74 | — |
| Plasticity No., cured (mm) | | | 2.78 | — |
| Film weight (mg/cm²) | | | 1.9 | 1.7 |
| Peel adhesion | Initial | 20 min. | 354 | 358 |
| (g/cm) | | 24 hr. | 421 | 382 |
| | 7 days roll | 20 min. | 134 | 74 |
| | storage @ | 24 hr. | 242 | 130 |
| | 70°C. | | | |
| Shrinkage, 13 × | Machine direction | | — | 1.05% |
| 13 cm. vinyl/steel, | | | | |
| 7 days @70°C. | Cross direction | | — | 0.55% |

It can readily be seen that remarkable improvements in the properties of hot melt pressure-sensitive adhesives can be made by irradiating them with EB and UV, but especially with EB.

Work very similar to the above irradiation of hot melt adhesives is described in a patent by Skoultchi and Davis.[34] Compositions, such as copolymers of vinyl acetate/2-ethylhexyl acrylate or vinyl acetate/ethyl acrylate/octyl acrylate, of Williams' plasticity numbers between 1.2–5.5 are applied as hot melts or as solutions and then are irradiated in the presence of polynuclear quinones, such as 1, 2−,and 1,4-naphthoquinone, phenanthraquinone and others. Improvements in adhesive properties are noted, for example, a 180° peel adhesion of 257 g/cm and shear adhesion of more than 48 hr. (versus peel adhesion of 313 g/cm and hold strength of 12 hr.). The work from the same laboratory which was patented earlier[35] uses similar polymers (e.g., vinyl acetate/octyl acrylate), but with a copolymerized acrylic photoinitiator which forms an integral part of the polymer. The photoinitiators of this work are prepared from hydroxyl or carboxyl groups carrying benzophenones or anthraquinones with glycidyl acrylate or methacrylate. Improvements in adhesive properties obtained are similar to those given in Reference (34). Thus, peel adhesion of 1390 g/cm and 12 hr shear adhesion is obtained after irradiation of 8 min with a UV lamp (275 watts) of a polymer which before irradiation had peel adhesion of 1474 g/cm and a shear adhesion of 4.5 min.

A further variation of the copolymerized photoinitiator approach is presented in the use of the reaction products of polychlorophenyl glycidyl ethers (such as the pentachlorophenyl ether) and acrylic or methacrylic acids.[36] Such polymerizable photoinitiators are copolymerized with the usual adhesive-producing monomers, such as methyl acrylate/2-ethylhexyl acrylate, vinyl acetate and others, to produce polymers which are applied as hot melts and irradiated by UV. One of the advantages of these compositions is said to be the constancy of adhesive properties over a wide range of cure conditions. In other words, they are difficult to overcure with the resulting loss of adhesive properties. That the chlorophenyl compounds are indeed active photoinitiators is shown by omitting them from a polymer (in which case the adhesive properties are essentially unchanged on irradiation).

Other work with polymerizable photosensitizers[37] includes the use of benzoin acrylate which is copolymerized with a variety of monomers, such as 2-ethylhexyl acrylate, butyl acrylate, tetrahydrofurfuryl acrylate, acrylamide, acrylic acid, and others, and then irradiated for varying periods of time. Voluminous adhesive data are given to show improvements in adhesive properties on irradiation. An earlier patent assigned to the same company also uses benzoin acrylate as a copolymerized constitutent in the copolymer.[38]

Other work to improve the adhesive properties of acrylic pressure sensitive adhesive polymers includes co-curing of an acrylic copolymer by EB, together with the usual additives, with polyfunctional acrylates or methacrylates.[39]

Similar improvements can be obtained in an ethylene/vinyl acetate copolymer on co-curing it by EB with polyol acrylates or methacrylates and with other additives.[40,41]

A departure from the work described is to introduce a reactive acrylic bond into an acrylic copolymer by means of the reaction of glycidyl methacrylate with the carboxyl groups on the polymer backbone and to cure the resulting "activated" copolymer.[42] Thus a copolymer from 2-ethylhexyl acrylate, ethyl acrylate and methacrylic acid, post-reacted with glycidyl methacrylate and cured by UV light gave the 180° peel adhesion of 846–984 g/cm.

Improvement of pressure-sensitive adhesive properties can also be obtained with compositions based on polybutadienes and modified polybutadienes. Thus, a patent[43] describes EB curing (at about 2 Mrads) of a mixture of an acrylic acid modified polybutadiene and nitrile rubber, together with 2-ethylhexyl acrylate and acrylic acid, as well as other additives. The properties of the improved adhesive included peel adhesion of 412 g/cm and shear adhesion of 40° for more than 3 hr. Two other patents[44,45] describe similar compositions from polybutadiene, nitrile rubber, acrylic acid and other additives, such as phenol-terpene resin, which on irradiation with EB at about 10 Mrads gives improved tackiness and creep resistance. Natural rubber and styrene/butadiene rubber, together with the additives, can also be used to prepare pressure-sensitive adhesives by EB.[46,47] Similarly, the adhesive properties of isobutylene rubber adhesives with a tackifier resin can be improved by irradiating them with UV in the absence of oxygen.[48]

A-B-A block copolymers of a conjugated diene and styrene can also be used to prepare pressure-sensitive adhesives with improved properties. Kraton rubber, in conjunction with additives and polyfunctional acrylates such as trimethylolpropane triacrylate, can be used to produce improved adhesives by curing in EB (in an inert atmosphere) or in UV in the presence of the usual photoinitiators.[49]

## CONCLUSION

In spite of the considerable amount of work which has been published and in spite of the interest this evolving technology has engendered, there is no published information that radiation-cured pressure-sensitive adhesives are commercial. It is probable, however, that radiation curing processes are being used by companies that manufacture adhesives without any public announcements. At any rate, this developing field is far from being exhausted. Nevertheless, many problems still await solutions before these adhesives will achieve full significance.

The problem of application of these new materials, especially of the polymeric type, is far from being a trivial one. An analysis of the cost of these materials compared with the conventional ones must be made, and it must, of course, include the cost of the energy required for curing and capital expenditures for equipment. The future availability of solvents for conventional adhesives and possibly their cost increase must be taken into account, as well as the present (and future) pollution legislation. However, the most important factor in commercialization of these materials may likely be to adapt them to a specialty area where their performance advantages over those of the conventional adhesives will be so desirable that the user will be willing to pay the increased price.

## REFERENCES

1. Steuben, K. C. *Adhesives Age*, **20**(6): 16 (June 1977).
2. Pappas, S. P. ed. *UV Curing: Science and Technology*, Technology Marketing Corp., Stamford, Conn., 1978.
3. Oster, G., and Yang, N. L. "Photopolymerization of Vinyl Monomers," *Chem. Rev.* **68**: 125 (1968).
4. Calvert, J. A., and Pitts, J. N. *Photochemistry*, New York: John Wiley and Sons, 1966.
5. Charlesby, A. *Atomic Radiation and Polymers*, Oxford: Pergamon Press, 1960.
6. Tawn, A. R. H. *J. Oil Col. Chem. Assoc.* **51**: 782 (1968).
7. Ledwith, A., Russell, P. J., and Sultcliffe, L. H. *J. Chem. Soc., Perkin Transact.* **2**: 1925 (1972).
8. Ledwith, A. *J. Oil Col. Chem. Assoc.* **59**: 157 (1976).
9. Sandner, M. R., and Osborn, C. L. *Tetrahedron Letters*, 415 (1974).
10. Sandner, M. R., Osborn, C. L., and Trecker, D. J. *J. Polym. Sci.*, **10**: 3173 (1972).
11. Sandner, M. R., and Osborn, C. L. U. S. Patent 3,715,923 (February 6, 1973) (assigned to Union Carbide Corp.).
12. Barber, R. C., Bettacchi, R. J., Lundsager, B., and Wood, L. L. U. S. Patent 3,920,877 (November 18, 1975) (assigned to W. R. Grace and Co.).
13. A description of light sources is given by V. D. McGinnis in *UV Curing: Science and Technology*, Technology Marketing Corp., Stamford, Conn., 1978.
14. Ramler, W. J. Proceedings of the First International Radiation Curing Conference, Puerto Rico, 1976.
15. Nablo, S. V., Quintal, B. S., and Fussa, A. D. *J. Paint Technol.* **46**: 51 (1974).
16. Nablo, S. V., *et al.*, "Electron Beam, Processor Technology," in *Non-Polluting Coatings and Coatings Processes*, (Gardon, J. L., and Prane, J. W. eds.) New York: Plenum Press, 1973.
17. Hoffman, C. R. *Radiation Phys. Chem.* **9**: 131 (1977).
18. Linear Kathode EB Processor, Product Bulletin, PPG Industries, Inc.
19. UV Dosimetry Systems, Product Bulletin, PPG Industries, Inc.
20. Fukukawa, S., Shimomura, K., Ijichi, I., Yoshikawa, N., and Murakami, T. Ger. 2,134,468 (January 13, 1972) (assigned to Nitto Electric Industrial Co., Ltd.); also compare Yoshi-

kawa, N., Shimitzu, Y., and Sunagawa, M. Jap. Kokai 75 136,328 (October 29, 1975) (assigned to Nitto Electric Industrial Co., Ltd.).
21. Nishizaki, T., Nishimura, K., Okazaki, H., and Yamano, K. Jap. Kokai 75 64,329 (May 31, 1975) (assigned to Sekisui Chemical Co., Ltd.).
22. Dowbenko, R., and Christenson, R. M. U. S. Patent 3,897,295 (July 29, 1975) (assigned to PPG Industries).
23. Dowbenko, R., Christenson, R. M., Anderson, C. C., and Maska, R. *Chem. Technol.*, 539 (1974).
24. Steuben, K. C., Azrak, R. G., and Patrylow, M. F., Ger. 2,715,043 (October 13, 1977) (assigned to Union Carbide Corp.).
25. Bibost, P. Ger. 1,594,193 (April 15, 1971) (assigned to Novacel SA).
26. Steuben, K. C., Azrak, R. G., and Patrylow, M. F. Ger. 2,813,544 (December 10, 1978) (assigned to Union Carbide Corp.); also Steuben, K. C., Azrak, R. G., and Patrylow, M. F. U. S. Patent 4,165,266 (August 21, 1979) (assigned to Union Carbide Corp.).
27. Steuben, K. C. U. S. Patent 4,111,769 (September 5, 1978) (assigned to Union Carbide Corp.).
28. Barzynski, H., Marx, M., Storck, G., Druscheke, W., and Spoor, H. Ger. 2,357,486 (May 22, 1975) (assigned to BASF AG).
29. Yoshikawa, N., Shimizu, Y., and Sunagawa, M. Jap. Kokai 75 36,328 (October 29, 1975) (assigned to Nitto Electric Industrial Co., Ltd.).
30. Brookman, R. S., Grib, S., and Pearson, D. S. U. S. Patent 3,661,618 (May 9, 1972) (assigned to Firestone Tire and Rubber Co.).
31. Hendricks, J. O. U. S. Patent 2,956,904 (October 18, 1960) (assigned to Minnesota Mining and Manufacturing Corp.).
32. Aubrey, D. W. British Patent 866,003 (January 3, 1962) (assigned to Adhesive Tapes, Ltd.).
33. Christenson, R. M., and Anderson, C. C. U. S. Patent 2,131,059 (December 28, 1972) (assigned to PPG Industries), see also Reference 23.
34. Skoultchi, M. M., and Davis, I. J. U. S. Patent 4,069,123 (January 17, 1978) (assigned to National Starch and Chemical Corp.).
35. Skoultchi, M. M. and Davis, I. J. Ger. 2,411,169 (September 12, 1974) (assigned to National Starch and Chemical Corp.).
36. Pastor, S. D., and Skoultchi, M. M. U. S. Patent 4,052,527 (October 4, 1977) (assigned to National Starch and Chemical Corp.).
37. Guse, G., Lukat, E., and Schulte, D., Ger. 2,743,979 (April 5, 1979) (assigned to Beiersdorf AG).
38. Guse, G., Lukat, E., Jaucken, P., Lende, W., and Pietsch, H. Ger. 2,443,414 (March 25, 1976) (assigned to Beiersdorf AG).
39. Gleichenhagen, P. and Karmann, W. Ger. 2,455,133 (May 26, 1976) (assigned to Beiersdorf AG).
40. Karmann, W., and Gleichenhagen, P. Ger. 2,350,030 (April 17, 1975) (assigned to Beiersdorf AG).
41. Yoshikawa, N., Kamano, T., and Sunagawa, M. Jap. Kokai 51 046,331 (April 20, 1976) (assigned to Nitto Electric Industrial Co.).
42. Uehara, K., Murakoshi, S., and Hisamtsu, H. Jap. Kokai 74 05,145 (January 17, 1974) (assigned to Dainippon Ink and Chemicals, Inc.).
43. Sasaki, T., Araki, K., Kawaguchi, T., and Ishiyama, H. Jap. Kokai 76 26,940 (March 5, 1976) (assigned to Japan Atomic Energy Research Institute).
44. Sasaki, T., Araki, K., Kawaguchi, T., and Ishiyama, H. Jap. Kokai 76 26,941 (March 5, 1976) (assigned to Japan Atomic Energy Research Institute).
45. Sasaki, T., Araki, K., Kawaguchi, T., and Ishiyama, H. Jap. Kokai 76 26,942 (March 5, 1976) (assigned to Japan Atomic Energy Research Institute).

46. Fukukawa, T., Shimomura, T., Yoshikawa, S., and Sunagawa, M. Jap. Kokai 74 037,692 (December 4, 1975) (assigned to Nitto Electric Industrial Co.).
47. Nakata, S., Honi, H., and Ohnishi, K. Jap. 74 029,613 (August 6, 1974) (assigned to Sekisui Chemical Co., Ltd.).
48. Kasper, A. A., U.S. Patent 3,328,914 (June 27, 1967) (assigned to Kendall Co.).
49. Hansen, D. R., and St. Clair, D. J. U. S. Patent 4,133,731 (January 9, 1979) (assigned to Shell Chemical Co.).

# Name Index


Abbott, N., 199
Albrecht, A. H., 382
Alexander, B. B., 153, 167, 192, 218, 327, 369
Alliger, G., 297
Alner, D. J., 77
Altenau, A. G., 188
Anderson, C. C., 328, 603
Anderson, E., 425
Andrade, E. N., 94, 98
Andrews, E. H., 150
Anthony, I. B., 425
Apikos, D., 297
Araki, K., 603
Aranyl, C., 328
Askadskii, A. A., 150
Aubrey, D. W., 64, 65, 69, 77, 150, 203, 219, 311, 322, 329, 603
Ault, 403
Azrak, R. G., 603

Baatz, J. C., 329, 585
Bafford, R. A., 503, 532
Bailey, F. E., 329
Bailey, J., 150
Bailin, M. M., 418
Barber, R. C., 602
Bartell, C., 14, 372, 382, 404, 405, 411, 412, 413, 418
Bartenev, G. L., 126, 150
Barzynski, H., 603
Bates, A., 548, 557
Bates, R. J., 48
Bauer, R. F., 48
Bauer, W., 327
Beatty, J. R., 48
Beiersdorf, P., 3
Bemmels, C. W., 23, 206, 219, 437
Benson, T. D., 462
Berejka, A. J., 297
Bergstedt, M. A., 327
Bettacchi, R. J., 602
Bey, A. E., 403
Bibost, P., 603
Bikerman, J. J., 48, 65, 73, 74, 77, 112, 150
Billmeyer, F. W., 188
Birchwood, 418
Blackford, B. B., 219, 425
Blackley, D. C., 584
Blickensderfer, P. S., 548, 556
Blum, A., 372, 382
Boltzmann, 110
Bond, H. M., 329
Bonnslaeger, S. R., 407, 418
Booth, G. L., 503, 532, 564, 573
Boranian, A. G., 375, 382
Borthwick, J. B., 572
Bradford, E. B., 542, 556
Bradstreet, J. A., 383
Braeunling, E., 16, 99
Bright, W. M., 54, 76
Brinker, K. C., 369
Brock, M. J., 188
Brookman, R. S., 328, 603
Brooks, B. A., 327
Brown, F. W., 329
Brown, G. L., 542, 556
Brown, R. C., 437
Brunt, N., 48
Buckley, D. J., 296
Bueche, F., 110, 150
Bull, R. F., 48, 195, 219
Busse, W. F., 110, 150
Butler, G. L., 21, 189

Caimi, R. J., 375, 382
Callan, J. E., 297
Calvert, J. A., 602
Campbell, J. K., 403
Cantor, H. A., 327
Capeland, F. S., 425
Carlson, D. W., 188
Ceresa, R. J., 275
Chan, H. K., 63, 76
Chandra, 403
Chang, F. S. C., 48, 74, 77
Charlesby, A., 602
Christenson, R. M., 328, 603
Clark, F., 76
Cluett, S. L., 413, 418
Cochrone, W. D., 497
Coffman, A. M., 328
Coker, G. T., 369
Colburn, A. P., 556
Collins, W. C., 376, 382, 414, 418
Colquhoun, J. A., 403
Corey, A. E., 329, 585
Counsell, P. J. C., 199, 200, 219
Cox, E. R., 297
Craver, J. K., 275
Cristmas, H. F., 378, 383
Crocker, G. J., 48, 219, 327, 376, 382
Crone, J. W., Jr., 378, 383
Crozier, R. N., 497
Currie, C. C., 497
Curts, H. A., 323, 328

Dabroski, W. C., 372, 382, 432, 437
Da Costa, N. M., 343
Dahl, R., 24
Dahlquist, C. A., 22, 23, 24, 42, 48, 51, 54, 56, 69, 74, 76, 77, 78, 98, 195, 199, 219, 329, 372, 377, 382
Dalibor, H., 328
Dann, J. R., 48
Dannenberg, H. J., 57, 76
Das, S. K., 328
Davis, D. S., 383, 418
Davis, I. J., 327, 328, 329, 600, 603
Day, H. H., 3, 23, 189, 218
De Bastiani, N. P., 14, 463
De Bruyne, N. A., 70, 77, 150
De Groot von Arx, E., 24
De Jager, D., 220, 572
Demmig, H. W., 377, 382
Dennett, F. L., 403
Deryagin, B. V., 58, 76
De Walt, C., 48
Dexter, J. F., 352
D'Ianni, J. D., 343
Dickard, L. R., 497
Diefenbach, W. T., 53, 76
Dillion, P. W., 556
Dillon, R. E., 542, 556
Dixon, G. M., 382
Doehnert, D. F., 327
Dotbauer, B., 328
Douck, M., 329
Dow, J., 48, 343
Dowbenko, R., 328, 586, 603
Drew, R. G., 23, 98, 189, 218, 418, 497
Druscheke, W., 603
Duke, A. J., 67, 76
Dunkel, W. L., 296
Dunlap, I. R., 405, 409, 413, 418

Eaton, R. F., 551, 557

Eby, L. T., 297
Egan, F., 77, 163, 164, 167
Eger, L. W., 411, 418
Engel, J. J., 542, 456
Engel, E. W., 411, 418
Erickson, D. E., 48
Estes, P. W., 383
Eustis, W., 210, 219
Evans, J. L., 375, 382
Ewart, R. H., 577, 585
Ewins, E. E., Jr., 220
Excelsior, 418

Fallwell, W. F., 579, 585
Feige, H. I., 382
Feigl, 168
Fellgett, P. B., 171, 188
Ferry, J. D., 95, 98, 150
Fesslar, R. R., 462
Fey, M. D., 384
Fieser, 168
Filer, T. D., 585
Finnegan, L. P., 406, 411, 418
Fisher, A. A., 425
Fisher, C. H., 329
Fitch, J. C., 24
Fitzwater, R. N., 542, 556
Follette, W. J., 532
Forbes, W. G., 48
Forese, J. L., 437
Forrestal, L. J., 322, 329
Fortry, G., 84
Fournier, A. A., 328
Fowells, E. W., 532
Fox, T. G., 313, 330
Freeston, W. D., Jr., 510, 511, 532
Fries, J. A., 329
Fujiki, T., 328
Fukuda, Y., 328
Fukukawa, S., 328, 602, 604
Fukuzawa, K., 14, 25, 65, 77, 150, 152, 153, 155, 157, 167, 426
Funk, C. S., 383
Furtner, V., 382
Fussa, A. D., 602

Gamo, M., 150
Gander, R. J., 327
Gardner, D. M., 328, 329
Gardner, T. A., 546, 550, 551, 556, 557
Gardon, J. L., 40, 48, 57, 65, 68, 74, 76, 77, 577, 585, 602
Geffken, C. F., 49
Gehman, D. R., 501, 532
Gent, A. N., 55, 56, 76
Gershberg, D., 578, 585
Gerstel, M. H., 15, 24
Gillespie, R. H., 33, 34, 35, 48
Ginosatis, S. M., 322, 329
Given, D. A., 329
Gleichenhagen, P., 603
Gobran, R., 329
Gogos, C. G., 532
Goland, 113
Golden, T., 422, 425
Good, R. J., 69, 77
Goodman, 425
Goodwin, J. T., 352
Gordon, D. J., 403
Gordon, J. I., 150
Goun, 405, 411
Grant, O. W., 532
Gray, R. A., 48, 98, 275
Greene, B. W., 585
Greenman, F. G., 418
Greensmith, H. W., 56, 76
Grenoble, M. E., 403
Grib, S., 328, 603
Grimes, W. D., 18, 438
Grolito, C. J., 418
Grossman, R. F., 377, 383
Guerin, J. D., 329, 585
Gul, J. E., 150
Gunner, L. P., 297
Guse, G., 328, 603
Gustafson, K. H., 418
Gutfreud, K., 328
Gutte, R., 328
Guttman, A. L., 297

Hagenweiler, K., 343
Ham, G. E., 327, 584
Hamm, F. H., 422, 425
Hammer, I. P., 297
Hammond, F. H., Jr., 32, 36, 44, 48, 98
Hansen, C. M., 537, 540, 556
Haplin, I. C., 150
Harianawala, A., 297
Harkins, W. D., 576, 577, 585
Harlan, J. T., Jr., 98
Harris, G. M., 19, 22, 297, 450, 462
Hasen, D. R., 604
Hata, T., 150
Hauber, R., 343
Hauser, M., 329
Hauser, W. R., 437
Hawrylewicz, E. J., 328
Hayes, M. W., 188
Hechtl, W., 382
Hechtman, J. F., 404, 407, 413, 417, 418
Hendricks, J. O., 22, 23, 24, 51, 54, 56, 76, 209, 219, 376, 382, 603
Heyse, W. T., 418
Higgins, J. J., 22, 276, 297
Higuchi, 153, 154, 167
Hildebrand, 41
Hino, K., 48, 556
Hinterwaldner, R., 562, 572
Hisamatsu, H., 603
Hittmair, P., 382
Hoch, C. W., 199, 219
Hock, C. W., 44, 48
Hockemeyer, F., 375, 382
Hodgson, M. E., 343
Hofman, E., 343
Hoffman, C. R., 602
Hokama, T., 562, 573
Holden, G., 98
Holt, F. W., Jr., 497
Holtz, A., 373, 382
Honi, H., 604
Horback, W. B., 327
Horn, M. B., 328
Hougen, O. A., 536, 537, 556
Houseman, G., 572
Houwink, R., 76, 150
Howard, G. J., 63, 76
Huber, E. W., 497
Huntsberger, J. R., 42, 48, 54, 76
Hurst, A. R., 375, 382
Hutton, T. W., 329, 585
Hwang, S. S., 511, 532

Ijichi, I., 328, 602
Ikeda, M., 77
Iler, R. K., 383
Imoto, T., 150
Ishiyama, H., 603
Ito, T., 48, 77, 541, 556
Iyengar, Y., 48

Jaffe, H. H., 188
Jagisch, F. C., 22, 276, 297
Jahn, R. G., 369
Jaucken, P., 603
Jayne, J. E., 404, 418
Jenkins, M. C., 556
Jepson, M. D., 552, 557
Johns, R. E., 556, 557
Johnson, E. G., 70, 77
Johnston, J., 63, 65, 66, 67, 77
Jones, F. B., 48
Jouwersma, C. J., 74, 76, 77, 150
Jubilee, B. D., 327
Jurrens, L. D., 573

Kaelble, D. H., 48, 68, 70, 71, 72, 73, 74, 75, 76, 77, 150, 422, 425
Kalleberg, M. O., 24, 330
Kamagata, K., 48
Kamano, T., 603
Kambe, H., 48
Kargin, V. A., 150
Karman, W., 603
Karr, T. J., 297
Kasper, A. A., 604
Kawaguchi, T., 603
Kazanceva, V. V., 150
Keck, F. L., 329
Keller, A., 188
Kellgren, W., 418
Kelly, G. E., 497
Kemp, M. H., 425
Kendall, D. N., 168
Kest, D. O., 24
Kimura, M., 328
Kirk, 188
Kirkland, J. J., 188

Kisbany, F. N., 207, 219
Klepetar, M., 209, 219
Kligman, A. M., 425
Kline, M. M., 343
Knapp, E. C., 327
Knibbs, R. W., 48, 200, 201, 202, 219
Knoepfel, H., 328, 329
Knutson, A. T., 330
Kolpe, V. V., 98
Komerska, J., 19, 478
Kono, T., 556
Korcz, W. H., 220
Kordzinski, S., 328
Korpman, R., 21, 24, 573
Kosaka, H., 48
Kosaka, T., 155, 167
Kosaka, Y., 328
Kosta, G., 532
Kothari, G. U., 573
Kozima, K., 150
Kraus, G., 43, 48, 83, 98, 275
Kremer, C. J., 297
Krotova, N. A., 76
Kucera, C. R., 209, 219
Kuczynski, W. J., 327
Kugler, J. H., 327
Kurulova, N. M., 76
Kwok, M. C. T., 343

Lader, W., 327
Lagani, A., 297
Lambert, R. J., 24
Langley, P. G., 48, 200, 201, 202, 219
Lannus, A., 57, 76
Lauck, J. E., 369
Lavanchy, P., 24, 377, 383
Lawton, D. W., 550, 556, 557
Lazor, T. J., 501, 532
Ledwith, A., 602
Lehman, G. W., 328
Lende, W., 603
Levine, I. W., 23
Lewis, A. F., 329
Lewis, C., 199, 219
Lewis, P. W., 403
Litz, R. J., 560, 572
Loshaek, S., 313, 330
Louth, G. D., 188
Lucas, A., 425
Lukat, E., 603
Lundsager, B., 602
Luongo, J. P., 188
Luskin, L. S., 327, 575, 584
Lydick, H. K., 327

Malinskii, YU. M., 150
Malofsky, B. M., 329
Mandelkern, L., 188
Mange, E. A. O., 462
Mao, T. J., 307, 322, 323, 329
Marchessault, R. G., 330
Mark, H., 316, 329
Marples, R. R., 425
Marrs, O. L., 48
Marsella, L. J., 548, 557
Marsh, G. R., 552, 557
Marschall, 418
Marshall, W. R., 536, 537, 556
Martin, J. B., 383
Marwitz, H., 382
Marx, M., 603
Maska, R., 328, 603
Massen, 157, 167
Matheson, L. A., 542, 556
Matsukawa, H., 153, 157, 167
McAdams, W. H., 545, 556
McCauley, H. J., 536, 537, 556
McFadden, R. S., 188
McGarry, A., 377, 383, 418
McGinnis, V. D., 602
McGregor, R. R., 352
McKenna, L. W., 328, 329
McLaren, D. D., 54, 76
McLeod, L. A., 48
Mendelsohn, M. A., 330
Merrill, D. F., 344, 352
Mestetsky, T. S., 374, 382
Meyer, E. J., 429, 437
Meyers, R., 520, 522, 532, 575, 584
Michelson, A. A., 188
Middleman, S., 512, 521, 532
Mihalik, R., 76, 319, 329
Miller, J. J., 520, 532, 585
Miller, R. G., 188
Miller, V. A., 406, 411, 418
Mills, T. L., 579, 585
Milutin, I. C., 382
Miranda, T. J., 314, 330
Morgan, B. D., 437
Moriguchi, H., 48, 541, 556
Morris, V., 418
Morton, M., 188, 296
Mosher, R. H., 383, 418
Mowdood, S. K., 329
Moyer, H. C., 297
Mueller, 21
Mueller, A., 328, 329
Mueller, H. W., 331
Murakami, T., 328, 602
Murakoshi, S., 603
Mushel, L. A., 518, 532
Musser, F. M., 573

Nablo, C. V., 602
Nachtsheim, H. G., 429, 437
Nakamura, T., 150
Nakata, S., 604
Nathan, J. B., Jr., 161, 162, 163, 167
Nemoto, T., 155, 167
Neu, R. F., 296
Newman, D. J., 541, 556
Newman, N. S., 327
Nichols, L. D., 380, 383
Nielsen, L. E., 275
Nishimura, K., 603
Nishizaki, T., 603
Noll, W., 403
Nolle, A. W., 96, 98
Nonaka, J., 151
Nuessle, A. C., 329
Nukatsuka, H., 49, 77
Nunn, C. J., 541, 556

O'Brien, J. T., 372, 382
O'Driscoll, K. F., 584
Ohnishi, K., 604
Okazaki, H., 603
Oliver, J. K., 541, 556
Orchin, M., 188
O'Malley, W. J., 351, 352
Orrill, G. R., 210, 219
Osborn, C. L., 602
Oster, G., 602
Other, 188
Overly, W. F., 556, 557

Page, D. H., 409, 418
Pagel, K. J., 556, 557
Pagendarm, R., 548, 557
Palermo, F. C., 573
Pappas, S. P., 602
Park, V. K., 562, 573
Parry, S. A., 48
Pastor, S. D., 329, 603
Paterson, S. D., 297
Patrick, R. L., 77
Patrylow, M. F., 603
Paulavičius, R. B., 53, 76
Pearson, D. S., 328, 603
Peckmann, H. V., 327
Pekarskas, V., 151
Perry, W. M., 585
Peterson, R. L., 323, 328
Petke, F. D., 297
Petrich, R. P., 55, 56, 76
Picard, L. E., 418
Pietsch, H. G., 323, 328, 603
Pike, C. O., 378, 383
Pitts, J. N., 602
Plummer, A. P., 330
Pohlhausen, E. Z., 556
Pomazak, E., 382
Popa, L., 327
Porsche, J. D., 382
Powers, P. O., 573
Prane, J. W., 602
Price, S. J., 161, 162, 163, 167
Prokopenko, V. V., 150

Quintal, B. S., 602

Rajeckas, V., 54, 65, 109, 150, 151
Ramler, W. J., 602
Ransaw, H. C., 188
Reegen, S. L., 307, 322, 323, 329
Regel, V. R., 150
Rehberg, C. E., 329
Rehnelt, K., 377, 382
Reinhard, H., 328

Reip, H., 437
Reissner, 113
Reppe, 331, 332
Reylek, R. S., 72, 75, 77
Reynolds, W. F., 383
Rhyne, J. B., 369
Richardson, C. A., 556, 557
Rienau, J. H., 570, 573
Rifi, M. R., 561, 572
Ritchie, P. F., 48
Ritson, D. D., 383
Rochow, E. G., 352
Roe, C. P., 585
Roehm, O., 327
Rogers, S. S., 188
Rolih, R. J., 382
Rollman, K. W., 48, 98, 275
Rothrock, C. W., Jr., 369
Rowland, 403
Russell, P. J., 602
Russel, T. E., 561, 573
Rutzler, J. E., 76, 316, 329
Rygg, S., 382

Sahler, W., 343
Saito, M., 328
Salditt, F., 343, 425
Salisbury, 403
Šalkauskas, M. J., 53, 76
Salomon, G., 76
Samour, C. M., 219, 327, 328, 585
Sanderson, F. T., 329, 501, 532
Sandman, I., 188
Sandner, M. R., 602
Sarkanen, K., 418
Sasaki, T., 603
Satas, D., 13, 14, 21, 50, 53, 54, 76, 77, 298, 319, 327, 329, 370, 419, 425, 426, 498, 533, 558, 574
Sauders, F. L., 585
Savage, R. I., 76
Saville, R. W., 329
Scales, 425
Scardiglia, F., 562, 573
Scharr, C. H., 425
Scheinbart, E. L., 297
Schilling, R. W., 497
Schlademan, J. A., 353
Schlauch, W. F., 375, 382
Schmidt, G. A., 329
Schneider, G. B., 521, 532
Scholz, K., 343
Schonhorn, H., 41, 48
Schroeder, G., 343
Schulte, D., 603
Schurb, F. A., 375, 382
Screnock, J. J., 437
Seiler, C. J., 54, 76
Settineri, R. A., 372, 382
Seymour, D. E., 343, 421, 425
Sharpe, L. H., 41, 48
Sheetz, D. P., 542, 556, 579, 585
Shecut, W. H., 3, 23, 189, 218
Shenfeld, R. S., 562, 573
Sherriff, M., 48, 200, 201, 202, 219
Sherwood, T. K., 536, 556
Shimizu, Y., 603
Shimomura, K., 602, 604
Shimomura, T., 328
Shoraka, F., 69, 77
Silver, S. F., 328, 329, 585
Sirota, J., 327, 328, 497
Sjothun, I., 297
Skeist, I., 188, 275
Skoultchi, M. M., 329, 600, 603
Slonimskii, N. L., 150
Slutsker, A. I., 150
Slutsker, A. I., 150
Smith, A. G., 24
Smith, A. L., 188
Smith, D. J., 573
Smith, R. M., 383, 418
Smith, T. E., 462
Smith, T. L., 150
Smith, W. C., 297
Smith, W. V., 577, 585
Snow, A. M., Jr., 24
Snyder, L. R., 188
Sohl, W. E., 382
Solomon, G., 150
Sonnichsen, H. M., 413, 418
Sorell, H. P., 343, 375, 382
Sparks, G. J., 73, 76, 77
Spitz, D. A., 548, 556
Spoor, H., 603
Stannett, V., 418
St. Clair, D. J., 220, 604
Stedry, P. J., 150
Stenzel, G., 328
Stepanov, Y. E., 150
Stephenson, A. R., 425
Steuben, K. C., 343, 597, 602, 603
Stickle, R., Jr., 329
Storck, G., 603
Strahan, T. W., 327
Striker, A. M., 497
Stucker, N. E., 22, 276, 297
Stowe, R. H., 219
Sultcliffe, L. H., 602
Sunagawa, M., 603, 604
Suzuki, M., 541, 556

Tadmor, Z., 532
Tager, A. A., 150
Takeuchi, T., 556
Tamburro, A. J., 497
Tarowski, H. L., 556
Tawn, A. R. H., 602
Taylor, P. C., 515, 532
Tess, R. W., 275
Thomas, A. G., 76
Thomas, R. M., 296
Thomas, R. T., 297
Thommen, E. K., 418
Tierney, H. J., 24, 418, 497
Tikeda, M., 49
Tikhonov, A. T., 150
Timm, W. C., 375, 382
Tobolsky, H., 150
Tomashevskii, E., 150
Tomita, J., 327, 329
Towers, 425
Toy, W. W., 327
Toyama, M., 42, 44, 48, 49, 68, 77, 153, 154, 167, 541, 556
Trecker, D. J., 602
Trotter, J. R., 297
Tsuge, S., 556
Tsuruta, T., 584
Turner, C. E., 24
Tyran, L. W., 297

Uehara, K., 603
Uemura, M., 328
Uffner, M. W., 49
Ulrich, E. W., 98, 305, 327, 497
Uraya, T., 328
Urjil, A. J., 327
Urzhumtsev, Y., 150, 151

Van Arsdell, W. B., 537, 556
Vanderhoff, J. W., 542, 556
Vassel, B., 382
Vievering, W. A., 88
Voit, A., 49
Volkersen, 113
Vona, J. A., 327
Voyutskii, S. S., 76, 150

Wake, W. C., 49, 329
Waldman, H. L., 24
Wall, J. S., 328
Wallace, D. A., 15, 24
Wallace, R. C., 573
Watanabe, 153, 154, 167
Watson, C., 558
Webber, C. S., 23, 24, 377, 378, 383
Weber, C. D., 403
Weidner, C. L., 150, 377, 383, 418, 497
Weiss, H. L., 503, 532
Weiss, P., 556
Welding, G. N., 150, 311, 329
Wendlandt, W. W., 188
Werdner, C. L., 418
Weschler, J. R., 411, 412, 418
Wetzel, F. H., JR., 36, 44, 49, 96, 153, 167, 192, 193, 199, 202, 218, 219, 369
Weymon, H. P., 497
Wherry, R. W., 369
Whitby, G. S., 296
Whitehouse, R. S., 199, 200, 219
Widmer, H., 343
Willeboordse, F. G., 551, 557
Williams, P. L., 382
Willis, H. A., 188
Willis, J. M., 585
Wilson, G. J., 297
Wilson, J. E., 384
Wilson, P. H., 15, 24

Winslow, L. E., 328, 585
Wislicenus, I., 343
Wittock, C., 219
Wolf, H., 585
Womack, H. G., 573
Wong, T., 150, 311, 329
Wood, L. L., 602
Wyart, J. W., 327

Yamada, F., 328
Yamano, K., 603
Yang, N. L., 602
Yap, W., 74, 77
Yocum, R. H., 579
Yoshikawa, N., 328, 602, 603, 604
Youman, 157, 167
Young, J. E., 375, 382
Young, K. C., 327

Zang, D. H., 327
Zapp, R. L., 296
Zdanowski, R. E., 57, 76, 328, 329, 585
Zelinski, R. P., 98
Zenk, R. E., 329
Zhurkov, 109, 110
Zigman, A. R., 328, 585
Zink, S. C., 532
Zisman, W. A., 40, 41, 42, 49
Zuev, Y. S., 150

# Subject Index

Italics indicate a major source of information


Abietic acid, 355
Accelerators, rubber cure 190, 206
Acrylic polymers
  acrylonitrile comonomer, 60–61, 317–319
  adhesives, 2, 16, 21, 59, 60, 92, 160–161, 163–166, *298–330*, 342–343, 419, 421, 424–425, 432, 480, 593–601
  aging of, 160–161, 163–166
  carboxylic groups, 305, 314, 319, 322
  compliance of, 319
  crosslinking of, 311–315, 322
  glass transition temperature of, 306
  monomers, 267, 299–305
  paper saturants, 410
  polar groups, 316
  reactive, 326
  release coatings, 372, 374, 375, 378
  tack, 306–309
Acrylonitrile-butadiene elastomer, 176, 407, 410, 414
Acrylonitrile comonomer. *See* Acrylic polymers
Activation energy of failure, 128, 137
Adherend, effect of stiffness, 120
Adhesion
  wetting of surface, 68–69
  work of, 40
Adhesive-cohesive failure transition, 50, 53–56, 61, 62, 64, 308–310
Adhesive failure, 50, 54, 55, 65, 111, 112, 119, 122, 128, 130, 153, 154
Adhesives
  acrylic. *See* Acrylic polymers
  block copolymer, 2, 22, *221–275*, 294, 414, 481, 561, 562, 601
  butadiene-acrylonitrile, 22
  butyl rubber, 22, *281–297*, 451
  compounding, 21, 502–503
  delayed action, 486–488
  EVA, 23, 368, 559–561
  floor tiles, 15, 291, 292, 561
  hot melt, *253–262*, 290, 292–294, 338, 342, 360, 480–481, 599, 600
  label, 289, 290, 291
  latex, 502–503
  medical, 290–291, 339, 341, 342, *419–425*
  natural rubber, 3, 16, 22, 44–47, 79–80, 85, 207–208, 368, 414, 415, 419, 421, 423, 480, 601
  polychloroprene latex, 23
  polyvinyl acetate, 601
  polyvinyl ether, 21, 84–85, 92, *331–343*, 419, 597
  polyvinyl pyrrolidone, 23
  repulpable, 19, 323, 339
  sales volume, 2
  SBR, 2, 21–22, 46, 234, 368, 480, 601
  silicone, 161–164, *344–352*
  syrups, 595–598
  thermosetting, 22, 215–216
Aging
  acrylic adhesives. *See* Acrylic polymers
  natural rubber, 153, 155–158, 160, 166, 217
  polyvinyl ether, 335–337
  silicone adhesives, 161–164, 352
  tapes, *152–167*, 217, 446
  Aliphatic-aromatic resins, 360–361
  Aliphatic resins, 230, 359–361
  Allyl polymerization, 315, 594, 598
  Aluminum foil tape, 9, 17–18
  Aluminum powder, 210
  Amide groups-crosslinking, 314–315, 322
  Amines-release agents, 378
  Amino silane cure, 349

Amorphous polypropylene, 3, 23, 283, 289, 292, 368, 561–562
Analogies for strength prognosis, 141–148
Antioxidants, 155–158, 178, 184, 195–196, 235–236, 278, 279, 283, 335–337, 480, 481
Antiozonants, 235, 237
Appliance industry tapes, 19
Aromatic-aliphatic resins. *See* Aliphatic-aromatic resins
Aromatic hydrocarbon resins, 231, 358–359
Arch dryers, 551–553
Atactic polypropylene. *See* Amorphous polypropylene
Automotive
  decals, 445, 448
  masking tape, 414–415, 439–440
  side molding tapes, 438–439
  striping films, 446–447
  tops, 448–449
  wood grain films, 439, 442–448
  tapes, 414–415, 438–449
Aziridine crosslinking, 315

Backing effect on peel, 65–67
Backing identification, 172–175
Backsize coating, 370, 377–379
Bailey's criterion, 130–135, 141, 149
Binding, ionic, 62–64, 319–322
Bitumen, 342
Blisters, coating, 546–547
Block copolymers, 2, 22, *221–275*, 294, 414, 481, 561, 562, 601
  blends, 234
  compounding, 224, 225, 228, 238–243
  degradation, 258–262
  hot melts, 258–262
  manufacturing, 243–244
  plasticizer effect, 232–234
  properties, 224–227
  solubility, 244–253
  solvent effect, 227–229
  structure, 222–223
  UV crosslinking, 267–268
  UV resistance, 274
Bond coats, 198, 283, 350, 370, 413–414
Bond failure
  activation energy, 128, 137
  adhesive. *See* Adhesive failure
  cohesive. *See* Cohesive failure
Bonding, secondary, 315–322
Bonds
  chemical, 110, 112, 126, 129
  intermolecular, 110–112, 128–129
Bond strength
  adhesive thickness, effect of, 121, 122
  bond width, effect of, 112
  deformation rate, effect of, 137–139
  prognosis of, 141–148
  static load, dependence, 123–125, 127–130, 136, 146
  temperature, effect of, 122–125, 127–130, 137–141, 144–146, 149
  varying load, dependence, 130–137
Bromo butyl rubber. *See* Butyl rubber
Building construction tapes, 14–15
Butadiene-acrylonitrile. *See* Adhesives
Butadiene-acrylonitrile rubber. *See* Acrylonitrile-butadiene elastomer
Butyl rubber, *286–297*
  adhesives. *See* Adhesives, butyl rubber
  latex, 280, 288
  tack, 282

Capillary flow, 537, 538, 542
Carbon black, 210
Carboxylated SBR, 411–413
Carboxylic groups. *See* Acrylic polymers
Carpet laying tapes, 214, 561
Carton sealing tapes, 211–212, 437
Cathodic protection. *See* Pipe wrap tapes
Cavitation, 55
Cellophane tape, 4, 7, 13, 159–160, 189–190, 372, 430, 434
Cellulose acetate tape, 7, 372, 375
Chain branching determination, 181
Chemical bonds. *See* Bonds, chemical
Chlorinated PVC adhesives, 12
Chloro butyl rubber, 277, 279–281, 283–285, 291
Chrome complexes, 375, 378–380.
Chromotography, 177, 180
Cleavage, chain, 335
Cloth tapes. *See* Fabric tapes
Clupak kraft paper tapes, 431–432
Coating
  air knife, 504, 513–515
  blades, construction of, 505–507
  calender, 419, 480–481, *527–532*, 565
  extrusion, 568–569

flexible blade, 510–513
floating knife, 507, 509
gravure direct, 398–399, 526–528
gravure, offset, 399, 526–527, 529
hot melt, *558–573*
inverted knife, 508, 509
kiss-roll, 522–523
knife-over-blanket, 507–508
knife-over-roll, 503–508
microtransfer, 525
pressure roll, 523–524
reverse roll, 400, 504, 518–523, 566–568
roll coaters, hot melt, 563, 565–568
roll coating mechanism, 524–525
roll-over-roll, 524–525
shear rates, 502
silicone, 391, 392, 397–402
slot orifice, 504, 563–565
thixotropy, 488–500
trailing blade. *See* Coating flexible blade
transfer, 398, 572
wire-wound rod, 399, 504, 516–517
Coatings
backsize. *See* Backsize coatings
dilatant, 498–500
Newtonian, 498–500
pseudoplastic, 498–499
release. *See* Release coatings
Cohesive failure, 50, 53–56, 61, 62, 64, 65, 68, 111, 118–120, 122, 126, 128, 130, 153, 154, 158–160, 204, 306, 308–310
Compliance, 42–43, 80–81, 83, 94, 95, 319
Compounding. *See* Adhesives compounding
Compression modulus, 82
Conductive adhesives, 210
Conductivity, electrical, 209–210
Constant rate period. *See* Drying
Contact angle, 40, 41, 69
Contour diagrams, 239–241, 263, 271, 272
Copolymerization, acrylates, 305–307
Copper foil tapes, 9, 18
Corona discharge, 373, 375
Coumarone-indene resins, 227, 231, 359–360
Creep, 78–79
compliance, 79–81, 94, 97
modulus, 143
resistance, 208–209, 317, 320–321, 324, 336, 353, 365–366, 369
shear, 95
tensile, 94
Critical relative humidity, 544
Critical surface tension, 12–14, 40–42, 68–69
Crosslinking, 90–94, 182–183, 203–209, 213, 265, 267–268, 279–281, 283–285, 311–315, 348–349, 351
accelerators. *See* Accelerators, rubber cure
acrylic monomers for, 267
agents, 208–209, 284
allylic double bonds, 315
amide groups. *See* Amide groups-crosslinking
amino silane. *See* Amino silane cure
aziridine. *See* Aziridine crosslinking
determination, 182–183
functional groups for, 279–280, 313–315, 322–324
ionic, 314
isocyanate, 209, 283, 315
paper saturant, 412–413
peroxide, 265, 348
polyamine, 208
quinoid, 284
radiation, 91, 209, 267–268, *584–604*
release coatings, 485
sulfur systems, 265, 284
titanate esters, 208
UV radiation, 267–268
Crystallinity
degree of, 181–182
X-ray analysis, 181, 183
Curing. *See* Crosslinking
Curling. *See* Drying, curling
Curometer, Wallace Shawbury, 205

Decals, 445, 448, 496
Defects, self-healing, 130–131
Deformation rate. *See* Bond strength
Degradation. *See* Block Copolymers
Degree of polymerization, 179, 334, 341
Delamination resistance, 405, 406, 411, 412, 416, 417
Delayed action adhesives. *See* Adhesives
Die cutting, 398, 475, 476
Dielectric
breakdown, 17, 18, 100, 101, 104–106, 164–165
constant, 103–104
dissipation factor, 103, 105
loss, 104
Differential thermal analysis, 169

Diffraction, X-ray. *See* Crystallinity
Diffusion, solvent. *See* Drying
Diisocyanates. *See* Crosslinking, isocyanate
Dilatancy. *See* Coatings, dilatant
Dipentene, 362
Dispensing of tapes, 435–436
Disproportionation of rosin, 356
Dissipation factor, dielectric. *See* Dielectric
Dissolving, 198, 244–251, 285, 288
Double-faced tapes, 11–12, 19, 214, 215, 449
Drafting tapes, 468, 470
Dryers
  airfoil, 554–556
  arch, 551–553
  convection, 549–556
  conveyor, 552
  floater, 552–553
Drying, 252, 253, *533–557*
  blisters, 546–547
  capillary flow, 537, 538, 542
  cascading, 549
  constant rate period, 539
  convection, 544–546
  critical relative humidity, 544
  curling, web 547–548
  curves, 538–540
  falling rate period, 536, 539, 540
  free volume theory, 540
  heat transfer, 544–546
  infrared, 546
  latexes, 542–544
  rate of, 533, 538–540, 542, 549–551
  solvent diffusion, 536, 537, 540–542
  solvent retention, 146, 538, 540–542
Duct tapes, 264–266
Dynamic shear storage modulus, 82

Edge tear, 411, 417
Elasticity, modulus of, 269
Elastomers, identification 175–176
Electrical tapes, 3, 7, 14–18, 54, 163–166, 215–216, 293–295
  aging of, 163–166
  conductivity, 209–210
  dielectric. *See* Dielectric
  electrical properties, 17, 18, *99–108*
  electrolytic corrosion, 102–103
  flammability, 107
  friction, 16
  harness, 99
  insulation resistance, 103
  International Specifications, 107
  loss index, 103
  natural rubber, 215–216
  *n*-minute test voltage, 101
  OEM, 15–16, 18
  paper, 17, 107, 417
  temperature, effect of, 17, 18, 104–106
  transformer, 99
  volume resistivity, 103
  weight loss, 106
  withstand voltage, 101
Electron beam generators, 592
Electron beam radiation, 209, 586–589, 593, 595–596, 598–601
Electrostatic charge, 57–59
Elongation, tape backings, 405, 413–414, 417
Embossible tapes, 497
Emulsifiers, 577–580
Emulsion polymerization, 575–578
Emulsions
  natural rubber, 210–211, 217, 417
  tackifying resins, 210–211
Epoxy film tapes, 18
Epoxy groups, crosslinking, 314–315
Esterification of rosin, 356
Ethylene glycol dimethacrylate, 313
Ethylene vinyl acetate copolymers (EVA), 23, 234, 289, 294, 368, 559–561
Explosive limit, 549
Extrusion coaters. *See* Coaters.

Fabric tapes, 5, 6, 15–18, 27, 60, 156, 158, 160, 350–351, 420, 428, 432–433
Factice, 424
Failure. *See* Bond failure
Falling rate period. *See* Drying
Fillers, 177–178, 184, 197, 234, 282, 325, 480
Film tapes, 7–8, 15, 17, 27, 426–431, 434–435
  cellophane, 430, 434
  cellulose acetate, 7, 372, 375
  epoxy, 18
  fluorocarbon, 8, 14, 16–18, 164, 351
  graphic art. *See* Graphic art tapes
  polyester, 7–8, 15–17, 65–67, 101, 107, 166, 263
  polyethylene, 8, 15, 16, 215, 373
  polyimide, 8, 16, 17
  polypropylene, 7, 14, 17, 212
  polyvinyl chloride, plasticized, 7, 15, 107, 215

polyvinyl chloride, unplasticized, 7, 14, 211–212
polyvinyl fluoride, 14, 17
solar control, 17
tetrafluoroethylene, 8, 16, 18, 164
wood grain, 439, 442–448
Flammability, 107
Flame retardants, 326
Flame tests, 169
Flatback paper tape, 6, 415–416
Floating knife. *See* Coating
Floor tile adhesives. *See* Adhesives
Floor tile, release agents for, 375
Fluorescent tape, 467
Fluorocarbon complexes. *See* Release
Foam sealants, 11, 14
Foam tapes, 10–12, 14, 19, 449
Foil tapes, 8–9, 15, 17–18, 60–62, 65, 210, 427
aluminum, 9, 17, 18
copper, 9, 18
lead, 9
Free radical polymerization, 588–589
Friction tape. *See* Electrical tapes
Functional groups. *See* Crosslinking

**G**el permeation chromatography. *See* Chromatography
Glass cloth tapes, 6, 15, 17, 350–351
Glass transition temperature, 43, 69, 79, 169, 182, 183, 200–202, 231, 242, 267, 270–272, 299, 305–307, 317, 365
Graphic art tapes, 13–14, *463–477*
blockout films 464–465, 474
color films, 470–472
drafting applique, 474–475
imagable, 497
masking film, 472–474
photographic, 470
printing plate mounting, 214
projectable color film, 465–466, 472
shading and pattern films, 472, 474
stripping, 464–465
Gravure coating. *See* Coating, gravure.
Gum rosin, 354

**H**arness tape. *See* Electrical tape
Heat of peeling, 69
Heat reactive resins, 361–362. *Also see* Crosslinking
Heat resistance, 139–141, 203–207
Heat transfer. *See* Drying
Holding power. *See* Creep resistance
Hot melt
adhesives. *See* Adhesives, hot melt
coaters. *See* Coating, hot melt
coating. *See* Coating, hot melt
degradation, 258–262
mixing, 254–258, 261–262
rheology, 255–257
roll coaters, 563, 565–568
Hydrocarbon oils, 232–233
Hydrogenation of rosin, 356
Hydrogen bonds, 60–61, 317
Hydroxyl groups, 314

**I**dentification
acrylonitrile-butadiene elastomer, 176
antioxidants, 178, 184
backing materials, 172–175
chain branching, 181
crosslinking, 182–183
differential thermal analysis, 169
elastomers, 175–176
fillers, 177–178
flame tests, 169
plasticizers, 177
resins, 176–177
Imagable tapes, 497
Industrial tapes, volume, Japan, 28
Infrared absorption spectroscopy, *169–171*, 173–179, 181–187
Infrared drying. *See* Drying
Inhibitors, UV, 237–238
Initiators, polymerization. *See* Polymerization
Insulation resistance. *See* Electrical tapes
Interliner, release. *See* Release liners
Intermolecular bonds. *See* Bonds
Inverted knife coater, 508, 509
Ionic binding, 62–64, 319–322
Ionic crosslinking, 314
Irradiation. *See* Electron beam, UV crosslinking
Isocyanates, 209, 283, 315

**J**apanese paper tape, 433–434
Japanese industrial tapes, 28, 29
Joint wrap tapes, 457–458

**K**eil test, 382, 402, 486

Kiss roll coater, 522–523
Knife coating, 505–507, 510–513
Knife-over-blanket, 507–508
Knife-over-roll, 503–508
Kraft paper tapes, 6, 17, 25, 160
*k*-value, 334, 341

Labels, 19, 20, 217–218, 339–340, *478–497*
  address, 491, 492
  adhesives, 289–291
  applicators, 481–483, 485, 486, 489
  bottle, 492–493
  delayed action, 478
  die-cutting of, 398
  freezer, 268–269, 273, 290, 479
  graphic arts, 493. *Also see* Graphic art tapes
  heat sealable, 478
  inventory, 489–490
  Labelon, 490–491
  library, 494
  manufacturers of, 20–22, 26
  office, 489–490
  packing list, 492
  permanent, 264, 479
  pricing, 488–489
  production volume, 2, 29
  remoistenable, 478
  removable, 290, 291, 479, 489
  shipping, 491, 492
  stock, 27, 29, 559
  tape cassette and reel, 490
  transparent, 490
  wire marking, 494
Lanolin, 21–22, 47
Lap shear strength, 346, 347, 351
Laser Raman spectroscopy, 168
Latexes
  adhesives, 502–503
  butyl rubber, 280, 288
  compounding, 502–503
  drying, 542–544
  natural rubber, 210–211, 217, 417
Lettering, dry transfer, 476–477
Limonene. *See* Dipentene
Liquid chromatography, 177
Lithographic blockout tape, 464–465
Lithographic masking tape, 213
Long line tapes. *See* Pipe wrap tapes
Loop tack tester, 381–382
Loss modulus, 83
Loss index, electrical, 103
Lower explosive limit, 549

Manufacturers, labels, 20–22, 26
Manufacturers, tapes, 20, 26, 29–31
Masking tapes, 190, 203, 212–214, 470
  automotive, 414–415, 439–440
  creped, 6, 405–407, 413
  colored, 407
  curing, paper saturant, 412–413
  delamination resistance, 405–406, 411, 412, 416, 417
  lithographic, 213
  paint, 3, 405, 413–415
  platers, 350
Mass spectroscopy, 178
Mastication, 191
Mayer rod. *See* Coating, wire-wound rod
Medical adhesives. *See* Adhesives, medical
Melting, hot melt, 571
Microtransfer coating, 525
Migration, plasticizer, 340
Migration, resin, 407
Milling, 191, 197–198
Mixing, hot melt, 254, 257–258
Modulus
  compression, 82
  creep, 143
  dynamic shear storage, 82
  elasticity, 269
  loss, 83
Molecular weight, 54, 62, 85–88, 92, 94, 180, 307–311, 333, 334, 387, 392, 592–593
  determination, 179–180
  determination, distribution of, 180–181
  distribution, 311–312
  number average, 180
  viscosity average, 180
  weight average, 180
Mounting tapes, 10–11
Monomers, acrylic, 299–305
Multiple internal reflection, 171–172

Name plates, 495
Natural rubber, 2, 3, 16, 22, 45–47, 79–80, 85, 153, 155–158, 160, 166, *189–219*, 368, 414, 415, 419, 421, 423, 480, 601
  adhesives. *See* Adhesives, natural rubber
  aging, 153, 155–158, 160, 166, 217

electrical tapes, 215–216
heat resistance, 203–207
milling, 197–198
pale crepe, 191
reclaimed, 160
SBR, blends with, 197
smoked sheet, 191
types of, 191
Neoprene. *See* Polychloroprene
Newtonian flow, 498–500
Nomex tapes, 17, 18
Nonocclusive tapes, 420, 422–423
Nonwoven fabric tapes, 8, 16–18, 420, 422
Nuclear magnetic resonance, 178

OEM tapes, 15–16, 18
Office tapes, 13–14
Oleoresin, 354
Oligomeric surfactants, 579–581
Oligomers, 326
Oscillating peel force, 56–59
Oscillatory methods, 96–97
Osmometry, 180
Oxygen absorption, 153–154

Packaging tapes, 7, 14, 211–212, 415–416, 426–437
Paint masking tapes, 3, 372, 405, 413–415. *Also see* Masking tapes
Pale crepe rubber, 191
Paper saturants, 410–411
acrylic, 410
curing of, 412–413
SBR, 410
Neoprene, 410–411
Paper, saturated, 433
delamination resistance of, 405–406, 411–412, 416, 417
Paper, saturating, 408–409
Paper saturation, 407–408
Paper tapes, 6, 17, 27, 263, 427, 431–432
colored, 417
electrical, 17, 107, 417
elongation of, 405, 413–414
flatback, 6, 415–416
Japanese paper, 433–434
kraft, 6, 17, 25, 160
printable, 417
repulpable, 19
rope stock, 6, 416
saturated, 3, 6, *404–418*
tensile strength, 404–405, 415–416
Paraquat, 354
Peel adhesion. *See* Peel resistance
Peel force. *See* Peel resistance
Peeling, heat of, 69
Peel resistance, *50–77*, 81–82, 153–154, 273, 308–311, 353, 365, 369
acrylic adhesives, 317–318, 320, 322–325
angle, effect of, 68
backing, effect of, 65–67
crosslinking, effect of, 204–205
oscillating force, 51, 56–59, 69, 82, 306, 311
peel rate, effect of, 54, 308–309, 318
polyvinyl ether adhesives, 336
resins, effect of, 194, 203
silicone adhesives, 346, 347
stick-slip, 51, 56, 69, 82, 306, 311
temperature, effect of, 54, 61–63, 204, 310, 318
thickness, effect of, 63–65, 74, 338–339
unwind, 53, 57
Peel strength. *See* Peel resistance
Peel tests, 33, 35–36, 51–53, 57, 65–66, 70–75
Peptizers, 198
Peroxide curing, 265
Petrolatum, 292
Petroleum-based resins, 153, 192, 195–195, 283, 354, 356–362
Phenolformaldehyde resins, 206–208, 213
Phenolic resins, 91, 206, 267, 284, 415, 417
Photographic tape, 470
Photoinitiators, 267, 588–591, 596, 597, 600
Pigments, 197, 282
Pimaric acid, 355
Pinene resins, 23, 79, 192, 199–201, 208, 209, 362, 363
Pipe wrap tapes, 19, 22, 293, 295, *450–462*
Plaster, surgical, 157, 159–160, 189
Plasticizers, 43, 184, 196, 232–234, 239–240, 283, 325
acrylic adhesives, 325
hydrocarbon oils, 232–233
identification of, 177
migration of, 340
natural rubber, 196
Platers masking tape, 350
Polar groups, 316
Polishing bar, 518
Polyacrylates. *See* Acrylic polymers

Polyamide waxes, 373, 379
Polyamines, 208
Polybutadiene, 562, 601
Polybutene, 283, 292, 419
Polybutylene, 562
Polychloroprene, 23, 117, 410–411
Polyethylene film tape, 8, 15, 16, 215, 373
Polyisobutylene, 22, 157, *276–297*, 419, 424
Polyisoprene, 2, 85–86, 197, 368
Polymerizable surfactants, 579
Polymerization
  acrylic, *574–585*
  allyl, 594, 598
  degree of, 179, 334, 341
  emulsion, 575–578
  equipment, 582–584
  free radical, 588–589
  initiators, 267, 332, 333, 576, 580–582, 588–591, 596, 597
  photoinitiators, 267, 588–591, 596, 597
  rosin, 356
  solution, 575–576
Polypropylene
  atactic, 3, 23, 283, 289, 292, 368, 561–562
  crystalline, 562
  film tapes, 426–429
Polystyrene resins, 195, 359
Polyurethane adhesives, 23, 161, 163
Polyvinyl acetate adhesives, 601
Polyvinyl ethers
  adhesives, 21, 84–85, 92, 331–343, 419, 597
  aging, 335–337
  chain cleavage, 335
  peel resistance, 336
  solubility, 334–335
Polyvinyl pyrrolidone adhesives, 23, 323–324
Porosity, tape backing, 406–407
Pressure roll coater, 523–524
Pressure test, release coatings, 381
Primers, 198, 283, 350, 370, 413–414
Printed circuits, 351
Printed graphic tapes, 467–469
Printing plate mounting tapes, 214
Porous tapes, 406–407, 422–423
Probe tack. *See* Tack
Protective tapes, 216–217
Pseudoplasticity, 498–499
Pyrolysis, 176, 177

Quinoid, 284

Radiation. *See* Crosslinking, radiation; Electron beam radiation; UV radiation
Reactive polymers, 326
Reclaimed rubber, 160
Reinforced tapes, 7, 9–10, 208, 427, 434
Relaxation stress, 117–118, 122, 141
Relaxation time index, 117
Release coatings 61–62, 198, *370–383*, 413, 484–486
  acrylic, 372, 374, 375, 378
  amine, 378
  branched alkyl, 375–378
  chrome complexes, 375, 378–380
  crosslinking of, 485
  floor tile, 375
  fluorocarbon complexes, 375
  polyamide wax, 373, 379
  polymer, 371–373
  pressure test for, 381
  printable, 372
  silicone. *See* Silicone release coatings
  wax, 373, 379, 429
Release liners, 12, 19, 393–394, 481–484
Repulpable adhesives, 19, 323, 339
Repulpable tapes, 19
Resins, 43–46, 190–195, 199–202, 206, 207, 227–231, 239–241, 270–272, 283, 296, 324–325, *353–369*, 480
  aging of, 152–155
  aliphatic hydrocarbon, 230, 359–361
  aromatic hydrocarbon, 231, 358–359
  coumarone-indene, 227, 231, 359–360
  dipentene, 362
  disproportionation of rosin, 356
  emulsified, 210–211
  esterification of rosin, 356
  gum rosin, 354
  heat reactive, 361–362
  hydrocarbon, 153, 192, 194–195, 230–231, *354–362*
  hydrogenation of rosin, 356
  identification, 176–177
  migration, 407
  oleoresin, 354
  peel resistance, effect of, 194, 203
  petroleum based. *See* Resins hydrocarbon
  phenolformaldehyde, 206–208, 213
  phenolic, 91, 206, 267, 284, 415, 417
  pinene, 23, 79, 192, 199–201, 208, 209, 362, 363

polymerization of rosin, 356
polystryrene, 195, 359
polyterpenes, 45, 153, 192, 283, 354, 362–363, 415
rosin and derivatives, 3, 152–154, 190–191, 200–201, 230, 283, 354–356, 414, 419, 471
tall oil rosin, 354
terpene-phenolic, 230, 283
Reverse roll coating. *See* Coating
Rheology, melt, 255–257
Roll coating, 520–525, 563, 565–568
Rolling ball tack, 32–35
Rolling cylinder tack, 32
Rope stock paper tape, 6, 416
Rotating drum tack, 32
Rubber
accelerators, 190, 206
bromo butyl, 280
butadiene acrylonitrile, 407, 410, 414
butyl, 276–297
chloro butyl, 277, 279–281, 283–285, 291
natural. *See* Natural rubber
pale crepe, 191
reclaimed natural, 160
tape, 66

Saturants. *See* Paper saturants
Saturated paper. *See* Paper, saturated
Saturating paper. *See* Paper, saturating
Saturation, paper. *See* Paper saturation
SBR adhesives, 2, 21–22, 46, 197, 234, 368, 411–413, 480, 601
Scanning electron microscopy, 169
Sealants, 11, 14
Secondary bonding, 315–322
Shear
creep, 95
modulus, 82
rates, coating, 502
strength, 118–119, 155, 158–160, 194–195, 310, 346, 347, 351
test, 158
Shoe tapes, 18
Side molding tapes, 438–439
Side chain substitution, 182
Signs, pressure-sensitive, 494–496
Silanes, 283
Silica, 197
Silicone adhesives, *344–352*
aging, 161–164
peel resistance of, 346–347
primer, 350
shear strength, 346, 347, 351
tape, 13, 18
Silicone release coatings, 346, 373–375, *384–403*
coating of, 391, 392, 397–400
crosslinking of, 385–390, 395–397
emulsion, 388–392, 394–395
solvent borne, 388–389, 391–394
solventless, 388–392, 395–397
Skin irritation, 421–422, 424
Slitting, 9, 13, 570
Slot orifice, 504
Smoked sheet, 191
Solar control tapes, 17
Solubility parameter, 41, 179, 245–246
Solution polymerization, 575–576
Solution viscosity, 285–287
Solvent
diffusion, 536, 537, 540–542
retention, 146, 538, 540–542
sensitivity to, 407
Solvents, 198, 227–229, 243–249, 252–253, 285–289
Spectroscopy
infrared, *169–171*, 173–179, 181–187
laser Raman, 168
mass, 178
Splicing tapes, 19, 295
Splitting temperature, 62, 308–309
Stearate, 47
Stearate, 47
Stearic acid, 278
Stick-slip peel force. *See* Peel resistance
Stiffness, backing, 120, 406
Strength
kinetic concept, 110
long term, 110
prognosis, 112, 122, *141–148*
Stress concentration, 111, 113–118, 122
Stress distribution, 70–73, 76, 114–115
Striping films, 446–447
Stripping tape, 464–465
Styrene-isobutylene, 562
Subsequent adhesion, 382, 402
Sulfur cure, 207, 265, 284
Surface energy, 40, 385
Surface protective tapes, 14, 416–417, 433, 496

Surface tension, 40–41
  critical, 40–42, 68–69
Surface wetting, 68–69
Surfactants. *See also* Emulsifiers
  oligomeric, 579–581
Surgical adhesives. *See* Adhesives, medical
Surgical plaster, 3, 157, 159–160, 189
Surgical tape, 3, 5, 13, 25, 54, *419–425*
Syrup, adhesive, 595–598

Tack, *32–49*, 79–83, 198–200, 202, 269, 271, 273, 289, 353, 364–367, 369
  acrylic adhesives, 306–309, 317, 324, 325
  butyl rubber, 282
  natural rubber, 192–195
  peel tests, 33, 35–36
  polyvinyl ether, 336
  probe, 33, 36–40, 44, 45, 317, 324, 325
  rolling ball, 32, 33–35
  rolling cylinder, 32
  silicone adhesives, 346–348
  testing, 33–39
  two-phase system, 198–200, 202
Tackifying resins. *See* Resins
Tall oil rosin, 354
Tapes. *See also* Appliance industry, automotive, building construction, carpet laying, carton sealing, double-faced, drafting, duct, electrical, embossible, fabric, film, fluorescent, foil, foam, graphic art, manufacturers, masking, nonwoven fabric, packaging, paper, pipe wrap, protective, reinforced, rubber, shoe, solar protection, splicing, surgical, transfer, weatherable
  dispensing of, 435–436
  production volume, 2, 28
  water vapor transmission in, 15
Tear resistance, 405, 411, 416, 417
Telescoping, 83–84, 371, 570
Temperature
  glass transition. *See* Glass transition temperature
  peel resistance, 54, 61–63, 204, 310, 318
Temperature-concentration analogy, 146–147
Temperature limits of electrical tapes, 17, 18, 104–106
Temperature-time analogy, 54, 141–148, 310
Tensile creep, 94
Tensile strength, 404–405, 415–416
Terpene-phenolic resins. *See* Resins
Transfer lettering, 476–477
Testing
  flame, 169
  Keil, 382, 402, 486
  loop tack tester, 381–382
  peel, 51–53, 57, 65–66, 70–75
  shear, 158
Thermal analysis, 169
Thermal degradation, 571
Thermoplastic rubber. *See* Block copolymer adhesives
Thermosetting adhesives, 22, 215–216
Thickening, 501–502
Thixotropy, 498–500
Time-temperature analogy. *See* Temperature-time analogy
Titanate esters, 208
Titanium dioxide, 480
T-peel, 51–52, 157
Trailing blade coater, 504, 509–513
Transfer coating, 398, 572
Transfer tapes 12, 13, 19
Turpentine, 354
Two-faced tapes. *See* Double-faced tapes

Unwind force, 352, 373
UV absorption spectroscopy, 168
UV crosslinking, 267–268
UV inhibitors, 237–238
UV irradiation of adhesives, 586–589, 592–594, 596–601
UV irradiation equipment, 591

Vinyl ether monomers, 332–333, 338–342
Vinyl pyrrolidone polymers, 23, 323–324
Viscometry, 502–503
Viscosity, 85, 88–90, 96, 285–287
Volume resistivity. *See* Electrical tapes
Vulcanizing agents. *See* Crosslinking

Wallace Shawbury curometer, 205
Wall coverings, 15
Water vapor transmission, 15
Waxes, 47, 292, 373, 379, 429, 559–560
Weatherable tapes, 272–274
Wetting, 370
Winding tension, 570
Wire-wound rod coating, 399, 504, 516, 517
WLF equation, 79, 141, 145
Wood grain films, 439, 442–448
Work of adhesion, 40

X-ray analysis, 181, 183

Zinc oxide, 168, 197, 204, 282–284, 480